Green Energy and Technology

Climate change, environmental impact and the limited natural resources urge scientific research and novel technical solutions. The monograph series Green Energy and Technology serves as a publishing platform for scientific and technological approaches to "green"—i.e. environmentally friendly and sustainable—technologies. While a focus lies on energy and power supply, it also covers "green" solutions in industrial engineering and engineering design. Green Energy and Technology addresses researchers, advanced students, technical consultants as well as decision makers in industries and politics. Hence, the level of presentation spans from instructional to highly technical.

Indexed in Scopus.

Indexed in Ei Compendex.

Ioan M. Ciumasu

Eco-cities

Scenarios for Innovation and Sustainability

Springer

Ioan M. Ciumasu
University of Paris-Saclay and University
of Versailles Saint-Quentin-en-Yvelines
Versailles, France

INTRAS Institute
Versailles, France

ISSN 1865-3529 ISSN 1865-3537 (electronic)
Green Energy and Technology
ISBN 978-3-031-57273-9 ISBN 978-3-319-14702-4 (eBook)
https://doi.org/10.1007/978-3-319-14702-4

This Springer imprint is published by the registered company Springer Nature Switzerland AG
The registered company address is: Gewerbestrasse 11, 6330 Cham, Switzerland

Paper in this product is recyclable.

Preface

Even as the cover title opens a wide horizon, the sub-title indicates a practical scope: generating options in concrete eco-city projects. The initial idea of this volume came from editor Anthony Doyle who invited me to consider making a book proposition that would develop further the method from a research paper using a Delphi-based approach to generate eco-city scenarios [1]. As so often with issue-driven works, this book took a decade to generate, so it incorporates the knowledge and the experiences that have been accumulating during all this time. Some of those include my own works within science, epistemology, sustainability, eco-innovation management and science-business relations (also outlined in a chapter of another Springer book, focused on smart cities and societies [2]). But more importantly, the book comes out in a world that is a lot different. Back then, long-term consequences of the 2008–2010 financial crisis loomed in the dark. Today, we are witnessing a shift: aggravating climate changes and the broad unsustainability crisis manifested through sets of sets of interconnected eco-socio-economic issues and emergencies, and an insufficient response capacity that persists in spite of thick specialist expertise existing in most areas. Ever since the end of the Cold War, true worldwide shocks were thought to be unlikely because global financial-industrial interdependencies had become the norm and geopolitical-economic negotiations emerged as a default approach. But for those who study complex dynamic systems (by sheer necessity in problem solving-oriented research), such cascading events do not surprise: they simply signal that some criticality thresholds (i.e., tipping points between alternative internal equilibrium configurations of those systems) have been reached. Humanity needs options and postponing again isn't one of them anymore.

Many books and papers touch on cities-and-sustainability. But few, if any, do more than informing about some of the problems and eventually proposing laundry lists of 'should' and 'must' advices or making sets of narrow-cover specialist suggestions that leave (knowledge gaps and/or) operational integration towards real solutions to some undefined 'others', or naively presume that integration comes by itself. Meanwhile, real-world crises keep multiplying and aggravating, and the still-rare (difficult but right) holistic projects afield keep lacking operational instruments.

This book needed time precisely because it took on the challenge of solving the methodological puzzle of necessary conditions that precede gap-bridging between what we know and what we do. In order to be an honest author, I wanted to develop a minimal working base of hitherto-missing integrative knowledge-action models, and thus truly assist real-world eco-city projects in spite of the traditional divisions between academic disciplines and between science, business and government.

Based on direct experience in pioneering projects, in-depth literature reviews, and generalized models, this book offers a way for working together productively.

Versailles, France
December 2022

Ioan M. Ciumasu

References

1. Ciumasu IM (2013a) Dynamic decision trees for building resilience into future eco-cities. Technol Forecast Soc Change 80(9):1804–1814. https://doi.org/10.1016/j.techfore.2012.12.010
2. Ciumasu IM (2023a) Smart solutions for sustainability: RDI for urban and societal transitions requires cross-sectoral experimentation platforms. In: Da Silva Portela CF (ed) Sustainable, innovative and intelligent societies and cities. Springer, Cham, CH. https://doi.org/10.1007/978-3-031-30514-6_10

Main Idea in 100 Words

Teaser Nutshell

The world is not some big-but-ultimately-predictable clockwork. It is a complex dynamic system, i.e., it includes not only order but also thresholds, tensions, unpredictable paths, and surprises. Today, the multiplication of not-just-local crises shows that we have already entered a deep, systemic phase transition to a new long-term structural and functional equilibrium—for better or worse. Like all societal transformations in the past, this one too may last for decades, it will be predominantly disordered, and its end outcome will remain uncertain for a long time. This book offers a practice-oriented tool for learning what matters and staying on course towards a great future.

Contents

Module 1
Definitions and Introductions

1.1 Premises

As a civilized society, we need to become operationally aware that *Homo sapiens* is part of nature and that Humanity depends on the integrity and the proper functioning of our planet's natural bio-physico-chemical processes. At present time, we have lost and need to reestablish a long-term balance with those. But (a) competent action requires preparation and (b) cities, as communities of people, are key to it. This gives the practical reason of the book: to close the gap between what we know and what we do collectively.

This volume is not some vain attempt to capture "all there is to know" on this hard topic or to lure you with "silver bullet" solutions. The book comes from an extended effort beyond armchair 'guiding principles'—at the bleeding edge of pioneering projects that cut across the traditional boundaries dividing academia, business and government, and it offers a hands-on method for efficient orientation, organization, and management eco-city projects. As I am first and foremost a scientist, the book relies firmly on research and science, academic references being used steadily throughout the text. But the book also regards every city as a living space-and-time, a unique combination of attributes and potentials that are uniting minds and destinies, a place with a character and soul of its own: "cities, like dreams, are made of desires and fears, even if the thread of their discourse is secret, their rules are absurd, their perspectives deceitful, and everything conceals something else" [69].

While acknowledging contexts and respecting identities, I started from within my own 'experience as horizon' [328] and took the operational perspective of a project manager seeking to connect science and society through the 'imperative of problem solving' [225]. I gave myself the task of making the right distinctions as justified by experience and knowledge at hand, and I aimed at distilling a set of knowledge-action models for those of us who undertake to solve the 'riddles of planning' [111] without naivety [102, 153] and to navigate between the Scylla and Charybdis of eco-city projects.

I. M. Ciumasu, *Eco-cities*, Green Energy and Technology,
https://doi.org/10.1007/978-3-319-14702-4_1

This book includes three autonomous but complementary modules that can be read and re-read separately. The first introduces the central notions grounding any eco-city scenario and project work. The second describes a procedure for generating coherent and evidence-based sets of scenarios as a basis for eco-city programs (i.e., sets/strings of interrelated projects that are needed for generating relevant solutions for cities). The third presents a set of generalized models that I deemed necessary in the preparation and the course of eco-city projects: main ideas for managers, plus technical details, and some small mathematical 'plug-ins' for experts.

Throughout the book, whenever a technical concept is mentioned for the first time, it is explained in plain words right away. But I encourage you to look it up and to maintain a healthy critical spirit at all times: the professional literatures and networks are brimming with work results that wait to be integrated into new solutions.

I hope that this book will serve as a conversation starter and as a common think-and-do platform for teams that include both non-technical and technical people.

1.2 What Is Sustainable Development?

The Challenge

There are at least two ways to concisely define sustainable development. The first is the general statement in the well-known Brundtland Report of the United Nations' World Commission on Environment and Development [440]: the kind of development that satisfies the needs of the present generation without compromising the possibility of future generations to satisfy theirs. This definition follows the prior emergence of key notions like 'ecology' [179], 'ecosystem' [403, 404], and 'ecosystem ecology' [299]; an understanding that we must protect the long-term socioeconomic value of nature, and related optimistic versus pessimistic takes [52, 75, 165, 170, 183, 233, 265, 308].

The second is directly through the applied systems perspective on sustainability, which points out that any economic system is a sub-system of a social system, itself being a sub-system of a natural/bio-physical system [167, 169, 439]. This nested systems perspective and model links straightforwardly with sustainability studies and with evidence from research, development and innovation (RDI). Crucially, it helps us visualize and understand correctly the unsustainability crisis as an unresolved conflict of priorities between humans and nature (Fig. 1.1), manifested across 'coupled human and natural systems' (CHANS) (Fig. 1.2), i.e., all functional interdependencies between human socio-economic systems and nature as a whole [147, 239, 240].

Because survival is the real top priority of people, human concerns begin with the economic issues, continue with the social issues and end up with the environmental issues. This order largely overlaps short-, medium-, and long-terms, except for emergencies and imminent threats, which trigger the "default crisis management" mode. So, the strength of support for any policy follows this order: economy now > society

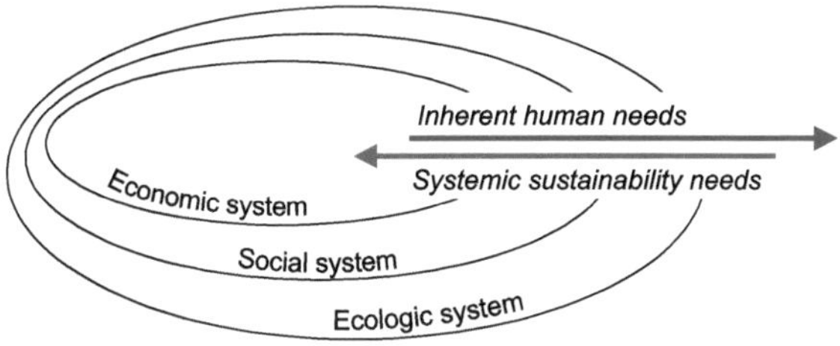

Fig. 1.1 The nested inclusion relation between economic, social, and natural systems. The arrows pointing in opposite directions illustrate the unsustainability crisis as an unresolved conflict of priorities identifiable in humans versus natural systems. Adapted after [99, 103, 167, 169]

Fig. 1.2 General representation of virtual CHANS case studies (sustainability issues) mapped on the systemic perspective model. Human systems can have various sizes: from a few people to a city to humanity itself. Black double arrows denote interdependencies (couplings)

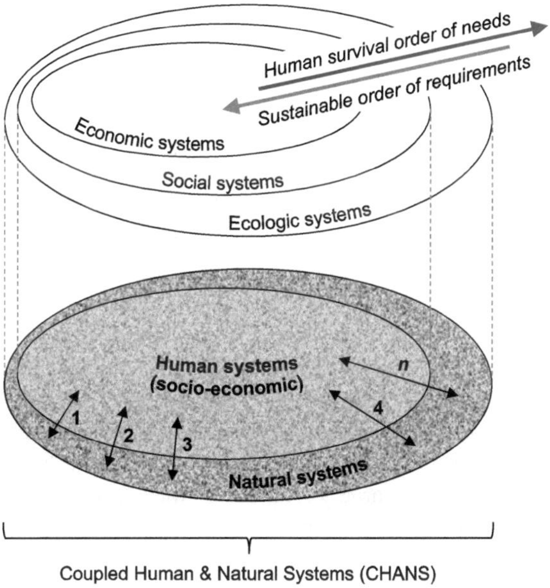

later > environment later on. But, to be compatible with sustainable development goals, proposed solutions must verify the objective space of possibilities allowed by the natural life-support systems ergo must follow the opposite order of priorities: environment as start basis > society (as next condition) > economy as end outcome.

From an epistemic perspective (also including information science and knowledge organization), the category mapping in Fig. 1.2 can serve as an operational start basis for purpose-driven knowledge integration across disciplines and professions, i.e., knowledge-action models. Seen through the lens of CHANS, "sustainability can be defined operationally as a feasible, desirable set of flows (material, currency, information, energy, individuals, etc.) that can be maintained despite internal changes and changes in the environment" and "sustainable development can be defined as the process by which CHANS can be moved towards sustainability" [257].

In theory, the nested relation between systems opens the way to concrete scenarios and operations [99, 103], in ways that respect both ecological complexity across scales (e.g., [398]) and humanity's aspirations. But can we? Will we? Earth existed without humans for most of its time and will continue to exist if humans disappear, but humans cannot exist without our planet's life-support system. We trust human ingenuity [334], but, even if we colonize other planets, the Earth's "Biosphere 1" (or Technobiosphere [414]) remains necessary and irreplaceable [114, 292].

Importantly, the concept of sustainability as-we-know-it has only gained social traction in recent decades, especially after the United Nations Conference on Environment and Development (UNCED; the so-called Earth Summit in Rio de Janeiro in 1992) [174, 364], but with two important observations. The first is that the core principles of sustainability, namely that (a) we must look beyond short-term interest and that (b) we must insure intergenerational fairness, are in fact part of common-sense wisdom since the dawns of human civilization. One could make a long list of human practices that embody de facto the principle of anticipation and avoidance of collapse by overuse or overexploitation—from the domestication of plants and animals and the multi-annual (long-term) management of those resources since Neolithic, fast-forward to modern day psychology which explains how small children learn (on their individual path to maturity) to refrain from immediate gratification in order to obtain a superior benefit later.

And, the core meaning we now give to the notion of sustainability has been documented for centuries. In 1713, Hans Carl von Carlowitz, a government-appointed manager of natural resources (mines and the forests as wood source for burning in mines) in German Saxony, used the word '*Nachhaltigkeit*' (the German word for Sustainability) with the same meaning that we have today. His book *Sylvicultura Economica* (i.e., forest economics)—which you probably noticed was published 63 years before 1776, the year when (1) Adam Smith's book *The Wealth of Nations* was published (largely regarded as the beginning of modern economics), (2) James Watt built his new steam engine (a key technological event in the Industrial Revolution), and (3) the American Declaration of Independence was made (in a revolution started as an objection to taxation without representation)—can be read today without difficulty and many parts of it are as relevant today as they were then. This is already remarkable. What is even more, when he refers to sustainability, von Carlowitz actually cites other authors (many of which he knew from his professional travels across Europe) that were using this idea for the management of natural resources on the continent for centuries before his time, for example the management of forests in France [173, 373]. And, the history of human civilization in every world region is replete with examples that illustrate the importance of the long-term view and the necessity of understanding the big picture.

In short, the notion that the socio-economic and overall development of Humanity needs to be sustainable (so that we do not undermine ourselves) is really a matter of common sense rediscovered across different sectors of activity, even if this comes in different forms and shapes, after some long and winding roads of history.

'Planetary Boundaries Framework' analyses based on the 'Earth System' approach to modeling of anthropogenic impacts show that 6 out of 9 main boundaries are transgressed (biogeochemical flows; freshwater change; land system change; biosphere integrity; climate change; novel entities (anthropogenic materials))—meaning the loss of 'safe operating space for humanity' (the 'still OK' others are: stratospheric ozone depletion; atmospheric aerosol loading; ocean acidification) [348].

In Principle, Is there a Logical Way Out of this Conundrum? Yes

The unsustainability crisis may appear unsolvable, leading to our "self-organized extinction" [170]. The particular cases (local solutions) when a depletion of local common resource pools have been avoided [9, 308, 309, 310, 311, 312] might be too peculiar to repeat in most other local contexts and across scales; this path is probably only a small part of the story. But an analytic regard at the nested systems model reveals a general knowledge-action model, i.e., a logic direction to follow. Namely, if the two conflicting ordered sets of priorities are projected against the series of phases and steps that are known from the overall process of problem resolution in complex situations [99, 375] it becomes obvious that the two series pertain to two distinct sub-processes inside the full-length cycle of problem resolution (Fig. 1.3).

In other words, the two arrows/ordered sets of priorities (1) occur at different moments of the evolution of the typical in-project conversation taking place in those projects that apply a truly systemic approach (i.e., projects that are complex-enough to cover the entire spectrum of basic domains of expertise), and, therefore, (2) they are actually in a relation of operational complementarity, not reciprocal exclusion. Respectively, they describe the requirements of two functionally distinct halves of a full cycle of problem definition and resolution: the "economy first" set of priorities expresses the order in which different types of concerns need to be assimilated during a holistic, "all-inclusive" definition of a problem faced by people, while "natural environment first" set of priorities shows the order in which knowledge and expertise from different domains must be considered by any project or program seeking to generate evidence-based solutions that can be, objectively speaking, sustainable on the long term [97].

The abstract model above can also be formulated as a set of principles (Box 1.1).

Box 1.1 Basic principles for solving the humans/nature conflict of priorities

- Short-term economic incentives do determine human priorities and the public business agendas
- Longer-term prosperity and economic-business competitiveness do depend specifically on those solutions which increase system sustainability
- Holistic approaches enable new action models. Therefore, short-term/economic priorities of people must be recognized as top incentives to act. This provides the necessary public basis for mobilizing professional networks and projects

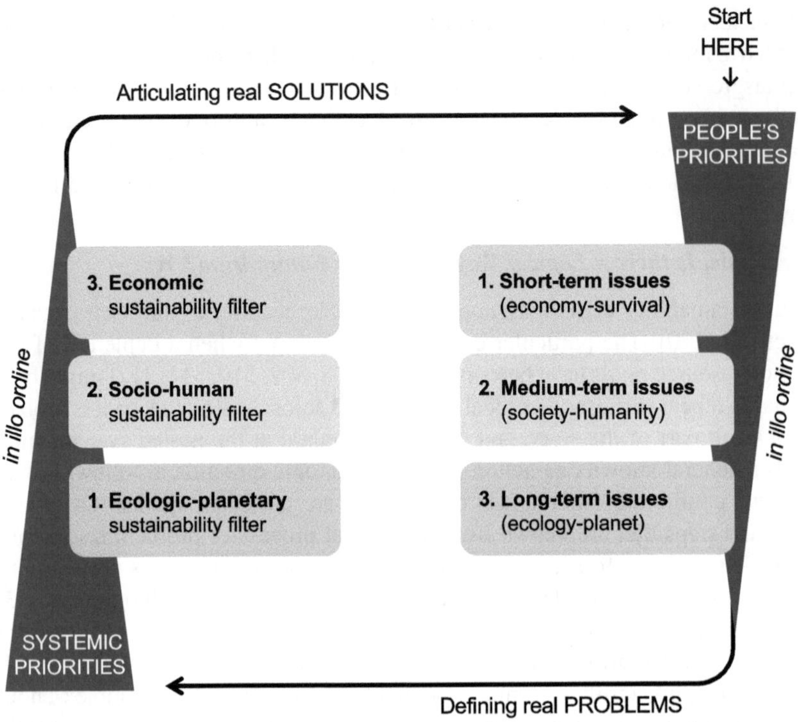

Fig. 1.3 Resolution cycle for the conflict of priorities between human livelihood and sustainable development. The two large arrows chasing each other explicate the conflict of priorities from Fig. 1.1 through projection on different but complementary planes of knowledge representation inside a full iterative (cyclical) process of problem resolution and updating. Numbered boxes indicate steps derived from the nested inclusion relation between economic, social, and natural systems, operationalized as knowledge filters, i.e., lists of domain-specific expert-determined indicators that must be met ("passed") along the process in their order of application inside the two cycle halves. The Latin expression *in illo ordine* means 'in that order'

This model along with its underlying principles represents the start-basis for the methods and scenarios that are being presented in the rest of the book.

Why Hasn't this Solution of Principle Been Applied Yet? ... or, has It?

The short answer is that some progresses are being made, but breakthroughs take time. This systemic nested inclusion relation between nature, society, and economy is not new. To use a beautiful and famous example, Elinor Ostrom (1933–2012), a pioneer of sustainable solution developments and recipient of the Nobel Prize for economic sciences (in 2009), already used in her own work the practice-oriented vocabulary of nested complex dynamic systems and CHANS, such as 'nested sets', 'nested ecosystems' and 'nested enterprises' (e.g., [312]), expressions not unrelated to the so-called 'problem of embeddedness' of economy [172]. Originally trained as a political scientist, she learned whatever it was necessary to better understand the problem and articulate appropriate solutions, thus using sociology, economics, anthropology, and mathematics and donated her Nobel Prize money to the lab she helped establish at her university for the study of local communities in their relation with their natural resources (e.g., [19, 309]).

But, even today, this vocabulary and the systemic perspective on sustainability as a whole is still new—even to the relatively small number of people that know about it (this is the best of cases); or it is simply ignored (this is the case in most fields outside sustainability studies and in society at large).

(As a telling example between brackets, some people, in particular those who see sustainability not as problem to solve but as a fashion to join or/and exploit, still use the logically incorrect representation of sustainability as a small area of "triple overlap" between environment, society, and economy (Fig. 1.4)—as if some part of the economy could exist outside social systems, and, respectively some part of human society could exist outside nature. Such a superficial (non-critical; often opportunistic) regard on sustainable development promotes falsehood and undermines functional science-society relations, thus creating more harm than good).

We Need Adequate Knowledge Representation and Integration

Far from being anodyne, the difference between a correct and an incorrect model is fundamental because the respective sets of definitions of categories and the logical relations between them are semantically and mathematically non-identical. Adequate and correct representations of real systems make the difference between success and

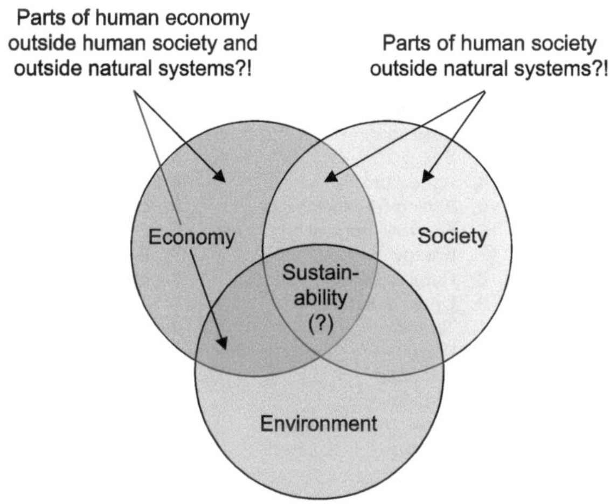

Fig. 1.4 **Incorrect representation of sustainability**. This now-outdated model had the merit that it recognized the problem of unsustainability, but then it used a vague definition (the initial, necessary compromise lowest common denominator between competing political views in the 90's). Science-wise, it fails to see humans as part of nature and tries to imagine the solution ("sustainability") by using the very same mindset that created the problem in the first place (i.e., incongruous representations of reality coming from previously separate fields) combined with the scientifically naive idea that nature is the environment "out there" instead of a physical reality everywhere

failure. Admittedly, the comparatively incorrect model shown above is a technically correct illustration of the mistaken views that characterize unsustainable development because it shows well the problem of knowledge fragmentation. This issue is important because misrepresentations may induce people to think that sustainability is a "nice-to-have" (not a "must") and will prevent them from seeing the "world of options" behind the problem [312]. Worse, such mistakes can trigger adversarial attitudes and scientifically unjustified beliefs, e.g., that sustainability was an impractical concept to be dismissed as an eccentricity or luxury.

Science works with proofs: it provides a basis for coherent action but also evolves with time. Bibliometric analyses (i.e., the study of trends in science reports) show that the field of 'sustainability studies' is not monolithic. It is actually a constellation of topics that reflect a multitude of perspectives that also changes in time: as our needs evolve and knowledge accumulates, new questions are being asked and thus science moves to a next logical stage. Figure 1.5 shows that the root label 'system' only emerged in the sustainability studies landscape by 2014, probably due to a growing need for operational integration of all the knowledge that has been accumulating.

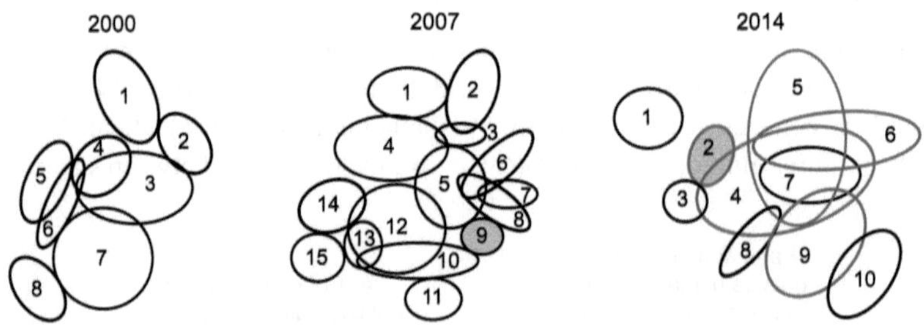

1. Business
2. Tourism
3. Sustainable development
4. Fishery
5. Forestry (tropical rain forest)
6. Energy
7. Agriculture
8. Forestry (biodiversity)

Basic concerns about limited resources

1. Forestry (agroforestry)
2. Business
3. Soil
4. Agriculture
5. Ecological economics
6. Forestry (tropical rain forest)
7. Energy
8. Rural sociology
9. Urban planning
10. Tourism
11. Water
12. Fisheries
13. Wildlife
14. Forestry (biodiversity)
15. Health

Problematization / breakdown into sub-problems

1. Governmental debt & fiscal policy
2. Urban & transport systems
3. Competitive advantage of firms
4. Economy & business systems
5. Environmental systems
6. Energy systems
7. Water resources
8. Education & sustainable human development
9. Fishery & forestry systems
10. Health

Knowledge integration driven by operational requirements

Fig. 1.5 Evolution of the main themes in the literature on sustainability studies. Ellipses reflect the size (volume) and distances (and overlaps) between sub-domains identified in research publications. Blue outlines mark the presence of the category 'system'; rosy fills signals 'urban'. *Source* Redrawn based on the bibliometric method (citation network analysis; topological clustering of terms identified in the titles of published sustainability studies) used by Kajikawa et al. [215, 216]

In this analysis, one notices the central place of the year 2007—in many respects, a peak of activity. Interestingly, 2007 is also when sustainability studies begun to be officially recognized as an established field: a historic moment in time when it ceased being just an eclectic area of shared concerns across (and with contributions from) traditional disciplines and actually gained "a room of its own" in academic institutions, notably in terms of dedicated funding [106]. Of course, that recognition does not imply that the integration of knowledge across (older or/and newer) fields had been achieved. It only shows (a) that this recognition came after efforts to problematize sustainable development in specific real-word contexts (after a decade-long period of societal introspection and intellectual redefinition of the relation between Humanity and its home planet) and (b) that operational integration (driven by concrete problem-solving) had become the next logical phase.

One also observes that 2007 is the year when 'urban' emerged as a core designation within the broad picture of sustainability studies. Arguably, this is more than just a coincidence, and the next sub-section of this book will focus on the place of cities in the whole challenge of sustainability. But before doing that, one last important question must be asked in this introduction: *How do the 'human and natural system couplings' (HANSC) from Fig. 1.2 translate into operational requirements?*

As explained below, all real systems (natural, human, or man-made) are complex dynamic systems (CDS), and agent-based modeling (ABM) is the usual way to represent them. In short, a core answer to the question above, and a starting point for further elaborations, is provided in Fig. 1.6.

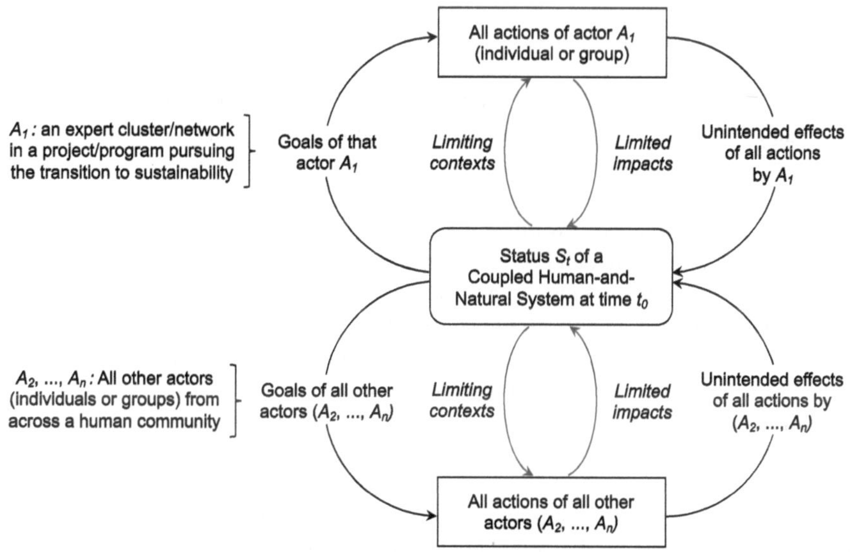

Fig. 1.6 General ABM model of the coupled feed-back loops involved in CDS: an abstract representation of the sum of sums of interdependencies, constraints, and uncertainties connecting actors and CHANS overall or chosen target HANSC inside CHANS. *Source* Redrawn based on [230, 396]

In practice, CDS are called just 'complex systems' or just 'systems', as near all real systems are complex and dynamic (bar the simplest) (Fig. 1.7); or the term 'complex adaptive systems' (CAS; CDS that are able to change based on learning from experience) is used instead (see, e.g., [76]). Examples come from all domains: the biosphere, the ecosystems, the colonies of social animals (insects, birds, mammals, etc.), the immune system, the brain, the developing embryo and the living cell itself; social organization of people (cultures, political parties, communities); stock markets, manufacturing, businesses, supply chains, and innovation systems (for a review, see, e.g., [384], also [270]); cities (e.g., [150]), urban health and wellbeing (e.g., [161, 241, 358, 411]), and so forth.

This complex causality means that realistic (ergo sustainable) solutions can only (1) come from a good understanding of the target systems and (2) be multi-purpose. The general problem-resolution cycle in Fig. 1.3 already reflects this reality.

At the wider societal level, we need to reach the point where most of us understand that we all live in CDS, not in some deterministic world-as-a-clock "machine". The mechanistic predictions based on Newtonian physics are but particular cases of the physical world and only apply in the "easy world" between certain boundaries. This is why experimental physics works by isolating some phenomena (eliminating all the rest) in laboratory. This reductionist approach helps us in the search for mechanisms (when and where they exist), but it always comes together with specifications about the boundaries within which every mechanism exists. When more parameters from the real-world are added, human planning quickly loses realism. Instead, we must understand and operate with the attributes of the CDS (Table 1.1).

To paraphrase a famous quote, preparing for the future is 1% planning and 99% learning. That is, we must build capacity to learn and to act based on learned lessons.

Indeed, the whole discussion about successful eco-city projects is about everyday relentless development of competence to carry out such projects and thus becoming better and better at it—in short, learning-by-doing and doing-by-learning [91, 92]. And, the same applies to sustainability transitions in society in general [163, 244, 245, 268].

Surely, learning our way towards a goal does not mean that everyone must become an expert in natural, social, and economic complex systems. Instead, a truly holistic project (1) must include enough expertise from different fields (even if that means very many people and organizations) so that the project team as-a-whole can cover all needs and (2) must find/develop enough common grounds by openly discussing their differences and agreeing on a minimum core of shared values and knowledge. This learning means also that we must work with nature, not against it. This starts with facing complexity instead of pretending it isn't there just because it's difficult, and continues with building the means for addressing that complexity—through a perpetual exercise of cultivating visibility (via realist scenario set iterations) while integrating knowledge as it comes, and as developed, in practice via disciplined experimentation ([96]; details discussed in Chap. 3, Sect. 3.8; see also the discussion at the end of Sect. 1.3 about cities as loci of experimentation).

These being said, it becomes clear that objective systemic 'complexity', a demonstrable physical reality, must not be confused (like it is often the case in daily life)

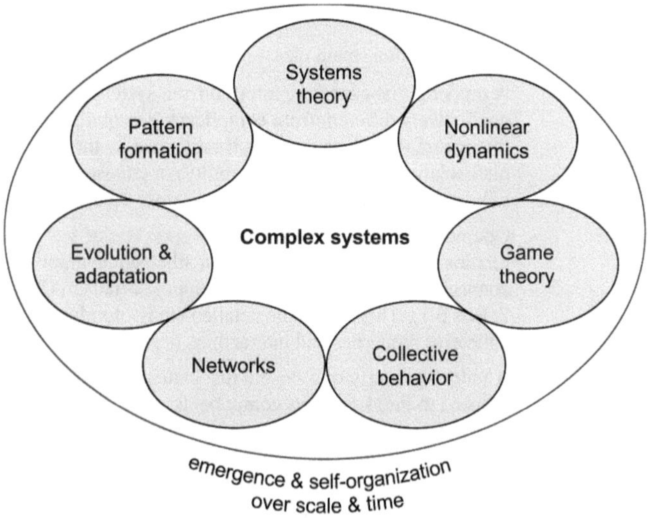

Systems theory	Feedback loops, entropy, homeostasis, autopoiesis, information, computation, cybernetics, self-reference, sense-making, complexity measurement
Pattern formation	Partial differential equations, fractals, dissipative structures, reaction-diffusion, self-replication, percolation, cellular automata, geomorphology, spatial ecology
Nonlinear dynamics	Ordinary differential equations, time series analysis, phase space, attractors, iterative maps, stability analysis, chaos, bifurcations, population dynamics
Evolution & adaptation	Evolutionary computation, artificial neural networks, machine learning, natural computation, viable system theory, artificial intelligence, genetic algorithms
Game theory	Bounded rationality, rational vs. irrational decisions, social dilemmas, n-person Prisoner's Dilemma, evolutionary game theory, networked games
Networks	Graph theory, scale-free networks, small-world effect, social network analysis, community identification, motifs, dynamic/adaptive networks, systems biology
Collective behavior	Agent-based modeling, collective intelligence, swarm behavior, ant colony optimization, synchronization, self-organized criticality, phase transitions

Fig. 1.7 A basic cloud-of-clouds model of (a selection of) topics in the study of complex systems. *Source* Modified after [67]

with perceived situational 'complicatedness'—a managerial contingency. In short, complexity and complicatedness are in fact not interchangeable (unless everybody in the conversation is known to be familiar with the difference between the two).

In business environments, there are heuristic tools for orientation in daily works and help with making the right distinctions, at least at a basic "sense-making" level. For example, the Cynefin framework use 'complicated' and 'complex' as labels for two successive stages on a gradient between what is 'simple' ('obvious' or 'clear') and what is identified as 'chaotic' situations (e.g., [286, 390]). Although this kind of tool is not real science-based knowledge management methods, it has the merit that

Table 1.1 Key features of complex systems

Attribute	Short definition (main idea)
Complexity	A property that cannot be inferred from system components; this is qualitatively different from complicatedness/multiplicity. Models that ignore this property (even if motivated by the need to gain mathematical-operational tractability) or choose to treat it as noise will inevitably be inaccurate and therefore useless in the real-world
Network character	Operational representability as networks: sets of sets of objects (and relations between them) coded as mathematical graphs (dots/vertices connected by links/edges; more technical details in Chap. 3, Sect. 3. 2, Box 3.1). This allows the detailed study of individual and collective properties and interactions between systems components
Emergent properties and behaviors	A unique property of a system that cannot be explained as a (or reduced to the) sum of its components, as it does not exist in those components, and only appears when these components interact as a system. Sometimes, they can be modelled/visualized as network effects
Self-organization	Order resulted as emergent property (unforeseen or unplanned); called 'self-organization'/'spontaneous order' in natural/social sciences
Nonlinear dynamics	Thresholds, steady-states between them & less ordered ("chaotic") transitions between those; feedback loops, inherent uncertainty, 'power law' (fractal) patterns; self-organized criticality, i.e., non-manifest accumulation of tension behind equilibrium thresholds followed by sudden re-equilibration events, e.g., "the straw that broke the back of the camel"

it can contribute to some extent (at least at the beginning, for starting conversations and some exploratory works) in the managing the uncertainties of planning in various areas that involve science and society alike, from ecosystems [34] to infrastructure safety [194].

1.3 What Is a City?

The Different Perspectives

What is a city? The simple definition from human geography is: a large human settlement (smaller ones being towns, villages, and hamlets). Urban geography studies urban processes, thus bringing together aspects from the biophysical, social, and economic systems. Different fields of study have different perspectives, reflecting their particular questions of interest (Table 1.2), but ultimately, a city is its people.

In this sense, some historic details stand out as cultural references for the present, but a holistic perspective is a logical next step in our quest to truly understand what is a city. This is not just a matter of intellectual curiosity: looking deep into this matter is to look deep into the soul of Humanity to see its past, present and future. To this

Table 1.2 Domain-specific perspectives upon cities (a non-exhaustive list)

From the view point of:	A city is ...
Physics	A complex system in space–time; flows of matter-energy/information
Engineering	A built environment, with all related infrastructures and appliances
Biology; ecology	A colony of species *Homo sapiens*[a]; a disturbance of local ecosystems
Chemistry; geology	A source/sink in biogeochemical cycles; a feature of *Anthropocene*[b]
Physical geography	A form of landscape modification, covering about 3% of Earth's land
History; archaeology	A location of events and material records thereof; a culture and power hub
Business; lifestyle	A hub of knowledge, learning and innovation; an engine of progress
Economics	A place of production, exchange and consumption of goods and services
Sociology	A social construct, a place of high diversity and density of interactions
Politics	A constituency, a spectrum of organizational preferences and choices

[a]In the usual terms of anatomical modernity and behavioral complexity, humans begun to exist approximately 300,000 years ago although the technical discussion is more nuanced (for a recent summary, see [368]). On a geological time-scale, this means:
... 0.18% of the history of mammals (which emerged ca. 168 million years ago);
... 0.05% of the existence of animals on Earth (earliest evidence from 580 million years ago);
... 0.008% of the existence of life on Earth (earliest evidence from 3.7 billion years ago);
... 0.0066% of the existence of Planet Earth (which formed 4.54 billion years ago);
... and knowing that less than 1% of all species that ever existed on Earth are still alive
[b]Because humans started to be a geological force with impacts on the planet's geosphere, hydrosphere, biosphere/ecosphere, and atmosphere, and generate an anthroposphere (or 'technosphere', i.e., that part of the environment that is made or modified by humans for their habitation and activities, including all artifacts, e.g., artificial satellites), geologists are debating whether we are in a distinct geological epoch (sub-division of geological time) called 'Anthropocene' [117, 141, 297, 395]. To date, an official decision has not been made, but 'pro' arguments keep accumulating. In any case, current evidence suggests (a) that behavioral changes connected to lithic technology goes back to 3.3 million years ago, at the origin of genus *Homo*, i.e., before the emergence of species *Homo sapiens* 2.8 million years ago [60, 184, 435], see also [16, 189, 204, 277, 376] and (b) that human exploratory behavior and "learner-driven innovation" was at the origin of lithic cultures already [445]. All in all, because (1) tools and tool-users co-evolve in a virtuous circle of success and capability and (2) as tools are known from hominids preceding *Homo sapiens*, one can logically argue that tools made us human [289]

aim, I will briefly evoke three moments in the existence of humanity before going on to the other managerial aspects that matter for current and future projects.

Firstly, after the defeat of the Greeks at Thermopylae in 480 BCE (by the Persians led by Darius, "the King of kings") and in the expectation of a decisive confrontation which they could only hope to win by strategically retreating on land and engaging the enemy at sea, Themistocles persuaded the Athenians to abandon their old city, essentially because "a city is not its stones but its people". Afterwards they saw their city burned to the ground but they also won the war and returned to rebuild Athens to an even greater glory and as a major socio-economic, cultural and political hub. This episode reminds us that a city is much more than what meets the eye.

Secondly, cities actually appeared in the history of humanity when a great number of people came together, i.e., gravitated around and concentrated at a given location, this spatio-temporal movement generating (a) complex forms of organization which were necessary to insure co-existence in a crowded area and (b) a sophistication of life which was not possible in scattered agricultural populations and which further promoted demographic growth and attracted other people. In this historical process, candidates to the status of proto-cities making the transition from rural to urban (or as "social experiments"; presumably different from a city-proper by a lack of planning and centralized rule) are known *inter alia* from:

- the Middle East, with settlements starting since ca. 9000 BCE but distinguishable as proto-cities especially since ca. 4000 BCE, notably the Ancient Jericho in Palestine (Tell es-Sultan; in Hebrew and Arabic, 'tell' means mound or hill), Brak (Tell Brak) and perhaps also the low-density, special purpose (manufacturing) settlement of Khirbat al-Fakhar, both in North-Eastern Syria, and Çatalhöyük in Central Anatolia. All of these predate the indisputably urban settlement of Uruk, from ca. 3100 BCE in Southern Irak (see, e.g., [262, 296, 421]).
- and Central Europe (the Cucuteni-Trypillia culture, ca 5500–2750 BCE overall, with large settlements known from the Middle Period of this culture, ca. 4000–3500 BCE) (see, e.g., [20, 77, 130, 273, 397, 412], see also [441]).

For the purpose of this book, I will illustrate the following short discussion mostly with details from the latter region (because I am familiar with it, as a native).

While the debate is still ongoing about which large settlement(s) from the dawns of human urbanity can be considered essentially a city or only pre-urban, a comparative ("traditional") disciplinary versus ("modern") holistic perspectives is interesting and informative for the fundamental question "*What is a city?*" that needs to be asked by the manager(s) of any project aiming at improving our cities today.

From the disciplinary historical and archeological perspective (also included in Table 1.2), the main tentative cut-off criteria tend to relate to whether or not the researcher can distinguish in the remains (or in the technical reconstitutions) of such settlements the presence of any form of 'division of labor' and especially 'social stratification', as these two aspects are regarded as proxy indicators of a complex civilization.

Through these lenses, the urban-or-not decision relies almost exclusively on the interpretation of the material evidence discovered so far at the study site and every gap in the used data, information and knowledge can hamper the recognition of an urban character in large settlements. However, understanding the beginnings of urban life requires that the (perfectly legitimate) disciplinary perspective mentioned here can and must be complemented by other disciplines and view angles, in a holistic concert of science. Concretely, from a current complex systems perspective, it is physically impossible that very large numbers of people (often up to 5, 10, 20 thousand inhabitants or more) known from the Neolithic and the transitional Chalcolithic a.k.a. (Eneolithic or Copper Age, i.e., agriculturalists with some influence from hunter-gatherers,) could actually co-exist without a 'next-level' complexity of social organization, regardless of how "peaceful" these populations might have been. (And the Cucuteni-Trypillia culture is thought to have been peaceful, because of a lack of archaeological evidence of abundant remains of weapons and warfare). On the contrary, based on our current understanding of Humanity, the archaeological evidence of the long-time coexistence of such a huge number of Neolithic-Chalcolithic people at one single site simply indicate either that (i) some inexplicable lack of evidence (some 'unknown unknown') completely distorted our image about warfare in that civilization or that (ii) warfare was not needed much because people had other ways to deal with each other, i.e., a functional complex socio-economic structure.

Technically, any student of the 'urban phenomenon' that would choose to ignore the simple observation summarized above (and easily verifiable in the literature and in real life) would not practice science but merely some naive form of empiricism (for a discussion about the Scientific Methods, see the summary provided in Sect. 3.7 and the references therein). In the following paragraphs, I will try to just briefly show why, starting from existing archaeological evidence about proto-cities, placed in context and in the big picture. I do this because I think this kind of exercise is useful in two ways: in the true study of cities (as a basis for improving cities), there is nothing healthier and more necessary than questioning given assumptions; and, in so doing, we actually arrive at some of the core questions and methodologies that can help us do what we preach: the holistic approach.

Let us start with the obvious observation that some arguments against the urban character of large Neolithic settlements often point to certain key factors may have actually favored a peaceful coexistence: (a) contrary to some present myths, those people (at least in the Cucuteni-Trypillia culture used here as lead example) actually experienced an abundance of food in their natural environment (not scarcity), so there was no need for a cut-throat battle for it, and (b) abundant archaeological evidence exists which can be interpreted as proof that those were matriarchal societies (which are presumed to be less prone to war), and, moreover, weapon remains are not particularly abundant in these archeological sites. However, these arguments (even if accepted as complete, which is not a given) might perhaps be enough to explain peaceful coexistence in the small-size settlements of that culture, also knowing that this culture included settlements of all sizes, but such explanations do not hold for large and densely populated settlements: there, we have a very different discussion.

We already know that people from Neolithic already had general anatomical and psychological modernity (these features have already been achieved about 50,000 years ago, and here we are talking about people from 5000 to 10,000 years ago). Given this simple fact, assuming that life in a settlement of 10,000 individuals was essentially the same as life in a settlement of few tens of individuals would be naive (and therefore such an argument would be flawed from the beginning). Quite on the contrary, the fact that such high concentrations of people were possible at all is in itself a hard proof under our eyes that people in such large settlements had already achieved a certain, different level of social organization. In addition, not all Neolithic sites had the same level of connectivity with other sites, the most connected ones (in a regional network of sites) having the conditions for more sophisticated social life than the sites that would be more marginal in (lesser connectivity) with a given Neolithic culture. Realistically, these factors cannot be assumed as irrelevant.

On top of that, recent studies using remote sensing technologies are already bringing material evidence for the presence of economic and social status differences between inhabitants, along the internal structure and organization of the settlement.

For example, here is a very telling extract from a study using remote sensing technologies that enabled full-picture study of a newly discovered Cucuteni small site in Romania [20]:

> This internal village structure linked with the topography of the site, with well-defined buildings and open spaces, pathways to the fortifications, fortified elements, etc., involved an impressive amount of labor which clearly emphasizes the existence of a well-established social organization, following the rules established within Cucuteni communities. The results of this study, together with those obtained by our team at other Cucutenian sites such as Ripiceni–Holm, Brăteşti–Dealul Chicera, Fulgeriş–La Trei Cireşi, Hândreşti–Dăiceni, etc., with their varied typology of documented internal organizations, shows how they anticipate the later evolution of the mega-sites and giant settlements seen in the Republic of Moldova and Ukraine.
>
> The existence in our case study of a fortified enclosure with semi-circular ditches and palisades in the core area, and the presence of exterior habitations surrounded by three narrow curvilinear ditches, shows a clear distinction that should be made between defensive elements and ones with a symbolic purpose. In other words, our study documents a novel situation that could provide clarifications regarding the functionality of these construction works, a topic that divides the scientific community and offers occasional ungrounded opinions that are intensely debated in the professional literature. (…) In view of these results, the use of non-invasive investigations in researching Eneolithic sites belonging to the Cucuteni-Trypillia cultural complex now seems mandatory. They permit creation of integrated research results that can significantly contribute to prioritizing the main directions of future work and a reassessment of archaeological excavation strategies.

Then, a second study on a large Cucuteni-Tripillia mega-site in Ukraine (called Maidanetske) [327] indicates significant difference in social and economic status between the dwellings located in different structures of the same settlement (on top of the many interesting details known from that culture, like a rich material and spiritual life and the deliberate burning of the houses at the end of a cycle of 75–80 years and new houses ae being built on top of those). Here are two snippets from that study:

> About 2300 house remains are visible in the magnetic map (...). The magnetic anomalies of the burned buildings are arranged along concentric ellipses around an inner vacant space. A corridor, which is free of houses, divides the settlement into an inner and outer part. The houses that are directly located along this corridor belong to the so-called "ring corridor". Inside this free space are the megastructures, buildings with a different architecture than houses. North, outside of the outermost ellipse, remains of an older settlement are visible.
>
> The buildings are arranged along concentric ellipses around an inner vacant space and a vacant ring. Buildings along this ring corridor have increased total magnetic moments. The total magnetic moment indicates the remaining building material and therefore the architecture. Lastly, architecture can reflect economic or social status. Consequently, the increased magnetic moment of buildings along the ring corridor indicates a higher economic or social status.

This echoes an earlier one [199], which argues based on available evidence (difference between building types and how these evolve in time) that these mega-settlements look like they might have harbored precisely the transformation of social organization from more egalitarian to more centralized and more socially stratified, with potential social consequences in the complex dynamics in the subsequent eras.

Furthermore, even the great similarity between houses (which is sometimes hastily interpreted as a sufficient proof of an egalitarian society), when it occurs, is not necessarily an indication of similarity in social status: it is also a matter of efficiency based on the construction technology available at the time and is to be expected if there is a division of labor where specialized builders lead the constructions (the same way, specialized potters do the pottery at the high level of refinement and extent known from this culture): serial production of buildings by (or under the supervision of) specialized builders which use some general templates (for basic shape and spatial patterns) to facilitate their work and then add or not more features and/or details depending on context (today, we call it "customization").

More broadly, and as argued elsewhere by Blanke and Walmsley [46], in order to properly understand systemic features of a city in given culture (specifically, the resilience and evolution capacity of the cities in the region known as 'Decapolis', i.e., "Ten Cities" in Greek, in southern Levant), archaeological evidence must not be interpreted strictly within the narrow confines of one discipline but needs to be corroborated with evidence from other fields of investigation.

For instance, in our example from Central Europe, we also know that one of the precursors of the Cucuteni-Trypillia culture mentioned above was the Vinča culture (or Vincea in Romanian), which is well-known for the *Tărtăria Tablets* (ca. 5300 BC; present day Romania). These objects display the world's first known and undisputed example of a Proto-writing System, i.e., visible marks communicating limited information, using ideographic and symbolic systems (in this case, along symbolic traditions from early Neolithic).

Although Cucuteni-Trypillia culture itself is sometimes described by archeologists as "static" because of its remarkable continuity, it shows abundant proofs of abstract thinking and great material achievements. One example is the emergence in this culture of the earliest forms of potter's wheel, and the wheeled toys (miniature models of animals and cups) discovered in this culture indicate the clear knowledge that objects can be pulled on wheels, which further suggests this mechanism was used in daily practice (although no evidence conserved well-enough to be discovered in modern times—most likely because these objects were made of wood, and we know that the local climate does not favor the conservation of wood artifacts). The very same is true about textiles: their existence is known in this culture from imprints on pottery, but their perishable nature made that they didn't conserve. And the pottery itself (one of the most important type of evidence from Cucuteni-Trypillia settlements) is one of great complexity of shapes and chromatics. Interestingly, some of the substances/materials used in this pottery are rare and can only be obtained from very few places across this entire area (of over 300,000 square kilometers) and needed to be brought from neighboring cultures further away.

When taken together and placed in a holistic context, such details indicate not only a remarkable diversity of ideas and practices but also specialized skills, which logically implies a division of labor and trade across large geographic areas. Basically, we talk about a complex society which cannot be artificially reduced to local rurality and agricultural subsistence—even though agriculture did (inevitably) play a big role in the sustenance of all known proto-cities and 'first true cities'.

Indeed, the high homogeneity known from the Cucuteni-Trypillia culture is to be expected if local settlements across the entire region are regarded as nodes that are communicating with each other in a network of settlements spanning the culture and further communicating with neighboring cultures. And there is no reason whatsoever to presume that they did not communicate; quite the opposite is true: there is evidence of communication capacity. We know that individuals from hunter-gatherer societies were capable of long-rage walking covering very long distances (hundreds of kilometers) in a relatively short period of time (weeks and months), i.e., humans can do that. Also, the Cucuteni-Trypillia culture was also close to the area where horses have been domesticated (in fact there is evidence of horse remains in this culture but it is not clear whether they were already domestic or just wild horses hunted for food) and we know that they were already using wheels. And, we are talking about a civilization that lasted for thousands of years. In short, communication and interactions (between the settlements located inside a regional civilization) must be recognized as a factor.

Once this recognition is made, any simple network analysis (based on current, well-established methods; see Table 1.3 for a brief introduction) or just a mental exercise using the principles of network analysis predicts right away that (1) some settlements are more central than others and in time they will gradually become more influential (i.e., they will gain a 'higher centrality' within the network), which will further stimulate their growth and influence, and (2) the most influential settlements will serve as cultural diffusion centers facilitating (and speeding up) exchanges and the spread of innovations across the entire culture. In other words, the simple network mechanics results in (1) a "horizontal" gain in homogeneity across the network and (2) a "vertical" gain in heterogeneity inside the strongest nodes of the network. Additionally, each settlement anywhere anytime is a network of individuals interacting with each other; in time, some individuals (or groups of tightly linked individuals, called "cliques" in network analysis) which are more central (i.e., have the highest numbers of connections) will inevitably become more influential because their centrality makes them looked-for because they serve as "shortcuts" between any two individuals in that network of person-to-person interactions. And, logically, the larger a settlement is/becomes, the more structured it will become internally, which makes inevitable role specializations and division of labor between individuals and between groups (i.e., social differentiation).

In this sense, another aspect was observed in the literature precisely about Cucuteni-Trypillia mega-sites: larger population sizes logically imply higher collective computational abilities of these settlements [278]. This is important, because we also know that major innovations and societal changes (the Industrial Revolution from the eighteenth century included) happen in situations where there are many

Table 1.3 Some general approaches to modeling cities as whole systems or/and units

Model type	Main idea
Adaptive cycle	A repeatable sequence of phases ("reorganization, exploitation, conservation, release"; or "transformation, golden age, decay, crisis") as a basic pattern of change-and-stability in the evolution of a city in its context, usually represented as a Möbius band (or a large infinite sign) [129, 176, 202, 444]
Nested networks	These are mathematical graphs: sets of vertices/dots (agents) and edges (links), as briefly mentioned also in Table 1.1; for more about graphs, see also Chap. 3, Sect. 3.2, Box 1.9). Thus, a city can be modeled as a high-density subnetwork (cluster) within wider (regional and global) networks while itself including smaller networks. (e.g., [182, 229, 281, 282]). Many real-world interactions between groups of agents cannot be described only as sums of pairwise interactions, but mathematics give us further tools, e.g., higher-order interactions involving groups of 3 or more vertices: hypergraphs, which are graphs in which an edge can simultaneously join any number of vertices (so, explicit pairwise relations are not needed), and simplicial complexes which include units like vertices, edges and triangles (they are more constrained than the hypergraphs, providing an intermediate level of structuring) (for a discussion on mathematical and physical basics, see, e.g., [26, 27]) Given the huge amount of data and information involved, also relevant is the notion of 'smart city', which refers to computer-aided automation and optimization of functions and processes within cities, which is part of the discussion on eco-cities when reasonable [41, 47, 118, 276, 285, 300, 443], [209] So, in the systems perspective on sustainability described above, each sustainability filter (SF) is definable as a hypergraph, and different interactions within or between SFs can be represented using classical network graphs, simplicial complexes or hypergraphs; then, periodic processes and temporal dynamics can be represented as coupled oscillators
Urban metabolism (practice-oriented metaphor)	A bounded system model (with in-/outputs) which identifies and quantifies relevant fluxes of matter-energy (and therefore also information). This makes a city system describable by basic and applied science (physics, chemistry, biology/ecology) and mathematics as a basis for the further research on various questions of interest (from sociology, economics, engineering, etc.), and for generating knowledge-action models and solutions to multi-dimensional problems. Recently, a related notion integrates smart city ideals: smart urban metabolism [42, 43, 121, 380, 381]

human agents interacting with each other, i.e., an "*emerging social brain*" [405]. In other words, Cucuteni-Trypillia mega-sites and other proto-cities or first cities had a major role in the evolution of humanity to urbanity because they represented "places where things happen". For what this book is concerned, namely system-level urban transformations, this is probably the most important aspect of proto- and first cities.

Put together, the two dynamics mentioned above (internal and external to a given node-settlement), one easily sees how population size and social structuring go hand-in-hand, generating a "next level" type of social organization (see also [441]). Mathematically, one can easily notice about this reasoning that "it's elementary!" Then, in research practice, this same reasoning can help explain the the remarkably persistent cultural commonality (a shared local Hellenistic culture of Eastern Mediterranean) across the Decapolis network of cities studied by Blankey and Walmsley [46] mentioned above, and their common hinterland.

Finally for this second aspect, the point with all this discussion (and in light of the different debates going on in the literature) is not whether a certain ancient human settlement is to be declared "urban" or not. After all, the more we go back in history towards the origin of cities, the harder it is to find abundant evidence (and the more tempted we are to apply present time (and methodological contingent) criteria to those very different contexts) (for a discussion about the issue, see, e.g., [162]). The point (from a system, operational, and holistic perspective, as taken in this book) is that the becoming of a city is about a process of transformation that is led by both internal and external forces in a complex nonlinear dynamic that can see situations of long-term stability and (relatively) fast change at historical scales. It is less about "officially" pinpointing a specific moment-and-location and more about understanding the process of evolution of human settlements in their context because this is what makes a city (and any human settlement) what it is. In this sense, this discussion made here is more in line with the *Functional Model* used in modern archaeology for (1) understanding a city in relation to its hinterland (and context in general) and (2) "to study urban sites from the point of view of evolving social networks" ([274], see also, [44, 388, 433]). For a modern social network analysis of ancient proto-cities see ([156, 157], for the networks of settlements in pre-Roman Central Italy) and ([259], for a case study of internal networks inside a proto-city—Çatalhöyük, Turkey). For reviews of social network analysis (and related complexity science) methods in archaeology, see [61, 269]. For a modern Big Picture perspective of the evolution of cities ("*Where do cities come from and Where are they going to? Modelling past and present agglomerations to understand urban ways of life*"); see [158].

For concrete illustration of the point made above with the example of Cucuteni-Trypillia 'mega-sites', I find it particularly useful to put here an extract from the conclusions from [157] in their study of the evolution of a regional network of settlements which resulted in the emergence of the city of Rome as dominant in its region (*Latium Vetus*) and beyond:

> This work has enabled the definition of a unified index combining all the useful information from infrastructure network topology in a compact expression that provides a single criterion to rank the settlements we have represented as the nodes of a system of connections. (...).

> As already shown in previous work, centrality indexes calculated on the river network perform well only for the most ancient times (...). In this new analysis we can confirm that probably river connections, at least at an intra-regional scale, were more relevant during the Bronze Age than in the Iron Age. Indeed, in this last phase the river network, as a graph with a specific topology, seems to be unable to add anything valuable to the information we can extract from the topology of the road network. This result may be related to the decreasing importance of fluvial connections that are no longer a deciding factor.
>
> From a Network Science point of view, it is probably the low adaptability of the river communication infrastructure that determined the decrease of its importance. At an initial stage, villages having good fluvial connections could have possibly had better chances to develop into larger settlements, but at that point they would have needed even better connections in order to flourish both politically and economically. The network of river routes is an almost static object. Societies and settlements were changing at a regional scale and, in order to fulfill the needs of emerging proto-cities, it was surely easier to adapt existing terrestrial routes. We may suppose that, even if at the beginning there is not a high correlation between the settlement size and the centrality indexes, the sites which grew the most were those whose position was central in both networks.
>
> The greater importance of terrestrial routes in the Early Iron Age and subsequent periods also seems to confirm the importance of the foundation of secondary settlements in between primary settlements during the later Earlier Iron Age. The foundation of these settlements in fact certainly improved the communication network and boosted the development of primary settlements. This interesting hypothesis will be further tested in future work.
>
> The coherence of the results of the network analysis with a convincing historical interpretation confirms the validity of the network approach and opens up new research perspectives. For example a comparison between the Etruscan system and the Latin system might help explain why in the end the smaller but more compact and hierarchical region of Rome prevailed over its neighbors.
>
> It also shows that it is important to tailor the methodology to a given specific historical question. Existing indexes and measures can be useful in an exploratory phase of the research but subsequently it is important to develop a new methodology that builds on already existing tools but adapts them to extract the largest possible amount of information from a given dataset with particular structure and features.
>
> Therefore a double deductive approach is required: specific research questions and working hypotheses have to be translated into Network Science terms. Then the dataset has to be inquired properly and finally the information and the results obtained have to be interpreted into historical narrative.

Thirdly, the city-as-we-know it emerged in the last 200 years and was shaped by the changes that came about with the industrial revolution.

Table 1.2 is of course not meant to (and could not) be exhaustive; it only covers a few general domains. A long list of other angles can be added to reflect the array of specialized fields existing today, plus an endless number of research topics, knowing that each topic can itself amalgamate any number of disciplinary perspectives. In this context, one should also remember that modern concerns about the nature of cities are as old as modern cities themselves, and the literature since 1900 to present is replete with works reflecting those, e.g., from the very early "the definition of a city" [256] and "what is a city?" ([279], see also [377]) to the more specialty-driven inquiries in contemporary efforts to understand multi-dimensionality of a city (and the underlying complexity), e.g., "spatial definitions of the city: four perspectives"

[318] and "the eco-city: ten key transport and planning dimensions for sustainable city development" [222].

These being said, cities are complex dynamic systems par excellence (e.g., [28, 39, 43, 63, 307, 360]). Moreover, through their 'hybrid natural and mad-made' character [330], cities are a prime example of CHANS and constitute a unique 'natural experiment' through which humanity develops new features and possibilities, and the 'urbanizing' process itself is key [13].

Most importantly, from a system (and network analysis) perspective, a city has a double, local-and-global, character: it bears a set of distinct features that reflect its history and dependence on the local biophysical systems in which it is embedded, but at the same time it acts as a unique node in a global network of cities ([91, 101]; see also [38, 222, 307, 336]) (Fig. 1.8).

In the earnest, we are in a period of reflection and experimentation which requires an effort to find the useful commons [377]. The ideal of an 'urban science' is to reach the level of "an integrated theory from the first cities to sustainable metropolises" [243]. And, regardless of all disciplinary lenses, philosophical perspectives or methodological commitments, the complex challenges faced by cities today bring to the fore both "the tension between specificity (or difference) and universality in conceptualizing the urban and in drawing together empirical evidence" and the need for ramping up work "for identifying key interventions that could shift cities onto a more sustainable path" [317].

This has important consequences: being defined by unique sets of flows (of people, matter-energy and information) that reflect local and global realities, a city is both a "development machine" for its hinterlands ([45, 272], see also [122, 195, 417]) and a logical best trade-off between scales that matter for sustainability transitions: a city is small enough to keep projects managerially doable (in spite of the great diversity of actors from all domains and fields of expertise), while also being (still) large-enough to be relevant for the complex system dynamics of the world as a whole [96, 97].

The sheer fact that cities are CDS systems makes them a node linking individual and institutional levels with wider regional and global networks of cities: "cities as systems within systems of cities" [37]. Thus, cities comprise "networks of composite relationships among their internal components and inter-laced networks that cities have with one another" so that "considering a city as a complex adaptive entity implies an integration of inner processes into macro processes on the level of systems of cities" ([355], see also [21, 84, 336]).

In the context of sustainability-aiming efforts, this unique basic property of cities also suggests that the city is an adequate unit for generating and monitoring the real progress in the transition to sustainable development [93,96].

How Can we Hope to Realistically Manage Such City-Scale Projects?

Premises for project and innovation management in general

From the discussion in the previous book sub-sections, it follows that:

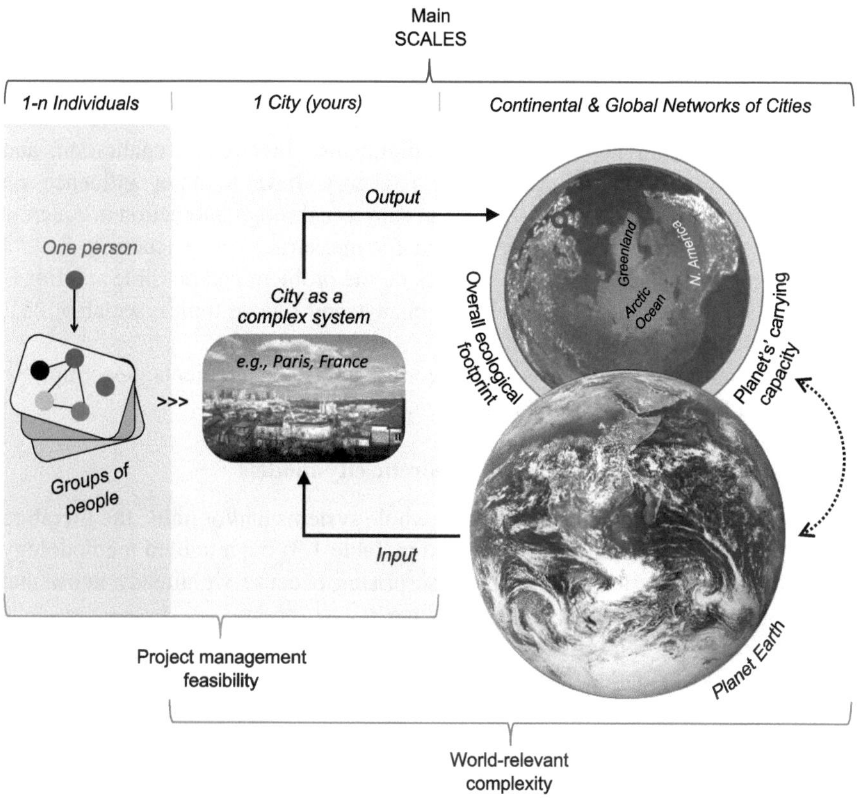

Fig. 1.8 Cities are "windows to the complexity of the world" and a still manageable scale in complex projects seeking to generate solutions for sustainability. *Source* Original, using a photo by the author (*La Défense* business district of Paris, as seen from the train coming from the neighboring, historic city of Versailles) and two modified public domain images: one by the National Oceanic and Atmospheric Administration (NOAA), with light density at night indicating the presence of the technosphere (as inspired by Turner [414]); and the 'Blue Marble' from NASA, taken in 1972, during Apollo 17; the quotation is based on [366, 367]

- City-focused projects may use any of the many field-specific perspectives evoked above or combination thereof. This virtually gives a huge epistemic and managerial diversity between and within projects (even if pruned by real-world limitations). In daily practice, this also means that hoping to find a sufficiently comprehensive range of expertise at one single location is unrealistic;
- City-scale projects undertaking holistic approaches will require interaction within networks of experts from across a diversity of institutional, socio-economic and cultural contexts (even within the same region of the Earth or/and similar levels of socio-economic development);

As per current RDI theory and practice, but also in view of policy developments, this overall reality directly translates into some necessary complementarities:

- between two types of organization: clusters and networks [171, 399];
- between universities, business, and government, i.e., the already 'classic' triple-helix model ([135, 232, 236, 354], for a related discussion about this model, see also [235]);
- between the different world views (individualist, hierarchist, egalitarian, and fatalist), which, according to Cultural Theory, have a major influence on personal preferences and choices but can converge through interaction in concrete projects—constitute a reference system for measuring social learning, i.e., "a convergence of stakeholder perspectives on the problem and possible solutions" [124, 430] (for applications to climate change and related topics, see also [231, 263, 294, 303, 321, 434]);
- between numerous other sets of categories describing gradients, spectrums or circles of factors as identified or identifiable in real systems.

The cognitive start basis offered by theoretic city-models

In terms of abstract modeling of cities as whole systems and/or units, the literature shows different philosophies (summarized in Table 1.3) but a unified methodology is lacking. This situation is in fact not surprising because we already know that knowledge integration is part of the challenge.

Nevertheless, we can see that those approaches are complementary. The first one (the adaptive cycle) marks the need to understand the city as a whole picture in space and time, and the imperative of understanding the city as a CDS. It stemmed from the Panarchy Theory of intrinsic dynamics of the real-world as a system of systems and recognizes that CDS do not (and cannot) "obey" human rules, top-down hierarchical "imperatives" or planning ideals but must instead be "danced with" along cycles of change and stability [10, 176, 202, 444].

This model connects with the debates about man-nature relations (couplings) that preceded both the Brundtland report and the current systemic perspective on sustainable development. It is useful in the study of cities and the conceptualizations of holistic attributes like urban resilience [11, 74, 224, 264, 266, 355, 379] but also for modeling the dynamics of all systems and CHANS that are functionally linked to urban sustainability more generally, e.g., from natural vegetation succession and the dynamics of ecosystems [62, 398] to innovation systems [54] and supply chain management (SCM) [444]. For example, in the field of management, it is known that precisely the moment a private company is most successful is also the moment when it is most vulnerable to complacency and when thinking about failure is necessary (whether this knowledge is applied or not in each company is another matter).

For an institution, this cyclicity means that its own role and impacts in society are subject to temporal effects: a new company may not be well known at a time when it is the most successful in value creation, and, by the time it has become known (it has established a brand), its capacity may have already be outdated. And this applies to all organizations: to use the words of Mihai Eminescu, a late nineteenth century poet, "*It was there when not seen, today we see it and is not*" (original text in Romanian: "*Era pe când nu s-a zărit, azi o vedem şi nu e*", in '*La steaua*' ('*To the star*', 1886), whereby he used the then-already-consecrated notion of light-years to allude to long-term cycles governing human life and society (since 1851, the notion of 'light year' was used as a unit for measuring cosmic distances and their time differentials; the exact speed of light was also known since 1883).

The broader point is that such time-lags and other temporal effects are another source of complexity that must be accounted for in human systems. Some of those can be grasped intuitively and are assimilated by the common culture, which means that they can be easily communicated and used by human communities in their deliberations about challenges and options. But most require more abstract analysis and modeling, first to be detected and then to be factored in by potential solutions—especially those connected to profound and detailed features of the systems but actually having major consequences on the system as a whole. For illustration, one can also make an analogy with human health: certain illnesses can be seen right away (with the naked eye), but most require more detailed analyses with modern technologies and methods to first diagnose then to cure.

In terms of real-world cities, an imperfect but easy-to-understand illustration is represented by the Californian city of San Francisco and its wider neighborhood (The Silicon Valley). Together, they have been a world's innovation powerhouse before most of the world knew about it, and, by the time cities around the world started to try to emulate it (attempting to create their own "Valleys"), this urban hub was already losing its sheen, many companies and people moving on to other places (e.g., Huston, Texas).

At the same time, this cyclicity tells us that a city's time of greatest struggle is also the time when solutions are found, laying the ground for a successful revival. Such was the case in the 90's when the cities (and the countries) from the entire eastern flank of the (current) European Union went through a necessary and painful transition from the failed communist rule and command economy (abusively imposed by the Soviet Union in the context following WWII) to an open society and market economy (and resulting in them joining the EU in 2004 and 2007). The remarkable characteristic of this region—from Tallin (Estonia) and Gdansk (Poland) to Prague (Czechia), Jassy (Romania), and Varna (Bulgaria)—is that of a unique combination of factors (notably rich cultures, extensive industrial traditions, a well-educated human capital, membership in and direct financial/logistic support from the EU, and a renewed dynamism) which create a special and globally unique potential for generating solutions for urban and general sustainability.

Furthermore, the adaptive cycle model acknowledges the reality of CDS depicted in Figs. 1.6, 1.7 and 1.8 and translates it into general practice: it says that cities are not some passive and malleable products of some "demiurgic" will of some individuals but possess a complex systemic character of their own—a set of systemic attributes which must be understood and respected. Relatedly, it also points out that historical ups and downs are natural phases in the nonlinear dynamics of cities as CDS.

In short, the adaptive cycle model is one (though not the only) good start for discussing *urban trajectories that are physically possible*. What needs to happen in a project is a good understanding of the fundamentals relations between what is possible and what is not.

A full array of abstract models of the city would fall outside the scope of this book which is dedicated to scenarios and related project management. However, the third module of this book touches, inter alia, on abstract modelling issues that are related to the systemic perspective on eco-cities and highlights the fact that any "smart city" approach (from computer sciences to industry) must build on the holistic foundation provided by the systemic perspective on sustainability.

Grounds provided by the current knowledge: the fundamental laws of physics

Technically, thermodynamics is positioned to be the most suited discipline for the study of complex systems because it is both the most fundamental field of science (quasi-all known physical phenomena and processes in the known universe can be explained in terms of thermodynamics (e.g., [449]) and it is probably the most structured discipline [32]. This is reflected in the wide application of thermodynamics to different fields, starting with the most obvious:

1. life itself, as a phenomenon, is a manifestation of the second law of thermodynamics [79, 374];
2. thermodynamics provides the core of basic ecology and of its first applications in landscape ecology (see [119, 168, 198, 221, 290, 335,382);
3. energetically and informationally, cities are dissipative structures ([58, 150, 319, 344], later on, see Chap. 3, Sect. 3.4, Figs. 3.10, 3.11 and Table 3.2); and
4. basic system analysis in sciences and engineering is grounded in thermodynamics and the mathematical tools from statistical mechanics (see, e.g., [1, 2], but the literature out there is immense and many relevant references can be found).

Illustrations from everyday life can be found everywhere because energy and entropy (and, relatedly, the basic notion of information) are at the origin of everything and pervade all research areas, in both natural and social sciences (and humanities, although this is usually less visible because the methods tend to be very different) (e.g., [63, 350, 407] and the references therein).

One lead example (because it merges all perspectives) is the field of management: thermodynamics is the ground zero and starting line for a fundamental, all-encompassing theory of decision-making, either directly or indirectly through the applied theory behind our understanding of complex dynamic systems and how they connect with the relation between possible and impossible, as well as between knowns and unknowns (see e.g., [97, 305, 394] see also the knowledge-action matrix in Chap. 3, Fig. 3.39), through the process of search and discovery, and the related phenomena represented by cognition itself, then technology, behavior, habits, culture, institutions, learning, innovation, and so on.

With regard to cities, beside the aspects mentioned above (and which influence cities in many ways), thermodynamics is being applied in high-quality urban planning ([151, 432], see also [350, 391]), and as related to what has been called "the complexity turn in planning" more generally, knowing that cities are participants in and beneficiaries of global supply chains [384], and also keeping in mind the importance of nature-based solutions as a systemic concept (not just "greening") in any form of urban planning (e.g., [136]). Of course, this is not meant to (and cannot) imply that a city is the sole result of planning; quite on the contrary, this use of the most fundamental laws of physics is a form of insuring objectivity in the pursuit of human aspirations (and, of course, studying and using the laws of nature is an obvious way of working with nature not against it).

Relatedly, thermodynamics is also applied in architecture and the so-called 'urban design' [51, 55, 63, 238, 350, 425]. Importantly, this notion is not to be confused with (or taken as a justification for using) the naive idea that cities can be "produced by design", i.e., contradicting reality and the established fact that cities are complex dynamic systems. Worryingly, such expressions can be seen in the literature (especially from disciplines which are not primarily scientific) and are still circulating "freely" in non-evidence-based policy circles in some countries; but, fortunately, this knowledge-action gap is in principle easy to mend by cooperation and practical learning. For the purpose of this book, I would only highlight that talking about "designing an eco-city" or "building an eco-city" only reflects a lack of familiarity with the topic of eco-city beyond the immediate interest or type of narrow specialism. For the reasons already summarized in this book (and as other discussions that will follow will also illustrate) one cannot "build" a city (let alone an eco-city) because a city is not equal to a number of buildings and appliances. And you don't "design" a city, because a city is not some serial product you manufacture and sell. But, yes, one can design and build some small parts that can become part of a city (and this kind of semantic clarifications must be made as soon as possible in eco-city projects).

Finally, numerous other interesting perspectives have been proposed in the academic literature, with various degrees of influence but, overall, benefitting from less attention than they would deserve. In Box 1.2, I only mention a few examples for brevity reasons, and I invite you to explore more of them in the literature.

Relatedly, different issues that are known intuitively are also modellable (in terms of physics and related complexity sciences) for better study and understanding, e.g., the inter-connected issues of 'metropolisation gradients' (in the wider discussion of rural–urban gradients and size and types of human settlements) as reflected by scaling laws (e.g., [149, 333, 359]), resilience at the level of entire regions (not just a city or metropolitan area, but a city with its hinterlands or a network of cities and their common hinterlands, and from different perspectives, starting with economics) [23, 320, 400], 'the fractal nature of resilience' as reflected in the interactions across 'micro', 'meso', and 'macro' scales [35] and by the scaling laws mentioned above, all within the long-hauled efforts to understand resilience and sustainability within a systems perspective (e.g., [68, 148]).

Perhaps less obvious (or less intuitive) but certainly not less important (nor less thermodynamic in nature) are the dynamic aspects of these complexities and of a city as a complex dynamic system (interacting with other complex dynamic systems). For example, a particularly interesting and informative phenomenon is the 'counter-urbanization' that was observed in most of the developed world since the 70's, i.e., "a movement from a state of more concentration to a state of less concentration" caused by "growing disamenities", i.e., premiums that "have to be paid to do business in larger urban agglomerations, where the social costs and environmental burdens of urban living are greater" combined with the fact that "new technologies are compressing time and space and accelerating change, lessening the need for face-to-face contact because of instantaneous electronic communications in the new age of the computer", the result being (1) a reduction in the "traditional" patterns of pressures on the local environment and (2) the emergence of 'polycentric' global networks [38]. Further, this is how the cited author sums it up: "*One way to characterize the resulting urban systems is as dynamic networks. No longer are vertical hierarchies arranged regionally into heartlands and hinterlands; instead, the "cores" are centers of creativity and entrepreneurial activity wherever they may be located, and are linked into transnational networks.*"

Box 1.2 Thermodynamic perspectives employed for understanding hard issues in different fields of study and issues of daily life

- understanding industrial revolutions as results of network effects of scientific discoveries, technological innovations, and information processing inside networked individuals, groups of people, and between hubs of industrial activity [405];
- cognitive models such as the "entropic citizenship behavior and sustainability in urban organizations" [110];
- the notion of "entropic disorder" as a research approach in environmental sociology [260, 261];
- system of systems models as CDS-based frameworks for modern governance for future sustainable societies (e.g., [68]);
- numerous studies in the field of ecology (which includes thermodynamics as part of its fundamentals anyways) make explicit reference to it when discussion topics in sub-fields like urban ecology and human ecology, and related to concrete topics like urban sprawl and other phenomena at the interaction between urbans systems and their bio-geophysical environment (e.g., [5–7, 99, 102, 161]);
- the related studies regarding CHANS (already discussed and referenced in this book so far, starting with Sect. 1.2);
- and the use of entropy and the basic laws of physics to advance the field of economics in general and with respect to humans-nature relations [15, 160, 164, 165, 353, 436]

The bottom-line is that the Second Law of Thermodynamics sits at the center of our scientific basis for the development of urban systems. In thermodynamic terms, the notion of *urban entropy* is entropy applied to cities as systems and refers to the amount of disorder in a system. It is also definable as 'information content' or 'the tendency of a system to descend into chaos'. For separate introductions to thermodynamics for cities, see (e.g., [63, 319]); for further discussions that extend to related aspects to urban and general complex systems, see (e.g., [28, 29, 58, 150, 331, 344, 345, 402]).

Now, as entropy is at the core of the study of all CDS, entropy-based models also exist in the study of cities as CDS. This is a first-rank technical support for management: the purpose of such models is to clarify (as soon as possible in the course of a project) what might be possible at all, and what can already be identified as a dead-end and a waste of resources. The basic entropy-based model in Fig. 1.9 is a simplest abstract representation of this idea (the possible and impossible "state-spaces" also connects to the notion of status of a CDS in Fig. 1.6).

One immediate, direct connection between entropy and nature-based solutions for sustainable cities and society is the notion of "low-entropy city" (here abbreviated as LEC in reference to its total system entropy): an urban system which imports lesser amounts of natural resources and exports less entropy (in the form of waste and pollution) into the environment. This is the opposite to a "high-entropy city", i.e., the "traditional" city which imports lots of natural resources and exports lots of waste. In everyday language, LEC is a low-disorder city, and HEC is a high-disorder city. This means that, for a city, undergoing a transformation from HEC to a LEC is not only a way of becoming less chaotic, more refined, and less resource-intensive

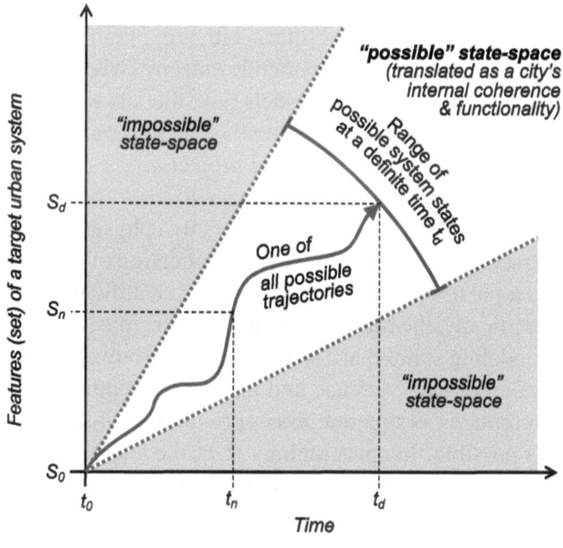

Fig. 1.9 Evolution of a city as a complex system in the realm of the possible (order vs. chaos), as defined in physics and information science, i.e., urban entropy. In this illustrative idealized model (an adapted, basic 'path-space model with increasing time horizon', here abbreviated 'P-SMITH'), a city's evolving set of features is represented as successive sets S_0, S_n, and S_d denoting 'initial', 'intermediate' and 'desired (target)' at different moments t_0, t_n, and t_d in the city's life-time. A state-space is the sum of all possible states in which a system can be, as reflected by all potential combinations of features (without discussing yet the detailed dynamics between those features). For simplicity, the borders between possible and impossible state-spaces are shown as straight lines (not implying deterministic outcomes). *Source* Redrawn starting from [150]; see also [29, 293, 319, 446]

and less dependent on those high amounts; it is also a way for that city to grow its capacity and expand its horizon of possibilities for the future.

Physically speaking, a straightforward way of doing this is to avoid artificial features that are unnecessary, i.e., by incorporating nature-based solutions as much as possible, thus doing less forcing of the natural biogeochemical cycles [319]. Here, a word of warning is necessary. For some people (especially those unfamiliar with scientific language, the built environment and man-made things in general might sound like "order" or "high order" while nature itself might sound (or be preconceived) like "disorder". However, man-made things can be high-order or low-order and nature also knows both high-order and low-order. The desirable situation is therefore defined as a combination of the high-order (or low-entropy) man-made things and high-order (or low-entropy) natural things. Put differently, a low-entropy city (or low-entropy system in general) is made from a combination of "what nature does best" and "what humans do best". Thus, traditional cities have a higher entropy than they could have because they include (too) many mad-made high-entropy components and (too) few low-entropy natural components (Table 1.4).

In a nutshell, becoming a LEC is part of a city's journey towards sustainability. For projects, this means (1) that urban sustainability models must incorporate systemic entropy (estimations) as an operational filter—either explicitly (when feasible) or at least intuitively and (2) that estimating systemic entropy (when possible) is a way to distinguish good sustainability-aiming models (i.e., models which stand a chance in the real world) from bad models (i.e., dead-ends which, if pursued, are a sure waste of resources which are taken away from good projects).

At this point, another observation is necessary to avoid potential confusions. In the paragraphs above and Table 1.4, entropy refers to a physical reality described by basic science (thermodynamics). But one other application of entropy is in the study of how science (as a phenomenon) works and how scientific research itself is to be made efficient. Specifically, the Maximum Entropy Principle (MEP) which sits at the core of scientific modeling states that a scientific model (built by researchers in order to be tested against empirical evidence and previously acquired knowledge), must agree with data but remains as structureless as possible. In this, "maximum entropy" means: as loose as possible. In epistemology (i.e., the study of science itself), this requirement of parsimony is defined as "maximum noncommittal with regard to

Table 1.4 Main situations of interaction between natural and human systems of high/low entropy

Principal entropy levels in cities as hybrid manmade and natural systems		Entropy of NATURAL systems and components	
		High	Low
Entropy of MANMADE systems and components	**High**	**High-Entropy City (HEC)**	Natural systems degraded by heavy human activity (ecological footprint)
	Low	Human systems restricted by unfavorable natural conditions	**Low-Entropy City (LEC)**

Bold words indicate the key situations

missing information" [97, 109, 211, 212, 383]. In the rest of the book, I will mention again these aspects when some clarifications are needed (see especially Chap. 3, Sect. 3.7, where I summarize the scientific method in relation to knowledge, and Sect. 3.8, where I discuss a stepwise model for developing solutions, e.g., a policy, a technology or anything else).

First consequences of fundamental science in projects and their management

To put the basic model in Fig. 1.9 and the notion of entropy in the broader perspective of humans-nature relations, the modern civilization of the twentieth century was driven by aspirational mottos like "the sky is the limit", concretely materialized by cities through famous 'skyscraper' buildings and unique 'city skylines', and even in space—by building the International Space Station, visiting the Moon and sending robotic probes on Mars (with the real perspective of building permanent human settlements on Moon and Mars) and further exploring the outer space.

This is about a deep human aspiration to know, one that is embodied by culture and popular ideas such as "to boldly go where no one has gone before" (a famous motto of *Star Trek: The Next Generation*, a cinematic science-fiction series that uses imagination and dramatic scenarios to explore the limits and the potential futures of humanity [24, 126, 314, 370]).

However, this upbeat "we can do anything" spirit does not imply that great achievements happen by doing "just about anything". For cities, fascinating constructions and skylines does not mean that buildings are all there is important about cities. What matters is the "unseen" process behind those great achievements, very often including great sacrifices. Instead of lingering on superficial perceptions, projects aiming at high goals have to meet some basic requirements (necessary but often insufficient conditions) for success.

These being said, the urban transitions to sustainability (and I am using the plural to highlight the multitude of identities and individual city endeavors involved in this shared aspiration) will definitely represent one of the greatest feats (or, given the complexity and the existential nature of it, even *the* greatest feat) of those local communities and of humanity as a whole. But this unprecedented challenge grew from all the complexity of human and natural systems and is not about knowing some key secret ("silver bullet") or knowing and controlling everything.

This formidable challenge is about us finding and using a realistic, open-ended dynamic equilibrium between structured-enough but not too-structured action (i.e., planning as much as possible but not more than that). Planning less than that means dangerous chaos; planning more than that means dangerous naivety (and then even greater chaos).

In engineering and management, optimization is the pursuit and protection of an optimal combination of factors (while undergoing/operating necessary change, typically trying to maximize desirable parameters and minimize undesirable ones). Optimization at system levels aims at avoiding sub-optimal (not enough) and the over-optimal (too much) situations, both with regard to single parameters and combinations thereof.

In many fields of science (from systems and evolutionary biology to cognitive psychology to sustainability and urban studies), finding a viable place to be between opposite and often hard to understand/predict dangers ("just right, not too hot and not too cold, not entirely certain and not entirely ambiguous") is often called the Goldilocks principle, after the name of character in a tale for children ("Goldilocks and the Three Bears") from English-speaking countries ([415], see also [81, 107, 349]). In science, where a complex dynamic system needs to probe and anticipate most favorable likelihood, the idea has also been called "the art of the probable" [166]. In RDI, a 'Goldilocks system' means "acceptable performance requiring not too much data (system not too expensive) to prove acceptability" [422]. Finally, in natural sciences and sustainability studies, a 'Goldilocks zone' can refer to the fact that planet Earth is not too close to but not too distant from the sun (and thus can be not to hot and not too cold for complex life to exist). On the Earth itself, it can refer, to a habitat that is particularly favorable for a high number of species (and thus a rich biodiversity) [115], or, in the context of climate changes, it can refer to the fact that since before the emergence of our own species, the concentration of carbon dioxide in the atmosphere was not too low and not too high (just right for a mild greenhouse warming effect), and, therefore, the excessive emissions of greenhouse gases (primarily carbon dioxide resulting from burning fossil fuels) represents a form of climate forcing that directly pushes the biosphere out of this equilibrium.

Before Discussing Eco-cities, Another Question Requires Attention: What Are Cities from a Crisis Management Viewpoint?

Taking a necessary step back to see the whole: a recourse to urban history

This book is meant to be a practical tool and a good occasion to stop for a moment from the daily rush, to look to the broad picture: in many senses, a city is a big place on a small planet. The literature contains different studies that offer to shed light on the big picture, and you might already know some of those.

In a spirit of dialogue, and in light of the discussion about thermodynamics above, I can also suggest two such papers, one focused on cities and aptly titled "Rethinking vulnerability in city-systems: a methodological proposal to assess urban entropy" [150], and one focused on decision-making based on the most realistic possible fundamentals and bearing an equally adequate title: "Thermodynamics as a theory of decision-making with information-processing costs" [305]. Each of them can make a good theoretical transition between these two book sections.

In the remaining of this subsection, I offer a third way, one that connects natural and social sciences and humanities around the observation that crises are moments when a city is being tested and therefore allows in-depth insights about what it is, how it works, what it can do, and what it needs. Put differently, the resilience issue is a cognitive door to all the other features that make a city be what it really is, and therefore a practical handle to open concrete conversations about urban futures. This is what I will do in this section. Later on (Sect. 3.4), I will use current knowledge and experience to identify and locate urban resilience in the operational context of all holistic (systemic) features possessed by a city aiming at sustainability.

First off, let me start from the reality that different people may not share the grand ideals discussed in university and policy circles, simply because they may not have the practical liberty or facility to engage in such conversations. For those of us who cogitate upon our place in the universe, yes, "our planet is a lonely speck in the vast and enveloping dark", "a lovely fragile little planet" which, photographed from the edge of our solar system, appears as a "pale blue dot" whose importance "lies only in what we make of it" ([361, 362], see also [287, 288]). For most people, however, the Earth is an overwhelmingly large abstraction (as opposed to their daily reality): not a single place but a long list of foreign places. Sure, people may understand in theory that this is "our dot" in the vastness of space, but a 'world crisis' is still something vague occurring somewhere else, beyond the powers of a mere mortal concerned with *hic et nunc* ('here and now') daily priorities of survival. For him/her, only those crises are actually real which involve well-defined events and effects taking place at locations and times that could touch him/her.

These being said, contrary to ideas like 'the world' and 'our common future' which transcend everyday life, a city is truly 'here and now' for anyone to see, to experience and, crucially, to appropriate. Therefore, cities have a uniquely important role that comes from their local-and-global character evoked earlier: they (a) connect us to each other and (b) provide us a concrete place-in-the-world. In short, "cities indeed are our emerging futures" [417]. On the long run, as already observed by many, not only has humanity been advanced in the past by cities as operational and civilizational hubs, but even the so-called "battle for sustainable development" will most likely be won or lost in cities (e.g., [14]).

To illustrate, I propose two examples, corresponding to two different-but-similar time perspectives: from past to present and from present to future. For the past-to-present perspective, to keep it concise, I will simply make recourse to archeology and its experts. For the long-term perspective, I will keep it short by bringing up a synthesis made by archeologists (Box 1.3), the main point, from a combined archaeological-and-resilience regard on the old cities being a convergence to the idea that cities and their fate (from the beginning to the end) are the final systemic outcome, emerging as a unique product of (i.e., more than the sum-total of individual) interactions between people in-and-around the city, and between those and the general natural and made-made environment of the city.

Box 1.3 Archaeology, resilience and sustainability. Illustration with snippets from [46]; see also [341–343]

> In archaeology, resilience theory serves as a potent companion to a 'longue durée' approach, given the ease by which resilience theory, as an explanatory agent of causes and consequences of change in communities and societies, can be applied universally in bridging historical and archaeological research. In breakthrough publications of the early 2000s, Charles Redman championed the role of resilience theory in defining and describing the evolution of cultural landscapes in archaeology, stressing the positive interplay between resilience theory and a longue durée approach in historical analysis. Resilience is, as Redman laid out in a 2014 publication, the ability of a system to

manage harmful events through change without breaking the essential qualities of that system—'function, structure, feedback capabilities, and therefore identity'—thereby safeguarding social consensus. Change, when resulting from the application of adaptive strategies, is as normal a state as stability, meaning stability is in no way a 'better' or more 'normal' condition to that of change. Overlapping with landscape studies, resilience theory has become firmly welded to environmental science and, within that, considerations on the ability of social-ecological systems to adapt to sudden and potentially catastrophic events derived from both natural and human causes

Classical cities were the product of cultural and political systems devised by the people who lived in and near them; resilience theory brings us ever closer to an intimate understanding of these community processes in the gestation and long history of those cities. Of universal value, the information generated by resilience theory has the capacity, validity and clarity to identify and evaluate degrees of success or failure in previous attempts to resolve historic social problems in ancient and medieval societies, modelling outcomes which can inform in the planning and implementation of effective behavioral responses in the race to ensure the endurance of the world's socio-ecological systems.

The relevance of resilience theory and sustainability science to understanding evolutionary processes in cities between antiquity and the Middle Ages is acute and jammed full of research possibilities. Redman, a practiced archaeologist and sustainability expert, has advocated for the intrinsic value of 'past experiences' as revealed in the archaeological record, with the past being nothing less than a vast 'laboratory for innovations'. This ever-expanding laboratory is not only a container for a historical archive of almost limitless human experience, but is also instructional by offering an original deep-history perspective on how to deal with the multifarious challenges of today. Here we see archaeology and the sciences going forward collaboratively, a two-way street of collective experience that illuminates not only the tangible 'hardware' of cities, but also their 'software'; that is, the urban cluster of human capital and the impetus cities can give to creativity and ideation. Yet the opportunity to build dynamic, cross-disciplinary interactions of mutual benefit have not yet eventuated as standard: 'just as resilience and sustainability practitioners have made poor use of history, those who do focus on the past and other cultures, such as historians, archaeologists, and anthropologists, have poorly framed their knowledge as useful to addressing today's challenges'. Achieving social value outcomes will rest on what questions are asked when change is considered; not just identifying the causes but also what archaeology might tell us about the mechanisms of community decision-making, what these change decisions were, and what alternatives were or could have been considered and why these options were not deemed feasible. From this more inquisitive and interpretive stance, practitioners of history, archaeology, and ecology would be well placed to 'make a well-understood past serve to create a more genuinely sustainable future and increase human resilience in the face of both gradual and sudden change'.

These being said, another extract from this work is worth for an epistemological insight, namely that sustainability studies (in general and urban sustainability studies in particular) not only benefits from contributions from all academic disciplines but the reciprocal is true: it has the potential to help academic research re-invent itself and pass to a next, superior level through theoretical and applied coordination on issues of common interest, e.g., via real life-inspired questions and operational concepts related to resilience and long-term (or *long durée*) perspective on cities as complex human systems.

For example, in a multi-disciplinary study of the evolution of the urban landscapes of a local network of classic Roman cities (Baysān/Scythopolis, Fiḥl/Pella and Jarash/Gerasa) in the Decapolis region (around Lake Tiberias in Southern Levant; currently split between Israel, Jordan and Syria) from late antiquity to earlier Islamic times (4th to tenth centuries), with a comparison of 'before' and 'after' the natural disaster (earthquake) from the year 749, [46] observe a trend which, in light of the current knowledge about human systems as CDS, is not at all surprising but illuminating from a multi-disciplinary viewpoint: before that big earthquake, those three urban societies were already striving towards a reset for a new era and therefore the shock generated by the earthquake did not create a vacuum but simply accelerated the transition to a new socio-economic organization, albeit according to their local details.

Further, this multi-century analysis provides a methodological insight about the value of multi-disciplinary regards, one that is of general interest for our society as a whole: there is a practical need for all research fields to have a critical regard upon themselves and move forward by reinforcing interaction across traditional divisions between disciplines. Here is the original sum-up by the authors:

> We contend that resilience theory provides a methodological structure to explore ideas more broadly and to compose new and contemporary narratives previously constrained by institutional partiality and academic reserve. It provides an uncompromisingly positive research model with which to produce rigorous results not only for archaeological delegations, but also as the basis of mandatory interaction with other fields of academic research including, but not limited to, environmental studies, urban sustainability, social inequity and economic disadvantage. New eyes, honed with data breadth and sound theory, have challenged old paradigms, and discovered fresh narratives with which to explain the past for the future.

Looking around to see the present: the context represented by climatic and broader environmental changes

As for the present-to-future perspective on cities, let us remind ourselves that a key issue within the unsustainability crisis is climate change: not only because it is a planet-scale phenomenon with direct impacts on people, but also because it is, by its fundamental nature, "the ultimate risk multiplier" [280], an aggravator of all the other problems straining people-nature relations: from various types of environmental degradation to their social and economic consequences (e.g., [8, 100, 155, 206–208, 423]).

Through climate lenses, a city is both a source of carbon dioxide and other greenhouse gases (GHG) which have causal contributions to this phenomenon, and a principal, multidimensional generator of viable solutions for climate change mitigation (and adaptation thereto), in a large cloud of themes and connections with sustainable development (Fig. 1.10). In this sense, the importance of cities is reflected in key numbers: the world's cities are responsible for about 90% of the global GDP and about 70% of the global GHG emissions ([391], the IPCC [206, 207] reports; see also [226]).

About cities, the most recent IPCC report on climate change mitigation [208] points out that "following a similar trend that appeared with *Agenda 21*, for which many cities adopted local plans, a growing number of cities have introduced Voluntary Local Reviews". These are similar to the Voluntary National Reviews (VNRs), except that they emphasize local actions and needs with respect to SDGs [306]. By 2019, 150 countries had already developed national urban plans (at least nominally), but less than half were in the implementation phase—and, of course, the success of those is another discussion [416].

As a highlight related to the required systemic transformations, I would mention that the same IPCC report [208] observes about cities (in their section C.6): "Urban areas can create opportunities to increase resource efficiency and significantly reduce GHG emissions through the systemic transition of infrastructure and urban form through low-emission development pathways towards net-zero emissions. Ambitious mitigation efforts for established, rapidly growing and emerging cities will encompass (1) reducing or changing energy and material consumption, (2) electrification, and (3) enhancing carbon uptake and storage in the urban environment. Cities can achieve net-zero emissions, but only if emissions are reduced within and outside of their administrative boundaries through supply chains, which will have beneficial cascading effects across other sectors (very high confidence)" (and the text is followed by cross references to internal sections of that same IPCC report giving more details, i.e., in: 8.2, 8.3, 8.4, 8.5, 8.6, Figs. 8.21, 13.2).

At this point, one terminological observation may be necessary. In practice and in some parts of the academic literature (especially but not only in social sciences and humanities), the term 'trade-off' seems to be often loaded with negative connotation, and a synonym with 'compromise' (another term which needs not but is often used with similarly negative semantic loads), basically implying some loss or a situation of "giving up on high standards". I do not subscribe to this semantics, and I actually use the bare meaning from science proper, where a trade-off refers to an objective equilibrium, and a sheer result of a combination of vectors (forces) of different sizes and directions. (Then, whether this result is desirable or not is a different matter and depends on what is being discussed).

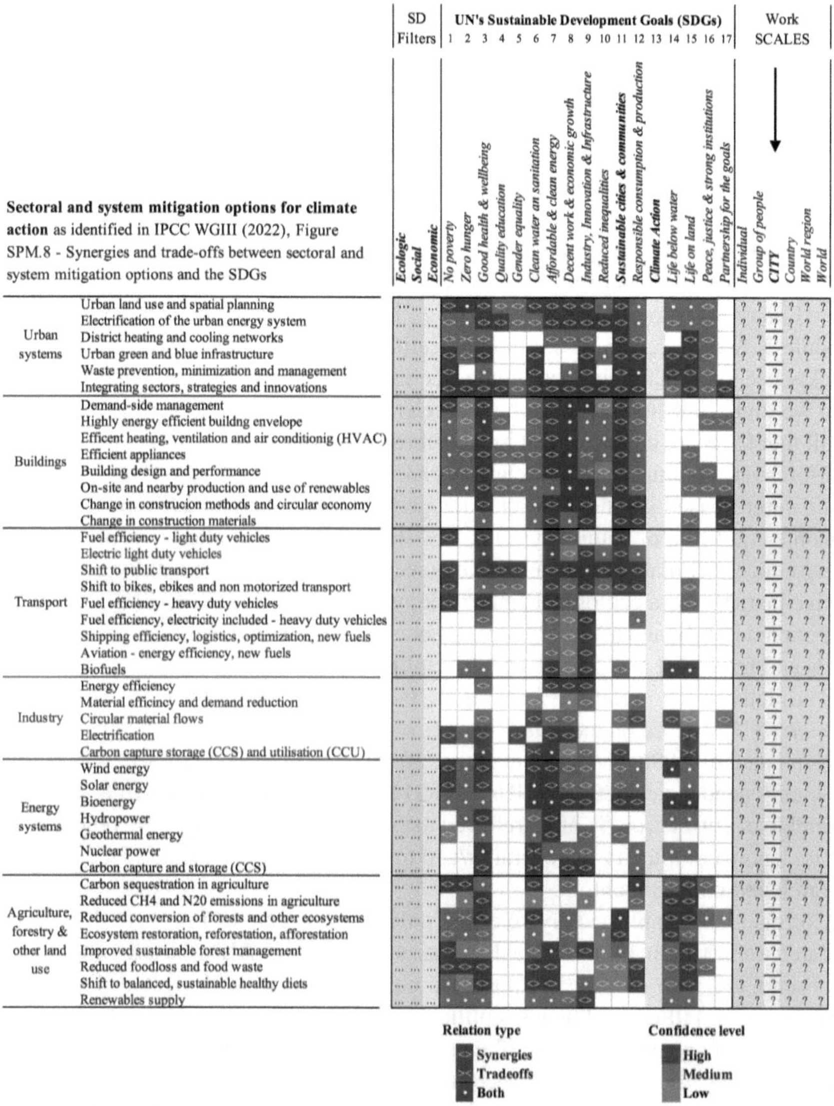

Fig. 1.10 Synopsis of outcomes of interactions inside complex systems. This is an illustration with the synergies and trade-offs that are known to exist between the mitigation of climate changes and the sustainability goals, as seen through sectoral topics. Each cell denotes an ongoing research space. Empty cells = current knowledge gaps ("unassessed yet"); "…" = technical lists (e.g., indices, indicators, parameters); "?" = knowledge-action options to be determined. Red color is used to highlight the text lines that explicitly refer to cities. *Source* Modified based on the lists of categories identified by the IPCC (Figure SPM.8 inside [208]; see also IPCC-WGII [207])

As pointed out by IPCC in their recent reports and as reflected in Fig. 1.10, current evidence shows that, objectively speaking, we need to find the relevant trade-offs in order to avoid both the worsening of the environmental degradation and the social (collective) paralysis, and the consolidation of a vicious circle between the two. For example, in the case of climate change, this is the rationale behind recent propositions that is necessary to use some technical trade-offs between the short-term (20 years) and the long-term (100 years) objectives related to the Global Warming Potentials, i.e., GWP20 and GWP100 [298]. Similarly, this is why we need combined (and coordinated) sets of policy measures aiming at both mitigation of and adaptation to climate changes (e.g., [100, 208, 385], see also in this book Module 3, Sect. 3.2, Fig. 3.6). Finding useful trade-offs is not only desirable but it quickly becomes an urgency, especially if climate change will actually happen faster than we think [448]—in fact, such an acceleration of issues is no surprise when we recall that CDS are characterized by non-linear dynamics.

Moreover, the imperative of relevant trade-offs is also an implicit call for prioritizing nature-based solutions, as they integrate human planning aspirations and the dynamic equilibria of natural systems [98, 102, 186], and similar trade-off requirements concern climate change and health (see, e.g., [410]), which call for using integrative insights from ecological anthropology (e.g., [116, 152, 228, 304]), the study of 'coupled human and natural systems' models [239, 240 and other interdisciplinary fields and from past and present transdisciplinary projects in general.

The plethora of correlations from Fig. 1.10 is also captured by the UN's *New Urban Agenda* (NUA) based on the *Habitat* series of conferences (*H-I* in Vancouver, 1976; *H-II* in Istanbul, 1996; and *H-III* in Quito, 2016) (e.g., [73, 401, 418, 419, 426]), and the interdependencies between NUA, SDG11 (cities) and the other SDGs. Such an ensemble of correlations is often referred to as a 'nexus'—a metaphor people tend to use to denote complexity in general or/and to capture detailed abstract models from research and management.

Correlations established by science (indeed, each cell in Fig. 1.10 is a large research topic) help us make realistic assumptions for ABM-based simulations of urban CDS with empirically observed categories, e.g., with System Dynamics Analysis (SDA) illustrated by a simplified model of a disturbance (Fig. 1.11).

But, even if we make more simplifications by only permitting a limited number of main states (e.g., on/off) and compact rules of interaction between vertices (like cells in a grid), the resulting computational devices (cellular automata) and patterns still show unpredictable long-term outcomes and dependency on initial conditions (e.g., [12, 365, 386, 393]). Overall, system-level failures inside a local social organization can result from several related types of situation, as summarized in Table 1.5.

Also, the double local-and-global character of cities discussed above means that urban risks are at the core of "globally networked risks" (from natural disaster risks to financial and economic crises), which can be addressed through network analysis as briefly described above and as detailed in the extant literature [159, 193, 214, 326, 329].

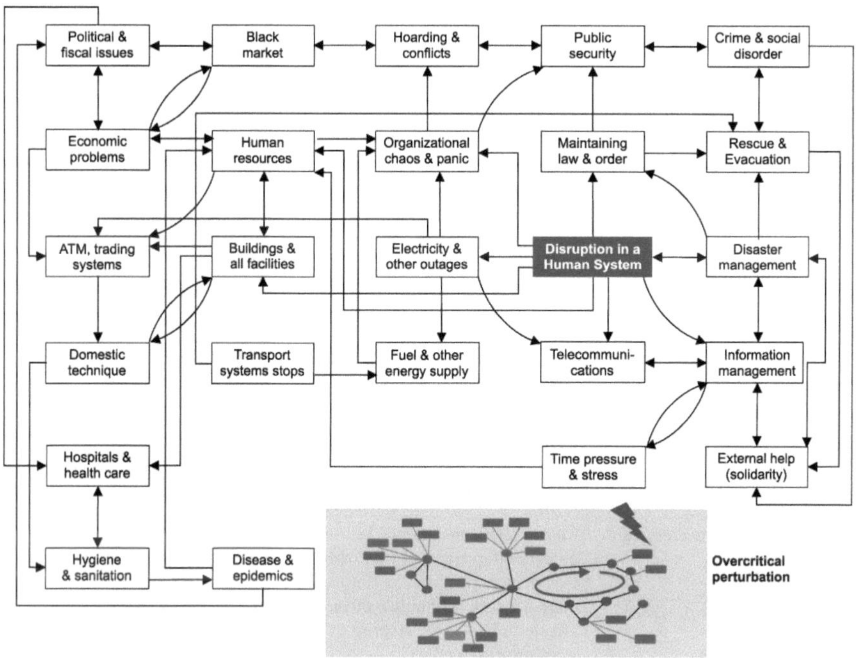

Fig. 1.11 Cascading effects of a crisis in a socio-economic system like a city: an SDA model. This illustration displays a still incomplete (for simplicity) set of interlinked causal loops between issues as seen from a crisis management viewpoint (on white background). In more detail, arrows can be labelled as positive (+) or negative (–) impacts, and the relation strengths are quantifiable as numerical (measured or estimated) weights that vary in time (and between chosen combinations of endogenous and exogenous factors). A perturbation becomes overcritical (disruption; in the grey field), if it is self-reinforcing via positive feedback loops ('vicious circles') and hits on sets of interdependent equilibria; this triggers a cascade of release (/re-equilibration) events: a number of components (or the whole system) switch from equilibrium states (shown in green) to new states (red). *Source* Redrawn based on [192], see also, e.g., [413]

This overall complexity must not be discouraging. The above-described relations between stability and instability in CDS relate to, and provide a scientific basis for, many other intuitive models known from the literature. They help us understand why cities tend to have unexpected trajectories in time, and how we can go about it. The core specifics of a knowledge-based stepwise approach to complexity will be discussed in the next modules of this book. In this section of the book, I only focus on the main ideas for starting the ball rolling in eco-city projects.

To give an illustrative example, SOC helps explain the fact that the 'release' in the adaptive cycle from Table 1.3 is sometimes (in the absence of some external trigger) manifested as a slow loss of function, but some other times (under the influence of some unexpected factor) is manifested as very sudden, chaotic and hard to control dynamics. Further, the concrete situations can be very diverse, and everyone can try and match these concepts and figure out what can explain different dynamics that

Table 1.5 Situations in which a complex dynamic system (e.g., local, national or broader) can find itself destabilized

Situation	Explanation of the idea, with examples from real life
Self-organized criticality (SOC)	Key features insuring system stability are being driven to a "critical point" (or "tipping point"), i.e., a systemic stability threshold beyond which the behavior of the system becomes intrinsically unstable (see, e.g., [59, 154, 371, 395]) In human systems, an example is the fall of communism in Central Europe (CE) in 1989. Previously, societal tensions accumulated and burst here and there without systemic effects until some spontaneous street protests started from a church (St. Nicholas) in Leipzig became "the drop that spilled the glass", triggering a chain of events that resulted in a multi-scale systemic shift Systemic tipping points are also studied in the Earth's climate systems (and biosphere) and their relation with human systems (notably via mitigation of and adaptation to climate changes), as reported by science and summed up by the International Panel on Climate Change [206, 207, 208] (see also Fig. 1.13 in Sect. 1.4); and in the field of management sciences, as part of the overall dealing with systemic complexity and uncertainties [305, 394]
Metastability + an overcritical perturbation	The system is metastable, i.e., robust to small perturbations (which quickly disappear over time), but this kind of stability is disrupted by an over-critical perturbation (for example a natural disaster), which has consequences on multiple other parts of the system (than those directly affected). Beside the surprise of the event itself, such a disruption can have surprising effects For example, an earthquake that caused a tsunami that had hit the Japanese coastal city of Fukushima on 11 March 2011, knocked on a nuclear disaster which entailed global political debates about nuclear energy, that resulted in civil nuclear energy being officially abandoned in Germany on 30 June 2011 (the last three nuclear reactors have stopped producing electricity on 15 April 2023—even as this book's manuscript was being finalized [25]). Similarly, the Chernobyl nuclear disaster in 1987 had led to Italy abandoning its nuclear power plants. At present times, the kinetic hostilities around the nuclear power plant in the town of Enerhodar (as part of the war that started on 24 February 2022 between Russia and Ukraine) have already generated global concerns about nuclear risks posed by armed conflicts
Metastability + co-occurrence of several small disturbances	The system is metastable, but several perturbations (in the network nodes or links) co-occur and their interaction is overcritical, triggering additional operational failures elsewhere in the system [65, 143, 203, 325, 424] This risk is particularly high when system nodes are linked by dependency links. As most disasters result from a combination of human errors and context [83, 120, 248, 316, 340], in today's networked world, the question is not whether but when such coincidences will happen; and how to manage them. In the unsustainability crisis, this notion is encapsulated by the idea of a looming "perfect storm": a convergence and reciprocal potentiation of unresolved systemic problems which, at a point, will leave us with no choice but witness a systemic collapse (and painful chains of hard consequences)—unless we get our act together and achieve a real transition to sustainable kinds of socioeconomic development before this has become too late [31, 146, 372]

Source: modified based on and the citations therein [192, 193]

we observe empirically or/and analytically in different cities. Here, the purpose is to simply introduce the reader to the different dynamics at play in urban systems and synthesize some practical conclusions for eco-city projects—not with the intention of being exhaustive, but with the twin purposes of (a) understanding the basic big picture and (b) getting eco-city dreams and projects started on a correct basis.

Discussing about city-scale projects: a selection of key insights

A first practical conclusion to be drawn from the technical insights selected above is that (1) every city must avoid the situations where its space of options is restricted, but that (2) disasters can have exactly that undesired effect because they undermine or/and erode the capacity of a city to function, and, in so doing, it restricts its horizon for future developments. However, (3) whether the city goes up or down after being hit by a disaster depends also on other attributes of the city, starting with how its people see themselves in the world—and this is not just rhetorics. In general, human societies engaged in civilizational development (and starting with cities as civilizational hubs) can be understood as emerging 'social brains': a form of self-organized swarm intelligence that generates solutions to identified problems [405]. And, cities themselves are—physically speaking—a form of manifestation of self-organization in nature [10], and, as such, are interconnected (coupled) with the biophysical systems around them (i.e., in the form of CHANS), this being why, to use the everyday language, we know since a long time that "systems of people and nature are not just social and ecological systems" [442], which means that addressing social and ecological systems as if they were "completely separate" (and because this appears easy from a traditional academic perspective and would spare us from the effort of learning beyond the comfort zone provided by narrow areas of specialized expertise) is (a) technically incorrect (in terms of our current knowledge) and (b) pragmatically not only counter-productive but also very dangerous in terms of our goals of generating real (not naively imagined) solutions for sustainability on Earth; if we fail, we don't have a planet B.

Further, the opposite of this necessary openness and learning-based progress is constituted by the negative attitudes towards knowledge that is perceived to be "external" to a given organization (the internal work environment). This is known (from practice and the literature) as the Not-Invented-Here (NIH) syndrome [17, 220]. Needless to say, this is a damaging phenomenon because it restricts productive interactions across relevant experts and actors in any organization, narrowing the horizon of that constituency of people and making it prone to internal vulnerability (tension buildup and release as described above in Table 1.5).

This relates to another managerial issue (disregard of the reality that knowledge is distributed and therefore can realistically be mobilized not "centralized"), known from corporate business but relevant for all large organizations: the Headquarters Knows Best (HKB) syndrome [53], and in general an attitude that overprioritizes "leadership" [427]: "the scalability of lessons learnt from these action networks is hampered by too strong a focus on leadership by the network administrators".

This attitude, in turn, is related to the double-sword issue of authority in public administration, manifested through "a natural fixation to focus on hierarchical notions of authority" which "is at odds with recent non-hierarchical aspects of modern-day cooperative roles and relations with respect to citizens that are meant to be serviced by such authority" [68] and may even undermine the epistemological and political foundations that are needed for the more enlightened perspectives on and forms of governance in future societies [137].

Thus, as resilience reflects how an urban system (and, the society in general) reacts with respect to risks and crises, and "urban resilience efforts must consider social and political forces" [138], addressing adequately the issue of urban resilience is paramount for finding ways forward in urban transformations towards better futures. This means that the current problems we see today with authoritarianism around the world might not be a mere contrariness of democracy (as it is usually perceived) but a deeper, systemic symptom of a tension between, on the one hand, an already outdated hierarchical form of social organization based on brute force, power and dominance abstracted as formal authority of central and centralizing political entities, and, on the other hand, truly cooperative forms of social organization that are emerging in the current era of internet (used as local and global infrastructure for efficient networked communication) and in which authority is not hierarchical but collective (or "collegial") and defined through the natural purpose-driven dynamics and equilibria at work in complex dynamic systems—a perspective that allows us to think about a "system of systems" type of governance [68] in which (a) individuals assume higher responsibilities and (b) the societal need for central-and-centralizing institutions is reduced to some flexible mechanisms that prioritize function over form and thus minimize the more "traditional" (and increasingly more problematic) features of public institutions like increasing over-burden, inefficiencies and frictions. In fact, some indications of the evolution of public administration in that direction is already visible with current phenomena like (1) digitalization of the public administration and the increasingly interactive forms of communication between public institutions and citizens (in a word, it is matter of e-governance [72, 123, 300, 378, 409]), as well as (2) a growing societal demand for competence, transparency, and accountability in the public administrations.

In addition, the social and political life (and the public life in general) is more and more taking place in and through online media, as these modern technologies and services overcome geographic (and temporal) distances and bring people closer together. The same is true about science, and expert-interactions in general, through the phenomenon of 'brain networking' from the fields on innovation and world development ([88–90, 95]; see also [48, 64, 80, 139, 234, 246, 284, 302], itself related to the 'Internet-of-People' (IoP) paradigm—a concept that only begins to reveal its potential in terms of expert networks and connecting local communities [96, 97, 451]).

In Module 3, Sect. 3.5, I present a concrete model for how cities can apply 'brain networking' to reach out and benefit from global expert networks. In here, I only make the point that current technology overcomes distances and no city is ever alone—quite on the contrary, a city can bring the global expertise to it through project-based involvement of experts from around the world, starting with its own Diaspora but not limited to it (in fact, Diaspora members can act as facilitators and catalyzers in such projects, and universities and other RDI institutions could facilitate the entire process, provided that local authorities and project managers convince those Diaspora members that they are serious and capable of meaningful co-work).

In other words, in expert networks, real work happens through distributed interactions triggered by commonly agreed goals, not by vertical commands (which, if applied will only chase experts away from projects). This can occur, for example, when the serious need for a sustainability transition of a city is expedited with "one-size-fits-all" (OSFA) or/and policies that are uncritically imported in given local (national or urban) contexts. As [387] puts it, "effective urban development and planning exercises must be highly context-sensitive. There cannot be one-size-fits-all policies. Disengaged, top-down planning approaches of the past have been shown not to work in numerous contexts". This approach is often used by national and local governments, especially in countries that are governed in a top-down fashion, and because those governments see the idea of "building new cities" as a "path-breaking" opportunity ([213], see also, e.g., [145]).

But more generally, the idea of "new cities" appears to many as a chance to "show leadership" despite the fact that they do not understand the networked and complex dynamics they are dealing with [and which are easily visible for anyone who studies networks and tries to find realistic ways to act within them (see, e.g., [30]). Typically, those would-be "leaders" are only motivated by some idea of positioning themselves socially (and understand "networks" just in some intuitive-but-inevitably superficial terms of public relations, such as 'networking' in the sense of "knowing many people" and "getting influential", i.e., without any real understanding of, or interest in, the science of it). It is important to understand that this observation is not some rant about non-scientists: incorrect understanding of terms and phenomena is a real problem for real practitioners. To illustrate, let me compare with another overused term that is directly relevant for our discussion here, namely "disruptive innovation"—not least because the terms of 'eco-innovation', 'eco-city', and other notions related to the transition to sustainability are often understood as, or closely dependent of 'disruptive innovations' or other 'disruptive' change. Here is what the father of the term (in the mid-nineties) and his colleagues that studied the evolution of this term in time (including the numerous mistakes and misuses) have to say about it [87]:

> A related issue is overuse of disruptive innovation/disruption as a synonym for any new threat (or substantial ongoing change) and underuse of disruptive innovation as a theoretical concept. Many popular writers invoke disruptive innovation to describe any new technology or startup that aims to shake up an industry and alter its competitive patterns; previously successful incumbents facing difficulties or going out of business are routinely said to have been disrupted [86]. Conflating disruptive innovation with any generic threat (and ignoring its more precise theoretical meaning) creates two potential risks. First, when the core ideas of prior work are obscured by indiscriminate use of its terminology, researchers will face difficulty building on and extending that work. This risk is especially pronounced in this case, given the widespread invocation of disruption-related terminology in academic journals, practitioner-oriented publications, and books in multiple disciplines. Second, practitioners who rely on incorrect or misleading renditions of disruptive-innovation theory may

be tempted to apply faulty ideas, reducing their chances of success. Given the contingent nature of disruption theory, applying a one-size-fits-all solution is a particularly egregious mistake.

In the same vein, one note from the literature [427] sums up the rather frequent problem of misplaced (and cognitively naive) leadership in terms of what happens in reality in 'action networks': "*their limitations relate to the difficulty in scaling up the knowledge generated. This is partly a consequence of too strong a focus on leadership within the networks—which itself is driven by the involved cities' eagerness to be seen as global leaders in low-carbon urban development. Perhaps less ambitious local action networks will, paradoxically, result in better scalable knowledge. This makes for a captivating question for future study*". In other words: learning and "transferring" knowledge in cities and related locally relevant organizations is not realistic if it is hastily understood as some sort of automatic general recipe, or as a topic of convenience for political maneuvers.

In a cognitively realistic perspective, the matter of urban sustainability is about real insights that can assist in what I call '*adequation*' to local specifics, and which involves studies, adaptations and choices made on the basis of a good understanding of the target context, of the problems and of the potentials. In absence of that, our world/the global network of cities will end-up as a mere collection of inadequate "solutions" that will worsen the initial problems and will threaten the common goal of sustainability of humanity on Earth.

In terms of the systemic perspective on sustainability, this means that a necessary condition to be met in the social sustainability filter is *adequation* (vs. *inadequation*) of what is decided to be implemented in view of attaining sustainability of humanity on Earth. That is, solutions in the transition to sustainability need not solve everything but must be logically compatible with the transition to sustainable development and then with the preservation of development sustainability (which is also to say that many of the solutions for the transition itself will have to be different that the solutions for maintaining sustainability because the two situations are different and their processes are inevitably different).

In all this, there is no safe 'trial and error' for us to play with, so a realistic approach must begin from an understanding the difference and the range (space of possibilities) between 'fail-safe' and 'safe-to-fail' [4].

In any case, a responsible approach is one based on the understanding that, innocent or deliberate, mistakes have consequences (which pull us down), while the existence of some mistakes is also inevitable (in other words, we must prevent them as much as possible but also expect them, and be prepared for the consequences).

Thus, the optimal approach is neither doing just about anything (to see what happens) nor doing nothing (to avoid mistakes), but one aiming at minimizing the losses (opportunity and transaction costs included). This means disciplined experimentation and requires the development of adequate methods and procedures [96]. In the last book module represented by Chap. 3, I describe the heart of it, i.e., a framework model consisting in progress along a scale of solution readiness. But right now, suffice it to say that, when it comes to urban systems, developing solutions requires us to develop a minimum of practically relevant abstractions, i.e., constant interaction between practice and theory through iterations of an 'Observation-Model-Experiment'-repeat cycle (alongside the literature and the experience from other projects/work clusters and networks). This allows us to become good-enough at 'experimental epistemology', i.e., the 'epistemology of the urban', and thus become good-enough at answering the question "*Through what categories, methods and cartographies should urban life be understood?*" [56, 113].

In the formal terms of operations research (which is a field of mathematics) and the related specialties in information research and computer science, answering this question involves a succession of three general steps: abstraction (identification of categories), modeling (understanding how those categories are coherent and relate with each other), and analysis (deducing further consequences of the model for the reality under study).

All in all, the details above show that the multitude of possible pathways along which cities can fall and rise again (bouncing back to a lower, similar or better level of potential and performance) depend on the initial and subsequently acquired properties of those urban systems. Ergo, to insure prosperity on medium- and long-term, a city as-a-community-of-people would be ill-advised to pursue "survival". Instead, it must deliberately aim high: active value-creation by constant self-evaluation, development of meaning and identity.

A local community (together with friends in the global network of cities, e.g., through twinning programs or/and thematic networks) needs (a) to cultivate and protect its space of options until it reaches sustainability and then (b) to continue to develop its overall ability to learn from own and others' experience. Sustainability can be lost again (with unfathomable consequences), but historical pathways of cities can be very instructive, if well-studied and understood: continual learning keeps the horizon fresh and the potential options close (for more discussions, see e.g., [223, 267, 363]).

Moreover, a crisis is not only an unpleasant situation. It is also an obvious occasion to finally address some systemic issues around which it would be hard (to impossible) to federate interests during "normal" times. In other words, crises can—if managed intelligently—serve as platforms to accomplish what has been called more optimistically (and practically) "bouncing forwards" [253], i.e., not just "bouncing back" to the old or previous equilibrium but making a leap to "a new state of balance and harmony" [190, 191].

Of course, one must keep in mind that the notion of resilience is a "double-edged sword. Because it is linked to so many aspects in human systems (and CHANS), the concept of resilience has acquired a "multitude of uses" [140] and this very fact turned the concept of 'resilience' too (like it happened with 'new cities' and 'disruptive innovation') into a "must-have" buzzword in discourses loaded with ambiguities that complicate the understanding of the term and make it hard to translate into concrete operations [66, 82, 301, 346, 357]. One of the authors cited above [190] summed it up like this: "*once revolutionary, now a buzzword, resilience is a concept that is difficult to transform into an integrated tool that stakeholders accept and adopt.*"

At this point, one observes that (1) the corruption of terms in all kind of 'battles of buzzwords' (for a discussion see, e.g., [196, 295]) is a recurrent problem and has become a research topic in the relations between science and society in general—precisely because this constitutes a supplementary problem and a major challenge complicating the entire process of evidence-based public policy. (In fact, this situation even obliges me to mention it again later in this book, with regard to some other issues). However, (2) the fact that a perfectly legitimate word such as 'resilience' happens to be overused and often abused does not mean that we should stop using the word: it only means that it must be consistently accompanied by proper contextualization and clarificatory definitions.

For our purpose here, let us remember that the understanding of the basic concept of resilience by different experts varies but it boils down to three ways of regarding resilience: (a) as a system property, (b) as a process, and (c) as an outcome [275]. Consequently, different cities/communities may regard one of these perspectives as "the" meaning of resilience (or a meaning that is taken as sufficient for them)—but, in general, the systemic view is dominant, since the other two are in fact logical results of the systemic dynamic of the cities [283].

Finally, one also observes, a truly resilient system is capable of both stability and adaptation-evolution [187, 406] and therefore the main management challenge of large projects in cities can be regarded as the challenge of insuring an ambidextrous capacity to combine exploitative and explorative activities [249], i.e., a city must transform itself without stopping to function as a city. Essentially, this is part of an operational philosophy of dynamic equilibrium that incorporate concepts like "city as experiment" [249], "ambiguity as opportunity" [250], "urban experimentation" [142], and "city as permanent experiment" [219], all within the broader goal of experimentation for sustainability (see e.g., [70, 369]).

Now, the richness of ideas and interpretations around resilience (as discussed above) did serve (and still do) as a unifier between different domains of knowledge—but not always in the desirable direction. As already observed in the literature [191], when faced with ambiguities or even contradictions between objectives and guidelines of resilience, the relevant stakeholders, experts, and actors are confronted with "grey areas" [131], i.e., a conceptual vagueness which favors a so-called "political reappropriation" [22, 33, 71, 134], which, because the notion of resilience has become a political and management imperative, it morphed into "a political and crowd-unifying tool" that can be used by incumbent actors for purposes other than the building of resilient systems, i.e., it can be actually diverted towards the consolidation of the status quo and towards the re-enforcement of those business-as-usual practices that undermine resilience [33].

Also important is that the literature on urban resilience points to a long list of "dimensions" (and relatedly, purpose- or topic-driven definitions) (e.g., [227]) that are similar to the list of perspectives or angles on the city given here in Table 1.2 at the beginning of this module section. In this sense, and considering all the complex effects discussed above, a city is also a concrete and useful window into the unsustainability crisis itself (and, in so doing, towards a space of potential options for addressing it).

Now we all understand that not everything can be anticipated to satisfactory levels from a resilience perspective, but what matters in such situations is the general capacity of a system to mobilize and reorganize itself. To pick a somewhat exotic example, geomagnetic storms—which are shock waves caused by solar winds, i.e., flares of charged particles (electrons, protons, and alpha particles) released from the corona (upper atmosphere) of the Sun, as they escape Sun's gravity because of the high energy-levels of fluctuations in the Sun's corona—can cause some radio and electrical blackouts at local/city levels for which it would be hard for the society to prepare against (e.g., [251]). These are extremely rare events and the response to such challenges are mostly technical, but what matters is whether or not the technical details fit into the overall system to generate a desirable outcome.

In terms of sheer network analysis of a system (as per Fig. 1.11), all of these mean that, by their nature, a crisis breaks some business-as-usual connections, and, in so doing, unlocks a space of options (for some new and better connections to be made) which did not exist before the crisis or would not exist (or would have demanded a wide agreement on widely unacceptable goals and efforts).

In brief, a city cannot possibly avoid all crises, but neither can a city afford to "waste a good crisis", because every crisis signals the potential location of a tipping point inside complex dynamic systems (e.g., [154, 207, 208, 371, 395]; relatedly, you will see also Fig. 1.13 in Sect. 1.4) and thus can be left develop into bigger, cascading crises, or it can be understood for what it actually is from a systems perspective: a functional window of opportunity inside existing systems to break away from the business-as-usual inertia and to make a leap forward on the path to a better institutional organization (starting with the definition of roles and actions, in a logic of societal evolution based on current knowledge and capacities, and preserving the previous civilizational achievements, starting with human rights and democratic government based on the respect of those human rights). Nor should a city "sit and wait" for crises to happen, as crises can push the city in undesired directions (unless the city really finds itself in a systemic lock-in and there is no other choice but to wait for the next crisis to open up some opportunity for change). Limiting our attention to "shocks and repair" would amount to choosing ignorance and deluding ourselves with ideas of returning to some "before" situations—a dead-end and a costly mistake (see also Chap. 3, Sect. 3.4 with a sub-section about the role of innovation).

Obviously, different city-windows (to the complexity of the world, as per Fig. 1.8; and to the complexity of the problems, as discussed above) do not give the same view or leverage, but they do offer a realistic access to it. They oblige the relevant actors to grasp the inherent difficulties of complex projects as quickly as possible and to identify those pathways (and to distill those processes) that enable coherent, productive conversations between all actor and stakeholder types ([78, 94, 210] and the remainder of this book).

1.4 What Is an Eco-city?

The Main Idea

The current literature on the matter ties cities to 'sustainable urbanism' aspirations (e.g., [43, 337, 351, 352]), but there is no generally agreed operational definition. Most importantly, the intrinsic systemic complexity is acknowledged but remains largely unaddressed (e.g., [91, 413]). Up to a point, this is understandable as these works come mostly from social scientists; natural sciences still representing a tiny minority in this effort. Of course, there are immense volumes of works that come from natural sciences and are directly relevant to the connection between cities and sustainability, but most of these works are being carried out in disconnection from the explicit effort for transitioning from cities to eco-cities.

More generally, as an epistemological observation, natural sciences seem to show a certain sluggishness in using holistic approaches, compared to social sciences, and I hurry to note that this is not necessarily a critique (and I am myself, *inter alia*, a natural scientist): we all know that different fields have different internal dynamics that must be understood and respected. For example, the traditional view on ecology and ecosystems takes the human presence off the table and studies how nature works in the absence of human "interference". On the contrary, social sciences have been at the front of adopting a holistic perspective on the relation between us and nature, not least because such epistemic issues fall in the domain of social sciences themselves to begin with, but also—as already discussed in the introduction of this book—because social sciences (which includes economics) respond to more urgent questions in the general order of priorities that people have. So, in a sense, the difference between the grand domains of study and learning reproduce the core patterns of interests of the human society and should be understood and managed accordingly.

However, in these disconnections and the lingering epistemic divisions, there is one very practical point of convergence which I will use here in the discussion about a common, operational work platform for holistic projects: social sciences too refer to S-shaped curves (describing typical transitions between CDS steady-states), as universal patterns of change in natural/human systems, especially in research fields concerned with innovation and world development (e.g., [85, 323, 324, 356, 437], for a detailed review of the basic physics behind it, see e.g., [305]).

Figure 1.12 and Table 1.6 show a synthetic model of current knowledge on cities and sustainability, as a basis for a practical-and-logical definition of the eco-city, one which I estimate to be able to defend based on my own experience in complex transdisciplinary RDI projects, entrepreneurship, policy-making and in managing graduate programs in triple-helix contexts bridging of natural and social sciences.

One important observation in Fig. 1.12 is that the cities that are currently estimated to be the most advanced eco-cities on the path to sustainability are still in the first half of the way and have not passed that critical systemic point of inflection; for this to happen, and thus to pass the current progress to a higher level, we need new and more potent methods (so, it is not a matter of 'political will' or any other will, it is a matter of intelligence, starting with knowledge integration and relevant RDI connecting science and society, and involving all major sectors in the triple-helix of academia, business and government).

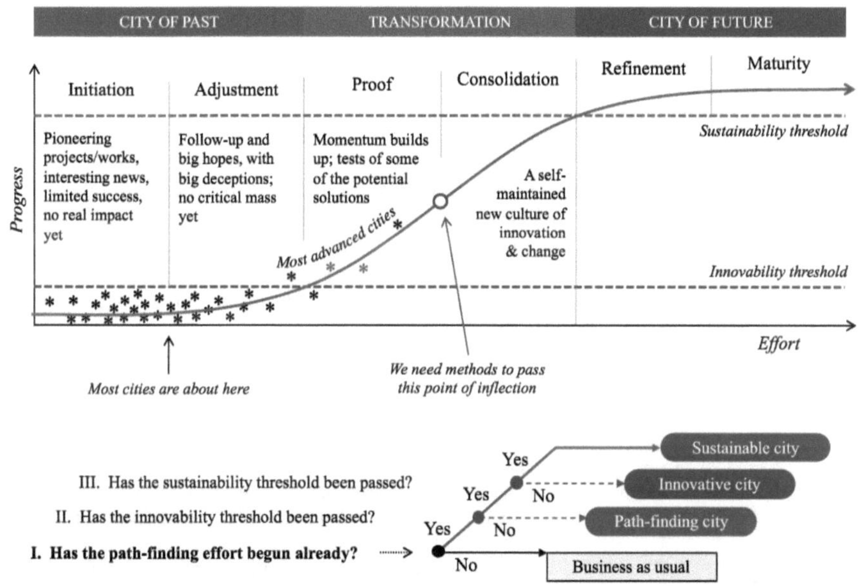

Fig. 1.12 General overview model of the non-linear (s-curve) evolution that can be expected for any city seeking sustainable development, and the fundamental scenarios derivable from it. The two plateaus (*City of Past* and *City of Future*) correspond to two different phases (steady-states), while the slope between them depicts a phase transition (i.e., systemic change). More detailed definitions of the stages along the curve are provided in Table 1.6. The decision tree and scenarios set at the bottom intuitively anticipates (and connects with) the scenario-generation method described in the next module. *Source* Original

Table 1.6 General steps (from 1 to 6) in evolution of a city towards sustainability

Phases and stages			Main characteristics of each situation
City of past	1	Initiation	Pioneer initiatives and projects have started here and there by small teams, with some initial resources, but this seems to have no major impact yet: there is no significant change in the business-as-usual Issues: interesting first insights but no big success; fragmentation of available knowledge; weak understanding of systemic complexity in the target city; insufficient experience with complex projects
	2	Adjustment	Follow-up projects are being pushed forward based on some initial lessons, notably a better recognition of city-system complexities. This leads to new great hopes and mobilization of new resources This is a critical time, with a mixture of exciting results and serious deceptions. The city suffers from insufficient momentum to escape inertia and to pass the innovability threshold, in spite of all efforts
Transformation	3	Proof	New endogenous developments or/and contexts generate a critical mass of actors wanting the city to change. A shared goal has been hashed out (not without controversy); new resources are provided Experiments are carried out to try and innovate how the city really works as a whole. Some solutions or/and components thereof are being proven. The possibility of change is clear; investments grow
	4	Consolidation	Change dynamics builds up via positive feedbacks between results, collective learning and investment in the target urban system, and novelty is being regarded predominantly as a (testable) advantage Efficient conciliation of known public and sustainability priorities. There is a sense that the city is in a process of "reinventing itself": a more capable community, better life quality and a reborn identity
City of future	5	Refinement	There is still room for improvement but the city already managed to find a basic equilibrium between the priorities of its people and the global requirements of sustainability (solutions, as per Fig. 1.3) Most local actors share a desire to keep this newly found equilibrium on the longer term, e.g., via cogent collaboration. Even the laggards are (at least) adjusting in reference to the new dynamics
	6	Maturity	Now the city essentially knows how to live sustainably in spite of eventual urgencies or surprises. So, it is more "relaxed" about the future, focusing on nurturing abilities and exploring possibilities Life goes on and the future is always open. As humanity advances, cities around the world learn from each other and continuously update their profile (role & status) within the global network of cities

Based on this synthesis, we can have a basic definition: *an eco-city is a city that has demonstrably commenced its evolution towards sustainability*. Consequently, an eco-city project is any project that materialize this commencement and evolution.

A minimal/weak interpretation of this definition would be that an eco-city is a city that has begun its own path-finding towards sustainability. A strong interpretation would be that an eco-city is a city that it has begun to transform itself on the path to its sustainability, i.e., it has already passed the innovability threshold, and thus it has proven its engagement-and-capacity to explore its potential sustainable futures. Technically, the ideal interpretation (according to which an eco-city has also passed the sustainability threshold) is not applicable because no city can become sustainable alone i.e., while the rest of humanity is not. Nonetheless, the ideal itself must remain in place as a collective goal for the global network of cities (as a necessary condition for the sustainability of humanity on Earth) as per Fig. 1.8.

In other words, the engagement of an individual city as an eco-city is a necessary (but not sufficient) condition for attaining sustainable development because of that double local and global character of every city. Beside 'doing their part' (or their "homework, if you will, in an expanded sense of this metaphor), the most advanced cities must also have enough influence to help other cities advance as eco-cities—because we are in this 'spaceship' boat together. Currently, our common goal is to have most or the bulk of cities in the global network of cities (representing most of the urban population of the Earth) pass the inflection point located right in the middle of the transformation phase from Fig. 1.12.

Brief Discussion on this Definition's Backgrounds and Implications

An eco-city is a force for good and a real generator-demonstrator of systemic solutions for the whole of Humanity

One notices that the general phase transition model in Fig. 1.12 shows the change from a lower to a higher steady-state, i.e., one with a higher level of complexity, which for us translates in a superior level of civilization. But it should be obvious that, technically, and based on the historical patterns of societal transitions in past civilizations, it is also possible that the new equilibrium situation resulting from a phase transition reaches a lower (not higher) level, which would translate in a worse (not better) case for a city and for Humanity as a whole. In some academic and tentative foresight circles intended to inform policy, these alternative possibilities are also known as forms of ("barbarization" through) "societal breakdown" [339, 205], and different wording of the possibility of a rough future often come from different specialized perspective but have the potential to be used in a more generalist sense. One example is the notion of "disorderly transition" that comes from the financial sector and primarily defined with a focus on the financial aspects of the inevitably complex ("chaotic") transition to sustainability [26, 27]. Another example is the notion of "missed opportunities", i.e., the problem represented by the fact that we already have a history of missing options for climate change mitigation (and adaptation) because of societal inertia and vested interests existing in those sectors of the economy that had (and some still have) more to lose from the transition to renewable sources of energy—which, given the central role of energy and climate in our socio-economic life, further translates in a loss of options for attaining the UN's SDG [206, 207, 208].

Nevertheless, in this discussion, different themes are actually converging rather than diverging, as especially visible in participatory approaches directly concerned with real-world situations and action targets (e.g., [3, 205]). Even if undesirable, such a breakdown is of course a real possibility; in fact, basic physics tell us the this is the objective default outcome (and base scenario), i.e., the inevitable result unless we get our act together. Since we failed all major opportunities so far (and we seem to do too little too late) this base scenario means, that we are heading towards disaster as we speak. Anyone having a basic understanding of complexity science and complex dynamic systems (as summarized in this book so far) can see that collapse is a logical end product of crossing series of tipping points (corresponding to systemic functional thresholds [371], see also [154]).

However, technically, the so-called "fatal synergies" between local effects of climate changes and other ecological phenomena (like local drought or cooling, harvest failures and food shortages, epidemic outbreaks, etc.) and effects inside human systems (e.g., violence within or between polities that are systemically vulnerable to disruption) may or may not lead to complete collapse—it depends on whether or not some sufficient adaptation pathways have been found by those human societies. Some vicious circles can be hard to break, like "from poverty to disaster and back" [180] and can only be addressed by systemic, multi-scale thinking.

And, cities have a key role to play in these interactions, both locally (e.g., through efficient mobilization to overcome crises in their hinterlands) and, through network of cities sharing certain functional common denominators, at wider scales across countries and continents (e.g., [429], see below at the discussion on networks of cities). This is why our actions require a rigorous evidence-based understanding of (past and potential, future) societal responses to climate changes (e.g., [128]). In that sense, we must take seriously the need to exercise realism and address the issues identified from the historical patterns in the interaction between humanity and nature, starting with the power of the "business as usual" which tends to be underestimated again and again.

In Fig. 1.13 I tried to illustrate the context in which cities themselves have to operate—as global events unfold and with respect to the UN's goal of limiting global climate change to 1.5 °C, knowing that national governments need to work closely with cities to be able to achieve their national GHG emission reduction targets.

Since many years already, the overall situation in which the global network of cities finds itself is describable by two key observations (at least for now, things are not improving) [391]: (1) "conventional urban planning approaches and capacity-building strategies to tackle increasing vulnerability to extreme events and growing demands for a transition to a low-carbon economy are proving inadequate as the need for large-scale transformation to achieve a 1.5 °C-warmer world becomes evident"; and (2) the so-called 'Global North' and 'Global South' have different spectra of goals and they operate differently with respect to both challenges and opportunities. In general, "traditional planning" is insufficient and needs to shift to the integration of our thinking about current realities and objectives.

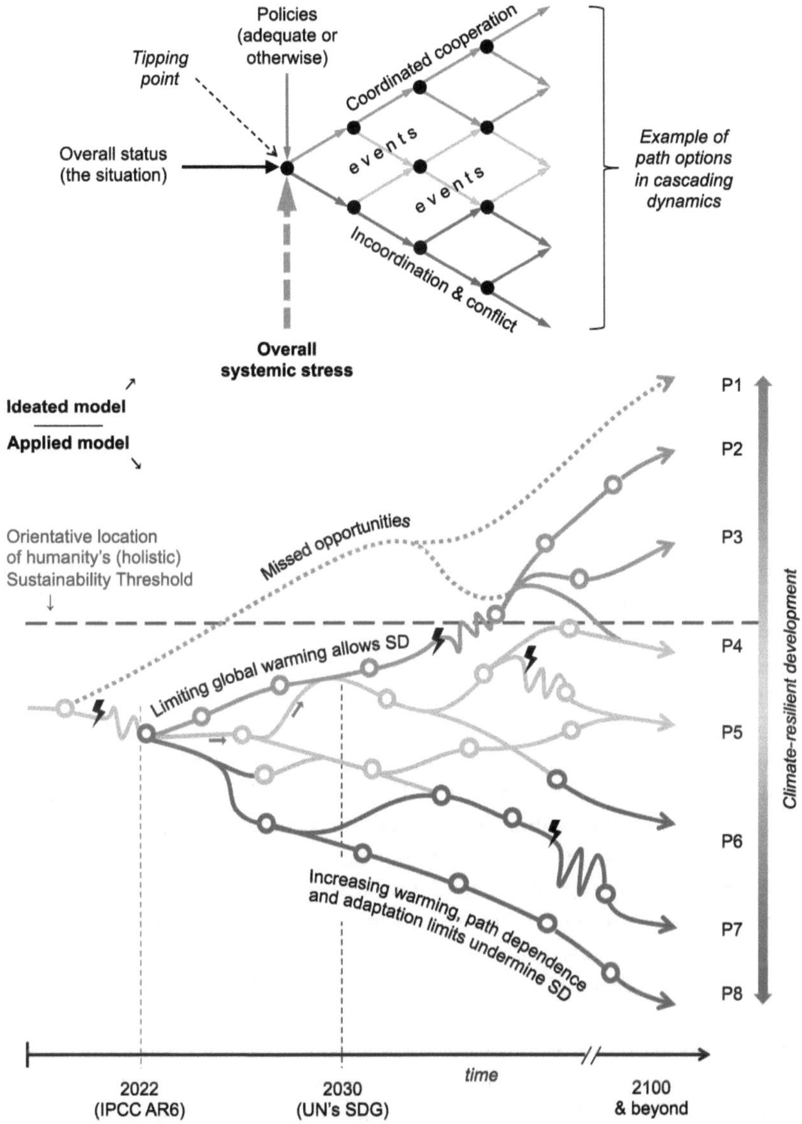

Fig. 1.13 A general model charting possible paths of Humanity (as per Fig. 1.9) depending on systemic tipping points (hard to anticipate, non-linear dynamics) (top) and an applied illustration (bottom) based on current knowledge on climate-sustainability relations. Black dots denote events that may cause systemic path change, some of which reflect tipping points. Colored circles denote potential opportunities for coordinated action. Lightening signs denote unexpected sources of instability. P1 to P8 denote paths along a virtual sustainability gradient. SD = Sustainable Development; SDG = SD Goals. Blue arrows suggest our current "best hopes" realistic trajectory. *Source* Original based on redrawing-modifications after [371] ('ideated') and [207] ('applied')

In sum, in order for us to raise to the challenge, all cities need to pass to a next-level. Figure 1.13 illustrates the paths that are opening in front of the global network of cities, and (even if some cities are in a better situation than others), because attaining sustainability is a global matter, individual cities have an obvious direct interest to help each other in their efforts to function as eco-cities as well. This can be done by cooperating at least on some issues even if there is inevitable competition on some other matters. Then, as those systemic solutions emerge in different contexts around the world', cities will have an ever-growing panoply of options that they can study and adapt for themselves or/and lessons and ideas for creating further solutions.

In this sense, I contend that the remark by Solecki et al. [391] that "these efforts must now shift to hyper-speed" can be achieved by the magnifying effect of learning across the global network of cities and that the notion of "brain networking" mentioned earlier and related to the notion of Internet-of-People (IoP) [97] can easily be adopted not only by individual experts and institutions (like Universities and other organizations involved in RDI). Within the logic of triple-helix projects involving universities, public authorities and the businesses), cities can deliberately apply a strategy of benefitting from global expert networks both directly and indirectly—as intermediated by universities and research institutions (and also by businesses, though their own expert networks and business networks). In this sense, I would call this approach Brain Networking for Eco-city Projects (BN-ECP if one needs an abbreviation). For a visual summary and model, see Chap. 3, Sect. 3.5).

To that aim, local authorities would be also well-advised to seek and participate in relevant regional and global alliances of cities and organizations aiming at helping cities cooperate, e.g., the C40 Cities Climate Leadership Group, the World Mayors Council on Climate Change (WMCCC), the Global Covenant of Mayors [112] and its regional/continental sub-alliances, and the International Council for Local Environmental Initiatives (ICLEI), the Council of European Municipalities and Regions (CEMR), and so forth. In spite of inherent contextual and organizational limitations, participation appears to be beneficial for learning and helping contributing experts developing further methods (for more discussions about the effectiveness of membership in these networks, see [188, 315], see also [49, 64, 197, 201, 217, 226, 242, 291, 392, 428]).

But, there is scope for other, local and/or topic-oriented alliances. A regional example is the long chain of coastal cities of West Africa (a quasi-continuous strip of urban area stretching for thousands of kilometers on the shore of the Atlantic and into the Gulf of Guinea) is emerging as one of the world's largest megalopolises. While this inevitably poses many challenges, it is also an opportunity for coordination and learning from each other for reciprocal benefit (not to forget, of course, that this great concentration of people also represents a huge potential common market, which, if materialized through coordination would be a magnet for international investors). And, similar situations can be identified around the world.

Such regional alliances of cities can also facilitate the mobilization of international expertise, not just by the usual institutions but also by direct participation of experts in concrete projects for addressing concrete issues, e.g., via the internet-based forms of collaboration, i.e., Brain Networking—to mobilize human resources from around the world, starting with (but not limited to) the skilled diasporas from those cities an countries (detailed method and discussion provided in Sect. 3.5; other coordination methods are described by the entire Chap. 3).

Further, to give a more topical example, many of the challenges faced by the cities along the west African coast are also faced by other cities around the continent. Of course, different cities can have very different sets of major challenges and priorities in different contexts. But, in the case of coastal cities, the specific sets of challenges faced by each city are compounded by a common phenomenon: the sea level rise that is caused by climate changes. For example, one planetary effect of climate changes is that polar ice melting will have concrete impacts on Sahel (sub-Saharan Africa) (see, e.g., [127]). In this sense, the last IPCC report ([207]—Table 9.8 therein) has identified especially 12 major cities: Abidjan (in Ivory Coast), Alexandria (in Egypt), Algiers (in Algeria), Cape Town, Casablanca (in Morocco), Dakar (in Senegal), Dar-es-Salaam (in Tanzania), Durban (in South Africa), Lagos (in Nigeria), Lomé (in Togo), Luanda (in Angola), and Maputo (in Mozambique). These and the other African cities sharing this challenge have a lot to gain by talking to each other and learning from each other. And the same applies to every other continent, and to every other shared challenge, from climate change to food, health, security and so on, as per current regional contexts. Relatedly, [207] includes a short case study that is worth studying, about the most vulnerable populations in this area (*Case Study 6.5: Climate-resilient Pathways in Informal Settlements in Cities in Sub-Saharan Africa*). For further studies on coastal cities around the world, see, e.g., [36, 127, 132, 207, 408, 431, 451].

In some cases, urban and rural communities have access to regional/continental joint programs, and country governments can facilitate the entire process of capacity-building and improvement via international agreements and national legislation, but the decisive factor is the people themselves. To pick an example from the European context that I know, rural communities in the Republic of Moldova (an official candidate country to joining the European Union) have been strongly affected by the economic changes caused by the war in Ukraine but can now benefit from a support program called *The European Village* financed by the European Commission (information about it can be easily found on internet; also the Diaspora of the country is more and more involved in the process of socio-economic development of all villages and cities, precisely because many of those people are living and working in the European Union already and so provide strong interactions at grassroot levels: individuals, enterprises, families, civic associations, events, and other projects). The key point is that, while geopolitical events served as trigger for certain changes, the positive local developments are possible because the country's government enjoys democratic (bottom-up) support, on the basis of which it is creating the legal basis and can take optimal decisions.

About the challenges of coordination

What in Fig. 1.13 refers to as 'potential opportunities for coordinated action' (the colored circles), IPCC [207] calls 'Arenas of Engagement', of which they suggest several types: community (i.e., cities and any other local human settlements); socio-cultural; political; ecological; knowledge + technology; and economic + financial. There are two important observations to make here. On the one hand, it is not enough to have some arenas and engagement: these interactions need to lead to some viable results (it cannot be just some tokenistic formalities that make everyone feel good as individuals while the systems in which they leave makes no real progress). On the other hand, solutions cannot be imposed (through "ambitious targets and decisive action" as one can often see written in public documents of intention that fail to say anything about "how" concretely can a problem be successfully addressed). In fact, the description between quotation marks above apply perfectly to totalitarian regimes, and totalitarianism is—by definition—not functional in complex dynamic system (and that is beside the moral issue that totalitarianism represents a denial of human rights). What may appear to the untrained person as "efficient action" on short term is only adding systemic problems that will burst later on in the form of cascading chains of events that will undo that system. All in all, it is just a logical observation that totalitarianism is not (and cannot be) compatible with sustainable development, so it cannot possibly be part of sustainability nor the transition thereto. On the contrary, what we understand based on current knowledge about systems and sustainability is that totalitarian regimes are themselves socially unsustainable, and necessarily unsustainable in the overall understanding of human sustainability. In short, we can expect that totalitarian societies will inevitably go down the red paths depicted in Fig. 1.13. Depending on how much they weigh on humanity as a whole, totalitarian societies can pull all of us down with them.

At the scale of the human society, solutions can only be generated with the participation of the people, not imposition of "solutions" on people. The only way to be realistic on the long run is we need to create a new paradigm of civilization based on true coordination between a large-enough number of people, communities/cities and countries. Given the distributed (not centralized) nature of knowledge, such a coordination implies that we manage to develop coherent policies, programs, and projects that pursue the set of conditions which allows productive works within clusters and across networks of experts and stakeholders. Relatedly, efforts are going on towards developing policy methods for adaptive plans that focus on different aspects reflective of these non-linear dynamics, e.g., identifying and using 'signpost signals' for monitoring critical changes (i.e., tipping points) in climate changes (e.g., [178]); using participative methods to identify "branching points" in "target-seeking scenarios" (e.g., [3]); integrating climate change adaptation plans with risk-based approaches—often in relation to the UN's *Sendai Framework for Disaster Risk Reduction* [420]—at city level, from the high-altitude cities in the Andes Mountains of South America faced with major landslide risks combined with socio-cultural and institutional challenges complicating environmental and economic issues [332] to the coastal communities in the Philippines requiring integrated approaches to maximize the use of limited resources [450]; and comparisons of cities all around the world (e.g., [125, 200]).

Current knowledge (as per academic literature, and as informed by practice in general, notably policy and business) indicates that many diverse and serious challenges persist in the local communities and organizations that are trying (and may already be legally obliged) to respond to climate changes (for example, see Box 1.4).

Box 1.4 Selected examples of challenges faced by local communities (and the related relevant organizations) when trying to respond to climate changes

- mismatches between support tools that were intended to help public and private administrators (but are too academic or/and heavy, or at least so perceived by the intended users in the public administration) and real needs of administrators (at least as perceived by them; which tend to need what they would regard as pragmatic support with how to address particular challenges, like conducting risk assessments to identify top priorities and doing cost–benefit analyses to decide between alternative potential action options, and to do so on a project basis within existing institutional and budgetary constraints)—an issue that actually starts with developed countries that already have more means than others to lead the way, like the UK and Germany (e.g., [105]);
- relatedly, mismatches between the high level of complexity of Integrated Assessment Models (IAMs) dealing with key, multidimensional and inherently complex topics (e.g., energy systems) and the capacity of real-world users to understand them represent another reason why very technical models remain unused: to date, their sheer complexity "has limited the wider understanding and influence of their underlying insights", on top of the fact that many such models exhibit a tendency to neglect most 'dynamic interdependencies', i.e., they lack 'dynamic realism' (see, e.g., [175] for a wider discussion), and this is something which, in the end, suggests that ignoring the models perceived as "too academic" is driven, at least to a certain extent, by a fundamentally correct intuition of the end users that "this is not what we need" (or "not good enough");
- gaps in even basic data in some regions of the world, a problem requiring significant capacity-building and international cooperation, especially in developing countries (e.g., [313]);
- or a lack of relevant data on target topics despite their popularity and benefits that are easily demonstrable in theory, e.g., the costs and advantages of green roofs and green walls (e.g., [237, 252, 258]), which shows that insufficient work is being done on the "how to do it" in concrete contexts;
- another fundamental aspect being the (all too-often unmet) requirement of a good interaction between citizens and local authorities, e.g., starting with

deliberations and choices about nature-based solutions, the design of urban public spaces, and in all eco- and sustainability- related goals (e.g., [57, 254, 347])

This persistence (and often even augmentation) of challenges points to a rather straightforward observation: each city must begin (and then continue) its own eco-city journey based on a proper identification of priorities. The next module will address the issue of how to do that under resource and time constraints. But the main points to underscore in here are (a) that the concrete power of the familiar "business-as-usual" is largely underestimated and (b) that all types of experts must try to bridge the gap between professions (although this is not easy, as it oftentimes requires changes in rules and institutional practices).

From an academic perspective at least, (in all likelihood) the 'normal' approach is to try to build the big picture by synthesis of all the knowledge that is available and then use that big picture as a backdrop and comparative reference for practical goals. This is a logical and mostly correct beginning, even if it is often underappreciated outside science and research institutions, because non-scientists tend to believe that science is already unitary, i.e., they tend to be unaware of its fragmentation and how it works (the issue of science in society will also be discussed in Chap. 3, Sect. 3.7). However, this underappreciation also hides a real point: the issue here is not that the 'sense-making' instinct of the scientists would be wrong (it isn't!). The issue is that the follow-up of this sense-making step seems indeed to be (or remain) too academic (and I can attest about this phenomenon myself, as a scientist and project manager observing other scientists in triplex-helix projects).

Surely, this problem can be clarified with a practice-inspired thought experiment, and the conclusion from it is worthwhile: it illustrates the problem with the famous problem of "silo-thinking" (failure to see the big picture, or "not seeing the forest for the trees") that is well-known in all professions. Namely, once they think they master a topic or subtopic, most scientists tend (as a reflex) to attempt to "guide" non-scientists through a lengthy series of steps leading to some "correct" result(s) or would even stretch out and basically try to educate everybody else (forgetting, in the process, that not everyone needs or wants to become a science expert). The logical and inevitable consequence of such attempts is that this "newly created science expert" person from outside science itself would have to garner expertise from all specialties (because there are scientists from virtually all domains who want to educate "the public" about their own specific domain). In other words, and to continue this logical exercise, each person outside science itself would have to become a sort of "super-scientist"—while the specialized "teachers" rarely dare (or care) to venture beyond their own field (which is a bigger problem in and by itself). This is of course laughable, and each scientist will realize it and laugh … once s/he reminds herself/himself that her/his own field of expertise in not the center of the universe. But there is a deeper point here: versions of this phenomenon occur in all professions, and specific examples can be found just as easily in the public administration and in businesses. In a more systematic perspective, these issues are known from management and psychology and are part of a long list of cognitive and behavioral biases that need to be considered in projects (In Chap. 3, I discuss in a bit more detail different aspects of this matter).

At this point I would note some other, related consequences that are relevant for projects, based on the example discussed above.

Those of us who are familiar with multi-disciplinary projects that cover different professions in the triple-helix of academia, business, and government are normally well aware of such field-specific ideas that cannot be applied when all details are put together. Field-specific (or profession-specific) ideas that ignore the broader picture lack sufficient consideration about professional constraints outside their domain, in this case in academia, naively insisting (for a while at least) on making a priority from the specific field they know about. In general, such ideas also assume unlimited resources and simply ignore basic managerial constraints like transaction costs (i.e., the real-world operational costs of implementation of a proposition), which would quickly reach astronomic dimensions, rendering such projects unfeasible. This kind of propositions also ignore the opportunity costs of implementing an idea, i.e., the main concern of an administrator or manager: within the realistic budget I have at hand, what is that I have to "sacrifice" (i.e., to not do) just in order to do this other proposition instead?

On the other hand, the same kind of practice-inspired thought-experiment can be run in the opposite way and will lead to a complementary set of problems (irony intended), and for the same root case: a natural tendency to use a certain functional oversimplification of everything we are not familiar with. A "purely non-scientific perspective" would be led by what seems "obvious" and "pragmatic" approach, for example by listing the necessary actions right away, so as to meet as soon as possible the professional requirements expected from them in their respective institutions and context, without much regard to underlying complexities—therefore running the risk of, and usually obtaining, results that are irrelevant or even detrimental the stated goals of their profession and institution. Often, such "results" are some spectacular short-term successes that do not address the stated long-term goal (and, in fact, are making the remaining work harder; and are squandering the resources that are left for that harder work that remains to be done).

Unfortunately, current institutional arrangements reinforce or even impose such biases, a most obvious example being the public administration in those countries where local/city authorities have in fact a very limited decisional power (because everything is decided from above). In such cases, the national government (and the mentality from which it is derived and which it engenders in a vicious circle) is part of the problem instead of being part of the solution, and the only way out is through the recuperation of freedoms and responsibilities of the local communities as a top priority. One can of course imagine (and reality often beats imagination) many other situations in which a good equilibrium is absent in the local community of people and must be recovered through re-balancing acts or processes.

The broader point is that, in all professions, both scientists and non-scientists tend to project their own way of doing things (and preferences) upon the rest of the world, thus oversimplifying reality in an ill-conceived attempt to force progress in a desired (and probably legitimate) direction. This is of course not new, but seems to keep happening anew—or, to use again the words of the poet, "All is old and new is all, what is good and what is broken … yours is to ask and reckon" (original: "*Toate-s vechi şi nouă toate, ce e rău şi ce e bine, tu te-ntreabă şi socoate*"; from the poem 'Gloss'/*Glosă,* by Mihai Eminescu).

This tendency of oversimplification comes from the combination of our deeply ingrained beliefs (based on what we know or think we know) and the psycho-social power of professional (and institutional) cultures which tend to re-enforce old practices and reject what does not conform to those. We know it under different names and in different languages, some popular descriptions being "the curse of the business-as usual" and "when all you have is a hammer, all you see around you is nails". Eco-city projects are of course not spared by it (not least because they need to also involve existing institutions), but, given the combination of very high complexity, difficulty and stakes, eco-city projects can easily fail when this problem is left unchecked.

Aside from the discussions in this book, you may also want to study the literature on Cultural Theory and Social Learning (already mentioned in Sect. 1.3) (see, e.g., 231, 263, 294, 303, 321, 430, 434), and on the cognitive (and related behavioral) biases known from project management (discussed in Chap. 3). Also, Cultural Theory can help us escape the narrow perspectives that often lead to a multiplicity of non-realistic and mutually exclusive versions of urban sustainability (or "sustopias", a shorthand for "sustainability utopias") and instead come to terms with the necessity of a critical reflection about urban sustainability, where the city is to be understood as a concrete place-in-the-world, one which "embraces creativity, critical practice, adaptation, and it places urban development and planning in a context of multiple spatial and temporal scales" [40], an old and re-invented "Cosmopolis" for a sustainable future [40, 101, 271].

Last but not least, an underlying point behind all this discussion is that we need to understand why people from other fields think the way they think, and that some misunderstanding and misevaluations can be resolved through dialogues, but also that some specificities must be respected for what they are. One example is simply the fact that the different disciplinary angles and the thematic or topical filters discussed in this module (e.g., the disciplines from Table 1.2, and the topics from Fig. 1.10) are a natural result of a division of labor between different domains of activity and are therefore complementary. Other examples of natural differences that must be recognized might be lesser known by those without direct experience in 'triple-helix' projects (or other complex RDI), but they are not less important. For example, a major difference in how people work is determined by the different work cycles in time: in science, the usual duration of a project is 3 years, while a major discovery or breakthrough takes about 7 years (on average; i.e., two successive projects and a bit), so work agendas are subsumed to this rhythm; while agendas in public administration is determined by a 4-year election cycles, and businesses in the private sector must obey annual and sub-annual evaluations (semestrial, trimestrial, monthly, and even more frequently).

To these, add a long list of issues that magnify the challenge of any project and can have ample effects in large and complex projects such as those pursuing eco-city agendas, and I will only briefly mention some as follows.

Firstly, for some people, learning per se is regarded as if it were a bad thing ("now you are learning?", automatically and mistakenly implying "not yet competent"). As a side observation, experienced scientists are fully aware of the limits of cognition in the face of the real-world complexity, as reflected in the popular expression "the more you study, the more you realize how little you know (compared to what is there to know and what remains to be discovered and understood through hard work)". This is why such experienced researchers and scholars often call themselves "students" (of this or that matter), but this can be confusing to non-academics—who might be thinking that "student" refers to some undergraduate student (i.e., the usual meaning in everyday life). So, let us not fall in that trap. Relatedly, let us not assume that students in universities are inexperienced: many are very experienced, and, in high-quality organizations, they are often involved in the same high-value work as the most experienced professionals and work with them for the betterment of our cities and societies. If we respect knowledge, as civilized people do, we must respect learning and studentship.

Secondly, some (or most) of the people who are very unfamiliar with science also have incorrect representations about it, for example thinking that science is simply about "precise measurement" (and usually implying that if a scientist talks about unknowns, uncertainties, probabilities and educated guesses based on available evidence, then s/he "doesn't really know" and "what kind of scientists are you, if you don't know such things?").

Thirdly, and related to the previous points, some may regard the notion of "theory" as being opposed to practice (and perhaps equivalent to "just an idea" anyone can have, or even a personal opinion or belief regardless of how justified that is, in gross disrespect of the fact that a scientific theory is hard won by generations of scientists, and often the product of personal sacrifices). Some may even refer hastily/dismissively to "theory" (for example, "this is too academic") as an opportunistic way to try facile excuses when they have none and just want to avoid supplementary but necessary work or to fence off scrutiny of their own work. Unfortunately, such practices are not uncommon, and reflect a failure to understand that modernity is the product of science, and therefore practice must be based on knowledge, e.g., professional abstract representation, i.e., theory.

This entire category of problems is even more difficult than these quick examples might suggest (for definitions of data, information and knowledge, see Chap. 3, Sects. 3.5 and 3.6). Here, I cut a long story short, by sharing with you an apparently simple observation that a more experienced colleague (sociologist Marina Fischer-Kowalski) made to me in a conversation we had 15 years ago in the break after her talk at a conference on sustainability and ecological economics in Nairobi (Kenya), and where my talks dealt with methods and scenarios in hard, holistic urban projects [103, 104]: "there is nothing more practical than a good theory".

Based on Fig. 1.13, where are we?

In this model from Fig. 1.13, one also notices that green path arrows are pointing upwards while yellow and red path arrows are pointing downwards (as per [207]—Figure SPM5 therein). This is one of the reasons why, in this figure, I explicitly located the sustainability thresholds in between. In the original model from IPCC cited above, an explicit sustainability threshold is not shown but is logically implied by the change in the orientation of the end arrows.

Because of systemic entropy (i.e., the tendency of disorder to increase over time), as time passes, all paths beneath the sustainability threshold will become red (which concretely means an amplification in time of the situation described in the first part of this module as the "tragedy of the commons"). Therefore, yellow translates as 'not ok and continually degrading' (or 'not yet red'): because the current overall situation of Humanity is unsustainable, the more we go on "eating up" the capacity of the Earth's ecosystems to regenerate themselves (by means that include but are not limited to the greenhouse gas emissions that heat up the planet), (1) the more the paths ahead of us will go down and (2) the harder it will be to reverse the downward trend.

These being said, and as I am correcting this manuscript (July 2023), it seems that we are heading towards missing the target of limiting global climate change to 1.5 °C (the limit agreed in the Paris Accord of 2015) and we are near to missing the 2 °C target as well because large-enough swift-and-intelligent action does not seem to be happening (see also [208]). This means that (a) we are already on the orange path signaled in Fig. 1.13 (applied model) with the small blue arrows, and (b) we are facing an imminent slide into the red domain from where it is realistically impossible to recover. It is therefore no surprise that the EU countries (and the European Commission), the USA and other developed countries have already understood that we now have to recuperate extremely fast the lost ground before it is too late. Of course, many developing countries have probably realized this situation too, but they tend to have lesser means to move equally decisive (and will therefor need to participate in joint global actions in the future).

As I am more familiar with the situation in Europe, I can say that EU looks decided to move "all in" and reduce GHG in an accelerated pace for the next few years, the purpose being to recover (by 2030) the chances for a green path as modelled in the 6th Assessment Report of the IPCC. But complex systems cannot be rushed at will, since this would bring other shocks, and we do not want a cure that is worse than the disease. The puzzle of correlations shown in Fig. 1.10 can only be resolved via competent action, not voluntarist or knee-jerk intervention on our socio-economic systems which evolved for millennia to become what they are.

To pick only one example, while our strategic goal to succeed in insuring at least a necessary minimum level of climate change mitigation, the twin goal of adaptation (to climate changes) cannot be realistically separated "at will" within our socio-economic systems: the two must be addressed simultaneously, the main operational aim being to generate synergies between them instead of miring humanity into a net of contradictions that will compromise both goals and will lock us into a downward spiral. As popular wisdom has it, "the road to hell is paved with good intentions".

Thus, the more so important becomes for us to recognize cities as necessary and key platforms for the development of concrete systemic solutions based on current knowledge. In a word, it is high time for eco-city projects and programs to drive our sustainability transition out of the dangerous place in which we find ourselves at the present time, and we need competent teams (both clusters and networks) to emerge and assemble themselves promptly for this task.

In finding viable paths, eco-city scenarios are knowledge-action organizers

In eco-city policies, programs and projects, the first imperative is the articulation of what is fundamentally feasible and what is not, within the limits of the possible, i.e., realistic scenarios. This means that developing useful eco-city scenarios essentially consists in finding the best trade-offs between objective constraints. This is the level of the content of eco-city projects (and programs and policies), and the next module discusses how to do that, concretely.

Then, obviously equally important for success in eco-city projects is the level of project management, and I cannot highlight this matter enough: only the non-experienced (or the false experts) would underestimate the fundamental and managerial challenges that come along with such complex projects (from epistemic, institutional and methodological mismatches to the hard issue of knowledge management across disciplines and through to finding the right operational routines). In short, without viable eco-city projects, there is no eco-city. This other level is addressed in the third module of this three-module book: details of a set of knowledge-action models that constitute (and support other) realistic and efficient pathways to urban (and wider) urban system sustainability. In the remainder of this book, I will only introduce and illustrate some key notions that are necessary for effective eco-city scenarios and projects.

A primordial operational aspect is both (a) logically deducible from the fact that we live in complex dynamic systems (as described in the first part of this module, i.e., Fig. 1.6 and the related discussions), and (b) very extensively documented from practice and experience (e.g., see again Fig. 1.10 and the related content), namely that trade-offs are necessary. And, this is not only a matter of negotiation and agreement between actors. Before any discussion about human interactions, 'trade-offs' is how nature works: they are physically unavoidable, and, trying to build scenarios without an evidence-based operational perspective would be pointless and, in the end, tends to be very costly.

Figure 1.14 illustrates this idea, based on the fundamental reality of what is known in science as the Pareto Frontier (or, sometimes, the 'Pareto Front'), which means that the best combination of desirable outcomes from a bundle of indicators (or/and parameters of interest) is bounded by the interactions between those factors, the total set of maximal combined outcomes being mathematically describable as a boundary line, the shape of which (i.e., the sum of the locations of all possible maximal outcomes) being variable and must be identified on a case-by-case basis: different methods are known and chosen as needed, coming from different fields that

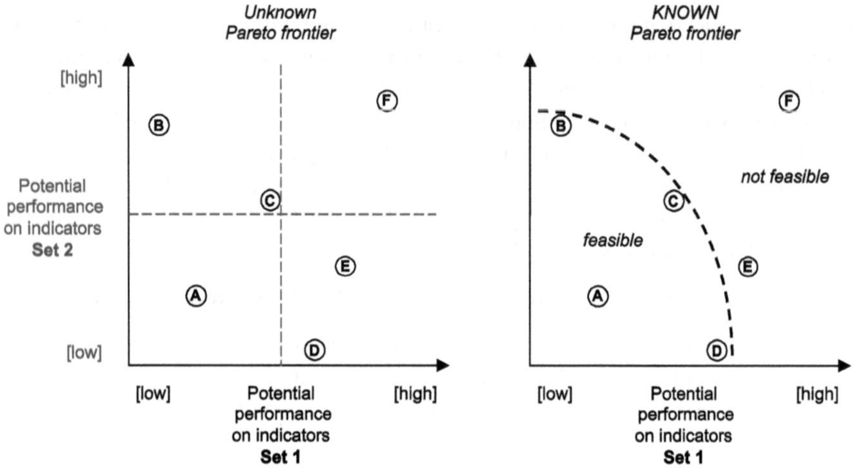

Fig. 1.14 Understanding the concrete value of eco-city scenarios without versus with the Pareto Frontier. When this limit is not known, all types of categories might seem equally achievable, and people may believe it is just a matter of planning. When this limit is known, the big picture of the scenarios looks very different, which shows how easy it is to slide into 'wishful thinking' when the matter (and reality itself) is understudied. *Source* Original, based on current well-established knowledge

use applied mathematics for optimization purposes, i.e., minimizing/maximizing some un-/desirable outcomes. The literature abounds in examples, especially (and by analytical and managerial necessity) in topics related to energy, smart-cities, transportation and land uses, which have a strongly multidimensional character (see, e.g., [18, 108, 133, 177, 181, 218, 247, 255, 322, 438, 447]).

Very often, it is assumed that any little change that can be identified as "green" is "ecological" and that "the more of those the better". The problem with this type of reasoning is that it rests on the assumption that there is a smooth continuum between the small "environmental" gesture and the big/systemic solutions for sustainability, which in fact means an assumption that the whole (system) is the sum of its parts. Complexity science (as introduced in the first part of this module and discussed all along) tells us instead that complex dynamic systems have non-linear behaviors. This means that, instead of assuming that we know the path, we need to actively search and confirm what path exist and what does not in spite of appearances; and that assuming is not equal to knowing. It is perfectly possible to do the wrong thing by focusing on very many small things hoping naively that this would lead to the big change. Complexity science tells us that it will not. Or, we can believe firmly that we can anticipate the exact path that is needed to "solve everything". Complexity science tells us that we cannot. In short, living in complex dynamic systems and being happy about it (because this is what makes human spirit virtually infinite and worth having) means that we have to do the hard path-finding work and we cannot cheat.

This and all the other insights discussed so far tell us (1) that eco-city scenarios must begin from concrete experience and be grounded in empirical knowledge, and at the same time, (2) that they must use a best-possible understanding of the theory (current knowledge) about how systems function. This makes that a (or even the) most natural way to approach scenario-generation for (and in) eco-city projects and programs is based on identifying and organizing priorities in the target city.

Importantly, in line with discussions on the illusion of one-size-fits-all, there is no "ready-to-use" scenario that can be directly transposed into a city of choice. Instead, relevant and useful scenarios can only be developed for the unique set of concrete situations existing in a real city/local community at a given moment in time. What I can do in this book is to provide some illustrations that convey the important points which need to be made in professional conversations, and which can then serve as starting points for meaningful generalizations (abstract models) and in-project works. This logical approach is also used, e.g., by IPCC, with the idea of scenario archetypes at world, nation and city levels (see e.g., [208]—Table 5.11 therein). In sustainability studies in general, scenario archetypes are useful because they help distill the main theme and trends (e.g., where issues converge or diverge), and thus can serve as common denominators for coordination and co-work across fields of expertise and for bringing different view angles into a common big picture (e.g., [50, 144, 185, 205, 389]). But even when using abstract scenario archetypes and models, it is necessary to rely on evidence and current knowledge to find those reliable ways to identify drivers, trajectories and the potential branching points of those potential trajectories (for further discussion see also [3]).

The next module of this book is focused on methodology.

References

1. Abreu EM, Neto JA (2022) Statistical approaches on the apparent horizon entropy and the generalized second law of thermodynamics. Phys Lett B 824. Article 136803. https://doi.org/10.1016/j.physletb.2021.136803
2. Ataman M, Hatzimanikatis V (2015) Heading in the right direction: thermodynamics-based network analysis and pathway engineering. Curr Opin Biotechnol 36:176–182. https://doi.org/10.1016/j.copbio.2015.08.021
3. Aguiar APD, Collste D, Harmackova ZV, Pereira L, Selomane O, Galafassi D, Van Vuuren D, Van Der Leeuw S (2020) Co-designing global target-seeking scenarios: a cross-scale participatory process for capturing multiple perspectives on pathways to sustainability. Glob Environ Chang 65:102198. https://doi.org/10.1016/j.gloenvcha.2020.102198
4. Ahern J (2011) From fail-safe to safe-to-fail: Sustainability and resilience in the new urban world. Landsc Urban Plan 100(4):341–343. https://doi.org/10.1016/j.landurbplan.2011.02.021
5. Alberti M, Marzluff JM (2004) Ecological resilience in urban ecosystems: linking urban patterns to human and ecological functions. Urban Ecosystems 7(3):241–265. https://doi.org/10.1023/B:UECO.0000044038.90173.c6
6. Alberti M, Marzluff JM, Shulenberger E, Bradley G, Ryan C, Zumbrunnen C (2003) Integrating humans into ecology: opportunities and challenges for studying urban ecosystems.

Bioscience 53(12):1169–1179. https://doi.org/10.1641/0006-3568(2003)053[1169:IHIEOA]2.0.CO;2
7. Alberti M, Marzluff JM, Shulenberger E, Bradley G, Ryan C, ZumBrunnen C (2008) Integrating Humans into ecology: opportunities and challenges for studying Urban ecosystems. In: Marzluff JM, Shulenberger E, Endlicher W, Simon U, ZumBrunnen C, Alberti M, Bradley G, Ryan C (eds) Urban Ecology. Springer, Cham, Switzerland, pp 143–160. ISBN: 978-0-387-73411-8
8. Aldy JE, Kotchen MJ, Stavins RN, Stock JH (2021) Keep climate policy focused on the social cost of carbon. Science 373(6557):850–852. https://doi.org/10.1126/science.abi7813
9. Aligica PD, Tarko V (2014) Institutional resilience and economic systems: lessons from Elinor Ostrom's work. Comp Econ Stud 56:52–76. https://doi.org/10.1057/ces.2013.29
10. Allen PM (2012) Cities and regions as self-organizing systems: models of complexity. Routledge, London. https://doi.org/10.4324/9780203990018
11. Amirzadeh M, Sobhaninia S, Sharifi A (2022) Urban resilience: a vague or an evolutionary concept? Sustain Cities and Soc. Article 103853. https://doi.org/10.1016/j.scs.2022.103853
12. Anderson T, Dragicevic S (2020) Representing complex evolving spatial networks: geographic network automata. ISPRS Int J Geo-Inf 9. Article 270. https://doi.org/10.3390/ijgi9040270
13. Andreev YV (1989) Urbanization as a phenomenon of social history. Oxf J Archaeol 8(2):167–177. https://doi.org/10.1111/j.1468-0092.1989.tb00198.x
14. Angelo H, Wachsmuth D (2020) Why does everyone think cities can save the planet? Urban Studies 57(11):2201–2221. https://doi.org/10.1177/0042098020919081
15. Annila A, Salthe S (2009) Economies evolve by energy dispersal. Entropy 11(4):606–633. https://doi.org/10.3390/e11040606
16. Antón SC, Middleton ER (2023) Making meaning from fragmentary fossils: early homo in the early to early middle pleistocene. J Human Evol 179. Article 103307. https://doi.org/10.1016/j.jhevol.2022.103307
17. Antons D, Piller FT (2015) Opening the black box of "not invented here": attitudes, decision biases, and behavioral consequences. Acad Manag Perspect 29(2):193–217. https://doi.org/10.5465/amp.2013.0091
18. Arent DJ, Bragg-Sitton SM, Miller DC, Tarka TJ, Engel-Cox JA, Boardman RD, Balash PC, Ruth MF, Cox J, Garfield DJ (2021) Multi-input, multi-output hybrid energy systems. Joule 5(1):47–58. https://doi.org/10.1016/j.joule.2020.11.004
19. Arrow KJ, Keohane RO, Levin SA (2012) Elinor Ostrom: an uncommon woman for the commons. Proc Natl Acad Sci 109(33):13135–13136. https://doi.org/10.1073/pnas.1210827109
20. Asăndulesei A (2017) Inside a Cucuteni settlement: remote sensing techniques for documenting an unexplored Eneolithic site from northeastern Romania. Remote Sens 9(1). Article 41. https://doi.org/10.3390/rs9010041
21. Asprone D, Manfredi G (2015) Linking disaster resilience and urban sustainability: a glocal approach for future cities. Disasters 39:s96–s111. https://doi.org/10.1111/disa.12106
22. Bahadur A, Tanner T (2014) Transformational resilience thinking: putting people, power and politics at the heart of urban climate resilience. Environ Urban 26:200–214. https://doi.org/10.1177/0956247814522154
23. Banai R (2020) Pandemic and the planning of resilient cities and regions. Cities 106. Article 102929. https://doi.org/10.1016/j.cities.2020.102929
24. Barrett D, Barrett M (2016) Star Trek: the human frontier. Routledge. London and New York. ISBN: 978-1-138-69959-5. https://doi.org/10.4324/9781315516493
25. BASE (2023) The nuclear phase-out in Germany. German Federal Office for the Safety of Nuclear Waste Management (Der Atomausstieg in Deutschland. Bundesamt fur Die Sicherheit der Nuklearen Ensorgung): https://www.base.bund.de/DE/themen/kt/ausstieg-atomkraft/ausstieg_node.html;jsessionid=030F992385D034AD4E9A5CDC9CC52033.internet952
26. Battiston F, Amico E, Barrat A, Bianconi G, Ferraz de Arruda G, Franceschiello B, Iacopini I, Kéfi S, Latora V, Moreno Y, Murray MM, Peixoto TP, Vaccarino F, Petri G (2021) The physics of higher-order interactions in complex systems. Nat Phys 17:1093–1098. https://doi.org/10.1038/s41567-021-01371-4

27. Battiston S, Monasterolo I, Riahi K, van Ruijven BJ (2021) Accounting for finance is key for climate mitigation pathways. Science 372:918–920. https://doi.org/10.1126/science.abf3877
28. Batty M (2009) Cities as complex systems: scaling, interaction, networks, dynamics and urban morphologies. UCL Working Papers Series, Paper 131. ISSN 1467-1298. https://discovery.ucl.ac.uk/id/eprint/15183/1/15183.pdf
29. Batty M, Morphet R, Masucci P, Stanilov K (2014) Entropy, complexity, and spatial information. J Geogr Syst 16:363–385. https://doi.org/10.1007/s10109-014-0202-2
30. Baumgartner M, Dell'Aglio D, Paulheim H, Bernstein A (2023) Towards the web of embeddings: integrating multiple knowledge graph embedding spaces with FedCoder. Web Semantics: science, services and agents on the World Wide Web, 75. Article 100741. https://doi.org/10.1016/j.websem.2022.100741
31. Beddington J (2009) Food, energy, water and the climate: a perfect storm of global events? In: Conference presentation given to the Sustainable Development UK Annual Conference, QEII Conference Centre, London, 19 March (2009). http://www.bis.gov.uk/assets/goscience/docs/p/perfect-storm-paper.pdf (archived; accessed 20 March 2020)
32. Bejan A, Errera MR (2016) Complexity, organization, evolution, and constructal law. J Appl Phys 119(7). Article 074901. https://doi.org/10.1063/1.4941554
33. Béné C, Mehta L, McGranahan G, Cannon T, Gupte J, Tanner T (2018) Resilience as a policy narrative: potentials and limits in the context of urban planning. Climate Dev 10(2):116–133. https://doi.org/10.1080/17565529.2017.1301868
34. Bentley C, Anandhi A (2020) Representing driver-response complexity in ecosystems using an improved conceptual model. Ecol Model 437:109320. https://doi.org/10.1016/j.ecolmodel.2020.109320
35. Bergström J, Dekker SWA (2014) Bridging the macro and the micro by considering the meso: reflections on the fractal nature of resilience. Ecol Soc 19. Article 22. https://doi.org/10.5751/ES-06956-190422
36. Berman M, Baztan J, Kofinas G, Vanderlinden JP, Chouinard O, Huctin JM, Kane A, Mazé C, Nikulkina I, Thomson K (2020) Adaptation to climate change in coastal communities: findings from seven sites on four continents. Clim Change 159:1–16. https://doi.org/10.1007/s10584-019-02571-x
37. Berry B (1964) Cities as systems within systems of cities. Papers Reg Sci 13(1):149–163. http://michael.blogweb.casa.ucl.ac.uk/files/2011/12/Berry-PPRSA-1964.pdf
38. Berry BJL (2008) Urbanization. In: Marzluff JM, Shulenberger E, Endlicher W, Alberti M, Bradley G, Ryan C, Simon U, ZumBrunnen C (eds) Urban Ecology: an international perspective on the interaction between humans and nature. Springer, Boston, MA, pp 25–48. https://doi.org/10.1007/978-0-387-73412-5_35
39. Bettencourt LM (2021) Introduction to urban science: evidence and theory of cities as complex systems. The MIT Press, Cambridge, Massachusetts. ISBN 9780262046008. https://mitpress.mit.edu/9780262046008/introduction-to-urban-science/
40. Beumer C (2017) Sustopia or cosmopolis? A critical reflection on the sustainable city. Sustainability 9(5). Article 845. https://doi.org/10.3390/su9050845
41. Bibri SE (2019) On the sustainability of smart and smarter cities in the era of big data: an interdisciplinary and transdisciplinary literature review. J Big Data 6(1):1–64. https://doi.org/10.1186/s40537-019-0182-7
42. Bibri SE (2022) Eco-districts and data-driven smart eco-cities: Emerging approaches to strategic planning by design and spatial scaling and evaluation by technology. Land Use Policy 113:105830. https://doi.org/10.1016/j.landusepol.2021.105830
43. Bibri SE, Krogstie J (2020) Smart eco-city strategies and solutions for sustainability: the cases of Royal Seaport, Stockholm, and Western Harbor, Malmö, Sweden. Urban Sci 4(1):11. https://doi.org/10.3390/urbansci4010011
44. Birch J, Hart JP (2018) Social networks and northern Iroquoian confederacy dynamics. Am Antiq 83(1):13–33. https://doi.org/10.1017/aaq.2017.59

45. Blake SE (2015) New York City's water supply: a case for a sustainable growth machine (Doctoral dissertation, Washington State University, Pullman, US-WA). https://rex.libraries.wsu.edu/view/pdfCoverPage?instCode=01ALLIANCE_WSU&filePid=13349961700001842&download=true
46. Blanke L, Walmsley A (2022) Resilient cities: renewal after disaster in three late antique towns of the east mediterranean. In: Martinez Jiménez J, Soulsby SO (eds) Remembering and forgetting the Ancient City. Oxbow Books, Oxford & Philadelphia, pp 69–108. Digital ISBN 978-1-78925-816-5. https://library.oapen.org/bitstream/handle/20.500.12657/54112/external_content.pdf?sequence=1#page=320
47. Bobbins K, Caprotti F, de Groot J, Pailman W, Moorlach M, Schloemann H, Densmore A, Finlay K, Fischat E, Siwali S, Links J (2023) Smart and disruptive infrastructures: re-building knowledge on the informal city. Urban Stud 61(1):165–179. https://doi.org/10.1177/00420980231172582
48. Bonilla K, Arrechea S, Velásquez Pérez LG (2022) Connecting scientists residing abroad: a review of converciencia as a practice to engage the Guatemalan scientific diaspora from 2005–2020. Front Res Metr Anal 7. Article 37. https://doi.org/10.3389/frma.2022.898496
49. Boon E, Wright SJ, Biesbroek R, Goosen H, Ludwig F (2022) Successful climate services for adaptation: what we know, don't know and need to know. Climate Serv 27. Article 100314. https://doi.org/10.1016/j.cliser.2022.100314
50. Boschetti F, Price J, Walker I (2016) Myths of the future and scenario archetypes. Technol Forecast Soc Chang 111:76–85. https://doi.org/10.1016/j.techfore.2016.06.009
51. Bostanci SH (2019) Critical thinking about urban studies linked with thermodynamic terms. In: IOP Conference Series: Materials Science and Engineering, vol 618, no. 1, pp 012078. IOP Publishing. https://doi.org/10.1088/1757-899X/618/1/012078
52. Boulding KE (1966) The economics of the coming spaceship earth. In: Jarrett H (ed) Environmental quality issues in a growing economy. Johns Hopkins University Press, Baltimore, Maryland, pp 3–14. https://doi.org/10.4324/9781315064147-2/
53. Bouquet C, Birkinshaw JM, Barsoux JL (2015) Fighting the headquarters knows best syndrome. Sloan Manage Rev 57(2):59–65. https://sloanreview.mit.edu/article/fighting-the-headquarters-knows-best-syndrome/
54. Boyer J (2020) Toward an evolutionary and sustainability perspective of the innovation ecosystem: revisiting the panarchy model. Sustainability 12(8):3232. https://doi.org/10.3390/su12083232
55. Braham WW (2016) Architecture and systems ecology thermodynamics principles of environmental building design, in three parts. Routledge, London & New York. eBook ISBN: 978-1-315-72772-1. http://refhub.elsevier.com/S0195-9255(19)30548-7/rf0055
56. Brenner N, Schmid C (2015) Towards a new epistemology of the urban? City 19(2–3):151–182. https://doi.org/10.1080/13604813.2015.1014712
57. Brink E, Wamsler C (2018) Collaborative governance for climate change adaptation: mapping citizen–municipality interactions. Environ Policy Gov 28(2):82–97. https://doi.org/10.1002/eet.1795
58. Bristow D, Kennedy C (2015) Why do cities grow? Insights from nonequilibrium thermodynamics at the urban and global scales. J Ind Ecol 19(2):211–221. https://doi.org/10.1111/jiec.12239
59. Brook BW, Ellis EC, Perring MP, Mackay AW, Blomqvist L (2013) Does the terrestrial biosphere have planetary tipping points? Trends Ecol Evol 28(7):396–401. https://doi.org/10.1016/j.tree.2013.01.016
60. Braun DR, Aldeias V, Archer W, Arrowsmith JR, Baraki N, Campisano CJ, Deino AL, DiMaggio EN, Dupont-Nivet G, Engda B, Feary DA (2019) Earliest known Oldowan artifacts at >2.58 Ma from Ledi-Geraru, Ethiopia, highlight early technological diversity. Proc Nat Acad Sci 116(24):11712–11717. https://doi.org/10.1073/pnas.18201771
61. Brughmans T, Hanson J, Mandich M, Romanowska I, Rubio-Campillo X, Carrignon S, Collins-Elliott S, Crawford K, Daems D, Fulminante F, De Haas T (2019) Formal modelling approaches to complexity science in Roman studies: a manifesto. Theoret Roman Archaeol J 2(1):1–19

62. Burkhard B, Fath BD, Müller F (2011) Adapting the adaptive cycle: Hypotheses on the development of ecosystem properties and services. Ecol Model 222:2878–2890. https://doi.org/10.1016/j.ecolmodel.2011.05.016
63. Butera FM, Palme M (2021) The city as a complex thermodynamic system. In: Palme, M, Salvati A (eds) Urban microclimate modelling for comfort and energy studies. Springer, Cham, pp 13–22. https://doi.org/10.1007/978-3-030-65421-4_2
64. Butler JR, Bergseng AM, Bohensky E, Pedde S, Aitkenhead M, Hamden R (2020) Adapting scenarios for climate adaptation: practitioners' perspectives on a popular planning method. Environ Sci Policy 104:13–19. https://doi.org/10.1016/j.envsci.2019.10.014
65. Buzna L, Peters K, Helbing D (2006) Modelling the dynamics of disaster spreading in networks. Physica A 363:132–140. https://doi.org/10.1016/j.physa.2006.01.059
66. Cai Y (2020) Renaissance of resilience: a buzzword or a new ideal? Manag Organ Rev 16(5):976–980. https://doi.org/10.1017/mor.2020.46
67. Calenbuhr V (2020) Complexity science in the context of policymaking. In: Šucha V, Sienkiewicz M (eds) Science for policy handbook. Elsevier, pp 118–127. https://doi.org/10.1016/B978-0-12-822596-7.00011-5
68. Calida BY, Jaradat RM, Abutabenjeh S, Keating CB (2016) Governance in systems of systems: a systems-based model. Int J Syst Syst Eng 7(4):235–257. https://doi.org/10.1504/IJSSE.2016.080313
69. Calvino I (1974) Invisible cities. First English edition: ISBN 0-15-145290-3. Original title: Le città invisibili, 1972. Giulio Einaudi, Turin. https://www.worldcat.org/oclc/914835
70. Caniglia G, Schäpke N, Lang DJ, Abson DJ, Luederitz C, Wiek A, Laubichler MD, Gralla F, von Wehrden H (2017) Experiments and evidence in sustainability science: a typology. J Clean Prod 169:39–47. https://doi.org/10.1016/j.jclepro.2017.05.164
71. Cannon T, Müller-Mahn D (2010) Vulnerability, resilience and development discourses in context of climate change. Nat Hazards 55:621–635. https://doi.org/10.1007/s11069-010-9499-4
72. Caprotti F (2018) Future cities: moving from technical to human needs. Palgrave Commun 4(1):1–4. https://doi.org/10.1057/s41599-018-0089-5
73. Caprotti F, Cowley R, Datta A, Broto VC, Gao E, Georgeson L, Herrick C, Odendaal N, Joss S (2017) The New Urban Agenda: key opportunities and challenges for policy and practice. Urban Res Pract 10(3):367–378. https://doi.org/10.1080/17535069.2016.1275618
74. Caputo S, Caserio M, Coles R, Jankovic L, Gaterell MR (2015) Urban resilience: two diverging interpretations. J Urbanism: Int Res Placemaking and Urban Sustainability 8(3):222–240. https://doi.org/10.1080/17549175.2014.990913
75. Carson R (1962) Silent spring. Houghton Mifflin Harcourt, Boston, MA. https://www.amazon.com/Silent-Spring-Rachel-Carson/dp/0618249060
76. Carmichael T, Hadžikadić M (2019) The fundamentals of complex adaptive systems. Complex adaptive systems: views from the physical, natural, and social sciences. In: Carmichael T, Collins AJ, Hadžikadić M (eds) Complex Adaptive Systems. Views from the Physical, Natural, and Social Sciences, pp 1–16. https://doi.org/10.1007/978-3-030-20309-2_1
77. Castellani V (2013) The lost world of old Europe: the Danube valley, 5000–2500 BC. https://doi.org/10.1080/10848770.2013.765105
78. Castle D, Culver K (2013) Getting to 'no': the method of contested exchange. Sci Publ Pol 40(1):34–42. https://doi.org/10.1093/scipol/scs118
79. Chaisson EJ (2015) Energy flows in low-entropy complex systems. Entropy 17(12):8007–8018. https://doi.org/10.3390/e17127857
80. Chávez JEG (2016) Institutional reform through innovative diasporas in the knowledge economy. Reforma institucional a través de diásporas innovadoras en la economía del conocimiento (Doctoral dissertation). Universidad Complutense de Madrid. https://hdl.handle.net/20.500.14352/26744
81. Chen Z (2020) Milestones: cognitive. In Benson JB (eds) Encyclopedia of infant and early childhood development (2nd edn). Elsevier. ISBN: 9780128165126; eBook ISBN: 9780128165119. https://shop.elsevier.com/books/encyclopedia-of-infant-and-early-childhood-development/benson/978-0-12-816512-6

82. Chelleri L, Baravikova A (2021) Understandings of urban resilience meanings and principles across Europe. Cities 108:102985. https://doi.org/10.1016/j.cities.2020.102985
83. Chiles JR (2002) Inviting disaster: lessons from the edge of technology. Harper, New York. ISBN 9780066620824
84. Chirisa I, Bandauko E (2015) African cities and the water-food-climate-energy nexus: an agenda for sustainability and resilience at a local level. Urban Forum 26:391–404. https://doi.org/10.1007/s12132-015-9256-6
85. Christensen CM (1997, 2011) The innovator's dilemma. Harper's Business. New York. ISBN 978-14221-8998-6
86. Christensen CM, Raynor M, McDonald R (2015) What is disruptive innovation? Harvard Bus Rev 93:44–53. https://hbr.org/2015/12/what-is-disruptive-innovation
87. Christensen CM, McDonald R, Altman EJ, Palmer JE (2018) Disruptive innovation: an intellectual history and directions for future research. J Manage Stud 55(7):1043–10781197. https://doi.org/10.1111/joms.12349
88. Ciumasu IM (2005) Environmental science and technology in Romania: the need and the possibilities of cooperation between the Romanian universities and the Romanian students and researchers at the European universities. Talk and book chapter. University in Society—UNISO 2005, 18–25 July. University of Caen Basse Normandie, Cherbourg. In Zaharia SE, Ivan MS (eds) UNISO 2005. University and labour market. ISBN 973-596-305-1, pp 309–312. https://www.ad-astra.ro/library/papers/Ciumasu_UNISO_2005_Diaspora_Plan.pdf
89. Ciumasu IM (2008) Brain networking—How to do it? Talk and book chapter. University in Society—UNISO 2008, 16–20 July. "Alexandru Ioan Cuza" University of Iasi, Romania. In Zaharia SE, Ivan MS (eds) UNISO 2008. Competence-based Higher Education: Challenges and Solutions, pp 197–203
90. Ciumasu IM (2010) Turning brain drain into brain networking. Sci Publ Pol 37(2):135–146. https://doi.org/10.3152/030234210X489572
91. Ciumasu IM (2013a) Dynamic decision trees for building resilience into future eco-cities. Technol Forecast Soc Chang 80(9):1804–1814. https://doi.org/10.1016/j.techfore.2012.12.010
92. Ciumasu IM (2013b) Is there a future for eco-cities? Prospective roles of Versailles Chantiers within the Greater Paris: a hub of multi-modal communication. Technical Report of the Econoving Integration Seminar (2011–2013). ECONOVING Chair for Generating Eco-Innovation, REEDS Center, University of Versailles Saint-Quentin-en-Yvelines. https://doi.org/10.13140/RG.2.1.3866.8240
93. Ciumasu IM (2014) Challenges of systemic innovation: cities as experimentation units? In: Deltas in Times of Climate Change II, Rotterdam, The Netherlands, 24–26 September, pp 278–279. http://www.climatedeltaconference2014.org/
94. Ciumasu IM (2018a) A coordination lattice model for building urban resilience. In: Proceedings of the 15th Information Systems for Crisis Response and Management (ISCRAM) conference, Rochester, NY, USA, 23–25 May, 419–427. http://idl.iscram.org/files/ioanmciumasu/2018/2119_IoanM.Ciumasu2018.pdf
95. Ciumasu IM (2018b) Networking brains: how Horizon Europe can foster excellence in central European science. Science|Business, 23 October 2018. https://sciencebusiness.net/framework-programmes/viewpoint/networking-brains-how-horizon-europe-can-foster-excellence-central
96. Ciumasu IM (2023a) Smart solutions for sustainability: RDI for urban and societal transitions requires cross-sectoral experimentation platforms. In: Da Silva Portela CF (ed) Sustainable, Innovative and Intelligent Societies and Cities. Springer, Cham, CH. https://doi.org/10.1007/978-3-031-30514-6_10
97. Ciumasu IM (2023b) Mapping Industry 4.0 onto eco-city transitions: a knowledge-action matrix. In: Da Silva Portela CF (ed) Sustainable, Innovative and Intelligent Societies and Cities. Springer, Cham, CH. https://doi.org/10.1007/978-3-031-30514-6_13
98. Ciumasu IM, Buzdugan L, Stefan N, Culver K (2012a) Ecosystems: planning. In: Jorgensen SE (ed) Encyclopedia of environmental management, Four Volume Set. CRC Press, pp 632–642. eISBN 9781351235860. https://doi.org/10.1081/E-EEM-120047256

99. Ciumasu IM, Costica M, Costica N, Neamtu M, Dirtu AC, De Alencastro LP, Buzdugan L, Andriesa R, Iconomu L, Stratu A, Popovici OA, Secu CV, Paveliuc-Olariu C, Dunca S, Stefan M, Lupu A, Stingaciu-Basu A, Netedu A, Dimitriu RI, Gavrilovici O, Talmaciu M, Borza M (2012b) Complex risks from old urban waste landfills—a sustainability perspective from Iasi, Romania. J Hazardous, Tox Radioact Waste 16(2):158–168. https://doi.org/10.1061/(ASCE)HZ.2153-5515.0000090
100. Ciumasu IM, Costica N (2010) Impacts of air pollution on the ecosystem and human health. Health and environmental impacts. In: Gurjar BR, Molina LT, Ojha CSP (eds) Air pollution health and environmental impacts. CRC Press (Taylor & Francis), pp 447–492. Boca Raton. eBook ISBN 9780429075599. https://doi.org/10.1201/EBK1439809624
101. Ciumasu IM, Culver K (2011a) Eco-cities in a globalized future—from Constantinopolis to "Cosmopolis"? In: The 9th international conference of the European society of ecological economics (ESEE), 14–17 June, Istanbul (2011). https://www.academia.edu/26227428/Eco_cities_in_a_globalized_future_from_Constantinopolis_to_Cosmopolis_
102. Ciumasu IM, Culver K, Costica M, Vanderlinden JP (2021) Ecosystems: planning and trade-offs. In: Fath BD, Jorgensen SE (eds) Handbook of environmental science, 2nd edn, vol 2. Managing biological and ecological systems. CRC Press, Boca Raton, Florida, pp 105–124. https://www.taylorfrancis.com/chapters/edit/10.1201/9780429346170-12/. https://doi.org/10.1201/9780429346170
103. Ciumasu IM, Lupu A, Costica N, Netedu A, Stratu A, Miftode V, Stefan N (2008c) Integrating social perception of nature with economic development expectations in a peri-urban area in Iasi, Romania. Talk and paper. In: ISEE 2008: applying ecological economics for social and environmental sustainability (10th biennial conference of the International Society for Ecological Eonomics), 7–11 August, Nairobi-UNEP
104. Ciumasu IM, Manolescu I, Costica N, Talmaciu M, Costica M, Nita V, Stefan N (2008b) Sustainable exploitation of ecosystems using scenarios of sustainable vs. unsustainable economic development: the Ciric River basin, Iasi, Romania. In: ISEE 2008: applying ecological economics for social and environmental sustainability (10th biennial conference of the International Society for Ecological Economics), 7–11 August, Nairobi-UNEP (2008).
105. Clar C, Steurer R (2018) Why popular support tools on climate change adaptation have difficulties in reaching local policy-makers: qualitative insights from the UK and Germany. Environ Pol Governance, 28:172–182. https://doi.org/10.1002/eet.1802
106. Clark WC (2007) Sustainability science: a room of its own. Proc Natl Acad Sci 104(6):1737–1738. https://doi.org/10.1073/pnas.0611291104
107. Clark WC, Harley AG (2019) Sustainability science: towards a synthesis. Sustainability science program. In: Kennedy school of government, sustainability science program working paper 2019–01. Harvard University, Cambridge, MA. Permanent link: http://nrs.harvard.edu/urn-3:HUL.InstRepos:42574531
108. Cocho-Bermejo A, Birgonul Z, Navarro-Mateu D (2018) Adaptive and morphogenetic city research laboratory. In: Computing for a better tomorrow—proceedings of the 36th international conference on education and research in computer aided architectural design in Europe (eCAADe), vol 2, pp 659–668 https://papers.cumincad.org/data/works/att/eCAADe_2018_volume2_screen_lowres_SCOPUS.pdf#page=673
109. Cofré R, Herzog R, Corcoran D, Rosas FE (2019) A comparison of the maximum entropy principle across biological spatial scales. Entropy 21:1009. https://doi.org/10.3390/e21101009
110. Coldwell D (2016) Entropic citizenship behavior and sustainability in urban organizations: towards a theoretical model. Entropy 18(12). https://doi.org/10.3390/e18120453
111. Colucci D, Sacco PL (2021). A glint of lights in the fog: invisible cities and the riddles of planning practice. J Plann Educ Res p.0739456X211027638. https://doi.org/10.1177/0739456X211027638
112. CoM (2023). The (global) Covenant of Mayors. Homepage https://www.globalcovenantofmayors.org (last accessed on JF 30 Sept 2023)

113. Cook M, Horne R, Potter S, Valdez AM (2019) Exploring the epistemicpolitics of urban niche experiments. In: Jensen Stissing J, Cashmore M, Späth P (eds) The politics of urban sustainability transitions: knowledge, power and governance. Abingdon, Oxon: Routledge, pp 125–147. https://files.core.ac.uk/pdf/86/161882888.pdf
114. Costanza R, D'Arge R, De Groot R, Farber S, Grasso M, Hannon B, Limburg K, Naeem S, O'Neill RV, Paruelo J, Raskin RG, Sutton P, Van Den Belt M (1997) The value of the world's ecosystem services and natural capital. Nature 387(6630):253–260. https://doi.org/10.1038/387253a0
115. Costello MJ, Chaudhary C (2017) Marine biodiversity, biogeography, deep-sea gradients, and conservation. Curr Biol 27:R511–R527. https://doi.org/10.1016/j.cub.2017.04.060
116. Crate SA (2011) Climate and culture: anthropology in the era of contemporary climate change. Annu Rev Anthropol 40(1):175–194. https://doi.org/10.1146/annurev.anthro.012809.104925
117. Crutzen PJ, Stoermer EF (2000) The anthropocene. IGBP Global Change Newsletter, 41:17–18. https://doi.org/10.1007/978-3-030-82202-6_2
118. Cugurullo F (2018) Exposing smart cities and eco-cities: Frankenstein Urbanism and the sustainability challenges of the experimental city. Environ Plann A: Economy Space 50(1):73–92. https://doi.org/10.1177/0308518X17738
119. Cushman SA (2015) Thermodynamics in landscape ecology: the importance of integrating measurement and modeling of landscape entropy. Landscape Ecol 30(1):7–10. https://doi.org/10.1007/s10980-014-0108-x
120. Daimon H, Miyamae R, Wang W (2023) Navigating theories of actions on disaster prevention: a systematic review on disaster research in Japan. Nat Hazard Rev 24(2):03123002. https://doi.org/10.1061/NHREFO.NHENG-1630
121. D'Amico G, Taddeo R, Shi L, Yigitcanlar T, Ioppolo G (2020) Ecological indicators of smart urban metabolism: a review of the literature on international standards. Ecol Ind 118:106808. https://doi.org/10.1016/j.ecolind.2020.106808
122. Daniels PW (2004) Urban challenges: the formal and informal economies in mega-cities. Cities 21(6):501–511. https://doi.org/10.1016/j.cities.2004.08.002
123. Dawes SS (2008) The evolution and continuing challenges of e-governance. Public Adm Rev 68:S86–S102. https://doi.org/10.1111/j.1540-6210.2008.00981.x
124. De Kraker J, Kroeze C, Kirschner P (2011) Computer models as social learning tools in participatory integrated assessment. Int J Agric Sustain 9(2):297–309. https://doi.org/10.1080/14735903.2011.582356
125. de Murieta ES, Galarraga I, Olazabal M (2021) How well do climate adaptation policies align with risk-based approaches? An assessment framework for cities. Cities 109:103018. https://doi.org/10.1016/j.cities.2020.103018
126. de Yarza C (2017) Star Trek–where no genre has gone before: application of mittell's television genre theory to the star trek series. J Multidisc Res 9(2):73–80. https://jmrpublication.org/wp-content/uploads/JMR9-2.pdf#page=75
127. Defrance D, Ramstein G, Charbit S, Vrac M, Famien AM, Sultan B, Swingedouw D, Dumas C, Gemenne F, Alvarez-Solas J, Vanderlinden JP (2017) Consequences of rapid ice sheet melting on the Sahelian population vul-nerability. Proc Natl Acad Sci 114(25):6533–6538. https://doi.org/10.1073/pnas.1619358114
128. Degroot D, Anchukaitis K, Bauch M, Burnham J, Carnegy F, Cui J, de Luna K, Guzowski P, Hambrecht G, Huhtamaa H, Izdebski A, Kleemann K, Moesswilde E, Neupane N, Newfield T, Pei Q, Xoplaki E, Zappia N (2021) Towards a rigorous understanding of societal responses to climate change. Nature 591:539–550. https://doi.org/10.1038/s41586-021-03190-2
129. Desmarais E (2023) Natural disasters and adaptive cycles: a case study of forced transformation from the 2011 Tohoku earthquake and tsunami. 岡山県立大学保健福学部紀要 29:123–129. https://doi.org/10.15009/00002447
130. Diachenko A, Menotti F (2017) Proto-cities or non-proto-cities? On the nature of Cucuteni-Trypillia mega-sites. J World Prehist 30:207–219. https://doi.org/10.1007/s10963-017-9105-8
131. Disse M, Johnson TG, Leandro J, Hartmann T (2020) Exploring the relation between flood risk management and flood resilience. Water Secur 9:100059. https://doi.org/10.1016/j.wasec.2020.100059

132. Doloisio N, Vanderlinden JP (2020) The perception of permafrost thaw in the Sakha Republic (Russia): narratives, culture and risk in the face of climate change. Polar Sci 26. Article 100589. https://doi.org/10.1016/j.polar.2020.100589
133. Doubleday K, Parker A, Hafiz F, Irwin B, Hancock S, Pless S, Hodge BM (2019) Toward a subhourly net zero energy district design through integrated building and distribution system modeling. J Renew Sustainab Energy 11(3). https://doi.org/10.1063/1.5093917
134. Duit A, Galaz V, Eckerberg K, Ebbesson J (2010) Governance, complexity, and resilience. Glob Environ Chang 20:363–368. https://doi.org/10.1016/j.gloenvcha.2010.04.006
135. Durán-Romero G, López AM, Beliaeva T, Ferasso M, Garonne C, Jones P (2020) Bridging the gap between circular economy and climate change mitigation policies through eco-innovations and quintuple helix model. Technol Forecast Soc Chang 160:120246. https://doi.org/10.1016/j.techfore.2020.120246
136. Dushkova D, Haase D (2020) Not simply green: nature-based solutions as a concept and practical approach for sustainability studies and planning agendas in cities. Land 9(1). Article 19. https://doi.org/10.3390/land9010019
137. Eagan JL (2007) Is authority in public administration separable from authoritarianism? Personalities, institutions, and resistance. Adm Theo Praxis 29(1):83–101. https://doi.org/10.1080/10841806.2007.11029562
138. Eakin H, Bojórquez-Tapia LA, Janssen MA, Georgescu M, Manuel-Navarrete D, Vivoni ER, Escalante AE, Baeza-Castro A, Mazari-Hiriart M, Lerner AM (2017) Urban resilience efforts must consider social and political forces. Proc Natl Acad Sci 114(2):186–189. https://doi.org/10.1073/pnas.162008111
139. Ekwo U (2011) Diaspora media and citizen engagement in the digital age: analyzing the impact of transnational journalism on democratization. In: Diaspora strategies: encouragement, evolution, and engagement. The Clinton Institute of American Studies, University College, Belfield, Dublin, Ireland, pp 9–11. https://www.mycmpi.org/wp-content/uploads/2012/07/Diaspora-Media-and-Citizen-Engagement-in-the-Digital-Age-Analyzing-the-Impact-of-Transnational-Journalism-on-Democratization.pdf
140. Emrich CT, Tobin GA (2018) Resilience: an introduction. In: Fuchs S, Thaler T (eds) Vulnerability and resilience to natural hazards. Cambridge University Press, pp 124–144. https://doi.org/10.1017/9781316651148
141. Erlandson JM, Braje TJ (2013) Archeology and the Anthropocene. Anthropocene 4:1–7. https://doi.org/10.1016/j.ancene.2014.05.003
142. Evans J (2016) Trials and tribulations: problematizing the city through/as urban experimentation. Geogr Compass 10(10):429–443. https://doi.org/10.1111/gec3.12280
143. Fan C, Jiang Y, Yang Y, Zhang C, Mostafavi A (2020) Crowd or hubs: information diffusion patterns in online social networks in disasters. Int J Disaster Risk Reduct 46:101498. https://doi.org/10.1016/j.ijdrr.2020.101498
144. Fergnani A, Song Z (2020) The six scenario archetypes framework: a systematic investigation of science fiction films set in the future. Futures 124:102645. https://doi.org/10.1016/j.futures.2020.102645
145. Ferguson J (2008) Global shadows: Africa in the neoliberal world order. Duke University Press, Durham, NC. https://doi.org/10.1017/S0022278X08003273
146. Fermeglia M, Lughi V, Pavan AM (2020) How to avoid the perfect storm: the role of energy and photovoltaics. MRS Energy Sustain 7:e34. https://doi.org/10.1557/mre.2020.36
147. Ferraro PJ, Sanchirico JN, Smith MD (2019) Causal inference in coupled human and natural systems. Proc Natl Acad Sci 116(12):5311–5318. https://doi.org/10.1073/pnas.1805563115
148. Fiksel J (2007) Sustainability and resilience: Toward a systems approach. IEEE Eng Manage Rev 35:5–15. https://doi.org/10.1109/EMR.2007.4296420
149. Finance O, Swerts E (2020) Scaling laws in urban geography. Linkages with urban theories, challenges and limitations. In: Pumain D (ed) Theories and models of urbanization. Lecture Notes in Morphogenesis. Springer, Cham, pp 67–96. https://doi.org/10.1007/978-3-030-36656-8_5

150. Fistola R, Gargiulo C, La Rocca RA (2020) Rethinking vulnerability in city-systems: a methodological proposal to assess "urban entropy." Environ Impact Assess Rev 85:106464. https://doi.org/10.1016/j.eiar.2020.106464
151. Fistola R, La Rocca RA (2014) The sustainable city and the Smart City: measuring the urban entropy first. In: Marchettini N, Brebbia CA, Pulselli R, Bastianoni S (eds) The Sustainable City volume I ISBN: 978-1-78466-023-9, volume II ISBN: 978-1-78466-024-6:537–548. WIT Press, Southampton and Boston. WIT transactions on Ecology and the Environment. https://doi.org/10.2495/SC140451
152. Fitzhugh B, Butler VL, Bovy KM, Etnier MA (2019) Human ecodynamics: a perspective for the study of long-term change in socioecological systems. J Archaeol Sci Rep 23:1077–1094. https://doi.org/10.1016/j.jasrep.2018.03.016
153. Flyvbjerg B (2021) Top ten behavioral biases in project management: an overview. Proj Manag J 52:531–546. https://doi.org/10.1177/87569728211049046
154. Franzke CL, Ciullo A, Gilmore EA, Matias DM, Nagabhatla N, Orlov A, Paterson SK, Scheffran J, Sillmann J (2022) Perspectives on tipping points in integrated models of the natural and human Earth system: cascading effects and telecoupling. Environ Res Lett 17(1):015004. https://doi.org/10.1088/1748-9326/ac42fd
155. Froese R, Schilling J (2019) The nexus of climate change, land use, and conflicts. Current Climate Change Reports 5:24–35. https://doi.org/10.1007/s40641-019-00122-1
156. Fulminante F (2012) Social network analysis and the emergence of Central Places: a case study from central Italy (Latium Vetus). Babesch 87:1–27. https://doi.org/10.17863/CAM.13080
157. Fulminante F, Lozano S, Prignano L (2016) Social network analysis and early Latin cities (Central Italy). In: Atterna P, Seubers J, Willemsen S (eds) Early states, territories and settlements in protohistoric central Italy: proceedings of a specialist conference at the Groningen Institute of Archaeology of the University of Groningen, vol 2, p 101. Barkhuis
158. Fulminante F (2021) Where do cities come from and where are they going to? Modelling past and present agglomerations to understand urban ways of life. Front Digital Humanities, 7. Article 633838. https://doi.org/10.3389/fdigh.2020.633838
159. Galaz V, Tallberg J, Boin A, Ituarte-Lima C, Hey E, Olsson P, Westley F (2017) Global governance dimensions of globally networked risks: the state of the art in social science research. Risk, Hazards and Crisis in Public Policy 8(1):4–27. https://doi.org/10.1002/rhc3.12108
160. Gao C, Gao C, Song K, Fang K (2020) Pathways towards regional circular economy evaluated using material flow analysis and system dynamics. Res Conservat Recycl 154, article 104527. https://doi.org/10.1016/j.resconrec.2019.104527
161. Gatzweiler FW, Reis S, Zhang Y, Jayasinghe S (2018) Lessons from complexity science for urban health and well-being. Cities & Health 1(2):210–223. https://doi.org/10.1080/23748834.2018.1448551
162. Gaydarska B (2016) The City Is Dead! Long Live the City! Nor Archaeol Rev 49(1):40–57. https://doi.org/10.1080/00293652.2016.1164749
163. Geels FW (2010) Ontologies, socio-technical transitions (to sustainability), and the multi-level perspective. Res Policy 39:495–510. https://doi.org/10.1016/j.respol.2010.01.022
164. Georgescu-Roegen N (1971) The entropy law and the economic process. Harvard University Press. https://doi.org/10.4159/harvard.9780674281653
165. Georgescu-Roegen N (1977) Inequality, limits and growth from a bioeconomic viewpoint. Rev Soc Econ 35(3):361–375. https://doi.org/10.1080/00346767700000041
166. Germain RN (2001) The art of the probable: system control in the adaptive immune system. Science 293:240–245. https://doi.org/10.1126/science.1062946
167. Giddings B, Hopwood B, O'Brien G (2002) Environment, economy and society: fitting them together into sustainable development. Sustain Dev 10(4):187–196. https://doi.org/10.1002/sd.199
168. Gkatsopoulos P (2017) A methodology for calculating cooling from vegetation evapotranspiration for use in urban space microclimate simulations. Procedia Environ Sci 38:477–484. https://doi.org/10.1016/j.proenv.2017.03.139

169. Gowdy J, Erickson JD (2005) The approach of ecological economics. Camb J Econ 29:207–222. https://doi.org/10.1093/cje/bei033
170. Gowdy J (2007) Avoiding self-organized extinction: Toward a co-evolutionary economics of sustainability. Int J Sustainable Dev World Ecol 14(1):27–36. https://doi.org/10.1080/13504500709469705
171. Granstrand O, Holgersson M (2020) Innovation ecosystems: a conceptual review and a new definition. Technovation 90:102098. https://doi.org/10.1016/j.technovation.2019.102098
172. Granovetter M (1985) Economic action and social structure: the problem of embeddedness. Am J Sociol 91:481–510. https://doi.org/10.1086/228311
173. Grober U (2007) Deep roots: a conceptual history of 'sustainable development' (Nachhaltigkeit). Wissenschaftszentrum Berlin für Sozialforschung. https://bibliothek.wzb.eu/pdf/2007/p07-002.pdf
174. Grubb M, Koch M, Thomson K, Sullivan F, Munson A (2013, 2019) The 'Earth Summit' agreements: a guide and assessment: an analysis of the Rio'92 UN Conference on Environment & Development, vol 9. Routledge, NY. First published in 1993
175. Grubb M, Wieners C, Yang P (2021) Modeling myths: on DICE and dynamic realism in integrated assessment models of climate change mitigation. Wiley Interdisciplinary Reviews: Climate Change, 12(3). Article e698. https://doi.org/10.1002/wcc.698
176. Gunderson L, Holling CS (2002) Panarchy: understanding transformations in human and natural systems. Island Press, Washington, DC, USA. ISBN 978-1-55963-857-9
177. Guo G, Dai B, Li D, Lin Y, Xu Z, Li B (2019) Data model analysis and integration technology based on electric power. J Signal Proc Syst 91:1249–1257. https://doi.org/10.1007/s11265-019-01468-3
178. Haasnoot M, van't Klooster S, Van Alphen J (2018) Designing a monitoring system to detect signals to adapt to uncertain climate change. Global Environ Change, 52:273–285. https://doi.org/10.1016/j.gloenvcha.2018.08.003
179. Haeckel E (1866) Generelle Morphologie der Organismen. Allgemeine Grundzüge der organischen Formen-Wissenschaft, mechanisch begründet durch die von C. Darwin reformirte Descendenz-Theorie, vol 2. https://www.biodiversitylibrary.org/bibliography/102933
180. Hallegatte S, Vogt-Schilb A, Rozenberg J, Bangalore M, Beaudet C (2020) From poverty to disaster and back: a review of the literature. Econom Disasters and Climate Change 4:223–247. https://doi.org/10.1007/s41885-020-00060-5
181. Handayanto RT, Tripathi NK, Kim SM, Guha S (2017) Achieving a sustainable urban form through land use optimisation: insights from Bekasi City's land-use plan (2010–2030). Sustainability 9(2):221. https://doi.org/10.3390/su9020221
182. Hannibal B, Meerow S, Woodruff SC, Roy M, Gilbertson PG, Matos M (2022) Who collaborates on urban resilience? An analysis of flood resilience planning networks in four coastal cities. J Urban Affairs (online first). https://doi.org/10.1080/07352166.2022.2137033
183. Hardin G (1968) The tragedy of the commons. Science 162(3859):1243–1248. https://doi.org/10.1126/science.162.3859.12
184. Harmand S, Lewis JE, Feibel CS, Lepre CJ, Prat S, Lenoble A, Boës X, Quinn RL, Brenet M, Arroyo A, Taylor N (2015) 3.3-million-year-old stone tools from Lomekwi 3, West Turkana, Kenya. Nature 521(7552):310–315. https://doi.org/10.1038/nature14464
185. Harrison PA, Harmáčková ZV, Karabulut AA, Brotons L, Cantele M, Claudet J, Dunford RW, Guisan A, Holman IP, Jacobs S, Kok K (2019) Synthesizing plausible futures for biodiversity and ecosystem services in Europe and Central Asia using scenario archetypes. Ecol Soc 24(2). https://doi.org/10.5751/ES-10818-240227
186. Hastings A, Abbott KC, Cuddington K, Francis T, Gellner G, Lai YC, Morozov A, Petrovskii S, Scranton K, Zeeman ML (2018) Transient phenomena in ecology. Science 361(6406):eaat6412. https://doi.org/10.1126/science.aat6412
187. Hegger DLT, Driessen PPJ, Wiering M, van Rijswick HFMW, Kundzewicz ZW, Matczak P, Crabbé A, Raadgever GT, Bakker MHN, Priest SJ, Larrue C, Ek K (2016) Toward more flood resilience: is a diversification of flood risk management strategies the way forward? Ecol Soc 21(4). Article 52. https://doi.org/10.5751/ES-08854-210452

188. Heikkinen M, Karimo A, Klein J, Juhola S, Ylä-Anttila T (2020) Transnational municipal networks and climate change adaptation: a study of 377 cities. J Clean Prod 257:120474. https://doi.org/10.1016/j.jclepro.2020.120474
189. Heinzelin JD, Clark JD, White T, Hart W, Renne P, WoldeGabriel G, Beyene Y, Vrba E (1999) Environment and behavior of 2.5-million-year-old Bouri hominids. Science 284(5414):625–629. https://doi.org/10.1126/science.284.5414.625
190. Heinzlef C, Robert B, Hémond Y, Serre D (2020) Operating urban resilience strategies to face climate change and associated risks: some advances from theory to application in Canada and France. Cities 104:102762. https://doi.org/10.1016/j.cities.2020.102762
191. Heinzlef C, Barroca B, Leone M, Serre D (2022) Urban resilience operationalization issues in climate risk management: a review. Int J Disaster Risk Reduct 102974. https://doi.org/10.1016/j.ijdrr.2022.102974
192. Helbing D (2012) Systemic risks in society and economics. In Social Self-organization (Springer, Berlin), pp 261–284. https://doi.org/10.1007/978-3-642-24004-1_14. Also available at: https://webarchiv.ethz.ch/soms/teaching/2010_crowds/material/systemic_risk_society_econ.pdf
193. Helbing D (2013) Globally networked risks and how to respond. Nature 497(7447):51–59. https://doi.org/10.1038/nature12047
194. Helmrich AM, Chester MV (2022) Reconciling complexity and deep uncertainty in infrastructure design for climate adaptation. Sustain Resilient Infrastruct 7(2):83–99. https://doi.org/10.1080/23789689.2019.1708179
195. Henderson JV (2010) Cities and development. J Reg Sci 50(1):515–540. https://doi.org/10.1111/j.1467-9787.2009.00636.x
196. Henry M, Schraven D, Bocken N, Frenken K, Hekkert M, Kirchherr J (2021) The battle of the buzzwords: a comparative review of the circular economy and the sharing economy concepts. Environ Innovat Soc Trans 38:1–21. https://doi.org/10.1016/j.eist.2020.10.008
197. Hernandez Y, Naumann G, Barbosa P (2020) Measuring the effectiveness of the Covenant of Mayors on the reporting of climate hazards by Municipalities. Heliyon 6(10). Article e05043. https://doi.org/10.1016/j.heliyon.2020.e05043
198. Ho MW (2013) Circular thermodynamics of organisms and sustainable systems. Systems 1(3):30–49. https://doi.org/10.3390/systems1030030
199. Hofmann R, Müller J, Shatilo L, Videiko M, Ohlrau R, Rud V, Burdo N, Dal Corso M, Dreibrodt S, Kirleis W (2019) Governing Tripolye: integrative architecture in Tripolye settlements. Plos One 14(9). Article e0222243. https://doi.org/10.1371/journal.pone.0222243
200. Hofmann SZ (2021) 100 Resilient Cities program and the role of the Sendai framework and disaster risk reduction for resilient cities. Progress in Disaster Sci 11:100189. https://doi.org/10.1016/j.pdisas.2021.100189
201. Hofstad H, Vedeld T, Agger A, Hanssen GS, Tønnesen A, Valencia S (2022) Cities as public agents: a typology of co-creational leadership for urban climate transformation. Earth Syst Governance 13. Article 100146. https://doi.org/10.1016/j.esg.2022.100146
202. Holling CS (1986) The resilience of terrestrial ecosystems: local surprise and global change. In: Clark WC, Munn RE (eds) Sustainable Development of the Biosphere, 14. Cambridge University Press, Cambridge, UK, pp 292–320
203. Huang W, Zhou B, Yu Y, Sun H, Xu P (2021) Using the disaster spreading theory to analyze the cascading failure of urban rail transit network. Reliab Eng Syst Saf 215:107825. https://doi.org/10.1016/j.ress.2021.107825
204. Hublin JJ (2015) Paleoanthropology: how old is the oldest human? Curr Biol 25(11):R453–R455. https://doi.org/10.1016/j.cub.2015.04.009
205. Hunt DV, Lombardi DR, Atkinson S, Barber AR, Barnes M, Boyko CT, Brown J, Bryson J, Butler D, Caputo S, Caserio M (2012) Scenario archetypes: Converging rather than diverging themes. Sustainability 4(4):740–772. https://doi.org/10.3390/su4040740
206. IPCC-WGI (2022) The Physical Science Basis. Working Group I Contribution to the Sixth Assessment Report. The International Panel for Climate Change. https://www.ipcc.ch/report/sixth-assessment-report-working-group-i/

207. IPCC-WGII (2022) Impacts, Adaptation and Vulnerability. Working Group II Contribution to the Sixth Assessment Report. The International Panel for Climate Change. https://www.ipcc.ch/report/sixth-assessment-report-working-group-ii/
208. IPCC-WGIII (2022) Mitigation of Climate Change. Working Group III Contribution to the Sixth Assessment Report. The International Panel for Climate Change. https://www.ipcc.ch/report/sixth-assessment-report-working-group-3/
209. Israilidis J, Odusanya K, Mazhar MU (2021) Exploring knowledge management perspectives in smart city research: a review and future research agenda. Int J Inf Manage 56.Article 101989. https://doi.org/10.1016/j.ijinfomgt.2019.07.015
210. Janssen M, Van Der Voort H, van Veenstra AF (2015) Failure of large transformation projects from the viewpoint of complex adaptive systems: management principles for dealing with project dynamics. Inf Syst Front 17(1):15–29. https://doi.org/10.1007/s10796-014-9511-8
211. Jaynes ET (1957a) Information theory and statistical mechanics I. Physical Reviews 106:620–630. https://doi.org/10.1103/PhysRev.106.620
212. Jaynes ET (1957b) Information theory and statistical mechanics II. Phys Rev 108:171–190. https://doi.org/10.1103/PhysRev.108.171
213. Jo HA (2019) Coordination vs. control: planning new cities as new economic engines in China. Master Thesis, Massachusetts Institute of Technology (MIT), Department of Urban Studies and Planning (DUSP). https://hdl.handle.net/1721.1/123961
214. Kachali H, Storsjö I, Haavisto I, Kovács G (2018) Inter-sectoral preparedness and mitigation for networked risks and cascading effects. Int J Disaster Risk Red 30:281–291. https://doi.org/10.1016/j.ijdrr.2018.01.029
215. Kajikawa Y, Ohno J, Takeda Y, Matsushima K, Komiyama H (2007) Creating an academic landscape of sustainability science: an analysis of the citation network. Sustain Sci 2:221–231. https://doi.org/10.1007/s11625-007-0027-8
216. Kajikawa Y, Saito O, Takeuchi K (2017) Academic landscape of 10 years of sustainability science. Sustain Sci 12:869–873. https://doi.org/10.1007/s11625-017-0477-6
217. Karhinen S, Peltomaa J, Riekkinen V, Saikku L (2021) Impact of a climate network: the role of intermediaries in local level climate action. Global Environ Change 67. Article 102225. https://doi.org/10.1016/j.gloenvcha.2021.102225
218. Karimi H, Ghadirifaraz B, Shetab Boushehri SN, Hosseininasab SM, Rafiei N (2022) Reducing traffic congestion and increasing sustainability in special urban areas through one-way traffic reconfiguration. Transportation 49:37–60 (2022). https://doi.org/10.1007/s11116-020-10162-4
219. Karvonen A (2018) The city of permanent experiments? In: Turnheim B, Kivimaa P, Berkhout F (eds) Innovating climate governance: moving beyond experiments. Cambridge University Press, Cambridge, UK, pp. 201–215. https://doi.org/10.1017/9781108277679.014.
220. Katz R, Allen T (1982) Investigating the Not Invented Here (NIH) syndrome: a look at the performance, tenure, and communication patterns of 50 R & D Project Groups. R&D Manage 12:7–20. https://doi.org/10.1111/j.1467-9310.1982.tb00478.x
221. Kay JJ, Schneider ED (1992) Thermodynamics and measures of ecological integrity. Ecological Indicators, vol 1. Springer, US, Boston, MA, pp 159–182
222. Kenworthy JR (2006) The eco-city: ten key transport and planning dimensions for sustainable city development. Environ Urban 18(1):67–85. https://doi.org/10.1177/0956247806063947
223. Khalili S, Harre M, Morley P (2018) A temporal social resilience framework of communities to disasters in Australia. Geoenviron Disasters 5:1–9. https://doi.org/10.1186/s40677-018-0114-4
224. Kim D, Lim U (2016) Urban resilience in climate change adaptation: a conceptual framework. Sustainability 8(4). Article 405. https://doi.org/10.3390/su8040405
225. Klein JT (2014) Discourses of transdisciplinarity: looking back to the future. Futures 63:68–74. https://doi.org/10.1016/j.futures.2014.08.008
226. Kona A, Monforti-Ferrario F, Bertoldi P, Baldi MG, Kakoulaki G, Vetters N, Thiel C, Melica G, Lo Vullo E, Sgobbi A, Ahlgren C (2021) Global Covenant of Mayors, a dataset of greenhouse gas emissions for 6200 cities in Europe and the Southern Mediterranean countries. Earth Syst Sci Data 13(7):3551–3564. https://doi.org/10.5194/essd-13-3551-2021

227. Kong L, Mu X, Hu G, Zhang Z (2022) The application of resilience theory in urban development: a literature review. Environ Sci Pollut Res 29:49651–49671. https://doi.org/10.1007/s11356-022-20891-x
228. Kottak CP (1999) The new ecological anthropology. Am Anthropol 101(1):23–35. https://doi.org/10.1525/aa.1999.101.1.23
229. Krätke S (2010) Regional knowledge networks: a network analysis approach to the interlinking of knowledge resources. Eur Urban Reg Stud 17(1):83–97. https://doi.org/10.1177/0969776409350794
230. Kutzschenbach MV, Schmid A, Schoenenberger L (2018) Using feedback systems thinking to explore theories of digital business for medtech companies. In Business Information Systems and Technology 4.0. Springer, Cham, pp 161–175. https://doi.org/10.1007/978-3-319-74322-6_11
231. Lacroix K, Gifford R (2018) Psychological barriers to energy conservation behavior: the role of worldviews and climate change risk perception. Environ Behav 50(7):749–780. https://doi.org/10.1177/0013916517715296
232. Lawton Smith H, Leydesdorff L (2014) The Triple Helix in the context of global change: dynamics and challenges. Prometheus 32(4):321–336. https://doi.org/10.1080/08109028.2014.972135
233. Levallois C (2010) Can de-growth be considered a policy option? A historical note on Nicholas Georgescu-Roegen and the Club of Rome. Ecol Econ 69:2271–2278. https://doi.org/10.1016/j.ecolecon.2010.06.020
234. Lewin AY, Zhong X (2013) The evolving diaspora of talent: a perspective on trends and implications for sourcing science and engineering work. J Int Manage 19(1):6–13. https://www.sciencedirect.com/science/article/pii/S107542531200066X
235. Leydesdorff L (2012) The Triple Helix, Quadruple Helix, …, and an N-Tuple of Helices: explanatory models for analyzing the knowledge-based economy? J Knowl Econ 3:25–35. https://doi.org/10.1007/s13132-011-0049-4
236. Leydesdorff L, Etzkowitz H (1996) Emergence of a Triple Helix of university—industry—government relations. Sci Publ Pol 23(5):279–286. https://doi.org/10.1093/spp/23.5.279
237. Liberalesso T, Cruz CO, Silva CM, Manso M (2020) Green infrastructure and public policies: an international review of green roofs and green walls incentives. Land Use Policy 96:104693. https://doi.org/10.1016/j.landusepol.2020.104693
238. Linxue LI, Simin TAO (2018) Towards thermodynamic architecture: research on systems-based design oriented by renewable energy. In: IOP conference series: Earth & Environmental Science, vol 146, No 1. IOP Publishing, p 012060. https://doi.org/10.1088/1755-1315/146/1/012060
239. Liu J, Dietz T, Carpenter SR, Alberti M, Folke C, Moran E, Pell AN, Deadman P, Kratz T, Lubchenco J, Ostrom E (2007) Complexity of coupled human and natural systems. Science 317:1513–1516. https://doi.org/10.1126/science.11440
240. Liu J, Dietz T, Carpenter SR, Taylor WW, Alberti M, Deadman P, Redman C, Pell A, Folke C, Ouyang Z, Lubchenco J (2021a) Coupled human and natural systems: The evolution and applications of an integrated framework. Ambio 50(10):1778–1783. https://doi.org/10.1007/s13280-020-01488-5
241. Liu J, Gatzweiler FW, Kumar M (2021b) An evolutionary complex systems perspective on urban health. Socio-Econom Plann Sci 75. Article 100815. https://doi.org/10.1016/j.seps.2020.100815
242. Liu M, Lo K (2021) A comparative review of urban climate governance in Chinese and western contexts. Urban Governance 1(2):81–88. https://doi.org/10.1016/j.ugj.2021.11.003
243. Lobo J, Alberti M, Allen-Dumas M, Arcaute E, Barthelemy M, Bojorquez Tapia LA, Brail S, Bettencourt L, Beukes A, Chen WQ, Florida R (2020) Urban science: integrated theory from the first cities to sustainable metropolises. Report submitted to the NSF on the Present State and Future of Urban Science. https://papers.ssrn.com/sol3/papers.cfm?abstract_id=3526940
244. Loorbach D, Frantzeskaki N, Avelino F (2017) Sustainability transitions research: transforming science and practice for societal change. Ann Rev Environ Res 42:599–626.https://doi.org/10.1146/annurev-environ-102014-021340

245. Loorbach D, Schwanen T, Doody BJ, Arnfalk P, Langeland O, Farstad E (2021) Transition governance for just, sustainable urban mobility: an experimental approach from Rotterdam, the Netherlands. J Urban Mobil 1:100009. https://doi.org/10.1016/j.urbmob.2021.100009
246. Londo DL (2020) The impact of international migration on immigrant's home country: exploring diaspora power and conflict. Am J Int Relat 5(1):14–25. https://doi.org/10.47672/ajir.508
247. Lozano-Garzon CA, Camelo M, Vilà P, Donoso Y (2015) A Multi-objective routing algorithm for wireless mesh network in a smart cities environment. J Networks 10(1):60–69
248. Madni AM, Jackson S (2009) Towards a conceptual framework for resilience engineering. IEEE Syst J 3(2):181–191. https://doi.org/10.1109/JSYST.2009.2017397
249. Majoor S (2015) Urban megaprojects in crisis? Ørestad Copenhagen revisited. Eur Plann Stud 23(12):2497–2515. https://doi.org/10.1080/09654313.2015.1014780
250. Majoor S (2018) Ambiguity as an opportunity: coping with change in Urban Megaprojects. In: Salet W (ed) The Routledge Handbook of Institutions and Planning in Action. Routledge, New York, pp 331–346. https://doi.org/10.4324/9781315111230
251. Mandea M, Chambodut A (2020) Geomagnetic field processes and their implications for space weather. Surv Geophys 41:1611–1627. https://doi.org/10.1007/s10712-020-09598-1
252. Manso M, Teotónio I, Silva CM, Cruz CO (2021) Green roof and green wall benefits and costs: a review of the quantitative evidence. Renew Sustain Energy Rev 135:110111. https://doi.org/10.1016/j.rser.2020.110111
253. Manyena B, O'Brien G, O'Keefe P, Rose J (2011) Disaster resilience: a bounce back or bounce forward ability? Local Environ: Int J Justice Sustain 16(5):417–424. https://doi.org/10.1080/13549839.2011.583049
254. Martins VWB, Rampasso IS, Anholon R, Quelhas OLG, Leal Filho W (2019) Knowledge management in the context of sustainability: literature review and opportunities for future research. J Clean Prod 229:489–500. https://doi.org/10.1016/j.jclepro.2019.04.354
255. Masoumi Z, van Genderen J (2023) Artificial Intelligence for sustainable development of smart cities and urban land-use management. Geo-spatial Inf Sci (published online, 20 April). https://doi.org/10.1080/10095020.2023.2184729
256. Maunier R (1910) The definition of the city. Am J Sociol 15(4):536–548. https://www.journals.uchicago.edu/doi/pdf/10.1086/211804
257. Mayer AL, Donovan RP, Pawlowski CW (2014) Information and entropy theory for the sustainability of coupled human and natural systems. Ecol Soc 19(3). https://www.jstor.org/stable/26269639
258. Mayrand F, Clergeau P (2018) Green roofs and green walls for biodiversity conservation: a contribution to urban connectivity? Sustainability 10(4):985. https://doi.org/10.3390/su10040985
259. Mazzucato C (2019) Socio-material archaeological networks at Çatalhöyük a community detection approach. Front Dig Humanit 6. Article 8. https://doi.org/10.3389/fdigh.2019.00008
260. Mckinney LA (2012) Entropic disorder: new frontiers in environmental sociology. Soc Perspect 55(2):295–317. https://doi.org/10.1525/sop.2012.55.2.295.296
261. McKinney L (2019) The entropy curse. In: Frey RS, Gellert PK, Dahms HF (eds) Ecologically unequal exchange: Environmental injustice in comparative and historical perspective, pp 143–165. https://doi.org/10.1007/978-3-319-89740-0_6
262. McMahon A (2021) Corinne Castel, Jan-Waalke Meyer and Philippe Quenet (eds.)–Circular Cities of Early Bronze Syria. Paléorient. Revue Pluridisciplinaire de Préhistoire et de Protohistoire de l'Asie du Sud-Ouest et de l'Asie Centrale 47(2):125–127. https://doi.org/10.4000/paleorient.1435
263. McNeeley SM, Lazrus H (2014) The cultural theory of risk for climate change adaptation. Weather, Climate, and Society 6(4):506–519. https://doi.org/10.1175/WCAS-D-13-00027.1
264. McPhearson T, Haase D, Kabisch N, Gren Å (2016) Advancing understanding of the complex nature of urban systems. Ecol Ind 70:566–573. https://doi.org/10.1016/j.ecolind.2016.03.054
265. Meadows DH, Meadows DL, Randers J, Behrens W (1972) The limits to growth. Universe Books, New York. Digitalized version available at: https://collections.dartmouth.edu/teitexts/meadows/diplomatic/meadows_ltg-diplomatic.html

266. Meerow S, Newell JP, Stults M (2016) Defining urban resilience: a review. Landsc Urban Plan 147:38–49. https://doi.org/10.1016/j.landurbplan.2015.11.011
267. Melendez A, Caballero-Russi D, Gutierrez Soto M, Giraldo LF (2022) Computational models of community resilience. Nat Hazards 111(2):1121–1152. https://doi.org/10.1007/s11069-021-05118-5
268. Mesjasz C (2016) Sustainability and complexity: a few lessons from modern systems thinking. In: Brauch HG, Oswald Spring U, Grin J, Scheffran J (eds) Handbook on sustainable transitions and sustainable peace. Hexagon series on human and environmental security and peace, vol 10. Springer, Switzerland, pp 421–450. https://doi.org/10.1007/978-3-319-43884-9
269. Mills BJ (2017) Social network analysis in archaeology. Annu Rev Anthropol 46:379–397. https://doi.org/10.1146/annurev-anthro-102116-041423
270. Mitchell M (2006) Complex systems: network thinking. Artif Intell 170:1194–1212.https://doi.org/10.1016/j.artint.2006.10.002
271. Mohammed SB (2022) Re-Thinking the global cosmopolis: an analysis of the UN-Habitat "city we need" policies in Helsinki and Sydney. Cogent Arts Humanit 9(1). Article 2132615. https://doi.org/10.1080/23311983.2022.2132615
272. Molotch H (1976) The city as a growth machine: toward a political economy of place. Am J Sociol 82(2):309–332. https://doi.org/10.1086/226311
273. Monah D (2005) Religie si arta in cultura Cucuteni [Religion and art in Cucuteni culture]. In: Dumitroaia G (ed) Primul muzeu Cucuteni din Romania [The first Cucuteni museum for Romania] (in Romanian). Bibliotheca Memoriae Antiquitatis XV:162–173. Editura Foton, Piatra-Neamţ, Romania
274. Monroe JC (2018) "Elephants for want of towns": archaeological perspectives on West African cities and their hinterlands. J Archaeol Res 26:387–446. https://doi.org/10.1007/s10814-017-9114-2
275. Moser S, Meerow S, Arnott J, Jack-Scott E (2019) The turbulent world of resilience: interpretations and themes for transdisciplinary dialogue. Clim Change 153(1–2):21–40. https://doi.org/10.1007/s10584-018-2358-0
276. Moustaka V, Vakali A, Anthopoulos LG (2018) A systematic review for smart city data analytics. ACM Comput Surveys 51(5). Article 103. https://doi.org/10.1145/3239566
277. Mulcahy NJ, Call J (2006) Apes save tools for future use. Science 312(5776):1038–1040. https://doi.org/10.1126/science.1125456
278. Müller J, Hofmann R, Shatilo M (2022) Tripolye mega-sites: "Collective Computational Abilities" of Prehistoric Proto-Urban Societies? J Soc Comput 3(1):75–90. https://doi.org/10.23919/JSC.2021.0034
279. Mumford L (1937) What is a City? (Republished in LeGates RT, Stout F (eds) The City Reader, 2011. https://www.taylorfrancis.com/chapters/edit/10.4324/9780203869260-21/city-architectural-record-1937-lewis-mumford). Also available as scans of the original paper: https://bpb-us-e1.wpmucdn.com/blog.uta.edu/dist/b/4296/files/2014/08/mumford_what_is_a_city_1937.pdf
280. Munasinghe M (2011) Addressing sustainable development and climate change together using sustainomics. Wiley Interdisciplinary Rev: Climate Change 2(1):7–18. https://doi.org/10.1002/wcc.86
281. Muñoz-Erickson TA (2014) Co-production of knowledge–action systems in urban sustainable governance: the KASA approach. Environ Sci Policy 37:182–191. https://doi.org/10.1016/j.envsci.2013.09.014
282. Muñoz-Erickson TA, Cutts BB, Larson EK, Darby KJ, Neff M, Wutich A, Bolin B (2010) Spanning boundaries in an Arizona watershed partnership: Information networks as tools for entrenchment or ties for collaboration? Ecol Soc 15(3). Article 22. http://www.ecologyandsociety.org/vol15/iss3/art22/10.1890/100035
283. Muñoz-Erickson TA, Meerow S, Hobbins R, Cook E, Iwaniec DM, Berbés-Blázquez M, Grimm NB, Barnett A, Cordero J, Gim C, Miller TR (2021) Beyond bouncing back? Comparing and contesting urban resilience frames in US and Latin American contexts. Landsc Urban Plan 214:104173. https://doi.org/10.1016/j.landurbplan.2021.104173

284. Murakami Y (2014) Influences of return migration on international collaborative research networks: cases of Japanese scientists returning from the US. J Technol Transf 39(4):616–634. https://doi.org/10.1007/s10961-013-9316-9
285. Myeong S, Kim Y, Ahn MJ (2020) Smart city strategies—technology push or culture pull? A case study exploration of Gimpo and Namyangju, South Korea. Smart Cities 4(1):41–53. https://doi.org/10.3390/smartcities4010003
286. Nachbagauer A (2021) Managing complexity in projects: extending the Cynefin framework. Project Leadership and Soc 2:100017. https://doi.org/10.1016/j.plas.2021.100017
287. NASA-JPL (1996) PIA00452: solar system portrait—Earth as 'Pale Blue Dot'. *Photojurnal*, PIA00452. Jet Propulsion Laboratory (JPL), National Aeronautics and Space Administration (NASA). https://photojournal.jpl.nasa.gov/catalog/PIA00452. Accessed: 30 March 2023
288. NASA-JPL (2020). PIA23645: Pale Blue Dot Revisited. Photojurnal, PIA23645. Jet Propulsion Laboratory (JPL), National Aeronautics and Space Administration (NASA). https://photojournal.jpl.nasa.gov/catalog/PIA23645. Accessed: 30 March 2023
289. Navarro J, Hancock PA (2023) Did tools create humans? Theor Issues Ergon Sci 24(2):206–232. https://doi.org/10.1080/1463922X.2022.2076954
290. Naveh Z (1987) Biocybernetic and thermodynamic perspectives of landscape functions and land use patterns. Landscape Ecol 1(2):75–83. https://doi.org/10.1007/BF00156229
291. Neij L, Heiskanen E (2021) Municipal climate mitigation policy and policy learning-A review. J Cleaner Prod 317. Article 128348. https://doi.org/10.1016/j.jclepro.2021.128348
292. Nelson M (2021) Biosphere 2's lessons about living on Earth and in space. Space: Sci Tech 2021. Article 8067539. https://doi.org/10.34133/2021/8067539
293. Netto VM, Meirelles J, Ribeiro FL (2017) Social interaction and the city: the effect of space on the reduction of entropy. Complexity. Article ID 6182503. https://doi.org/10.1155/2017/6182503
294. Newman TP, Nisbet EC, Nisbet MC (2018) Climate change, cultural cognition, and media effects: worldviews drive news selectivity, biased processing, and polarized attitudes. Publ Understanding Sci 27(8):985–1002. https://doi.org/10.1177/0963662518801170
295. Noss RF (1995) Ecological integrity and sustainability: buzzwords in conflict? In: Westra L, Lemons J (eds) Perspectives on ecological integrity. Environ Sci Technol Library, 5:60–76. Springer, Dordrecht. https://doi.org/10.1007/978-94-011-0451-7_5
296. Oates J, McMahon A, Karsgaard PA, Quntar S, Ur J (2007) Early Mesopotamian urbanism: a new view from the north. Antiquity 81(313):585–600. https://doi.org/10.1017/S0003598X00095600
297. O'Brien K (2021) Reflecting on the Anthropocene: the call for deeper transformations. Ambio 50(10):1793–1797. https://doi.org/10.1007/s13280-020-01468-9
298. Ocko IB, Hamburg SP, Jacob DJ, Keith DK, Keohane NO, Oppenheimer M, Roy-Mayhew JD, Schrag DP, Pacala SW (2017) Unmask temporal trade-offs in climate policy debates. Science 356(6337):492–493. https://doi.org/10.1126/science.aaj2350
299. Odum E, Odum HT (1953) Fundamentals of ecology. XV. B. Saunders Co, Philadelphia
300. Oliveira TA, Oliver M, Ramalhinho H (2020) Challenges for connecting citizens and smart cities: ICT, e-governance and blockchain. Sustainability 12(7):2926. https://doi.org/10.3390/su12072926
301. Omata N (2023) The role of developmental 'buzzwords' in the international refugee regime: Self-reliance, resilience, and economic inclusion. World Dev 167:106248. https://doi.org/10.1016/j.worlddev.2023.106248
302. Opute J (2018) Management of diaspora businesses: issues and learning. In: Hack-Polay D, Siwale J (eds) African Diaspora Direct Investment. Palgrave Studies of Entrepreneurship in Africa. Palgrave Macmillan, Cham. https://doi.org/10.1007/978-3-319-72047-0_9
303. O'Riordan T, Jordan A (1999) Institutions, climate change and cultural theory: towards a common analytical framework. Glob Environ Chang 9(2):81–93. https://doi.org/10.1016/S0959-3780(98)00030-2
304. Orlove BS (1980) Ecological anthropology. Ann Rev Anthropol 235–273. https://www.jstor.org/stable/2155736

305. Ortega PA, Braun DA (2013) Thermodynamics as a theory of decision-making with information-processing costs. Proc Roy Soc A: Math, Phys Eng Sci 469:20120683. https://doi.org/10.1098/rspa.2012.0683
306. Ortíz-Moya F, Saraff E, Kataoka Y, Fujino J (2021) State of the voluntary local reviews 2021: 40 From Reporting to Action. 1st ed. IGES, Kamiyamaguchi, 66 pp. https://catalog.lib.kyushu-u.ac.jp/opac_search/?lang=1&amode=9&opkey=B169756800074518&start=1&cmode=0&place=&srvce=0
307. Ortman SG, Lobo J, Smith ME (2020) Cities: complexity, theory and history. PLoS ONE 15(12):e0243621. https://doi.org/10.1371/journal.pone.0243621
308. Ostrom E (1990) Governing the commons: the evolution of institutions for collective action (political economy of institutions and decisions). Cambridge University Press, Cambridge UK. ISBN: 0-521-40599-8
309. Ostrom E (1999) Coping with tragedies of the commons. Annu Rev Polit Sci 2(1):493–535. https://doi.org/10.1146/annurev.polisci.2.1.493
310. Ostrom E, Burger J, Field CB, Norgaard RB, Policansky D (1999) Revisiting the commons: local lessons, global challenges. Science 284:278–282. https://doi.org/10.1126/science.284.5412.278
311. Ostrom E (2009a) A general framework for analyzing sustainability of social-ecological systems. Science 325(5939):419–422. https://doi.org/10.1126/science.1172133
312. Ostrom E (2009b) Beyond markets and states: polycentric governance of complex economic systems. Nobel Prize Lecture (The Sveriges Riksbank Prize in Economic Sciences in Memory of Alfred Nobel 2009), 8 December 2009 at Aula Magna, Stockholm University, Stockholm. https://www.nobelprize.org/prizes/economic-sciences/2009/ostrom/lecture/
313. Osuteye E, Johnson C, Brown D (2017) The data gap: an analysis of data availability on disaster losses in sub-Saharan African cities. Int J Disaster Risk Reduct 26:24–33. https://doi.org/10.1016/j.ijdrr.2017.09.026
314. Ott BL, Aoki E (2001) Popular imagination and identity politics: reading the future in Star Trek: the next generation. Western J Commun (Includes Commun Rep) 65(4):392–415. https://doi.org/10.1080/10570310109374718
315. Palermo V, Bertoldi P, Apostolou M, Kona A, Rivas S (2020) Assessment of climate change mitigation policies in 315 cities in the Covenant of Mayors initiative. Sustain Cities Soc 60:102258. https://doi.org/10.1016/j.scs.2020.102258
316. Papakonstantinou N, Van Bossuyt DL, Linnosmaa J, Hale B, O'Halloran B (2021) A zero trust hybrid security and safety risk analysis method. J Comput Inf Sci Eng 21(5). https://doi.org/10.1115/1.4050685
317. Parnell S, Robinson J (2017) The global urban: difference and complexity in urban studies and the science of cities. In: Handbook of Social Science. Routledge, London, pp 13-31. ISBN 9781032022567
318. Parr JB (2007) Spatial definitions of the city: four perspectives. Urban Studies 44(2):381–392. https://doi.org/10.1080/00420980601075059
319. Pelorosso R, Gobattoni F, Leone A (2017) The low-entropy city: a thermodynamic approach to reconnect urban systems with nature. Landsc Urban Plan 168:22–30. https://doi.org/10.1016/j.landurbplan.2017.10.002
320. Pendall R, Foster KA, Cowell M (2010) Resilience and regions: building understanding of the metaphor. Camb J Reg Econ Soc 3(1):71–84. https://doi.org/10.1093/cjres/rsp028
321. Pendergraft CA (1998) Human dimensions of climate change: cultural theory and collective action. Clim Change 39(4):643–666. https://doi.org/10.1023/A:1005323809980
322. Perera ATD, Coccolo S, Scartezzini JL, Mauree D (2018) Quantifying the impact of urban climate by extending the boundaries of urban energy system modeling. Appl Energy 222:847–860. https://doi.org/10.1016/j.apenergy.2018.04.004
323. Perez C (2010) Technological revolutions and techno-economic paradigms. Camb J Econ 34(1):185–202. https://doi.org/10.1093/cje/bep051
324. Perez C (2015) From long waves to great surges: continuing in the direction of Chris Freeman's 1997 lecture on Schumpeter's Business Cycles. Eur J Econom Soc Syst 27(1–2):9–80. https://econpapers.repec.org/article/risejessy/0005.htm

325. Pescaroli G (2018) Perceptions of cascading risk and interconnected failures in emergency planning: Implications for operational resilience and policy making. Int J Disaster Risk Reduct 30:269–280. https://doi.org/10.1016/j.ijdrr.2018.01.019
326. Pescaroli G, Nones M, Galbusera L, Alexander D (2018) Understanding and mitigating cascading crises in the global interconnected system. Int J Disaster Risk Reduct 30:159–163. https://doi.org/10.1016/j.ijdrr.2018.07.004
327. Pickartz N, Rabbel W, Rassmann K, Hofmann R, Ohlrau R, Thorwart M, Wilken D, Wunderlich T, Videiko M, Müller J (2022) Inverse filtering of magnetic prospection data—a gateway to the social structure of Cucuteni–Tripolye settlements? Rem Sens 14. Article 484. https://doi.org/10.3390/rs14030484
328. Pickering M (2004) Experience as horizon: Koselleck, expectation and historical time. Cult Stud 18(2–3):271–289. https://doi.org/10.1080/0950238042000201518
329. Porter MN, Mora MCET, Trevino MJP, Eugster J, Papamichalis T (2022) The systemic impact of debt Default in a Multilayered Global Network Model. Int Monetary Fund. https://www.elibrary.imf.org/view/journals/001/2022/171/article-A001-en.xml (Accessed 08 May 2023)
330. Portugali J (2016) What makes cities complex? In Complexity, Cognition, Urban Planning and Design: Post-proceedings of the 2nd Delft International Conference, pp 3–19. Springer International Publishing. https://doi.org/10.1007/978-3-319-32653-5_1
331. Prigogine I (1997) The End of Certainty: Time, Chaos and the New Laws of Nature. The Free Press, New York. ISBN 0684837056
332. Puente-Sotomayor F, Egas A, Teller J (2021) Land policies for landslide risk reduction in Andean cities. Habitat International, 107. article 102298. https://doi.org/10.1016/j.habitatint.2020.102298
333. Pumain D, Rozenblat C (2018) Two metropolisation gradients in the European system of cities revealed by scaling laws. Env Plann B: Urban Anal City Sci 46(9):1645–1662. https://doi.org/10.1177/2399808318785633
334. Pyka A, Bogner K, Urmetzer S (2019) Productivity slowdown, exhausted opportunities and the power of Human ingenuity—Schumpeter meets Georgescu-Roegen. J Open Innovat: Tech, Market, Compl 5(3). Article 39. https://doi.org/10.3390/joitmc5030039
335. Quattrochi DA, Luvall JK (2014) Thermal infrared remote sensing for analysis of landscape ecological processes: current insights and trends. In: Wang G (ed) Scale issues in remote sensing, Chapter 3, pp 34–60. https://doi.org/10.1002/9781118801628.ch03
336. Raimbault J (2020) Indirect evidence of network effects in a system of cities. Environ Plann B: Urban Analyt City Sci 47(1):138–155. https://doi.org/10.1177/2399808318774335
337. Rapoport E, Vernay AL (2011) Defining the eco-city: a discursive approach. Paper presented at the Management and Innovation for a Sustainable Built Environment conference, International Eco-cities Initiative, Amsterdam, The Netherlands. pp 1–15. http://resolver.tudelft.nl/uuid:87abcfee-e6fe-4e3a-86a3-7741a1a36642
338. Raskin PD (2005) Global scenarios: background review for the Millennium Ecosystem Assessment. Ecosystems, 8(2):133–142. https://doi.org/10.1007/s10021-004-0074-2.
339. Reason J (1990) Human error. Cambridge University Press, Cambridge, UK. https://doi.org/10.1017/CBO9781139062367
340. Redman CL (2005) Resilience theory in archaeology. Am Anthropol 107(1):70–77. https://doi.org/10.1525/aa.2005.107.1.070
341. Redman CL (2014) Should sustainability and resilience be combined or remain distinct pursuits? Ecol Soc 19(2). Article 37. https://doi.org/10.5751/ES-06390-190237
342. Redman CL, Kinzig AP (2003) Resilience of past landscapes: resilience theory, society, and the longue durée. Ecol and Soc 7(1). Article 14. http://www.consecol.org/vol7/iss1/art14.
343. Rees W (2012) Cities as dissipative structures: global change and the vulnerability of urban civilization. In: Weinstein MP, Turner RE (eds) Sustainability science: the emerging paradigm and the urban environment. Springer, London (UK). https://doi.org/10.1007/978-1-4614-3188-6_12
344. Rees W, Wackernagel M (1996) Urban ecological footprints: why the cities cannot be sustainable—and why they are a key to sustainability. Environ Impact Assess Rev 16:223–248. https://doi.org/10.1016/S0195-9255(96)00022-4

345. Reghezza-Zitt M, Rufat S, Djament-Tran G, Le Blanc A, Lhomme S (2012) What resilience is not: uses and abuses. Cybergeo. Article 621. https://doi.org/10.4000/cybergeo.25554
346. Ricart S, Berizzi C, Saurí D, Terlicher GN (2022) The social, political, and environmental dimensions in designing urban public space from a water management perspective: testing european experiences. Land 11(9):1575. https://doi.org/10.3390/land11091575
347. Richardson K, Steffen W, Lucht W, Bendtsen J, Cornell SE, Donges JF, Drüke M, Fetzer I, Bala G, von Bloh W, Feulner G (2023) Earth beyond six of nine Planetary Boundaries. Science Advances (published online: 13 September 2023). https://doi.org/10.1126/sciadv.adh2458
348. Robinson RT, Huppler AR (2017) The Goldilocks model of immune symbiosis with Mycobacteria and Candida colonizers. Cytokine 97:49–65. https://doi.org/10.1016/j.cyto.2017.05.015
349. Rondinel-Oviedo DR, Keena N (2023) Entropy and cities: a bibliographic analysis towards more circular and sustainable urban environments. Entropy 25(3):532. https://doi.org/10.3390/e25030532
350. Roseland M (1997) Dimensions of the eco-city. Cities 14(4):197–202. https://doi.org/10.1016/S0264-2751(97)00003-6
351. Roseland M, Devuyst D (2001) The eco-city approach to sustainable development in urban areas. In: How green is the city? Columbia University Press, pp 85–104. https://doi.org/10.7312/devu11802-009.
352. Rosser JB Jr (2021) Econophysics and the Entropic Foundations of Economics. Entropy 23:1286. https://doi.org/10.3390/e23101286
353. Rowan NJ, Casey O (2021) Empower Eco multiactor HUB: a triple helix 'academia-industry-authority' approach to creating and sharing potentially disruptive tools for addressing novel and emerging new Green Deal opportunities under a United Nations Sustainable Development Goals framework. Current Opin Environ Sci Health 21:100254. https://doi.org/10.1016/j.coesh.2021.100254
354. Rogov M, Rozenblat C (2018) Urban resilience discourse analysis: towards a multi-level approach to cities. Sustainability, 10. Article 4431. https://doi.org/10.3390/su10124431
355. Rotmans J, Kemp R, van Asselt M (2001) More evolution than revolution: transition management in public policy. Foresight 3(1):15–31. https://doi.org/10.1108/14636680110803003
356. Rufat S (2015) Critique of pure resilience. In: Reghezza-Zitt M, Rufat S (eds) Resilience Imperative. Elsevier, pp 201–228. https://doi.org/10.1016/B978-1-78548-051-5.50011-5
357. Rutter H, Savona N, Glonti K, Bibby J, Cummins S, Finegood DT, Greaves F, Harper L, Hawe P, Moore L, Petticrew M (2017) The need for a complex systems model of evidence for public health. The Lancet 390(10112):2602–2604. https://doi.org/10.1016/S0140-6736(17)31267-9
358. Rybski D, Arcaute E, Batty M (2019) Urban scaling laws. Environ Plann B: Urban Analyt City Sci 46(9):1605–1610. https://doi.org/10.1177/2399808319886125
359. Rybski D, González MC (2022) Cities as complex systems—collection overview. Plos One 17(2). Article e0262964. https://doi.org/10.1371/journal.pone.0262964
360. Sagan CE (1990) The Earth from the frontiers of the Solar System—The Pale, Blue Dot. Magazine, 9:11–12. https://news.google.com/newspapers?id=_upSAAAAIBAJ&pg=4800,1930437&dq=pale-blue-dot+-book&hl=en (Last accessed: 29 March 2023)
361. Sagan CE (1994) Pale Blue Dot: A Vision of the Human Future in Space. Random House, New York. ISBN 0-679-43841-6
362. Saja AA, Teo M, Goonetilleke A, Ziyath AM (2021) A critical review of social resilience properties and pathways in disaster management. Int J Disaster Risk Sci 12:790–804. https://doi.org/10.1007/s13753-021-00378-y
363. Saner R, Yiu L, Kingombe C (2019) The 2030 Agenda compared with six related international agreements: valuable resources for SDG implementation. Sustain Sci 14(6):1685–1716. https://doi.org/10.1007/s11625-019-00655-2
364. Santé I, García AM, Miranda D, Crecente R (2010) Cellular automata models for the simulation of real-world urban processes: a review and analysis. Landsc Urban Plan 96(2):108–122. https://doi.org/10.1016/j.landurbplan.2010.03.001

365. Sassen S (2012) Cities: a window into larger and smaller worlds. Eur Educat Res J 11(1):1–10. https://doi.org/10.2304/eerj.2012.11.1.1
366. Sassen S (2017) Beyond differences of race, religion, class: making urban subjects. Ethics of the Urban: The City and the Spaces of the Political. Harvard University Graduate School of Design. Zurich: Lars Müller Publishers
367. Scerri EM, Will M (2023) The revolution that still isn't: the origins of behavioral complexity in Homo sapiens. J Human Evol 179. Article 103358. https://doi.org/10.1016/j.jhevol.2023.103358
368. Schäpke N, Stelzer F, Caniglia G, Bergmann M, Wanner M, Singer-Brodowski M, Loorbach D, Olsson P, Baedeker C, Lang DJ (2018) Jointly experimenting for transformation?: shaping real-world laboratories by comparing them. GAIA 27:85–96. https://doi.org/10.14512/gaia.27.S1.16. Also available at: http://refhub.elsevier.com/S2214-6296(20)30299-1/h0405
369. Scharf MP, Roberts LD (1994) The interstellar relations of the federation: international law and Star Trek: The Next Generation. Facul Publ 867. Case Western Reserve University, School of Law, Case Research Paper Series in Legal Studies. Cleveland, Ohio. https://scholarlycommons.law.case.edu/faculty_publications/867
370. Scheffran J (2016) From a climate of complexity to sustainable peace: viability transformations and adaptive governance in the Anthropocene. In: Brauch HG, Oswald Spring U, Grin J, Scheffran J (eds) Handbook on sustainable transitions and sustainable peace, pp 305–346. Hexagon Series on Human and Environmental Security and Peace, Vol. 10. Springer, Switzerland. https://doi.org/10.1007/978-3-319-43884-9
371. Schlör H, Venghaus S, Fischer W, Märker C, Hake JF (2018) Deliberations about a perfect storm–The meaning of justice for food energy water-nexus (FEW-Nexus). J Environ Manage 220:16–29. https://doi.org/10.1016/j.jenvman.2018.04.097
372. Schmithüsen FJ (2013) Three hundred years of applied sustainability in forestry. Unasylva 64(240):3–11. https://doi.org/10.3929/ethz-a-009955604
373. Schneider ED, Kay JJ (1994) Life as a manifestation of the second law of thermodynamics. Math Comput Model 19(6–8):25–48. https://doi.org/10.1016/0895-7177(94)90188-0
374. Scholz RW, Spoerri A, Lang DJ (2009) Problem structuring in transitions: the case of Swiss waste management. Futures 41(3):171–181. https://doi.org/10.1016/j.futures.2008.09.013
375. Schwartz JH, Tattersall I (2010) Fossil evidence for the origin of Homo sapiens. Am J Phys Anthropol 143(S51):94–121. https://doi.org/10.1002/ajpa.21443
376. Scott AJ (2017) The constitution of the city. Economy, Society, and Urbanization in the Capitalist Era. Palgrave Macmillan (Springer Nature). https://doi.org/10.1007/978-3-319-61228-7
377. Seduikyte L, Gražulevičiūtė-Vileniškė I, Povilaitienė I, Fokaides PA, Lingė D (2023) Trends and interdisciplinarity integration in the development of the research in the fields of sustainable, healthy and digital buildings and cities. Buildings 13(7). Article 1764. https://doi.org/10.3390/buildings13071764
378. Sellberg MM, Wilkinson C, Peterson GD (2015) Resilience assessment: a useful approach to navigate urban sustainability challenges. Ecol 20(1). https://doi.org/10.5751/ES-07258-200143
379. Shahrokni H, Lazarevic D, Brandt N (2015a) Smart urban metabolism: towards a real-time understanding of the energy and material flows of a city and its citizens. J Urban Tech 22:65–86. https://doi.org/10.1080/10630732.2014.954899
380. Shahrokni H, Årman L, Lazarevic D, Nilsson A, Brandt N (2015b) Implementing smart urban metabolism in the Stockholm Royal Seaport: Smart city SRS. J Ind Ecol 19:917–929. https://doi.org/10.1111/jiec.12308
381. Sherwin WB, Prat i Fornells N (2019) The introduction of entropy and information methods to ecology by Ramon Margalef. Entropy 21. Article 794. https://doi.org/10.3390/e21080794
382. Shore J, Johnson R (1980) Axiomatic derivation of the principle of maximum entropy and the principle of minimum cross-entropy. IEEE Trans Inf Theo 26(1):26–37. https://doi.org/10.1109/TIT.1980.1056144

383. Skrimizea E, Haniotou H, Parra C (2019) On the 'complexity turn' in planning: an adaptive rationale to navigate spaces and times of uncertainty. Plann Theo 18:122–142. https://doi.org/10.1177/1473095218780515
384. Shindell D, Borgford-Parnell N, Brauer M, Haines A, Kuylenstierna JCI, Leonard SA, Ramanathan V, Ravishankara A, Amann M, Srivastava L (2017) A climate policy pathway for near- and long-term benefits. Science 356(6337):493–494. https://doi.org/10.1126/science.aak9521
385. Simmonds J, Gómez JA, Ledezma A (2020) The role of agent-based modeling and multi-agent systems in flood-based hydrological problems: a brief review. J Water Climate Change 11(4):1580–1602. https://doi.org/10.2166/wcc.2019.108
386. Simon D (ed) Rethinking sustainable cities. Accessible, green and fair. Policy Press, University of Bristol, 2016, Bristol, UK. ISBN: 978-1-4473-3284-8
387. Sindbæk SM (2022) Weak ties and strange attractors: anomalocivitas and the archaeology of urban origins. J Urban Archaeol 5:19–32. https://doi.org/10.1484/J.JUA.5.129841
388. Sitas N, Harmáčková ZV, Anticamara JA, Arneth A, Badola R, Biggs R, Blanchard R, Brotons L, Cantele M, Coetzer K, DasGupta R (2019) Exploring the usefulness of scenario archetypes in science-policy processes. Ecol Soc 24(3). https://doi.org/10.5751/ES-11039-240335
389. Snowden D, Rancati A (2021) Managing complexity (and chaos) in times of crisis. A field guide for decision makers inspired by the Cynefin framework. Publications Office of the European Union, Luxembourg. https://publications.jrc.ec.europa.eu/repository/handle/JRC123629
390. Solecki W, Rosenzweig C, Dhakal S, Roberts D, Barau AS, Schultz S, Ürge-Vorsatz D (2018) City transformations in a 1.5 °C warmer world. Nat Climate Change 8(3):177–181. https://doi.org/10.1038/s41558-018-0101-5
391. Sporchia F, Marchi M, Nocentini E, Marchettini N, Pulselli FM (2022) Sub-national scale initiatives for climate change mitigation: refining the approach to increase the effectiveness of the covenant of mayors. Sustainability 15(1). Article 125. https://doi.org/10.3390/su15010125
392. Stamellou E, Kalogeropoulos K, Stathopoulos N, Tsesmelis DE, Louka P, Apostolidis V, Tsatsaris A (2021) A GIS-cellular automata-based model for coupling urban sprawl and flood susceptibility assessment. Hydrology 8(4):159. https://doi.org/10.3390/hydrology8040159
393. Stanton MCB, Roelich K (2021) Decision making under deep uncertainties: a review of the applicability of methods in practice. Technol Forecast Soc Chang 171:120939. https://doi.org/10.1016/j.techfore.2021.120939
394. Steffen W, Leinfelder R, Zalasiewicz J, Waters CN, Williams M, Summerhayes C, Schellnhuber HJ (2016) Stratigraphic and Earth system approaches to defining the Anthropocene. Earth's Future 4(8):324–345. https://doi.org/10.1002/2016EF000379
395. Sterman JD (2000) Business dynamics: systems thinking and modeling for a complex world. McGraw-Hill Publishing Company, Boston, Massachusetts
396. Stevanović M (1997) The age of clay: the social dynamics of house destruction. J Anthropol Archaeol 16(4):334–395. https://doi.org/10.1006/jaar.1997.0310
397. Sundstrom SM, Allen CR (2019) The adaptive cycle: more than a metaphor. Ecol Complex 39:100767. https://doi.org/10.1016/j.ecocom.2019.100767
398. Suominen A, Seppänen M, Dedehayir O (2019) A bibliometric review on innovation systems and ecosystems: a research agenda. Eur J Innov Manag 22(2):335–360. https://doi.org/10.1108/EJIM-12-2017-0188
399. Swanstrom T, Chapple K, Immergluck D (2009) Regional resilience in the face of foreclosures: Evidence from six metropolitan areas. In: A report prepared for the MacArthur Foundation's building resilience regions project (Working Paper 2009–05). Institute of Urban and Regional Development: Berkley, CA, USA. https://cloudfront.escholarship.org/dist/prd/content/qt23s3q06x/qt23s3q06x.pdf. Last accessed on 17 Oct 2023
400. Sykes O, Demaziere C, Nurse A (2020) 'Green cities' as urban models: contributing to new urban agendas, but how? Introduction. Town Plann Rev 91(4):349–355. https://doi.org/10.3828/tpr.2020.20; and https://core.ac.uk/download/pdf/327070557.pdf

401. Tan X, Huang B, Batty M, Li J (2021) Urban spatial organization, multifractals, and evolutionary patterns in large cities. Ann Am Associa-tion of Geographers 111(5):1539–1558. https://doi.org/10.1080/24694452.2020.1823203
402. Tansley AG (1935) The use and abuse of vegetational concepts and terms. Ecology 16(3):284–307. https://doi.org/10.2307/1930070
403. Tansley AG (1939) British ecology during the past quarter-century: the plant community and the ecosystem. J Ecol 27(2):513–530. https://doi.org/10.2307/2256377
404. Tao Y, Sornette D, Lin L (2021) Emerging social brain: a collective self-motivated Boltzmann machine. Chaos, Solitons and Fractals 143. Article 110543. https://doi.org/10.1016/j.chaos.2020.110543
405. Tempels B, Hartmann T (2014) A co-evolving frontier between land and water: dilemmas of flexibility versus robustness in flood risk management. Water Int 39:872–883. https://doi.org/10.1080/02508060.2014.958797
406. Tiezzi EPB, Pulselli RM, Marchettini N, Tiezzi E (2008) Dissipative structures in nature and human systems. In: Design and Nature IV: Comparing Design in Nature with Science and Engineering, vol I, pp 293–299. WIT Press. https://doi.org/10.2495/DN080301
407. Touili N, Baztan J, Vanderlinden JP, Kane IO, Diaz-Simal P, Pietrantoni L (2014) Public perception of engineering-based coastal flooding and erosion risk mitigation options: lessons from three European coastal settings. Coast Eng 87:205–209. https://doi.org/10.1016/j.coastaleng.2014.01.004
408. Tomor Z, Meijer A, Michels A, Geertman S (2019) Smart governance for sustainable cities: findings from a systematic literature review. J Urban Technol 26(4):3–27. https://doi.org/10.1080/10630732.2019.1651178
409. Tonmoy FN, Cooke SM, Armstrong F, Rissik D (2020) From science to policy: development of a climate change adaptation plan for the health and wellbeing sector in Queensland, Australia. Environ Sci Pol 108:1–13. https://doi.org/10.1016/j.envsci.2020.03.005
410. Tozan Y, Ompad DC (2015) Complexity and dynamism from an urban health perspective: a rationale for a system dynamics approach. J Urban Health 92:490–501. https://doi.org/10.1007/s11524-015-9963-2
411. Tringham R (2005) Weaving house life and death into places: a blueprint for a hypermedia narrative. In: Bailey D, Whittle A, Cummings V (eds) (Un)settling the Neolithic. Oxbow Books, Oxford, UK, pp 98–111. https://www.ruthtringham.com/project/weaving-house-life-and-death-2005/
412. Tsolakis N, Anthopoulos L (2015) Eco-cities: an integrated system dynamics framework and a concise research taxonomy. Sustain Cities Soc 17:1–14. https://doi.org/10.1016/j.scs.2015.03.002
413. Turner DP (2011) Global vegetation monitoring: toward a sustainable technobiosphere. Front Ecol Environ 9(2):111–116. https://doi.org/10.1890/090171
414. Tversky B (2016) Lines: orderly and messy. In: Complexity, Cognition, Urban Planning and Design: Post-Proceedings of the 2nd Delft International Conference, pp. 237–250. Springer International Publishing. https://doi.org/10.1007/978-3-319-32653-5_13.
415. UN-GA (2019) Resolution adopted by the General Assembly on 15 October. 12, 2. The United Nations General Assembly. See also: "Political Declaration of the High-Level Political Forum on Sustainable Development Convened under the Auspices of the General Assembly: resolution/adopted by the General Assembly". United Nations Digital Library. https://digitallibrary.un.org/record/3833352?ln=en
416. UN-Habitat (2016) Urbanization and Development. Emerging Futures (World Cities Report 2016). The United Nations Human Settlements Programme (UN-Habitat), Nairobi. ISBN 9789211327083. https://unhabitat.org/world-cities-report-2016
417. UN-NUA (2017) New Urban Agenda English. United Nations. ISBN 9789211327311. https://habitat3.org/wp-content/uploads/NUA-English-With-Index-1.pdf
418. UN-NUA (2022) Progress in the implementation of the New Urban Agenda. Report of the Secretary-General. United Nations Digital Library. https://digitallibrary.un.org/record/1628008

419. UN-SFDRR (2015) Sendai Framework for Disaster Risk Reduction 2015–2030. United Nations, Geneva. https://www.undrr.org/publication/sendai-framework-disaster-risk-reduction-2015-2030#:~:text=The%20Sendai%20Framework%20for%20Disaster,Investing%20in%20disaster%20reduction%20for
420. Ur J (2017) The Birth of Cities in Ancient West Asia. In: Tsuneki A, Yamada S, Hisada, K (eds) Ancient West Asian Civilization. Springer, Singapore, pp 133–147. https://doi.org/10.1007/978-981-10-0554-1_9
421. US-GAO (2020) Technology readiness assessment guide. Best Practices for Evaluating the Readiness of Technology for Use in Acquisition Programs and Projects. Report GAO-20-48G. U.S. Government Accountability Office (GAO), Washington DC. https://apps.dtic.mil/sti/pdfs/AD1173994.pdf
422. US-SEPWC (1985) Greenhouse effect. Senate Environment and Public Works *Committee*, December 10, 1985. Video by C-SPAN: https://www.c-span.org/video/?125856-1/greenhouse-effect. Also, by YouTube: https://www.youtube.com/watch?v=1NOdTEIihFU
423. Valdez LD, Shekhtman L, La Rocca CE, Zhang X, Buldyrev SV, Trunfio PA, Braunstein LA, Havlin S (2020) Cascading failures in complex networks. J Complex Netw 8(2). Article p.cnaa013. https://doi.org/10.1093/comnet/cnaa013
424. Vallero DA, Brasier C (2008) Sustainable design. The science of Sustainability and Green Engineering. John Wiley & Sons, Inc., Hoboken, New Jersey. http://refhub.elsevier.com/S0195-9255(19)30548-7/rf0350
425. Valencia SC, Simon D, Croese S, Nordqvist J, Oloko M, Sharma T, Taylor Buck N, Versace I (2019) Adapting the sustainable development goals and the new urban Agenda to the city level: initial reflections from a comparative research project. Int J Urban Sustainable Dev 11(1):4–23. https://doi.org/10.1080/19463138.2019.1573172
426. Van der Heijden J (2016) Experimental governance for low-carbon buildings and cities: value and limits of local action networks. Cities 53:1–7. https://doi.org/10.1016/j.cities.2015.12.008
427. Van der Heijden J (2019) Studying urban climate governance: where to begin, what to look for, and how to make a meaningful contribution to scholarship and practice. Earth Syst Gov 1. Article 100005. https://doi.org/10.1016/j.esg.2019.100005
428. van der Heijden J, Patterson J, Juhola S, Wolfram M (2019) Special section: advancing the role of cities in climate governance–promise, limits, politics. J Environ Plann Manage 62(3):365–373. https://doi.org/10.1080/09640568.2018.1513832
429. Van der Wal M, De Kraker J, Offermans A, Kroeze C, Kirschner PA, Van Ittersum M (2014) Measuring social learning in participatory approaches to natural resource management. Environ Policy Gov 24(1):1–15. https://doi.org/10.1002/eet.1627
430. Vanderlinden JP, Baztan J, Touili N, Kane IO, Rulleau B, Simal PD, Pietrantoni L, Prati G, Zagonari F (2017) Coastal flooding, uncertainty and climate change: Science as a solution to (mis) perceptions? A qualitative enquiry in three coastal European settings. J Coastal Res 77:127–133. https://doi.org/10.2112/SI77-013.1
431. Vandevyvere H, Stremke S (2012) Urban planning for a renewable energy future: methodological challenges and opportunities from a design perspective. Sustainability 4(12):1309–1328. https://doi.org/10.3390/su4061309. https://doi.org/10.3390/su4061309
432. Vela RD (2019) Interlocking networks and the sacred landscape of Hellenistic northern Etruria: capturing social and geographic entanglement through social network analysis. Open Archaeology 5(1):505–518. https://doi.org/10.1515/opar-2019-0031
433. Verweij M, Ney S, Thompson M (2022) Cultural Theory's contributions to climate science: reply to Hansson. Eur J Philos Sci 12(2). Article 34. https://doi.org/10.1007/s13194-022-00464-y
434. Villmoare B, Kimbel WH, Seyoum C, Campisano CJ, DiMaggio EN, Rowan J, Braun DR, Arrowsmith JR, Reed KE (2015) Early Homo at 2.8 Ma from Ledi-Geraru, Afar, Ethiopia. Science 347(6228):1352–1355. https://doi.org/10.1126/science.aaa134
435. Von Schilling C, Straussfogel D. (2008). Introducing the research. Entropy debt: a link to sustainability? In 52nd Annual Meeting of the International Society for the Systems Sciences. https://journals.isss.org/index.php/proceedings52nd/issue/view/8. Also available at http://refhub.elsevier.com/S0195-9255(19)30548-7/rf0370

436. Voß JB, Bornemann B (2011) The politics of reflexive governance: challenges for designing adaptive management and transition management. Ecol Soc 16(2). Article 9. https://www.jstor.org/stable/26268901
437. Wang Z, Jiang L, Yang Y, Xiang T, Song J, Yan S (2015) Research on multi energy data transaction model based on vector evaluate genetic algorithm. In: 2015 International Symposium on Material, Energy and Environment Engineering (ISM3E 2015). Atlantis Press, pp 559–562. https://www.atlantis-press.com/proceedings/ism3e-15
438. Waring TM, Kline MA, Brooks JS, Goff SH, Gowdy J, Janssen MA, Smaldino PE, Jacquet J (2015) A multilevel evolutionary framework for sustainability analysis. Ecol Soc 20(2). Article 34. https://doi.org/10.5751/ES-07634-200234
439. WCED (1987). Our common future. Report of the World Commission on Environment and Development (WCED), to the General Assembly of the United Nations (04.08.1987). ("The Brundtland Report".) https://digitallibrary.un.org/record/139811?ln=en
440. Wengrow D (2015) Cities before the state in early Eurasia. Goody Lecture 2015, Max Planck Institute for Social Anthropology. https://discovery.ucl.ac.uk/id/eprint/1539997/1/2015_Goody_Lecture_RO151126_1_1.pdf
441. Westley F, Carpenter SR, Brock WA, Holling CS, Gunderson LH (2002) Why systems of people and nature are not just social and ecological systems. In: Gunderson LH, Holling CS (eds) Panarchy: understanding transformations in systems of humans and nature, pp 103–119. Washington: Island Press. http://hdl.handle.net/10919/65569
442. Wey WM, Peng TC (2021) Study on building a smart sustainable city assessment framework using Big Data and Analytic Network Process. J Urban Plann Dev 147(3):04021031. https://doi.org/10.1061/(ASCE)UP.1943-5444.0000704
443. Wieland A (2021) Dancing the supply chain: Toward transformative supply chain management. J Supply Chain Manag 57(1):58–73. https://doi.org/10.1111/jscm.12248
444. Wilkins J (2020) Learner-driven innovation in the stone tool technology of early Homo sapiens. Evol Human Sci 2. Article e40. https://doi.org/10.1017/ehs.2020.40
445. Wu C, Wang X, Jin J, Zhou Y, Bai X, Zhou L, Tong F, Zhang L, Cui Y (2023) Structure simulation and equilibrium evaluation analysis of regional water resources, society, economy and ecological environment complex system. Entropy 25(2):181. https://doi.org/10.3390/e25020181
446. Xie Y, Liu C, Chang S, Jiang B (2022) Urban sustainability: integrating socioeconomic and environmental data for multi-objective assessment. Sustainability, 14(15). Article 9142. https://doi.org/10.3390/su14159142
447. Xu Y, Ramanathan V, Victor DG (2018) Global warming will happen faster than we think. Nature 564:30–32. https://doi.org/10.1038/d41586-018-07586-5
448. Ying L (2015) Twin universes: universal laws of thermodynamics. Am J Mod Phys 4:1–4. https://doi.org/10.11648/j.ajmp.s.2015040101.11
449. Yoshioka N, Era M, Sasaki D (2021) Towards integration of climate disaster risk and waste management: a case study of urban and rural coastal communities in the Philippines. Sustainability 13(4):1624. https://doi.org/10.3390/su13041624
450. Ystgaard KF, Atzori L, Palma D, Heegaard PE, Bertheussen LE, Jensen MR, De Moor K (2023) Review of the theory, principles, and design requirements of human-centric Internet of Things (IoT). J Ambient Intell Humaniz Comput 14(3):2827–2859. https://doi.org/10.1007/s12652-023-04539-3
451. Zanuttigh B, Nicholls RJ, Vanderlinden J-P, Thompson RC (2014) Falk Burcharth H (eds) Coastal risk management in a changing climate. Butterworth-Heinemann

Module 2
Generation of Priority-Based Scenarios

2.1 Premises

How to Read Different Sections: Choose Whatever Order Suits You

In this module, I describe a set of generalized models that are borne in practice and further consolidated by abstract analysis and by being confronted with the extant literature and other practical experiences.

This and the following sections are roughly composed of two parts: one introducing the issue under discussion and the other describing the method of addressing it.

After the core Big Picture summarized in Module 1, some of you might want some context details before getting to an abstract model. Others will prefer to go directly to the model/method and to read the contextual details afterwards.

Being a scientist myself, it is rare that I read a scientific production in the novel-like order from start to end. Like virtually all my fellow researchers, I basically use (employ) each such expert production as a collection of specialized parts with different categories of specific content (literature review/introduction, methodology, results, discussions, conclusions, list of cited references, supplementary materials), which means that the order in which I read them really depends on what I want from that specific scientific report. Sometimes I start with the beginning (if it happens that I need familiarization or some details about that specific topic), but usually I go directly to the technical sections that interest me (Method, Results, or Conclusions) and I start reading from there; and I can finish reading the Abstract or the Method or the Discussions again.

While I hope this book reaches out to a great diversity of people (not just scientists but every person that is interested in eco-cities and eco-innovation), I wrote it having in mind the expectation that the reader will use it according to whatever diverse needs she or he might have.

This is also the reason why I have structured the book in autonomous, thematic modules (modules and sections; each with its own purpose, models, and explanations)

I. M. Ciumasu, *Eco-cities*, Green Energy and Technology,
https://doi.org/10.1007/978-3-319-14702-4_2

that are ordered as distinct (though related and cross-referenced) parts of the general storyline of the book.

Therefore, I fully encourage you to use whichever reading style suits you. With regards to the method and discussions in this module, I will be happy to answer the questions you may have (for example via LinkedIn or by email), I welcome all other thoughts you might want to share, and I will do my best to respond as fast as possible.

Context and Motivation—Acknowledging Complexity Is a Must

The differences in disciplinary perspectives on cities shown and discussed early on (Table 1.2, Module 1) warn us that, when we are dealing with complex issues, we cannot know everything but we need to know what we are doing—we need to study it properly and act accordingly. They also indicate that insufficient or inadequate integration of the components of a city (i.e., a situation with high entropy) inevitably results in disfunction and failures. When unresolved problems have been accumulating for a long time, the systemic breakdown will happen sooner rather than later. Ergo, the complexity of modern life which we are now used to (and many still take for granted) will be lost.

The point is that functional integration of available knowledge is incompatible with overreliance on planning, or "over-planning" (unrealistically detailed and rigid planning in and about complex dynamic systems). The same applies to overreliance on technology: sufficient integration is impossible if the requirements of the systemic perspective on sustainability (as discussed in the introduction of this book) are ignored. Unfortunately, this is exactly the case with some of technology-driven commercially motivated concepts. For example, the notion of Industry 4.0 presumes (with no scientific grounds, and rather hastily) that a socio-economic system can put "cyber" before sustainability [1]. For example, some ideas in "smart city" developments even go directly against sustainable development while pretending they help it. This is how [2] sums it up: "management logics of smart infrastructures attempt to program and transform vegetation and its ecologies into uniquely efficient and responsive urban organisms. In the process, these programs of efficiency have the potential to exacerbate extractive economies and social inequalities". Logically, such "eclipses of reason" result in "Frankenstein urbanism" [3]. (As a technical remark about those scientific grounds, please note that a summary on the scientific method is provided in Sect. 3.7 in the discussion about knowledge, information and data).

True city development projects are neither predictive clockworks nor chimeras. They reconcile in practice all approaches and insights coming from natural and social sciences. For example, unlike the management of natural systems, any "system of nature and people" also requires symbolic construction (i.e., meaning), because "human beings are constantly making sense of the events they perceive … thereby creating a 'virtual reality' around them" [4–9]. These "sensemaking" efforts (i.e., interpretive approaches) [10] aim to clarify how different objective realities and representations thereof "are embedded in each other across a panarchy", e.g., how "a green business model" interacts with "the climate crisis" in cognitive and operational terms [9].

In this context, eco-innovation, i.e., any innovation that helps with diminishing ecological footprints, is a process of multi-dimensional functional reconfiguration within urban (and other nested) systems, and thus reflects practical and epistemic convergences between older preoccupations that even preceded the establishment of sustainability studies as a field (the "problem of embeddedness" [11] included).

For this reason, recognizing complexity is a necessary condition for putting a city in a useful historical perspective, i.e., to understand its past trajectory as a start for unveiling its potential futures based on empirical evidence, and remembering that the information provided by historic sources themselves need to be corroborated with perspectives from other disciplines that deal with complex human systems in general, e.g., sociology and cultural anthropology, and natural sciences (e.g., [12]).

Further, as [13] observed when discussing cities as complex systems, "*a general framework regarding the social processes behind urbanization needs to account for empirical regularities that are common to both contemporary cities and past settlements known through archaeology and history*" and "*only by adopting an explicitly historical perspective can such fundamental structure be revealed*". In addition, evidence is only obtained through active study, and I put a strong emphasis on "active"—not in the sense that it would be fine to "overdo it" (it would not; efforts must be adequate to the task) but simply to keep in mind that "history shows but learning is up to us" [1], and this requires sustained multi-disciplinary and integrative efforts to use all available evidence and insight for a big picture understanding as a basis for competent generation of knowledge-action options.

Sure, there will always be some who care not if the world collapses, and simply think "After me, the deluge" (*Après moi, le déluge*). Realistic projects need to count them as part of the problem and act accordingly (to minimize all unwanted effects, because these people will not hesitate to contribute to causing the deluge). As a side note, I suppose such misanthropes do not bother to read books such as this one, but if they did, I would ask them to look in the mirror and tell to themselves what is that they see. Caring about the state of our cities/communities and planet/humanity is part of the meaning of our existence, and contributing to solving our common problems is part of what it means to be human (and is the real meaning of seen or unseen personal sacrifice).

Core Approach

This module describes the practice-borne method (a) whose core was originally published a decade ago [14], (b) motivated the writing of this book, and (c) is now being consolidated and refined for current needs.

The start idea is fairly simple: because resources tend to be scarce (time included), and because doing "everything at the same time" is an impossibility, every project must start from a competent identification of the current priorities in the relevant community of stakeholders and actors.

The reciprocal is equally straightforward: available/mobilizable resources must be allocated to those projects and programs that competently address the most important problems (and thus, can realistically yield the strongest positive results).

At this point, I would bring to the fore a practical summary made by Friedman [15] about the broad topic comprising foresight, forecasting, and scenarios, from technological and business issues to societal and geopolitical matters (for our purposes here, I just needed to replace the country name in the original text with a general denomination expression): "There are two keys to forecasting: a ruthless, objective understanding of imperatives, followed by an equal comprehension of constraints. It demands, in other words, an understanding of what [the specific human system of interest] must have and raises the question of why it doesn't have it. Forecasting is for the simple-minded. The most obvious things are the most useful".

In light of the discussions I made in the previous book module, I would also add that, when it comes to complex systems, the most obvious things are only obvious after a fair amount of delving into the details and doing the sense-making work. But this means neither that we should make things more complicated than they are, nor that we should act based on superficial appearances. It only means that, because the (objectively) most obvious things are the most important, we need to do the necessary homework to clarify what is "the most obvious" in light of available evidence.

Furthermore, in terms of operational knowledge-action relations, the most consequential issues awaiting resolution also tend to be the most urgent. This means not only that we should not waste time; equally important is to keep in mind at all times that we cannot afford to drain our resources by spending them on less-than-top priorities. This means that, yes, our most important resource is time but the imperative of acting on priorities means that we must "make time for the work that matters" [16]. Experience confirms this rationale but it also tells us that it is easier said than done: in the end, what we need is a set of good methods.

Starting from this managerial concern, this second module is focused on how to get the ball rolling in the right direction while it hops, bumps and jumps on a rough road (Box 2.1). It proposes a procedure for generating scenarios as a means to address the need for a local community of people (a city or any other human settlement as a whole) to develop a "shared reference story" about how they come to be, what they can be in the future and what they want to be.

This methodology is then complemented, in the third module, by a set of support models (that are general enough to serve for intuitive reasoning and be applicable in virtually all contexts).

Box 2.1 In complex projects, usual difficulties faced by foresight exercises are related to:

- a set of known and unknown (yet) obstacles;
- the exponentially growing complexity of systemic problems and of the viable solutions;
- the transaction costs, i.e., the cost of processing all connections that need to be considered, also known in mathematics, computer science and engineering as, or related to "algorithmic complexity"). They tend to grow

exponentially, i.e., they swell until they render a project unaffordable if not outright harmful;
- the hard choices in the allocation of resources to a project and not to another, also known in economics as "opportunity costs", i.e. what else you could have done with a given amount of money/resources in general;
- difficulties in managing the practical connections implied in the notion of "eco-city" and other paradigms like "smart city", "circular city", and so on, whose merits with respect to the goal of sustainable development are still debated (as needed) in expert and public agoras.

In other words, the aim is to make people be able to look and say "We want to be something "like this" or "like the other" general case and story line", and to provide a set of conceptual-managerial tools.

How to Organize Interactions Between Relevant Actors?

Basic Semantic Clarity is Not an Academic Luxury but an Operational Necessity

First of all, it is important to understand that the work on definitions is not primarily about academic finesse. It is an operational prerequisite for success in eco-city (as in all complex) projects. Similarly, responding to the challenges faced by cities is not about what is academic or non-academic: it is a matter for the whole of the society, and the best available knowledge needs to be employed, which requires genuine collaboration between all the actors across the triple-helix of academia, business, and government. Regardless from where we come as professionals or/and as citizens, we need to understand what is behind the terms that we use in projects in general and for eco-city-related matters in particular. Otherwise, the entire notion of eco-city risks to be misunderstood and perceived as "elitist" [17] and eco-city projects will stand on moving sands.

The following paragraphs make a brief (and inevitably incomplete) summary of the main points that eco-city project managers and participants need to have in mind from the beginning.

To start with governmental actors, the (occasional or more entrenched) tendency of local public administrators to ignore inputs coming from the other sectors, for example for reasons such as "too academic" (e.g., [18] or, for that matter "too business-like) is not automatically acceptable (just as it is professionally unacceptable for experts from Academia to ignore inputs from outside the university environment as "too non-academic" or similar reasons; and so on).

Quite on the contrary, because the public administration is employed by the rest of the society as a whole to act on our behalf and defend our common interests, they have a fundamental duty not just to be "receptive" but to actively seek inputs from all experts that are relevant for a given topic (because this is a necessary condition for fulfilling their job), and to make the due effort to understand those inputs as they come. To emulate here the reasoning of a famous expression about science in the new era of transitioning to sustainable development ("a new social contract for science and scientists" [19, 20]), in the above-mentioned context of the triple-helix for generating solutions for sustainability, and also given the already long-time discussions about business sector (who is already at least aware about its societal responsibility in this new era, via more or less perfect concepts like corporate social responsibility and the like), one can and should reasonably demand for a "new social contract for public administrators".

Now, if the underlying problem in public administration units is the (often demonstrable) lack of sufficient expertise to assimilate and use those external insights, then the right course of action for them is to signal that problem properly and then to find ways to address it, e.g., by acquiring/updating some basic expertise and by sustained interaction with academia (and the business environment). If this requires significant restructuring of existing institutional ways of doing things, then this is what must be done. After all, we are still relying on institutions that emerged in their current form during the times of mass industrialization of the late nineteenth century and were designed to meet the challenges of that era.

If such institutional updates fail, then we will find ourselves in a situation in which we are talking about change while refusing change ourselves, with a sort of absurdist discourse that says "let us make some changes on the condition that nothing changes", as in the famous twentieth century social derisions that we know from the "theater of the absurd" depicted by playwrights Eugene Ionesco and Samuel Beckett [21] (see also [22, 23]). This can only lead to the type of anti-human and undignified public administration known from the kind of totalitarian, bureaucratic worlds and societies described by George Orwell in his dystopian novels like "Animal Farm" and "Nineteen Eighty-Four" [24–26] and by Franz Kafka—"a rigidly inflexible and sterile bureaucratic universe" [27] as based on his experience in the Habsburg Monarchy (Austria-Hungary) at the beginning of the twentieth century [28], a cultural legacy that is usually referred to as "Kafkaesque" public administrations around the world, including also in Organization Studies and Sustainability Studies [29–32].

In fact, a regard at the matter from a modern twenty-first century systemic perspective, one can argue that the claim made by Max Weber over a century ago (first translated into English right after WWII) that bureaucracy is "necessary" in the modern society is outdated but unfortunately still adhered-to by those who lack a current understanding of the complex dynamic systems in which we live—which only underscores once more the importance of developing a stronger and more adequate relation between science and society. In light of the current knowledge and operational capacities (our collective intelligence, digitalization, constant learning and communication) and the incomparably higher added value of human activity, I contend that we should stop making the confusion between public administration and bureaucracy: public administration can clearly be done without apparatchiks while the notion of bureaucracy is only apt to describe corrupt or/and incompetent or/and otherwise abusive public administration.

Worse even, the perpetuation of this confusion is rather toxic since it suggests, with the backing of some historic name (Max Weber being widely regarded as one of the founders of the field of Sociology, and is one of the first names a student finds about in undergraduate sociology courses), that a corrupt form of public administration is necessary. This is not only factually wrong: it creates a notional and cultural ground for a fundamental distrust of citizens in their common institutions (which are then to be regarded as a necessary evil instead of a natural organization in human systems), it generates other unwelcome confusions (e.g., in the definition of notions like individual freedoms and responsibilities), and it instills reasons (motivations) for non-cooperative behavior in the social context of a person. In brief, it is an idea that directly undermines the very foundations of social sustainability (without which there is no sustainable development at all).

The message here is clear: this is about all and each of us; methods and institutions need to keep up with time by updating (and even dissolving and reforming) themselves to meet current (not past) challenges; individuals and institutions who will fail to do this will lose purpose, will become inadequate and fill face demise.

Current Experience and Knowledge Indicate that the "Triple-Helix" Model of Co-Work Known from Innovation Studies is a Solid Start Basis for Eco-City Projects

In a nutshell,

(a) public administrators must understand that all citizens need not become experts in administrative forms and regulations (some kind of "second class administrators");
(b) scientists must understand that not everyone must be re-educated to become "aware about" every single topic that scientists deem as a "must"—usually something from their own discipline regarded as the hub of all human knowledge (to seemingly produce a sort of "second class scientists" which will nevertheless have to be trained in all sciences, while scientists themselves can eventually remain comfortably in only one specialized domain and not be asked to step outside of it); while
(c) businesses must guard against overinvestment (in or out of a given company; because this ultimately results in value destruction instead of value creation) and always remember that people are neither perfect rational economic agents (whatever that may mean in theory) nor a commodity they can amass or dispense.

To be clear, the point with the summary above is not about the examples chosen to illustrate the point (those can change) but about a key common necessity: maintaining a healthy critical spirit at all times.

The same need for changing the business-as-usual (BAU) applies for private investments and business. And, of course, the BAU problem also exists in academia. This means that the triple-helix mentioned above is not only a necessary conceptual framework and mindset in which different actors must learn to interact for the benefit of the society as a whole. It can and it does manifest in the form of a convergence of interests of those who want to preserve the situation as it is (regardless of consequences) and a default, tacit agreement between BAU-favoring actors to avoid disturbing each other with uncomfortable questions about sustainability and the like.

Most often though, people simply survive at the limit of their powers (living off whatever edge they have as individuals or groups/institutions)—a phenomenon known in complexity science as "evolution to the edge of chaos". Thus, another and perhaps the most widespread issue is that, when a potential project or action "looks complicated" (meaning that potential risks weigh heavier than potential benefits), many people and teams start "dragging their feet", "kicking the can down the road", "swiping the dirt under the carpet". They will "brush away" the questions to which they have no adequate (or at least convenient) answer, and will limit themselves to doing strictly what there are being asked to do (or even refusing that, if they can), the end result being one or more (and simultaneous) types of lock-ins (Table 2.1) that prevent problem-resolution and advancement in the transition to sustainable development in cities, or, for that matter, in any institution and human system.

Also important is to have a practical big picture representation of different roles that different actors may have in the business of generating real solutions on the market (for a start, one could use Table 2.2) and to be able to compare how different approaches can help in the management of projects that aim at generating positive change (for a start, see Table 2.3).

2.2 A Delphi-Based Method and Procedure

Context: Addressing the Accumulation of Systemic Risks

Cognitive Issues and Human Resources in Eco-City Foresight and Projects

First of all, being informational, material, and commercial nodes within the world's network of interconnected cities (Fig. 1.8 in Module 1, Sect. 1.3) (see also [36–38]) makes cities natural creativity hubs [39, 40] and "growth machines" for their hinterlands [41, 42]. As [43] writes, "*cities create wealth, generate employment and drive human progress by harnessing the forces of agglomeration and industrialization*". And, in so doing, cities exert their direct (individual) and indirect (collective) pressure upon the environment, i.e., they consume from the carrying capacity of the Earth's ecosystems and thus generate their own environmental footprint (for basic ecological details, see, e.g., [44]).

As defined in Sect. 1.3 of this book, an eco-city is a city that is demonstrably engaged on a development path that is compatible with the sustainability of Humanity on earth—an operational definition based on both experience and current disciplinary, inter-disciplinary and trans-disciplinary understanding in the literature.

However, not everyone is in the position to comprehensively follow cutting-edge or integrative insights, and there is no deed for everyone to do that. Instead, real-world project teams (clusters and networks) have to meet two basic organizational requirements:

(a) to collectively cover a minimum of expertise to keep a project on a relevant and efficient course, and

Table 2.1 A summary of types and characteristics of lock-ins that are preserving and perpetuating unsustainable practices

Types of lock-in	Key characteristics
Infrastructural and technological	• Technologic, economic, and techno-economic forces sometimes lead to inertia, notably through built-in interdependencies in complex systems • Long life cycles, large necessary investments (often as megaprojects) and related sunken costs, certain long-lived (usually systemic) effects • Initial choices account for narrow motivations, often (but not always) private, which ignore (social and) societal costs and benefits • Random or/and unintentional effects affect final outcomes in all complex systems (as per Fig. 1.6 in Module 1, Sect. 1.2)
Institutional	• By definition, institutions are designed to provide sufficient stability of those modi operandi that were deemed desirable at a certain moment in time. When the context changes and new ways are necessary, that past stability becomes a drag (it morphs from advantage to a disadvantage) • Vested interests and/or entrenched habits: economic, social, and political actors with different types and levels of influence/power tend to seek a consolidation of a status quo when that favors their interests • Related to the above point are issues like over-optimization ("the better becomes the enemy of the good") and ambidexterity (the need to find the right balance—a dynamic equilibrium) between stability and flexibility, knowing that long periods of stability breeds functional rigidity (a lack of those challenges that would maintain flexibility) • Deliberate institutional capture by certain actors, i.e., intended beneficial outcome for some actors but detrimental to the rest of the society • Deliberate but misguided (hasted) choice in favor of one type of solution or another, often related to the well-known (but often unaddressed) problem (bad practice) of "picking winners" by public authorities, which inevitably results in either token (i.e., facade) "success", "results", and/or "role models", while in fact the original problem remains essentially unaddressed and the best possible (but inevitably impossible to see beforehand) outcome is being missed • Related to above is the idea of intervening to "correct the market" when in fact the problem is not with the market but the institutional actors that fail to develop adequate institutions or to understand the underlying complex dynamics between actors, factors, systems, and effects
Behavioral	• Recurrent choices by individuals, related to different psychological processes, and phenomena, e.g., individual motivations, biases, habits, or overall competence (or insufficiency thereof) • Single deliberate calculation and choice sliding into a chain of subjective and self-justificatory actions in the wrong direction • Social norms and pressures (and the decision to conform even when the right path is not to) • Complacency and failure to see that interruption of damaging habits (and strings of bad or unjustified decisions) is possible by learning to use some different methods

Source Modified (generalized and expanded from recent discussions on climate changes in past projects) after [33]

Table 2.2 The roles of different actors in sustainable business: a simplified, illustrative list, with a focus on investors in eco-innovation

Business actors	Roles
Venture capitals	• Skilled "sustainability investors" who can accept slower ROI (Return on Investment) if needed for higher social and environmental returns • Demonstrators of "successful sustainable business investment formats" or/and supporters thereof • Investors (and associations thereof) that are capable to inject money into the developing world to generate new value • Funders of unprecedented (or unusual) projects and programs • Makers of small but smart investments
Entrepreneurs and startups	• Challengers of current business models and incumbent businesses, to potentially disrupt the market (see also Fig. 3.41 and related text in Module 3, Sect. 3.10) • Filling the need for skilled "sustainable entrepreneurs"
Large companies	• Drivers of major changes, generator of cultural shifts through the power of the example, and promoter of sustainability-compatible trends • R&D or/and manufacturing partner or/and major customer • Investor in (resource provider for) start-ups (e.g., via open competitions, partnerships or/and shared projects)
The Government	• Enabler of sustainable innovation via legislation (e.g., regarding product end-of-life, internalization of environmental "externalities" such as natural resources and ecosystem goods and services) • Facilitating fruitful networking and interactions across constituencies and sectors • Leading and/or supporting public–private partnerships (PPPs) and other types of collaborative projects and financing
The wider public	• Generator and user of new and different financing models, e.g., crowd-funding and peer-to-peer lending, anything that can help the emergence of new and innovative sustainable businesses (some of which will turn out to have systemic effects)

Modified after [34]

(b) to be sufficiently efficient in integrating their total/collective knowledge into productive workflows, and ultimately into mature solutions to the problems they want to address.

Now, these two points may sound rather straightforward, but bringing all relevant actors and stakeholders (or at least a minimal "most of them") on the same page is a heavy task that tends to be substantially underestimated. This is a serious issue, and it happens largely because of two inter-related contributing causal factors:

(a) as a natural consequence of the typical divisions between knowledge fields (as well as between different levels of familiarity of different people with different fields and the relevant literature, and between different types of professional and personal contexts) and

Table 2.3 A list of managerial aspects to be considered when steering change: a kickstart (non-exhaustive) synthesis from two complementary approaches, Adaptive Management and Transition Management

Aspect	Details
Theoretical background	Relevant research domains, e.g., • Complexity science • Ecology (and all of the natural sciences contributing to it) • RDI • Resource management • Eco-innovation management
Application realm	Human systems and natural systems + combinations thereof, e.g., • the so-called "socio-ecological systems" (SES; "functionally or spatially-defined systems", e.g., river catchment areas, natural parks, etc.) • the so-called socio-technical systems (STS; "arrangements providing societal functions", e.g., energy provision, agriculture, transportation, education, scientific research, etc.) • filling the need for skilled "sustainable entrepreneurs"
Fundamental objectives	A combination of Change and Adaptation (e.g., as with the twin goals of mitigation of and adaptation to climate changes, but generalized to include all challenges related to sustainability, climate change being only one of the UN's global sustainability goals (as per Fig. 1.10 in Module 1)) As an observation, change is understood to consist of transformations of the current STS by managing the ongoing innovation processes in ways that favor those solutions that are compatible with the basic sustainability goals or/and otherwise help in the transition to sustainability
Grounds	The reality that we live in complex coevolving systems, and the state of our knowledge about CDS behaviors, which include systemic steady states and transitions (typically described by S-shaped patterns, as mentioned in Module 1, Sect. 1.4)
Governance	To be based on Experimentation & Learning (while maintaining essential functions so as to avoid systemic shocks). This involves, inter alia: • insuring good government practices that are compatible with democracy and human rights, starting with the fundamental principle of Consent-of-The-Governed (CoG), and the Whole-of-Government (WoG) approach to policy • using triple-helix projects (government-academia-business), where different types of actors bring different view angles, methods, and skills for addressing the common systemic problems of interest • and relatedly, developing platforms for systemic experimentation and the overall collaborative development of systemic solutions

Reconceptualization based on [35] and as per current knowledge and expertise

(b) because most people do not visualize clearly that the number of possible interactions between involved participants increases exponentially (not proportionally) with the number of participants. In short, as pioneer cross-field projects grow in size (as they need to, if they want to become relevant), the number of aspects and interactions to be managed simply "explodes"; this is why and when new and truly "next level" methods are necessary.

In Module 1, I already evoked the very serious issue of people adopting the wrong conceptualization of the notion of sustainable development, and the subjacent issue that the very fact that this confusion even happens points to other major issues in the relations between different knowledge fields and the general relation between science and society. Well, this very problem is true with many (if not most) concepts that are relevant for the topic of this book, starting even with the basic understanding of the notion of eco-city itself.

For example, the fact that a project team decides to use the operational (and dully-justified) definition of an eco-city as defined in this book (Module 1, Sect. 1.4) does not mean that all their collaborators (within the relevant community of actors and stakeholders) will adopt it automatically (or even at all). A specific type of work is needed (in every complex project involving many disciplines and vocabularies) to reach concrete, explicit agreement on terminology and on all the other details that matters for the project, starting with the goals of the project and what they actually mean and imply (see the details in the model and method in Module 3, Sect. 3.4).

For instance, a city can be studied in terms of the relation between its inhabitants and the surrounding natural ecosystems, and therefore some may understand urban sustainability in this narrow perspective. For example, Breuste and Qureshi define the "sustainable city" as "an ecologically healthy human settlement modeled on the self-sustaining resilient structure and function of natural ecosystems" and "an entity that includes its inhabitants and their ecological impacts" [45]. But, as I explained in Module 1, cities cannot be sustainable in isolation from their environment/hinterlands. Neither can a city exist in isolation nor as some "feudal" hubs of some "ecologically-autarchic" regions so to speak—and that is before even considering sustainable development goals. Then, cities that have engaged on a path to sustainability must have started to undergo a process of systemic transformation from a predominantly local identity (characterized by a weak influence in the global network of cities) to a double, local-and-global character (and identity) as participant in the global affairs of humanity [46]. This line of thought coheres with current discussions in the literature, as famously encapsulated in the expression (and apparent riddle) "why cities cannot be sustainable and why they are critical to sustainability" [47].

The problem with the definition of a "sustainable city" used above for illustration of a superficial understanding, and the rather obvious answer to the apparent riddle evoked above is the following: based on current knowledge that can be observed and mobilized from the multi- and trans-disciplinary terminology of sustainability studies (but also as intuitively understandable by non-scientists too), cities cannot be sustainable per se because they are simultaneously dependent on their local context (starting with the carrying capacity of the local ecosystems, typically expressed as "ecosystem goods and services" (EGS)) and on their overall capacity to access global resources through the possibilities/options for useful exchanges within the worldwide networks [48]. In somewhat more prosaic terms, the idea that a city can be sustainable within its administrative borders cannot be taken seriously, simply because, by vocation and definition, a city is a meeting place and a concentrator of resources (of all types) from and for the benefit of a wide area beyond its administrative borders.

To connect back with established vocabulary, one cannot expect a city to be a "development machine" for a region but at the same time expect it to be sustainable within its administrative limits.

All in all, this means that a local community (and a project constituency) must, before anything else, establish their collective priorities and then keep updating them. Absence of that, an eco-city project has no basis to start from.

Then, a related challenge is that a local community may have already started to acknowledge the imperative character of, and to engage on, a self-transformative path, but many issues require dedicated policies, starting with separate recognition in the first place. For example, everyone in the relevant community of actors and stakeholders may agree that they need to do collective foresight, and they may successfully accomplish some collective agreement on their priorities as a community. Nevertheless, the application of those insights into practical reality can easily turn out to be a very different (and much harder) challenge.

The problem starts from the way in which the collective foresight exercise is conducted. Very often (typically), the process of decision-making in a complex community of actors and stakeholders involves trade-offs and arrangements that tend to seek avoidance of conflict or even the avoidance of controversy at all (or, in the most extreme cases, even the avoidance of the admission that there is a problem or that there are important uncertainties that must be acknowledged and addressed properly, i.e., studied instead of being "boxed in" convenient clichés). As many of you and I can attest based on your/our own experience, such differences and issues are easily noted in those projects that involve the necessary triple-helix participation from across academia, business, and government. The inevitable consequence of such situations is that the actual results of the conversation are rather conservative and thus the advancement on the path towards sustainability is hindered, halted, or even reversed [49–51].

Transition management theory too describes such situations in terms of an abstract (qualitative) multi-level model: "niche" innovations (i.e., startup organizations or/ and projects) compete with each other while attempting to alter existing "regimes" which are by definition resistant to change except when change is absolutely required (usually during some crises) and very often not even at that moment (which is what leads to those crises which open new possibilities and some components of the regime are being replaced with some "newcomers"). This is because regimes are de facto alliances of incumbents which cooperate to elaborate and enforce rules that prevent new arrivals but also compete with each other to influence the deeper (and even broader), quasi-immutable contexts (called "landscapes") in which they operate. In other words, change happens when new needs require new solutions and postponement is not one of them anymore [52, 53], a fundamental level in the whole picture being the operational level [54]. Using this metaphor, a city can also be imagined as a multi-layered system with niche, regime, and landscape factors/vectors, and the accumulation of tensions in the phenomenon of self-organized criticality (tensions later bursting into crises) as discussed in the first book module can be understood as manifesting itself inter alia in the interactions between agents within and between organizational agents, niches, regimes and landscapes.

This representational model/theory can also link with other important ideas in the big picture before (and as part of the preparations for) operations, as it provides another angle at understanding how different facts fit together, e.g., that knowledge is mobilizable not centralizable, and that change in complex systems is more about evolution than revolution [54] but that this evolution may happen through alternations of slow and fast change (i.e., systemic steady states and transitions between steady states). And, long-term change happens through series of shifts that transform how the system works while it keeps working. The fact that some may call this or that shift a "revolution" (that "changed everything") is only an unstudied reference to some of those manifestations of a CDS that may be more apparent (and perhaps more spectacular) but are part of a more profound dynamics that requires competent study and knowledge (not voluntarist actions as this will inevitably create more harm than good). This is exactly what the s-curve pattern of change used in the previous module (Fig. 1.12) is all about. It is not enough to just want a change to happen, we need to know what we are doing.

Of course, when we do not know, the mature approach is not to avoid admission of ignorance (nor to point to simplistic "silver bullet" or incomplete/insufficient solutions but to engage in study and learning, and do the work that needs to be done. In the next module, I address the issue of known-and-available vs. unknown solutions. At the frontier of knowledge, as is the case with eco-cities and eco-innovation, this means competent research, development, and innovation (RDI), competent policy-making, and competent management—all of which in fact begin with competent prioritization (which again, must start somewhere sometimes, usually by addressing a target problem, and then gets better with practice).

As discussed in Module 1, Sect. 1.3, crises and emergencies are both symptoms of problems and major opportunities for change and improvement. This is important in the general situation because the ideal of sustainability (and the UN's Sustainable Development Goals mentioned in Fig. 1.10) is fundamental—a necessary condition for the future of humanity [44, 48, 49], and therefore requires important modifications in the socio-technical landscape evoked in the metaphor above. On the pathways to future sustainability, hardships and their manifestations are essentially commensurate with this ideal. In the knowledge-and-innovation-based economy and society of today, relevance and effectiveness of decision-making for urban transformation depend directly on an operationally-integrated understanding of the entire plethora of angles of view from research, business and public administration, i.e., many disciplines, each with their own distinct language, methods, stakes in (and takes of) cities, and eco-cities [55].

In the general operational model of complex dynamic systems (Fig. 1.6, Module 1), this fragmentation is a seriously limiting context and a usually underestimated challenge. Usually, it results in progress being stifled early-on in the course of eco-city projects and programs, after which it tends to be replaced with face-saving claims that recycle hard-to-contest clichés which are meant to create the impression in the community of stakeholders of that project that advances were being made when in fact the invoked advances had been given no real chance in practice (as too many unresolved to managerial challenges co-occurred, knowledge fragmentation being only one of them). This face-saving but societally-toxic practice may have different (though easily deducible) motivations across the triple-helix (e.g., commercial motivations in business, political or/and ideological motivations in public administration, intellectual appeal in academia), but the fundamental result is that the common interest invoked at the beginnings of those projects is being sacrificed by and thrown under the easier-to-do business-as-usual (BUS).

As a society, and in spite of periodic launches of grand new visions, we seem to be stuck in a situation in which, despite the existence of respectable efforts to bring together in projects the full range of expert views upon urban systems [56], our cities do not (yet) fulfill the widely-shared old hopes about sustainability in general—or at least diminishing the ecological footprint of cities [47, 48, 57].

Unaccommodatingly, the price tag of renewing urban infrastructures keeps rising in many cities, while national and local budgets face numerous economic shortcomings and legal restrictions in most countries (rich and poor). That is beside the fact that doing any major infrastructure work is virtually impossible, especially in the large cities of the developed world, and beside all the characteristics of megaprojects that make everything harder (e.g., [58]) (Now, theoretically, anything is possible, but complexity and related costs tend to make it unfeasible; instead what can be done at a minimum is updating or/and repurposing old infrastructure: still at very high prices but with lower overall budgets).

This situation exists since decades and has further aggravated after the financial crisis started in 2007–2008 and during the ensuing economic crisis that took place in 2010 and which continues to have ripple effects to this day. On top of that came the socio-economic and environmental shockwaves created by the decisions taken under (public emotional) pressure during the COVID-19 pandemic started in 2020 (by the way, this is a huge stand-alone topic within the broader topic of sustainability, and which must be mentioned when needed but will not be treated directly in this book), and the impacts of the war between Russia and Ukraine started in February 2022.

This series of crises and the related conundrum of increasing debts (which often translates in practice as decreasing freedom of action) have put governance systems in a situation of intrinsic vulnerability that hardens existing problems and pile-up tensions in the complex dynamic systems that are the cities (as described in the first part of this book, especially in this module). All these add on top of the environmental crisis that is unfolding since at least half a century and that is now accelerating (as already visible in the scientific syntheses provided by researchers, and as most visible in the field of climate changes [59–61].

In this picture, it becomes clear that (1) the emergent issue of urban sustainability goals and policies overlaps with that of risk management and (2) each city must find a sound way to prioritize its urgencies, so as to correctly identify, usefully articulate, and quickly address the top issue that await resolution.

Unfortunately, the number of natural disasters and the human and economic losses caused by natural disasters are increasing faster than the capacity to address those, although big differences exist between cities and countries, and the measured or estimated parameters depend a lot on purpose, context and methods, and therefore need to be well targeted and addressed for detailed understanding for projects meant to address the issue in concrete locations; also, vulnerability may decrease in some locations and increase in others, or/and according to different disaster sources and types of losses (e.g., [62–71]).

For example, extreme weather events are a relevant and frequent topic for cities in Europe, [72] and details can be checked in the European Severe Weather Database [73]. In fact, even as I am finishing the manuscript in this book in the second half of 2023, extreme weather events unfold in Europe: Italy just faced devastating floods, France is facing drought, and all this is part of the usual news.

Over a decade ago, in a review published during the 2010–2012 economic crisis and not long after year 2007 (a year marked by many natural disasters but also a turning point in Sustainability Studies and IPCC reporting on climate change effects, a report [74] has distilled four main critical needs of human society with respect to extreme weather: (1) reducing vulnerability; (2) enhancing adaptive capacity, which also includes a certain flexibility in decision-making; (3) improving the overall usability of scientific research and information in that decision making; (4) "understanding and addressing local causes of harm through participatory, community-based efforts formulated within the larger policy context".

At about the same time, a cross-disciplinary attempt to study how cities responded to climate change risks and hazards [75] suggested three areas of potentially fruitful overlap between disaster risk reduction (DRR) and climate change adaptation (CCA), or rather areas in which inter-disciplinary work was lacking—and arguably, continue to do so: (1) extreme event likelihood: hazards, risks and uncertainty; (2) event impact parameters: exposure, vulnerability, and equity; and (3) societal responses to extreme events: adaptive capacity and resilience.

Also, in an analysis on urban adaptation to climate changes, Fünfgeld [76] saw four major challenges for urban risk managers: (1) true understanding of all the scientific information about climate-related hazards and impacts; (2) true understanding of how socio-economic processes influence urban vulnerabilities; (3) operational integration of information concerning climate risk and vulnerability into concrete planning processes and development agendas; and (4) a dearth of suitable governance/management frameworks for managing climate-related risks.

Climate change is perhaps the most evident environmental change to which communities will have to respond [50], but one should not forget that climate is only one component of the human–environment relation, and cities need to avoid actions (i) that contradict (and even cancel) each other (unfortunately, this is what cities often did during this last decade (see, e.g., [77] and the last IPCC reports)) or/and (ii) that fail to take a participatory approach that includes citizens (e.g., [78]).

In short, since more than a decade, a local-and-global situation lingers in terms of urban disasters, which can still be described as a decade ago [14] "(a) city unsustainability, associated with (b) a fundamental incapacity of science as a whole to provide an integrated knowledge base for sustainability making".

Efforts to bridge disciplines have a long history and some results provide ground for encouragement, but breakthroughs are overdue. While knowledge integration is very much a "work in progress", we should also not forget that that plurality of views is NOT the problem, in fact it is a necessity. The problem is the lack of sufficient epistemic and operational middle-ground or/and methods and tools for tapping onto the potential for creative thinking and value creation enabled by that diversity of viewpoints. We need both: (a) to acknowledge that "pluralism that cannot be boiled down to a common language" and (b) to obtain tractable/useful "trade-offs between generating a homogenous language and a collaborative appeal; between creating a stable creole and a situated semantic plurality" [79], the goal being the classic practical collaboration in the triple-helix between academia, business, and government [80, 81] under the "imperative of problem solving" [82] underlying the notions like eco-innovation and eco-city [1, 14, 83, 84] and open innovation [85], and to do so in ways that are acceptable in and accepted by the society as a whole (e.g., [86]).

All in all, we do not have time to wait anymore. The decades-long lack of progress means that cities are steadily accumulating systemic vulnerabilities, problems not only multiply, but harden and become wicked—even more complex and even less tractable. Local crises are becoming perhaps more frequent (although this would be rather hard to quantify) and, more importantly, they are becoming more and more damaging.

At the scale of the global network of cities, the local crises were already multiplying at least a decade ago and are surely appearing to do so now, from war in Europe (and in Gaza, and potentially in the western Pacific) to related soaring energy prices and food crises in poor countries that have nothing to do with that war, and through to profound reconfigurations of global supply chains. In the not-so-long-ago past, this kind of perspective on the global affairs suggested that we were entering a time period of concatenated global crises [87]. At present times, what looked then as worrying theoretical possibility has morphed into something that is more like an indication of a systemic reality. What 10–15 years ago loomed like a "perfect storm" (see Table 1. 5 in the introduction about cities: Module 1, Sect. 1.3) [88–93], may have just started now with a "mild shower". The unabated accumulation of multi-dimensional global issues related to known topics such as population, resources (including national and local jurisdictions over land and waters), peace, environmental changes in general (and climate changes in particular; see Fig. 1.10 in the same Sect. 1.3) firmly calls into question the business-as-usual approaches to economic development and urban management, and raises the perspective of that "perfect storm" to an all-time high level.

Therefore, the purpose of pioneering, experimental projects and programs must be double-sided: the development of both resilience and managerial capacity for emergency situations [94]. What was already known a decade ago is now becoming "burning": "we not only need to provide technological and social innovations for sustainable development, but we need to innovate fast" [14]. In that paper from 10 years ago, I commented as follows: "in the community of experienced students of cities, question marks arise now whether—after decades of inertia and rather unfavorable political contexts—we are in the situation of finally making the big steps towards the sustainable city, a new type of community for the twenty-first century, or rather writing the obituary of the eco-city idea [95]". Today (as I am correcting this manuscript in May 2023), the answer seems to be: no, we didn't.

Further in that 2013 paper, I also wrote something which I need only repeat because it is just as true: "epistemic difficulties associated with the lack of a unified theory of sustainability (or, for that matter, a unified theory of urban systems) are not an exception in the current era of fast change and pervasive knowledge fragmentation. In addition, despite substantial advances in formulating global policy options in terms of global scenarios, cities are rarely able to translate this into meaningful scenarios at their scale and type of discussion. Local communities need to overcome concrete difficulties related to both downscaling and availability of reliable local data. These, together with the local biophysical and social-economic peculiarities, pose a major challenge for city managers and stakeholders [50, 96]. In this complex context, science is called upon to evolve towards multi-disciplinarity and time-efficiency [97]".

I will address this issue later in the book. But before doing that, we need to answer a self-evident question coming out from the note above: so, are we talking about a "lost decade"? Based on the non-linear model of progress used in Fig. 1.12 in the previous module, the short answer is: "not necessarily". In fact, it shows that humanity as a whole is still in that initial phase of painful searches that are very costly and time-consuming but the critical mass has not built yet. Importantly, that critical mass is not only about a necessary quantity of work/investment/commitment moving in the stated desirable direction, but is especially about the quality of it, i.e., the transformational aspect. In short, we are still at the stage of trying to find our way towards that commonly agreed goal, and this will be a necessary-but-insufficient condition. However, all insights gathered during this period of "high-investment and low-return" are potentially useful for the winning re-configurations that will unlock faster progress.

Operationally speaking, the scale and dynamics of qualitative change is not necessarily the same as those of quantitative change, and one or a few decades is still a short time for the history of humanity. Whether this specific time was spent in the right direction (or was just one of the many divagations that history is familiar with) will be known later on, probably after some other decades. In any case, reflexivity (and "reflexive governance" [35]) is a necessary condition for eco-city projects and programs, as opposed to just reacting to burning crises—provided that it does not swirl into sterile debates, i.e., they must happen within a knowledge-action framework of interaction between participants. Postmodern managers in general cannot fully rely on science for real-time integration and conceptualization of the situations, contexts, and problems with which they are faced; they must make choices here and now, and live with the consequences of those choices (e.g., see [98] for an in-depth introduction to the bolts and knits of reflexivity in management education and learning). This is in fact the condition for emancipation from the unsustainability crisis that tethers a community of people to the ground, and for "learning to fly" towards a better future. (As a side observation, in projects, the reflexivity condition is to be found in the "reflexivity cycle" of the DIKAR_process model in Module 3, Sect. 3.6).

To address this universal challenge of knowledge integration (within science itself but also between science and society, and across all domains), different procedures have been proposed, but here I only focus on those directly relevant for the fundamental knowledge-action model for solving the basic conflict of priorities described in the introduction to sustainability (Module 1, Sect. 1.2, Figs. 1.1, 1.2, 1. 3 and Box 1.1).

Starting from the issue of urban solid waste in Switzerland, Scholz et al. [99] have proposed a methodology called "problem structuring for complex transition" (SCT), which I have compared and combined with the method I developed together with my own collaborators in Central Europe (Romania, Switzerland and Germany included) for the same issue of urban solid waste: a set of sustainability filters derived from the systemic perspective on sustainability (as anticipated in Module 1 of this book) [100, 101] and mapped logically onto the succession of stages described by the Swiss team.

The result was a managerial mechanism that filters (evaluates) virtually any program or city re-/development trajectory, and shows which proposed paths are compatible with sustainable development and which are not (based on all available knowledge, i.e., technically operationalized as lists of indicators). This evaluation method can be applied to all envisaged or deduced scenarios or proposed solutions for a city or for city district, from a holistic, systemic perspective.

Supplementary Observations Before the Method Description

Importantly, both transdisciplinary approaches mentioned above are grounded in "participation and negotiation followed by decision-making", and not on some "comprehensive understanding (or logical synthesis)" of all available data and information, although technical data is part of it. In fact, an efficient use of science in and for urban communities implies (1) a realization that substantial postponing of difficult choices in order to "wait for a better understanding" of the matter, usually give an already serious problem time to aggravate, and, consequently, that (2) it is better to go on addressing the problem with prioritization-based action, even before in-depth scientific descriptions become available [97], and doing so in a way that is flexible enough to allow swift re-prioritization when new information requires it.

A general rationale applies in all cases, action towards sustainability is a matter of "problem solving-oriented research" and "purpose-driven knowledge management", which does not, in any way, imply leaving science aside. How much time and resources should be assigned to prioritization is a matter of managerial decision, but efficient prioritization is required. This is part of what the necessary "disciplined experimentation" [84] and concrete materialization of the "a new social contract for science and scientists" in the twenty-first century [19, 20]. In the case of eco-city projects, such experimental developments often take the form of a triple-helix of collaboration, whereby actors from universities, industries and governmental agencies join forces and thus assume a collective role of "experimental epistemologist" for the purpose of transition to sustainability, the idea being that we must be able to know "how cities think" [102] and, in the context of local community and citizen involvement, how policies, programs and projects can apply the concept of "democratic action as experimental problem-solving to the spatiality of the city" [103].

This experimental approach allows an ordered (at least less disordered), cautious, evidence-based, stepwise transition to sustainability through a succession of exploration–exploitation cycles, one that is known in transition management theory as "reflexivity" [84, 89, 104, 105], and one which insures organizational ambidexterity both at project level (so in triple-helix ensemble) and within the participating organizations, knowing that different types of actors have different propensities (e.g., universities will favor exploration, industries will favor exploitation, while government will favor whatever benefits social order).

All in all, and to re-state for the purpose of this module an observation made in the first part of the book, that the main obstacle that prevented a faster progress towards sustainable development (in spite of many pioneering advances during the last 1–2 decades) is the persisting gap between knowing and doing. This means that, the technical goal we need to attain before even having a chance to concretely contemplate sustainability per se is to reach that inflexion (mid)point on the S-shaped curve of the transition to sustainable development in Fig. 1.12 (Module 1, Sect. 1.4). This requires knowledge management approaches.

In the next subsection, I apply a prioritization-oriented deliberation-based method: dynamic decision trees. Originally developed based on the real-world developments and experiences in two mid-sized, historic European cities—Jassy (Iași; in Romania) and Versailles (in France), this aims to be a "simple, integrative but sharp decision-making tool for further testing towards enhancing city resilience and sustainability" [14].

Now, building realistic scenarios is a self-disciplined way to employ and expand knowledge in creative but practical ways [106, 107]. But this is easier said than done: we live in complex dynamic systems with inherent limitations, uncertainties and side effects, and we have to deal with many unknowns by using a heterogenous corpus of knowledge that is distributed and largely non-integrated. In particular, cities that are located in developing countries face the additional challenge that most relevant insights about city development are scattered (demanding lots of resources to find for practical use) and they mostly come from the western or/and developed countries (see., e.g., [108]). While this is yet another instance of the reality that knowledge is distributed, it also underscores how important it is to participate in collaboration networks: they are a main (and in many cases the principal or even the only) way to stay connected to what other cities do and how to potentially combine different insights from across the world and develop solutions that fit the unique context of each city.

Needless to say, the fact that richer countries produce more insights (because more resources are invested in all kind of projects) does not mean that cities in those countries are self-sufficient: they are not, and they also have the same challenge of finding and using relevant insights. In a word, each city has its own concrete puzzle to solve, and this puzzle cannot be solved in disconnection from the global network of cities, projects, and experts.

In cities, planning for sustainability inevitably requires an inclusive "debate and decide" process. Given all the complexities discussed in this book so far, this deliberation process cannot possibly be replaced by information and communication technology (ICT) but benefits from its massive support. In the era of smartphones and instant communication (a major landmark being the launch of the iPhone in July 2007), ICT also includes aspects that are now common place in discussions about smart cities, e.g., Machine Learning (ML) and Artificial Intelligence (AI), so the available ICT support can be very sophisticated. Nevertheless, this deliberation is a rather social exercise, not a computer-driven "predict and provide" process [109, 110], and enthusiasm about ICT must be properly calibrated (sometimes tempered and some other times cultivated). ICT support is indeed greatly valuable but is much less sophisticated than it appears at a superficial regard, for the simple reason that automation cannot replace real-world complexity, and is—instead—limited by it. In fact, ICT methods and tools are no miracle (in spite of being often advertised using exactly this word, especially in commercial settings that may try to elicit irrational desires), and those methods and tools themselves require a great deal of dedicated development for real-life solutions. Later in the book, I discuss this issue in more detail, using the case of the relation between the so-called Industry 4.0 and the transition to sustainability. For now, suffice it to say that science (i.e., hard evidence and proper justification through hard work) and related RDI projects remain our best guide.

Method: Delphi-Based Scenarios

General context

Building scenarios is a typical approach in prospective studies and was initially a military method for strategizing through developing realistic action options towards winning a war (e.g., [111]). Scenarios do not predict what happens; they envision potential situations that may unfold in the future, and have at least one of the following two (and complementary) purposes:

- a broad (and usually long-hauled) purpose of building scenario is to understand what drivers and patterns can and are likely to shape the future set of action opportunities and limitations:
- a narrow purpose (and normally meant for situations where there is no time for in-depth study) is to support contingency plans and emergency procedures to be followed as fast as possible when the situation demands it.

Relatedly one must avoid confusing scenarios (defining a practical space of options) with the strategy one applies to arrive at a desired outcome given that space of options [112]. As scenarios must be used in conjunction with other methods, the discussion here involves different concepts and ideas that project teams may need or want to study (Box 2.2).

Box 2.2 A selection of other concepts and ideas that may inform foresight and scenario-making exercises in eco-cities:

- "Scenario improvisation" (related to sensemaking) and "promoting articulated action" (related to the advantage of having and applying a stepwise procedure for "path-finding" in complex systems [113]);
- "Scenario-based strategizing" where behavioral biases are explicitly accounted for by developing scenarios and strategies separately and then seeing how they fit, based on a "matrix" of efficacy and robustness [114];
- Relatedly, scenario clustering and evaluations (depending on the purpose of the scenarios, also including probabilities) [111];
- Cognitive aspects and building social capital, to try and address urgencies faster than previously thought [115];
- Embedding (known) real options in (more complex) scenarios, as necessary in operations, and how to advance the learning skills of firms in uncertain environments [116];
- "Strategic conversations" and "openness disposition" [117] (this is of course related to the well-known concept and strategy of "open innovation)";
- Combining scenario and resource mapping approaches [118].

Very important to note is that the notion of "branching points" mentioned in the previous Module (Fig. 1.13, and [78]) is key for the trajectories that cities may and actually take (or not) in the course of their evolution. There are many ways to interpret branching as relations between situations and options in trajectories that develop in time.

Here, I will use a (arguably even the) most fundamental understanding, as related to cluster analysis: sense-making of available knowledge and evidence about a problem at hand by clustering together those aspects that fit together, and ranking them in terms of priorities. Figure 2.1 illustrates this idea starting from available evidence (data), i.e., top-down. The end result will be a set of scenarios (usually known under the name of "decision tree") that starts from the most important cluster on (i.e., in Fig. 2.1, this means a bottom-up direction).

This simple concept will be applied here in the structuring of the interaction between the participants in a scenario-generation workshop (normally experts, but, depending on the exact purpose, it can or/and should include citizens and any other type of actor that is relevant for insuring realism of the final set of scenarios). But it is worth noting that, at the core (or mathematically speaking), the concept is closely related to, and is known in the literature under, different forms and applications (and this is true since a long time, both and social science and natural sciences and engineering), notably "hierarchical clustering" (i.e., groupings based on weight or other parameter reflecting the levels of importance (e.g., [119]), ground-truthing, i.e., verifications aiming to insure that what is assumed in an given abstract modeled is supported by empirical data and observations "on the ground"), usually related to

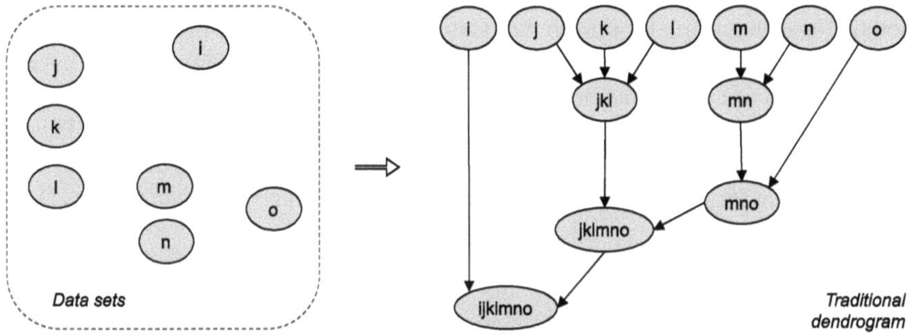

Fig. 2.1 The Delphi-based method for collective articulation and ranking of issues in multidisciplinary groups of experts is equivalent to traditional hierarchical clustering. Each bubble represents a data set. The dendrogram is also an example of a mathematical graph (for more about graphs, see Module 3, Sect. 3.2, Box 3.1). Black arrows represent transformations: interpretation of current data and structuring based on available information and knowledge (for details about the relation between data, information and knowledge see the section on the DIKAR_process model in Module 3, Sect. 3.6). Visual distances between bubbles intuitively roughly denote conceptual distances. *Source* original, based on current knowledge; for mathematical details on hierarchical clustering and decision trees, see (e.g., [125, 126])

optimization and the notions of entropy (introduced in the previous Module) (e.g., [120, 121]) and/or principal component analysis (PCA) (e.g., [122–124]).

In theory, if problems would be only multidimensional but not complex, i.e., non-linear dynamics of real-world CDS would not be involved (or in some very specific situations in engineered systems), priorities can be organized and ranked just by using weighted weights as per the Analytic Hierarchy Process (AHP)—and this is exactly what happens in certain functions in urban systems (e.g., mobility), or even to try and evaluate the level of complexity of different projects (e.g., [127–130]). AHP is a multi-criteria analysis (MCA) method to evaluate and prioritize alternatives by pairwise comparison of (selected) criteria, using classical matrices of relative importance of compared criteria such as this below:

$$A_{ij} = \begin{bmatrix} a_{11} & a_{12} & \dots & a_{1n} \\ a_{21} & a_{22} & \dots & a_{2n} \\ \vdots & \vdots & \ddots & \vdots \\ a_{n1} & a_{n2} & \dots & a_{nn} \end{bmatrix}$$

However, this is only the beginning, and its utility tends to decrease with increasing number of criteria (i.e., the situation quickly becomes "messy"), so such approaches need to be complemented with other methods.

A most (and perhaps the most) straightforward approach for scenario generation in "messy" contexts is based on purpose- and context-dependent versions of structured integration of expert opinions and is known under a generic name: the Delphi method (or approach). In essence, this has been characterized half a century ago as "a method for structuring a group communication process so that the process is effective in allowing a group of individuals, as a whole, to deal with a complex problem". And, "when viewed as communication processes, there are few areas of human endeavor which are not candidates for application of Delphi" ([131], see also [132]). This is what I used and adapted over a decade ago (quick details in Box 2.3) to develop an iterative procedure for generating versatile decision trees and eco-city scenario sets [14].

Box 2.3 A short history of how the Delphi-based method described in this book came to be.

During late 2011, while I was searching for a "clear-cut but versatile" method for collective decision-making in a triple-helix eco-innovation cluster (called Econoving) engaged in complex eco-city projects for urban developments in the city of Versailles [89], and being dissatisfied with the results of my search, I have started from this direct project experience and the most useful insights (that I found in the literature at the time) to develop a new method of scenario generation, one which recognizes from the onset and uses the practical (operational) complementarity between what I would call in an imperfect but quick summary as "the predominantly-qualitative approaches of social sciences and the predominantly-quantitative approach of natural sciences and engineering".

In the original method, I was thinking about priorities in terms of "pressures" (as a contraction of "pressing needs"). In discussions about disasters and resilience, this is the same with "risks", using—for illustration—the list of indicators that was already published for the city of Jassy (Iași), Romania [100, 133], combined with the results of a public local exercise and analysis about potential and desired (or not) futures (or details for the future) of the city [133].

Here, I present and discuss this method, with an emphasis on urban risk management. First, I give you an extremely short description of the main idea, and then I dedicate the rest of this section to providing you (based on experience and current literature) with enough structured material and explanation so that you can do it yourself.

Important to know or remember about the Delphi method is that it is easy to think that it's easy when you didn't practice it, and then it's easy to give up too soon under pressure (premature disappointment), i.e., before it becomes easy for you to apply it because you simply gained experience with it. However, once you "get the hang of it", the method is extremely efficient at hashing out a set of priorities and their ranking, especially after a few iterations that stabilize the result of work, and this is all you need for a valid kickstart of eco-city projects and programs.

Depending on available resources, purpose and level of skill of the organizer, one iteration method can be exercised in a half-a-day session or in a half-a-year.

In a nutshell, this Delphi-based scenario-building method consists of organizing and moderating a session of collective expert judgment and decision where a group of experts (well-chosen to be relevant and to cover all needed areas and types of expertise) discuss and vote to produce a ranked list of priorities, which list is then discussed again for verification, refinement, and consolidation (knowing that this part may last a very short time when the issue is easy and a long time when the issue is hard and requires extensive discussion before voting again), and then each priority is reformulated as a question that can be answered with "yes" or "no" (in practical terms, like "do this or not"). Then this set of questions are being answered until the end: if the answer is no, what does this means (how do we describe the situation when this question is being answered "no"); if the answer is "yes", then we (you) can go to the next question and so on. The result will be a set of situations (equal with the number of questions plus one), which are now characterized by the chain of answers given to those specific questions. Describe what that means in practice, based on available knowledge. For example, if the first question (corresponding to the top priority) was answered No, this automatically means "business as usual (BUS)", and you have to describe what this BUS entails. If the top priority was answered Yes and the next question was answered No, this represents another type of situation; and so on. Finally, chose an adequate name (descriptive but short: symbolic and easy to remember) for each type of situation. Each of these situations is technically identified by the path (yes-yes-…-no) to it. So: List of priorities, discussion and consolidated list, formulation of priorities as (Yes/No) questions, and taking stock of the set of possible end outcomes (what situations have been produced and how we got there).

Brief Important Observations Before Method Description

The first detail to remember is that the core method is developed in response to all the issues described in the book so far and is based on, and informed directly by practice. Therefore, it is flexible, iterative, and the more you do it, the better you do it, and the better its end-product will be for guiding and assisting decisions in practice.

The main advantage is that it allows constant operational orientation of action based on existing knowledge in spite of the multi-tiered fact that:

(a) Science in general is not some perfectly organized and coherent body of {data; information; knowledge} but an immense (and overwhelming) mosaic of disciplines and sub-disciplines (and the same is true across different professional fields within the triple-helix of academia, business, and government), each of those disciplinary domains having their own distinct traditions, methods, vocabularies, expectations, and work cycles;
(b) There is no unified theory of sustainability and no unified theory of a city; and this is actually a rather normal state of play. For instance, the field of physics is not characterized by a unified theory of everything (although this remains a general goal of it) but a set of theories addressing different foci of interest, and some of which contradict themselves (and uniting them is itself a research topic).
(c) Knowledge, in general, is, by its nature, distributed across very large and complex networks of experts, and can therefore be only "mobilized" via well-defined purposes, not "centralized" as many of those unfamiliar with the overall

workings of Science (and knowledge overall, as studied by epistemology and cognitive sciences in general) may imagine.

In other words, the method allows a team (a) to have a common perception about the target problem (and the world in general), and (b) to make sure that this perception is dynamic and hence current and thus allows near real-time expert response—which is especially important in situations when the context changes very fast, e.g., during some stages of a crisis situation (and I am not only talking about imminent or ongoing natural disasters but also about systemic disasters, like societal collapse, war, revolutions, or even positive development like mass mobilization for a noble cause or/and high-pace reconstruction after a profound societal challenge).

As a corollary, we need also remember that we humans (even experts) are, by and large, not very good at articulating the knowledge that we have. In this respect, the interactive method above helps individual experts (and really all participants in such exercises, including citizens) articulate better our own individual knowledge as part of the overall social exercise of articulating our collective knowledge in the form of structured lists, conceptual relations, technical tests, and operational representations and knowledge-action models.

Then, those yes/no questions are essentially policy questions, and when the answer is "yes", this means the city decides to have (as opposed to not have) a policy on that matter. So, the list of questions is essentially a list of issues to be answered in an expert-determined order of priorities (but can also be submitted to, or simply obtained by expert-informed popular vote by the citizens in the target local community, i.e., a vote held after a series of public workshop and debates connecting citizens and experts).

The issue of decision-making is a huge topic of research, and, from a project management perspective, is of particular interest in organization studies, psychology, mathematics, and information sciences (but this is by no means of a sufficient list).

In the discussion I make in the expanded description that follows, I use examples I know from practice and literature and I provide a series of general and analytic categories. This shows how the result can look like, while making the connection with the materials from the other book modules and allowing space for your own operational details to fit in.

Implicit in this procedure are at least two other major issues which I address in detail in some separate sections of the book: knowledge integration across expert areas that are very diverse and hard to integrate; and the policy-making process.

Then, the final module of the book is meant to assist the method with a detailed knowledge-action matrix for codifying activities and insuring an optimal balance of efficiency-and-flexibility in the kind of complex projects and programs that are required when a city wants to be an eco-city.

The Method Consists of Five Steps, as Follows (with Brief Explanations)

Step 1. Preliminary identification agreed (formulation) and prioritization (ranking) of the top p main (sources of) risk for the city.

These risks are to be understood (by the designated team; and by all relevant actors and stakeholders) to be "the most important short, mid and long-term stresses for which the city must prepare". For discussion inside the team (even if only for illustration, but usually not only) for a target city, all relevant information items can be piled up and can be imported from previous studies covering the target city (from lists of indicators to detailed conclusions from previous analyses) (Fig. 2.2).

In addition, such social network analyses of knowledge networks confirms and explicates (for in-depth analysis of patterns that can inform operational insights) the basic intuition (of anyone with some practical experience) that the networks of knowledge and expertise that exist in a city are only a small (local) part of the wider global networks that can provide input for local actors. For example, a recent study trying to answer the question "Who collaborates for urban resilience?" [136], mapped the most connected organizations in US cities like Baltimore, Boston, Fort Lauderdale and Seattle, and concluded that "resilience governance is multi-level, with city, regional, state, and federal government organizations present in all city networks".

This idea links back with and is compounded by (1) the existence of several relevant scales as shown in Fig. 1.10 (synopsis of outcomes of interactions inside complex systems) in the first module, (2) the relation between cities and their hinterlands, and (3) the operational requirement for integrative approaches and units. One

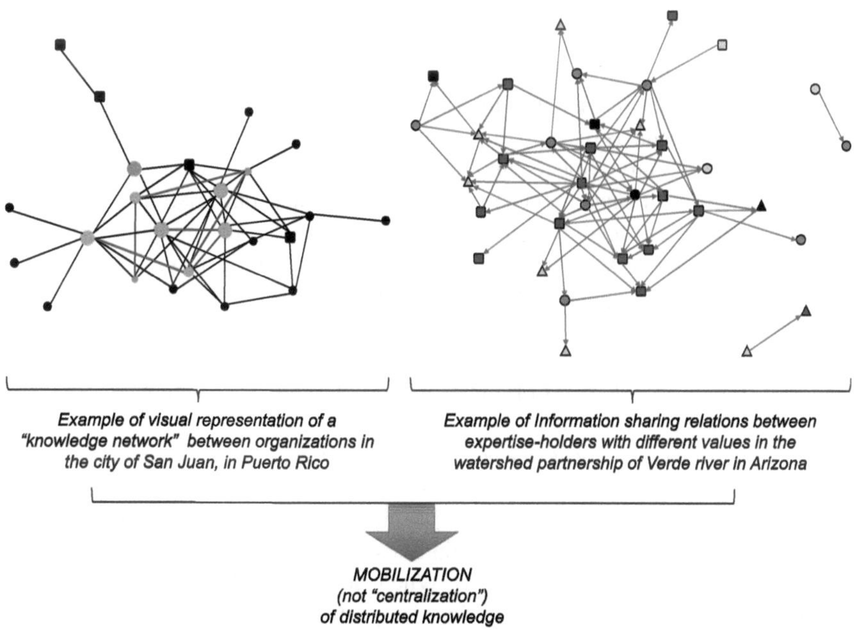

Fig. 2.2 General illustration of the phenomenon that knowledge is distributed (as consistent with practice, and as deducible from the reality that human and natural systems are complex dynamic systems), and is therefore mobilizable not centralizable. Dot shapes and colors indicate typological diversity. *Source* combined simplifying redrawing based on detailed research versions by ref. ([134, 135]; for more examples and discussions, see, e.g., [134, 137])

example worth mentioning for illustration purposes is the case of water pollution in cities: a river-basin approach is necessary in the study and management of the overall (waterborne, airborne, or land pollution contributing to final) water pollution upstream or downstream (e.g., [41, 100, 134, 138]). Another example is the case of atmospheric pollution and health risks: the use of extended regional and global scales are necessary in the study and management of the relation between economic development and pollution; a well-known pattern is the Environmental Kuznets Curve (EKC), i.e., the idea that pollution levels per capita increase up to a certain level of economic development of a country, after which it decreases (because of higher living standards, cleaner technologies begin to be demanded by emancipating people): this hypothesis is confirmed in some cases and infirmed in others, depending on the question being asked and the scale of observation being used. Notably, pollution in many cities in richer countries has decreased mostly because those polluting industries have actually moved to cities in the poorer countries while global pollution levels have actually increased (see, e.g., [139–143]).

In the broader discussions that exist in the literature on how to manage complex projects in large networks and clusters of experts, the distributed (non-centralized and non-centralizable but problem-driven mobilizable) character of knowledge is related to knowledge organization (KO) [144, 145] and knowledge management (KM) ([1, 146]; see also [147, 148]) in general, and to some more specific methodology issues like problem structuring in knowledge-action models, and knowledge representation (notably mind maps, concept maps and knowledge maps, which will be treated in Module 3, Sect. 3.3).

At this point in the book, it is enough to mention that the scenario sets that I am about to describe represent a particular (result- and process-oriented) case of knowledge mapping: one that is focused on streamlining expertise in spite of the inevitable complexities of eco-city projects, and taking the form of what is known in the literature (especially in management, operation research and information sciences) as "decision trees", the procedure that I'm about to describe representing (strictly technically speaking) an iterative decision-making algorithm for insuring relevance and promptness (and an adequate, context-determined trade-off between efficiency and flexibility).

First off, all knowledge mapping involves network analysis, i.e., how different categories (topics or/and agents or/and problems, etc.) relate to each other, how they rank in importance and so forth.

In terms of network analysis of a network of priorities, a high priority is equivalent to a highly "central priority", i.e., a top priority is a central priority and the topmost priority has the highest *Degree Centrality*. Similarly, a most influential agent (actor, institution, etc., any abstract agent) is the most important when it has the highest *Degree Centrality* in a network of agents. When the agent is a knowledge (and expertise) management method for integrating data, information and knowledge across fields or angles of study and practice (around a common topic of interest, like a city or eco-cities in general), that method or model is bound to have a high methodological centrality (i.e., importance) in multi-dimensional and complex projects (ergo influence in eco-city projects).

Mathematically, *Degree Centrality* is definable as:

$$D_c = \sum_j x_{ij}$$

where i denotes an actor and x_{ij} is the value of the connection between actors i and j. Also, x_{ij} is an entry (i, j) in the network's adjacency matrix (the table indicating which nodes are linked with which other nodes).

Then, relatedly, *Betweenness Centrality* is definable as:

$$B_c = \sum_{i<k} (g_{ijk}/g_{ik})$$

where g_{ijk} is the number of geodesics (i.e., shortest paths) connecting nodes i and j. For detailed discussions with applications, one can check the literature on network analysis applied to parameters that are relevant in cities (e.g., [136]).

In practice, ranking is the best done by elicitation and voting. In my own works, I did it as follows (I use narration because I estimate it to be the easiest to help the reader grasp the idea): a succession of two parts of a list-generation workshop, i.e., live. However, this can also be done through online communications (but it takes more time and much of the advantages of "thinking together" of a team is lost (for comparison see, e.g., the 2-round Delphi exercise done online by [149]).

On a whiteboard (or blackboard …) or directly on a computer with the computer screen projected on the wall (or any other screen that can be seen in real time by everyone in the room), I started a "preliminary list of issues, in any order (bullets) and in any start formulation, and had the team keep proposing issues (and refining listed formulations/bullets) until there is nothing new to propose in unrefined form. Then, we started consolidating the preliminary list, for example (and usually unavoidable phase of) merging the issues that can be merged (because many would overlap) and splitting the issues that can be split because the team estimates we should do so (because they require distinct discussions and tackling). Discussions can be messy, this is why the moderator needs to exercise this a number of times (s/he will become better in time) to get very good at handling and weighing all details as they come into the discussion and make decisions on the spot to separate the wheat from the chaff in order to keep the discussion both useful and streamlined. Of course, the moderator can also have helpers, e.g., someone helping with writing or just providing any needed support (logistic or/and conceptual; for example, the main moderator can be assisted by some people with complementary expertise, so that, in the end, all relevant knowledge fields are well covered). But, in the end, everyone in the room is supposed to help the moderator help everyone for distilling the list of topics.

As a side note, there is of course also the issue of conflict and conflict resolution—for that matter, there are other dedicated books. The main point here is that, of course, the team of experts participating should cover all relevant domains. Here, because we talk about sustainability, we already assume that everybody understood that we need all types of experts. If the prioritization is only done by economists or only by chemists, the results will not competently describe the situation of a city.

As a second side note, the intermediate purpose here is to produce a list of issues that are a bit similar to the cloud of topics in the whole field of sustainability studies as shown in Fig. 1.5 (Module 1, Sect. 1.2 of this book)—with the important difference that this is done for a specific city (not in general) and the purpose is operational: not only research but also the generation of viable pathways and solutions for specific problems along that pathway. But the big similarity is the purpose: the team taking this responsibility must make sure it does not forget anything important. This is why, when such an exercise is being done as part of an understanding with some local authorities, some people with experience in the management of the city must participate in the exercise (and must make sure it brings along all materials and information that are relevant for the city as a whole, not only for a sector).

Even if such an exercise is being done for a specific sector (for example, as a strategy to get the ball rolling), the team involved in this exercise needs to know how that specific sector is functionally inter-linked with the whole functioning of the city. As already discussed in this module, systemic problems tend to accumulate (and are particularly difficult) especially when interactions at the scale of the whole system are not being paid due attention.

This is not to say that problems cannot be broken down into subproblems: oftentimes, it can, and this must be done; instead, I only want to say that this must be done properly, based on knowledge. In the third module (chapter) of the book, relying on established knowledge and practice, I describe a model for allocating efforts in piecemeal approaches. But before that, it is necessary to have a good understanding of how to go about identifying the pieces and what matters in priority.

In the process, we might detect forgotten issues and see where we fit them (inside existing topics or as a new one). This exercise involves a mediator (a role which I fulfilled) that makes sure discussions go on and are efficient in distilling that preliminary (yet unranked) list of topics that must not forget anything important. Depending on the nature of the work and the preference of the team, these topics can be thought of as "problems" or "themes" or other useful categories; but in the end, they must be understood as pressures faced by the city. This is why I use "pressures" as the most general (i.e., the least epistemically committing but the most operationally relevant) formulation.

Step 2. Reality check of the shortlist made during the previous step.

This is done through a cross-analysis of pairs of risks (two-by-two matrix, basically a four-cell table), each risk having two possible states (*low* & *high*) on a low–high axis denoting an intensity gradient. The purpose of this step is to scan for potential contradictions or issues that have been left outstanding (or which require renewed attention) in the big picture of the city-as-a-whole. In other words, it is a form of "structured sensemaking" and is considered to be necessary to ensure correct cover of the target case study, as a basis for obtaining relevant and useful conclusions (Fig. 2.3).

The set of 12 situations represents a basic taxonomy covering all possible types of interactions between the selected top three pressures.

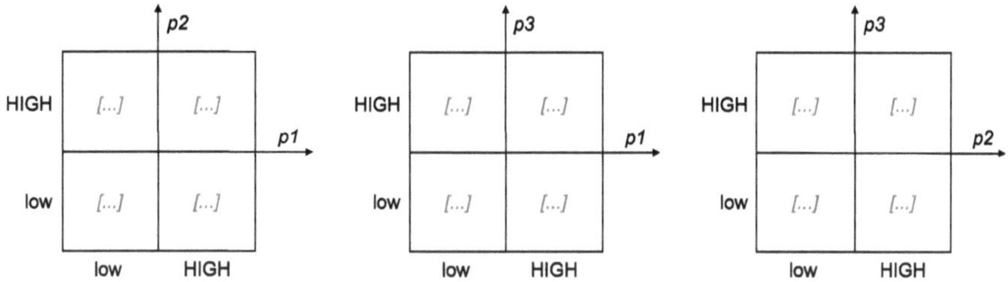

Fig. 2.3 Set of 4-cell visual matrices. Inside each sector, [...] denotes technical details and descriptions issued from expert workshops: they must write down all imaginable connections and issues as per current knowledge, with a particular care not forget anything important (if some aspects escape this step of "systematic scanning", the resulting list of priorities (or at least their formulations) risk to be later critiqued as not being well thought-through, and perhaps too weak to be relied upon). *Source* original

Each matrix includes four situations *S*, which may also be called "exploratory scenarios" (ES), since every *S* is an exploration of what each combination of low/high risk actually means in practice in the study context, based on all knowledge available to the analytic team performing the exercise.

The direct objective of this step is to verify whether all combinations of interaction between selected risks from the first step are relevant in the given context of the city of interest. If this does not verify, i.e. if at least one *S* does not make sense within the ensemble of details available from the study city, or if the exploratory situations revealed in the 4-cell matrix scenarios do not seem relevant for the study city) then it follows that the two risks in that particular pair should not be shortlisted together (i.e., they are incompatible/they exclude each other) in any strategy and sequence of actions, and, consequently, *Step 3* must be performed (iteratively, if needed). If/when they do verify, the procedure can be continued directly with *Step 4*.

For example, one can prepare the work session by making a review of available information, and then classify the situation of knowledge in each of the 12 boxes as *w, i or m.* Then, the team could focus specifically on re-verifying the *w* situations, completing the work (or advancing it as much as possible) in the *i* situations, and addressing the knowledge gaps represented by the *m* situation (or acknowledging that it is actually an empty set, and treat it as such). This step can result in deploying some models for problem-solving-oriented knowledge integration (see Module 3, all sections but especially Sect. 3.6), or generate questions for it, to be addressed according to known best practices.

This means that completing this step can very easily take a short time or a long time depending on how much relevant and actionable information is already available for the target city. If this takes more than a quick review, i.e., several or more work sessions, then the *m*, *i*, *w* codes (to keep the example above) can quickly indicate where the work stands at any moment in time (this might be needed to inform project propositions and related estimations of required work and time volumes).

The minimal action needed after *Step 2* (default *Step 3*) is to vote again on the ranking of the selected priorities from *Step 2* (by now, team members have an even

Table 2.4 Number of 4-cell matrices required for different lengths of the list of top priorities

if n	*then*	C_2^n
= 4	The number of matrices to work out	$C_2^4 = 6$
= 5		$C_2^5 = 10$
= 6		$C_2^6 = 15$
= 7		$C_2^7 = 21$
= 8		$C_2^8 = 28$
= 9		$C_2^9 = 36$
= 10		$C_2^{10} = 45$
= 11		$C_2^{11} = 55$
= 12		$C_2^{12} = 66$
= 13		$C_2^{13} = 78$
= 14		$C_2^{14} = 91$
= 15		$C_2^{15} = 105$

deeper understanding of the matter). The resulting ranking is the consolidated one (to be taken further in the procedure). But more work might be needed, as discussed below.

Other technical observations about this step are made at the end of the section and are important for understanding its rationale and necessary conditions for effective completion. In the next paragraphs, I continue with the rest of the general method so that you can see what this gives in terms of scenarios (how the sets of scenarios actually look like in the end).

Other observations important for this step

The number of matrices M that are necessary in this step is calculated to be the total number of combinations of n taken by k, i.e., the well-known C_k^n, where $k = 2$:

$$C_2^n = \frac{n!}{(n-2)!n!}$$

Therefore, the number of required 2 × 2 analytical matrices varies (Table 2.4).

In simple words, if the number of top priority pressures p selected for in-depth cross-analysis is 3 (as with the case discussed here), then the number of distinct pairs (and thus, the number of 2 × 2 matrices) is 3; and the total number of exploratory scenarios in the "scan" session becomes 12.

For managers (specifically, in terms of how much time and resources are needed for this step), it is important to remember how the number of cross-analysis matrices and exploratory scenarios increases with higher numbers of top priority pressures p selected for in-depth cross-analysis—as shown above. But the multiplication of the variable does not even stop here, because every ES will include some lists of detailed parameters, and every ES matrix and set of matrices represents a sensemaking effort from the part of the involved experts (both as individuals, in their own mind, and in terms of within-group communication and processing). So, doing a higher number of top priorities is not just a matter of multiplying the number of work hours: it also makes it more arduous.

Moreover, it is important to understand that this list of top priorities is a form of sampling the complexity of the studied urban system (or of a target problem/system in general), and the set of 2 × 2 analytical matrices derived from them constitute a set of windows that are deliberately open towards that complexity—to allow expert inspection, so that the experts can verify together and check the coherence of what they know collectively about that matter. By systematically matching the selected top priorities one-to-one, and by giving a range (low/high) of the potential interactions between these any two priorities, we basically seek to unveil the details of their interaction in real life: different ESs will evoke different expert insights and information items which together become the work content (and record) of each ES and matrix. This creates the occasion for a multidisciplinary team to corroborate what they know and to identify potential mismatches, gaps or needs—i.e., to refine, update, and consolidate what they know collectively about the studied system or/and problem. This deliberate, systematic elicitation of available understanding within a constituency of experts makes them pass, de facto, from a collection of individual understandings to a collective understanding on the study case.

Moreover, this explicit cross-analysis is not only an effective, disciplined method to "bring experts on the same page", which is a very important goal in and by itself. It is also a way to search for and to identify the relevant fragments of a potential big picture—and to put them on the same table (as a necessary preliminary condition for solving the puzzle).

The more the total number of ES increases, the more the completion of this step will require the experts to keep a higher number of variables in mind, and this becomes harder and less efficient. In some cases, it can be estimated judicious to deepen the study of priorities by including the first 4 or 5 (or more) priorities. But this requires correspondingly higher efforts (also bearing in mind that this will require a schedule of multiple meetings and it is required that the team composition is both large enough to be representative and to remain the same during the entire exercise). The basic case with three top priorities (as described above), represents a minimum that can be used in all situations (and, when resources are scarce, will be the only one applicable at all).

One must also be aware that each of the 12 ESs represents a unique case, with implications on insights from the collective experience of the analytical team. If, for some of the 12 cases, there is no practical content to be acknowledged, this means that those categories are empty and should be recorded as such (otherwise, they can be critiqued for being some potential situations that have been left unstudied, i.e., the work was superficial).

Step 3. Consolidation of the initial list of priorities.

Unless a simple default re-vote for consolidating the ranking suffices, the priority list of p must be modified, either (i) by replacing one or more of the initially identified top risks with the next ones on the long list of priorities, or (ii) by re-doing the first two steps to insure that eliminating one item from the list does not exclude some information that is too important to be excluded. Logically, after any such update, the four-cell tables must be re-done (re-run) as many times as necessary, i.e., until the priority shortlist is confirmed. A metaphor might help describe this step intuitively: the four-cell matrices serve as logical "digesters" of the pressures that have been identified intuitively and proposed (as preliminary formulation) during *Step 1* in the work group/team for further consideration down the procedure.

In that sense, a "good digestion" means that the reality check step was successful, and one can proceed to the next steps in the procedure. One important decision to be taken here concerns the size of the priority shortlist of p. As mentioned above, simple combinatorics indicate that the entire step becomes much more voluminous, ergo resource-consuming to analyze too large a number of top-of-the-list risks in a given exercise. Obviously, if four (or five) priority risks are being shortlisted (or five) for the "digestion" trial (instead of three), the number of matrices becomes $M(4, 2) = 6$, and the number of S becomes 24; or $M(5, 2) = 10$, and S becomes 40.

All in all, the chosen number of top priorities must reflect an equilibrium between two necessary conditions: the step of reality check with the four-cell matrices is necessary, but so is also a minimum number of pressures (or "risks"). This means that a reasonable short list would normally have to include three top pressures in most projects and 4 (or more) top pressures in very well-resourced projects (that can also afford the time).

Then, the final accomplishment of this step (achieving either confirmation of the concrete composition or change-and-rerun of step 2) is symbolized by capitalization: we stop talking about pressures p and start talking about pressures P.

As a methodological highlight, the procedure described so far (and as introduced a decade ago specifically for cities [14]) resembles the general-purpose model called Delphic Hierarchy Process (DHP) introduced 35 years ago by Khorramshahgol and Moustakis [128], which combines the Delphi approach and the Analytic Hierarchy Process (AHP; introduced a few years earlier by Saaty [129]) and which is the core of the full methodology that I am describing in this section. Of course, one can find in the literature some other applications of the Delphi family of methods alone or in combination with AHP (e.g., see [150, 151]).

Step 4. Conversion of priorities into operational questions.

In this short but key step (which should therefore be treated with utmost attention), confirmed-and-ranked risks P need to be transposed into an interrogative form, i.e., reformulated as (binary) decision-making questions to be answered collectively by the local community of stakeholders. Normally, if the identified priorities (and the reasons behind them) are already clear, this step should not pose particular problems. However, in the case of highly sensitive matters, all possible formulations should be considered and different versions can be then decided by vote if necessary. Also, if this step turns out to be very difficult for content-related reasons, then it might be the case that the work in the previous steps is actually not finished and needs some verifications (or completely redone if the issues are major).

Step 5. Observation of the scenarios resulted along the decision tree.

This final step essentially consists of the generation of scenarios by responding "yes/no" to the decision questions generated in *Step 4* in their order (the highest-ranked first) and identifying the "stop" situation, i.e., identifying each scenario as a history (succession) of "yes, yes … no" answers.

Since each scenario being the explicit result of a chain of decisions taken (go/not go answers given) to a "priority set of urban management questions", all scenarios generated along this decision tree acquire automatically an ontological description (i.e., they are coded by a chain of yes/no answers).

Given all steps, the result obtained in this last step is, essentially, the product of a procedure that can be iterated "automatically", i.e., a collective decision algorithm. Based on (and starting from) this explicated ontological identity, each scenario thus generated can be further studied in terms of its complete set of main implications in the real world, and can be refined as needed in operational practice, e.g., to develop policies, to base further managerial decisions, etc.

Last-but-definitely-not-least, it is important that a symbolic name be found for each scenario: one that effectively conveys the core idea of that scenario within the full set of scenarios. In the original project that led me to developing this method (in an eco-innovation cluster in the city of Versailles (France), details given in the next subsection), I used a context-dependent set of names, because the immediate purpose was the communication within the participants at the project. But then, for the purpose of analytical generalization of the method (using a case study from the city of Jassy, Romania, details in the same next section), I have chosen a typological set of denominations, as a way to convey the general relevance of the method beyond the original case study. I expect this to be true for future project too: in the particular case of each project, a context-inspired denomination would probably fit better, but the communication between contexts will require reference to a more general taxonomy. Because you have to start somewhere, I hope that this general typology can serve as a start case for you, and then you are of course free to generate and propose further typologies within your work projects.

General Important Observations About the Method

To insure success of the method, some conditions need to be met, as I summarize in the following lines based on my own experience with it and on the reports in the literature.

First, to insure maximum objectivity, the organizer and leader of the Delphi workshops must make sure that all experts included in the exercise have the same voice and there is no inhibition from the part of anyone involved, and the only and shared goal of the exercise is to identify the priorities in the target city as a human system. To that aim all power asymmetries, hidden agendas and "gaming" are eliminated from the outset in the interaction between participating experts. This is achieved by stating clearly from the beginning the goal, by explaining everything clearly and verifying that everyone understands and agrees with the rules of the method. The only kind of leadership in the room is that of the workshop organizer and his/her leadership is strictly tied to the commonly agreed goal of the exercise.

This also means that:

1. the leader must have sufficient professional skill and credibility in moderating professional conversations (otherwise the participants will not engage seriously)—which also include the full backing of the main stakeholder, i.e., the organization that wants these scenarios to be produced (which normally is the Local Authorities or/and the delegated organization, e.g., a University or another relevant organization) and
2. the participants must agree to act in good professional faith and accept the rule of equal voice, even if (or rather precisely because) a maximal diversity of views is necessary as long as they come with arguments and are based on evidence, relevant competence, and reason. If the leader of the exercise has any suspicion of the contrary, s/he must simply not include this or that expert in the list of participants.

Another aspect, already alluded to above, is that all conversations must be based on evidence, in line with the general principle of good management (of any business, be it private or public). Under this rule, promoting group agendas (of any kind) or taking political sides or advancing personal preferences is not allowed. Any potential conflict of interest should also be eliminated. (To give a somewhat entertaining example, diehard fans of a rival city should not be allowed to participate). The group of experts participating in the Delphi exercise can gather its members from around the world but at least enough of them should be familiar with the target city) and they are convened exclusively on behalf of the main stakeholder: the local community of people as a whole.

In general, the organizer (helped by all participants) should strive to insure the best conditions possible for an efficient conversation and to allow all participants to function at the maximum of their capacities as experts: equal and full access of everyone to all relevant information, ease of communication, and so on. Of course, one can make further observations in the same sense, but I think you already understood the idea: high professional standards and full objectivity. Details are then to be managed by the workshop organizer.

Importantly, no one is born with the skill in organizing and leading such exercises. So, several rounds might be necessary for the organizer to become very good at it, and for the participants to become very good at contributing to the collective decision in the workshop.

Finally, general consultations of the people in the target city—when necessary and feasible—can serve as a great start source of information for the convened Delphi team because they provide the basic information about what people want and they can take it from there: find the best options for maximizing applicability of the collective will of the people. Not only that, but it goes without saying that such exercises cannot ignore what people want, and the wider socio-political and institutional process leading to decision in the local community must include all forms of practical interactions with the local community, e.g., social surveys, "science-in-society" round tables, Questions & Answers session involving experts and citizens, *et cetera*. As shown with the examples in the next section, sometimes a public consultation (following a stage when the public had the chance to propose and a period of intense debates which insures that all interested members of the public had the chances to think about it and have a say) followed by a final public vote on a raw list of issues can settle the matter and distill the order of priorities to be subsequently addressed by the local authorities in collaboration with the competent institutions and experts.

Collateral Observations About the Method in the Context of Sustainability Aims

More than a "theory of everything" in sustainability, it is important that insights from unique pioneering projects to date and overall resource represented by the individual and collective experiences of experts from all relevant-albeit-very-different domains and sub-domains (of the triple-helix of science, business, and governance) is being used to a practical end. This means that the overall resource represented by experts finds an operational voice that is directly heard by decision-makers. This requires models of (human resource and) knowledge management that are capable to clearly elicit and integrate these experiences into tractable procedures and work results that can ground decision even before all scientific questions have been answered.

It is not unusual to hear in conferences and publications academics call for more research to be done in order to "better understand" an issue. But, to make a friendly collegial jest (I am also a researcher myself, *inter alia*), in the non-academic world people are fully aware that decisions must be made because the situation allows no delay. I wrote "friendly", because I know from casual (usually "incognito") people-to-people conversations on the streets (and in urban parks) that in the non-academic world, people do not understand how the "university world" works. It is not unusual to hear complaints like the following (citing from memory, from a man I talked to during a walk in the public Versailles Gardens, a few years ago, and whose specialty was in business, specifically logistics): "the problem with academics is that they don't understand anything". This is just one instance pointing to the fact that true experts (academics included) do listen to such complaints because it speaks loudly about the role of science in society (with the problems that need to be addressed in that relation), the importance of practicing science-society interactions, and the already-mentioned triple-helix approach in practical projects for urban evolution towards sustainability: when science and academics lose touch with the rest of society (and vice-versa), sustainable development is but an empty word.

Any observer of the literature can make their own reviews and they will notice that the Delphi method is particularly used (suited) for identifying and gauging/ranking challenges, i.e., issues with a substantial "unknowns" component.

In this sense, one may also notice that, *mutatis mutandis*, the logic behind this method is similar to that underlying the notion of "communicative action", in which people interact and coordinate their actions based on an agreed interpretation of the situation (e.g., [152]) and the related notion of "deliberative democracy", in which legitimacy of decision (in society in general) is gained by authentic prior deliberation (not merely the aggregation of the general preferences represented by voting; and where "authentic" deliberation means a deliberation that is free from distortion from unequal economic or/and political power or active lobby of interest groups) (e.g., [152–158]).

This highlights the importance of true deliberation in communities of people with a shared purpose (the common good), which further highlights the importance of cultivating the virtuous circle between trust and objective conversation as opposed to imposition, which places the method fully within the scope of the social sustainability as discussed in Module 1 (Figs. 1.1, 1.2 and 1.3).

Original and Generalized Examples

The method was developed in the Econoving industrial innovation cluster focused on eco-innovation and eco-city projects at the University of Versailles (UVSQ; now part of the Paris-Saclay University consortium) [83, 159]. This was a public–private funded project that included six universities from the Paris region, six main industrial partners (covering energy, construction, transport, water, and waste) and numerous smaller partners from the private sector, and governmental authorities. A main project of the Cluster was the transformation of a train station (*Versailles Chantiers*, a main transportation hub in the Paris region) and its urban surroundings according to urban sustainability goals, in view of using this station as a model for the other stations in France (at least).

I organized this interaction as Director of the *International Professional Master in Management of Eco-Innovation*, the world's first master program dedicated entirely to eco-innovation, in a context in which the Integration Seminar (which was part of the master program) fully overlapped with the cluster's work package dedicated to foresight (scenarios). Figures 2.4 and 2.5 give further visual details on these works and Fig. 2.6 shows an example of scenario set obtained with the Delphi-based method described above (with the participation of experts from all participating institutions as per our internal work organization). Important to mention is that the works in the Econoving cluster were only a part of the initial reflections and planning in different expert groups. The subsequent city district development was a completely different program, in which we did not participate directly as our job was already done: we provided SNCF (the main beneficiary) and through SNCF (and via other direct and indirect interactions) also the city of Versailles itself with expert inputs about potential options and rationales. However, I was able to watch personally the transformation of this train station and district and the city itself. A number of details come to mind, which have been discussed during the Integration Seminar (notably as propositions from students) and which I was able to observe in these ten years of urban evolution since 2014 (knowing of course that many of these ideas probably came towards the local authorities from multiple directions, not just from us, and this convergence was probably very convincing):

- the basic idea of increasing the surfaces of green areas in the city center (including also in Avenue de Sceaux between Versailles Chantiers and the Chateau, and in general strengthening this direct axis);
- the related idea of making more of "a city for people" (and less of a "city for cars", to keep a long story simple) while at the same time improving the public transportation system across the North–South axis which was under-developed for historic reasons;

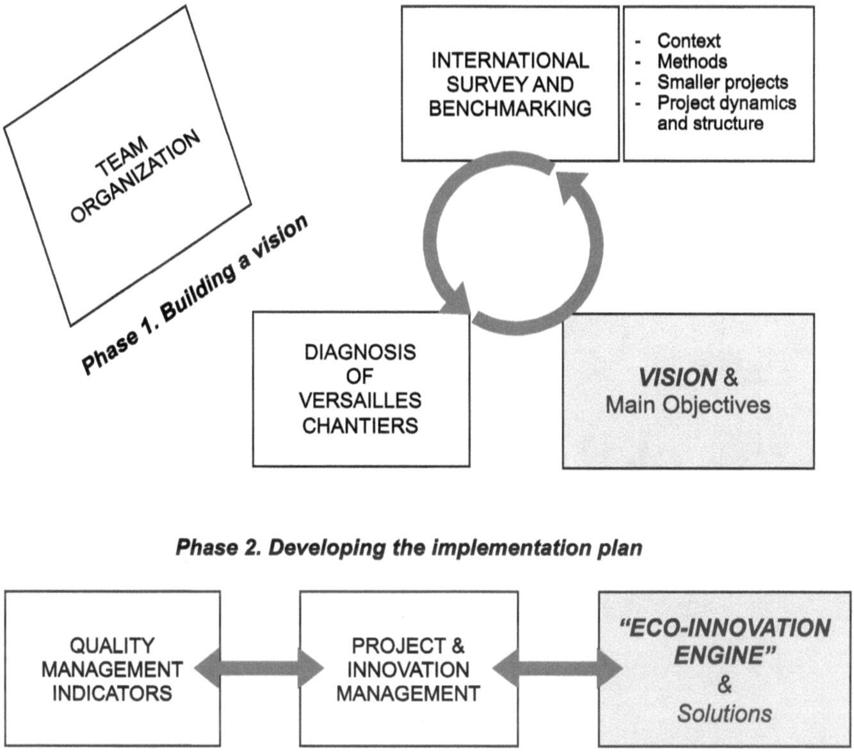

Fig. 2.4 Visual summary of the integration seminar work process in the Econoving cluster, focused on the redevelopment of Versailles Chantiers train station and city district. *Source* original (credit: I thank Rachel Guilloteau for help and practical inputs (from a project manager perspective) during the development and organization of this workflow, Keith Culver for the groundwork that insured many important academic, business, and organizational aspects until the summer of 2011, and Christelle Hue for logistic and organizational support during the entire project)

- the introduction of hydrogen-based buses already (on certain lines);
- experimentations with autonomous vehicles (driverless cars); using the historic streets and places (green areas included) literally as exhibition areas (see for example, Fig. 3.9 in Module 3, Sect. 3.3);
- the observation I made during the works, based on available documents that it would be logic to use as a main and key connection the passage under the bridge platform that is connecting the road directly with the first floor (one level up) of the station in the new circumstances of the re-modeling of the place;
- and other ideas which cannot be mentioned here for lack of space.

Perhaps less obvious to the hurried eye but very important to highlight is that the city itself built on the experience from Versailles Chantiers and pursues other urban redevelopment projects in other districts, notably in the district of *Gally* (reconversion of an old military site of 20 hectares near the ancient Royal domain) [160, 161].

Fig. 2.5 Illustrations of works. **Top half (indoors)**: snapshots from the Integration Seminar (2011–2013) focused on developing scenarios for Versailles Chantiers (details in the main text) (modified photos by Elodie Gonthier). **Bottom half (outdoors)**: A public info panel on a wall behind a bicycle parking about the new city district being developed there; physical works on the site (September 2018); the new public park of *Etangs Gobert* in the same city district (August, 2019), and the new extension of the station as part of a hub of multi-modal communication (to the right there is a large regional bus station: "La Gare Routière"; to the left there is the business quarter called *L'ilot Est* with office, conference and commercial spaces, and gardens (March 2020) (*photos* by the author)

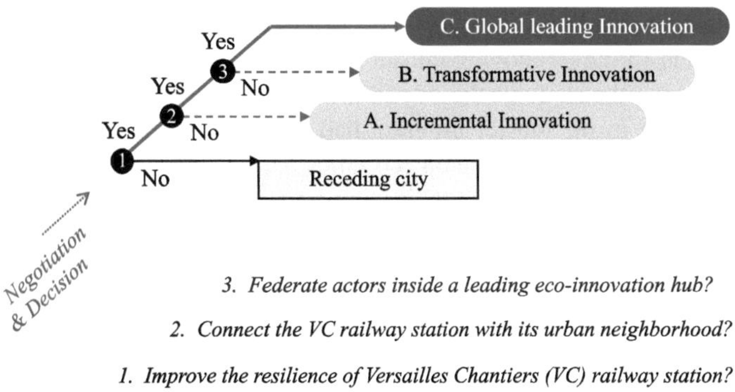

Fig. 2.6 Full set of scenarios formulated in terms on urban-city related eco-innovation starting from an urban hub (further details in Table 2.5)

Table 2.5 Main details of the set of scenarios developed in Versailles

Original question emerging from the econoving cluster's Integration Seminar* (listed from bottom up)	Decision path (in the decision tree algorithm)	Resulted scenarios: generic names (& type of intended content)**
3. Involve local authorities, citizens, stakeholders in creating a leading eco-innovation hub?	Yes; Yes; Yes	C. ***Global leading innovation*** (EGC 2050, i.e., a holistic regard on the city emulating (and as reflected in) the idea behind the well-known "*European Green Capital*" prize/label (work aim), combined with the rich history of the city, also including science, technology, innovation, and urban development
2. Connect the station with its neighborhood?	Yes; Yes; No	B. ***Transformative innovation*** (Re-development of the boulevard *Avenue des Sceaux* connecting two key poles of the city: Chateau de Versailles and Versailles Chantiers (VC); this avenue is one (the southernmost) of the famous "trident" set of three streets radiating from the chateaux roughly eastwards towards and through the town itself
1. Improve the resilience of Versailles Chantiers railway station?	Yes; No	A. ***Incremental innovation*** (Technology packages for updating the overall services and capacities, without essentially changing how the place works; basically just a strategy of keeping up with the latest technology)
	No	***Business as usual (BUS)***

* A work conducted in the *International Professional Master Program in Management of Eco-innovation* directed by the author inside the *Econoving* Cluster in a triple-helix (university-business-government) configuration; it overlapped the Work Package WP3-Anticipation of the cluster's Railway Station Program (in French "Programme Gares")

** As reflected in three distinct reports elaborated in a triple-helix exercise of consultation—the result of 2 years of work

And finally, from a broader project management perspective, it is worth mentioning that, in the Econoving cluster and graduate program, we used multiple forms of involving the students directly in the local and global works. For example, the same students in the master program had practiced reporting of the works of an international event (conference on eco-cities we organized in 2012), as a way to enlarge their horizon and increase their competence in the works in the cluster and for their own professional development overall. In other local-and-global events we focused on entrepreneurship, in the collaboration with our partners from industry. And the list could continue for much longer than there is space in this book section.

Shown in Fig. 2.5 (top half; indoors) is one of the three teams of students/experts of the first annual Integration Seminar iteration interacting with the in-project jury, moderated by me (as Director of the Econoving Master program) and supported by our project management staff. Credits for work contributions are due to:

- the 2011–2012 students of the Econoving Master program (Anna Aghvanyan, Adrian Carrillo, Krishna Chandran, Boubacar Coly, Fanny Courry, Maxim Doubrouski, Elodie Gonthier, Sirin Hamsho, Alejandra Hoyos, Xiaochao Huang, Collins Ihedioha, Natalia Kozdra, Anna Lomteva, Javad Mahmoodirad, Ravi Nakulan, Alexandra Ostrovskaya, Kevin Ramirez, Marcel Roquette, Dina Salakhova, Maria Stoica, Sandra Velasco);
- the 2012–2013 cohort of students of the same Master program (François Apicella, Gabriel Bermeo, Zoheir Haghighi, Ludmila Kobyakova, Renan Le Tohic, Habib Liamidi, Hélène Luciani, Diego Martinez, Mingyan Meng, Teodora A. Morar, Surya Muraleedharan-Nair, Julian Muriel, Marina Nikitina, Kalliopi Ntanou, Maria-Lucia Portocarrero, Cécile Tchoubaev);
- contributing experts: Jean-Louis Jourdan, Alain Adrianssens, Julien Pouget, Thierry Lafont, Yann Riffonneau, Benjamin L'Henoret, Vincent Delcourt, Fabienne Herlaut (from SNCF), Agnes Roux and team (Versailles city hall), Bernard Blez, Etienne Drouet, Olivier Bordelanne, Joelle Giton, Carole Lunot, Malia Belkacem, Aurélien Alfaré, Valentin Gavan, Roch Drozdowski, Valérie Bichler (GDF SUEZ), Laurent Schmitt, Regis Hourdouillie, Sebastien Lannez, Jean-Christophe Passelergue, Guillaume Foggia (ALSTOM GRID), Frédéric Cuq, Fabrice Nauleau, Remi Usquin, Sebastien Masclet (SAUR), Christophe Lombard, Jean-Philippe Vacher, Mélanie Shink (ITALCEMENTI Group), Dominique Agreche (X-Ange); Marc Dufau (OSEO); Olivier Dupont (Demeter); Sylvie Faucheux, Jean-Luc Vayssière, Martin O'Connor, Jean-Paul Vanderlinden, Daniel Vidal-Madjar, Sophie Godin-Beekmann, Philippe Bousquet, Caroline Gans-Combe, John Dean, Eberhard Falk, Krzysztof Kanawka, Bertrand Le Cun, Alain Bui, Soufian Ben Amor, Murat Ahat, Issiaka Traoré, René Audet, Sophie Némoz, Frédéric Louradour, Amélie Coulbaut-Lazzarini, Yann Hermans, Thor Jensen, Elizaveta Kuznetsova, Guillaume Bailly, Elaine Penny, Quentin Courtier, Claudie Lesage, Sylvie Faure, Delphine Paracchi and others (UVSQ); Filipa Lopes, Behnam Taidi, Dominique Pareau, Estelle Couallier (École Centrale Paris), Beat Bürgenmeier, Jean Simos (University of Geneva), Christophe Robillard (QUALISTEO), Suat Topsu (UVSQ and OLEDCOMM), Frédéric Chomé (Factor-X,

Brussels); Enrico Zio (École Centrale Paris; SUPELEC, Politenico di Milano); Patrick Boucher, Marc Petit (SUPELEC); John Gowdy (from Rensselaer Polytechnic Institute, Troy, New York); Joachim Spangenberg (Sustainable Europe Research Institute, Cologne, Germany), Richard Hudson, Maryline Fiaschi, Gail Edmundson and team (Science|Business, Brussels); Vivek Tandon (Aloe Private Equity, London & Paris); Nadia Hoyet (ENSA-Versailles); Rachael Ritchie (UVSQ; OECD); Ian Gillespie, James Philp and team (OECD, Paris); Anne-Christine Ayed (Tarkett, Luxembourg); Hélène Chepfer, Thanh-Tâm Lê (École Polytechnique, Paris; Climate KIC), Audrey Nivet (UPMC-IPSL; Climate KIC), Marjolaine Chiriaco (LATMOS-IPSL; Climate KIC); Jean-Pascal Dumas, Arnaud Comolet (SQY City Hall);
- the previous two generations of students of the Econoving's Master Program who worked (on the supervision of Keith Culver, the Econoving Chairholder during 2009–2011) on the neighboring city of Saint-Quentin-en-Yvelines (SQY), and thus greatly helped with the initial developments of the Integration Seminar methodology that was subsequently updated for Versailles, namely the 2010–2011 cohort (Thomas Aubron, Carine Bergez, Hélène Boga, Alexis Collet, Philippe Derouette, Yannick Djoko, Thomas Etesse, Sahba Farahi Jahromi, Jeffrey Fleishman, Alexandre Gobbo Fernandes, Isabelle Le Quellec-Brocard, Carolina Fernandez Maestri, Erica Otto, Eleanor Ramm, Nicola de Sanctis, Mona Katarina Rouot). And, a similar merit must be acknowledged for the 2009–2010 cohort (taken collectively) which I didn't know myself but was the group that kickstarted the Integration Seminar in its preliminary form.
- and many others who contributed in many ways and to whom I am deeply grateful.

From the very practical aspect of inter-cultural communication and cooperation, it is highly relevant (and beautifully symbolic) that Econoving students came from all around the world (Armenia, Belarus, Brazil, China, Columbia, Ecuador, France, Greece, India, Iran, Italy, Mexico, Nigeria, Poland, Romania, Russia, Salvador, Senegal, Sweden, Syria, UK, USA)—so much so that I was endearingly calling them "*The United Nations of Econoving*". In fact, if I would try to also count the countries of origin of the contributing experts, the list would be even larger (I don't even know it myself fully, but I know it also included *inter alia* Belgium, Canada, Germany, The Netherlands, Luxembourg, New Zeeland and Switzerland).

At this point, it is also fair to mention students from other Master programs who used in their practical (individual and team) projects under my supervision this Delphi method to study the perspectives of their chosen city of interest (along with other methods like the knowledge mapping described in this book in Sect. 3.3 and the rest of the Module 3), and especially the 2015–2020 cohorts from the Master program Adaptation to Climate Changes (in the course *Transition to Sustainability: Human & Technological Systems* that I held in that program). These programs too brought together students and teachers from different countries around the world.

The great diversity of backgrounds of all these people is intrinsically reflected in the collective work outcomes of the Econoving cluster and gives them truly global dimensions to this methodological description—precisely because this is about local communities from anywhere around the world, in a spirit of practical collaboration.

Yet another context detail is particularly relevant for grasping the value of these scenarios and related methods. Beside hosting the world's first master program dedicated entirely to Eco-Innovation (that alone is already a pioneering project), the Econoving cluster involved not just one or two private companies in close cooperation with the university (like it is usually the case with science-business projects), but six of them plus a wider network of smaller actors and it generated home-grown startups and different other initiatives (not discussed in this book). Given all these, it wasn't really a surprise that the cluster's master program climbed in the top 10 masters of France in the Section "Management of Innovation" [162] after only three years of existence. In short, behind these scenarios, there is a real story of a real pioneer program applying learning-by-doing and doing-by-learning.

Of course, there are many inspiring projects and programs out there around the world. The details above only serve to show that, cooperation and hard work lead to good results, and while projects come and go, the good lessons need to be learned. And, to end this summary in a sober note, the first lesson is that generating good scenarios is harder that might appear at the first thought (Table 2.5).

Among other things, this original case study illustrates well the need to find the right interplay between expert clusters and expert networks according to purpose.

On the one hand, the purpose of the foresight exercise had a local scope: exploring options through a minimal set of realistic scenarios (potential future updates) for the Versailles Chantiers train station in/and its urban surroundings (the city district Chantiers of the city of Versailles, the city of Versailles, the Greater Paris Region, in the context of the then on-going Grand Paris program of development of infrastructure and communications in the Paris Region) [163, 164], the country itself, and the EU. On the other hand, the purpose of a company involved in the project, for example, a startup company (which was co-incubated by the Econoving Cluster) and which was developing and a general-use solutions for energy consumption based on the NIALM (Non-Intrusive Appliance Load Monitoring) technology was—obviously—to gain market shares anywhere, and therefore it participated in the Railway Station program (and related projects of the Econoving cluster) because this was a platform for experimenting with different NIALM-related types of solutions for better meeting the demand on the national and international markets (some cheaper and in certain ways better solutions for monitoring and optimizing electricity consumption in buildings). Similarly, a bigger company, namely SNCF (the French Railways), had the direct interest of developing a new model of middle-size train stations that they would then be able to replicate throughout France and possibly across Europe by international partnerships.

The competing-but-complementary scenarios generated by us at the beginning of the process have then been critiqued and improved through weekly, monthly and yearly cycles, and by different teams. Research inputs have been provided as results of the research conducted by Econoving and its partners, including a group of PhD students and post-doctoral fellows, and concerned among others: models of energy flows in the target station [165], smart microgrids [166, 167], other complex engineered systems in the given urban context and in general [168], and use time-optimization for electrical vehicle round-the-clock [169].

Generalization Using a Different Context: The City of Jassy

A different and more general type of replication of the logic behind the method described in Versailles is based on the results of a previous public consultation that happened in Jassy (Iași), Romania, the European Union, which I mapped onto the procedure already developed in Versailles.

As also summarized in a study from a decade ago [14], a transparent process of public consultation was carried out by local authorities in the city of Jassy, which followed a chain of logical steps that maps (i.e., correspond directly, 1-to-1) onto steps 1–3 in the procedure described above in the case of Versailles Chantiers. For Jassy (city as a whole), the following particular list of principal drivers (pressures) and order of priority had emerged:

1. Level of governance capacity;
2. State of local ecosystem carrying capacity, with focus on climate/environmental changes;
3. Population/human resource dynamics (emigration in general, brain-drain in particular);
4. Economic aspiration and opportunity (entrepreneurial/job/economic);
5. Contribution by science and technology advances;
6. Emissions (pollutants released into air, water, soils);
7. Vulnerability to poverty/economic exclusion.

Three top pressures and a set of three matrices with a total of 12 exploratory scenarios (ES) are being identified for the three priorities, with details connecting them as per Fig. 2.3, and illustrated in the abstract template shown in the equation below and in the Table 2.6.

If you are not immediately concerned with the basic mathematical formalism, you can skip directly to Table 2.6 to get the main idea, as this brief formalism is provided to help the modeling-minded people keeping track of technical details. But if you do want to look at the mathematical formalism too, I have also explained it in words and hope this is helpful for you too.

Namely, any theoretical Exploratory Scenarios Set (ESS) coming from the combination of the priorities can be labeled with a name (A, B, C) and value (0 for invalid, 1 for valid, i.e., existent) in a matrix, i.e., exploratory set matrices (ESM):

$$ESM_{ij} = \begin{bmatrix} ESS_{11} & ESS_{12} & ESS_{13} \\ ESS_{21} & ESS_{22} & ESS_{23} \\ ESS_{31} & ESS_{32} & ESS_{33} \end{bmatrix} = \begin{bmatrix} 0 & \text{A} & \text{B} \\ \text{A} & 0 & \text{C} \\ \text{B} & \text{C} & 0 \end{bmatrix} = \begin{bmatrix} 0 & 1 & 1 \\ 0 & 0 & 1 \\ 0 & 0 & 0 \end{bmatrix}$$

Then, each of the three combination of priorities participating as a valid element (ESS) in ESM can further be expanded as a matrix where the gradient ends of each priority (Low; High) are assigned a value: the minimal default value (= 1) for Low

Table 2.6 Triple template identifying the three 2 × 2 matrices (here shown as a spreadsheet core)

Matrix A ($ESS_{12} \equiv ESS_{21}$)		*p1. Governance capacity*	
		Low	*High*
p2. State of ecosystems	*Poor*	EST-a	EST-b
	Good	EST-c	EST-d
Matrix B ($ESS_{13} \equiv ESS_{31}$)		*p1. Governance capacity*	
		Low	*High*
p3. Population dynamics	*Low*	EST-e	EST-f
	High	EST-g	EST-h
Matrix C ($ESS_{23} \equiv ESS_{32}$)		*p2. Population dynamics*	
		Low	*High*
p3. State of ecosystems	*Poor*	EST-x	EST-y
	Good	EST-z	EST-w

EST stands for Exploratory Scenario Type. Lower case letters are used as unique identifiers in the 12-set (only the non-zero types from the TES matrix). Each type includes an expandable list of details (data, information, knowledge, etc.), which can be generated in work sessions by using physical paperboard or electronic tables (or both). The equivalence sign ≡ indicates which ES sets in the combinatorial matrix ESM_{ij} above correspond to the same situation in practice

and a theoretically unknown, non-zero value *n*, which for analytical simplicity can be given a weight of 2 (arbitrarily meaning twice as big as the minimum; the purpose here is not to gauge the maximum end of the priorities per se but to distinguish it from the minimum end of the gradient, and thus be able to define a valid type of exploratory scenarios, within the sector of the space of options that is being studied); then details of how far that maximum goes can be specified and discussed in the concrete content that a team produces for that category, based on available data, information and knowledge.

This allows us to define the *Total-set of Exploratory Scenario types (TES)* as the Kronecker product of matrices of, respectively, pressure priorities and pressure gradients on an urban system (this type of product means that every value in the matrix on the left of the product symbol ⊗ is replaced with four values corresponding to the respective products of that value times each of the four values in the matrix on the right):

$$TES_{ij} = \begin{bmatrix} 0 & 1 & 1 \\ 0 & 0 & 1 \\ 0 & 0 & 0 \end{bmatrix} \otimes \begin{bmatrix} 1 & n \\ 1 & n \end{bmatrix} = \begin{bmatrix} \begin{pmatrix} 0 & 0 \\ 0 & 0 \end{pmatrix} & \begin{pmatrix} 1 & n \\ 1 & n \end{pmatrix} & \begin{pmatrix} 1 & n \\ 1 & n \end{pmatrix} \\ \begin{pmatrix} 0 & 0 \\ 0 & 0 \end{pmatrix} & \begin{pmatrix} 0 & 0 \\ 0 & 0 \end{pmatrix} & \begin{pmatrix} 1 & n \\ 1 & n \end{pmatrix} \\ \begin{pmatrix} 0 & 0 \\ 0 & 0 \end{pmatrix} & \begin{pmatrix} 0 & 0 \\ 0 & 0 \end{pmatrix} & \begin{pmatrix} 0 & 0 \\ 0 & 0 \end{pmatrix} \end{bmatrix}$$

$$
= \begin{bmatrix} 0 & 0 & 1 & n & 1 & n \\ 0 & 0 & 1 & n & 1 & n \\ 0 & 0 & 0 & 0 & 1 & n \\ 0 & 0 & 0 & 0 & 1 & n \\ 0 & 0 & 0 & 0 & 0 & 0 \\ 0 & 0 & 0 & 0 & 0 & 0 \end{bmatrix}
$$

where $n, n \neq 0, n \geq 1$.

Or, if you prefer, you can apply $n = 2$ (as suggested above), i.e.,

$$
TES_{ij} = \begin{bmatrix} 0 & 1 & 1 \\ 0 & 0 & 1 \\ 0 & 0 & 0 \end{bmatrix} \otimes \begin{bmatrix} 1 & 2 \\ 1 & 2 \end{bmatrix} = \begin{bmatrix} 0 & 0 & 1 & 2 & 1 & 2 \\ 0 & 0 & 1 & 2 & 1 & 2 \\ 0 & 0 & 0 & 0 & 1 & 2 \\ 0 & 0 & 0 & 0 & 1 & 2 \\ 0 & 0 & 0 & 0 & 0 & 0 \\ 0 & 0 & 0 & 0 & 0 & 0 \end{bmatrix}
$$

Alternatively, one can represent this set of situations by storing category labels:

$$
TES_{ij} = \begin{bmatrix} 0 & \mathrm{A} & \mathrm{B} \\ 0 & 0 & \mathrm{C} \\ 0 & 0 & 0 \end{bmatrix} \otimes \begin{bmatrix} L & H \\ L & H \end{bmatrix} = \begin{bmatrix} 0 & 0 & \boldsymbol{a} & \boldsymbol{b} & \boldsymbol{e} & \boldsymbol{f} \\ 0 & 0 & \boldsymbol{c} & \boldsymbol{d} & \boldsymbol{g} & \boldsymbol{h} \\ 0 & 0 & 0 & 0 & \boldsymbol{x} & \boldsymbol{y} \\ 0 & 0 & 0 & 0 & \boldsymbol{z} & \boldsymbol{w} \\ 0 & 0 & 0 & 0 & 0 & 0 \\ 0 & 0 & 0 & 0 & 0 & 0 \end{bmatrix}
$$

Or, you can represent it as a table, as shown (Table 2.6).

At this point, it is important to remember that all urban systems are simultaneously multi-dimensional and complex. This entails three levels of attention that are required from us. Firstly, it is useful to have a good, systematic representation of the multi-dimensionality. This can be done most efficiently, by using the formalism provided by the Analytic Hierarchy Process (AHP): the literature provides sufficient examples of application in cities, usually on specific topics, and often in combination with Delphi methods (e.g., [105, 127–130, 170–174]), but, naturally, this list of applications also includes a broader spectrum of societal issues (e.g., [175, 176]). As already mentioned previously (at the end of the description of Step 3), a combination of the Delphi approach and the AHP logic is basically what I have also done so far in the description above.

Then, the very fact that AHP is known to be used in combination with the Delphi approach (since shortly after its introduction) tells us that AHP covers well the issue of multi-dimensionality but one needs something else to cover the non-linear behavior of cities as complex systems. The latter constitutes the second level of attention (and has the characteristics of CDS as discussed so far in the book).

And thirdly, one needs to combine the two levels in an operationally valid and efficient way. This is to say that care must be taken, when using AHP (and other similar methods) not to let understand that addressing multi-dimensionality is the same as (or enough as a method for) addressing complexity. In other words, it is important to have a clear method for organizing the work in a way that accounts efficiently for multi-dimensionality, for example using an AHP matrix for seven priorities, e.g.,

$$A_{ij} = \begin{bmatrix} ES_{11} & ES_{12} & \cdots & ES_{17} \\ ES_{21} & ES_{22} & & ES_{27} \\ \vdots & & \ddots & \vdots \\ ES_{71} & ES_{72} & \cdots & ES_{77} \end{bmatrix}$$

but this is only the preparation for the much harder work of addressing CDS dynamics. And this must be done without implying an urban system is multi-dimensional without being complex.

Then, to continue the procedure with the Jassy case study, the concise conclusion of the "reality check" step, all 12 exploratory scenarios are found to be relevant: they describe true (valid) potential situations for the city of Jassy (the minute observations and comments with local details are not shown, for brevity purposes; instead, such details must be filled in by the reader for their city of interest).

Thus, the initial selection of main types of risks (categories of pressures, in their formulation) is acknowledged (confirmed) as realistic and therefore maintained as such (However, in the case of the reader's city of interest, this might not always be the case, which means that the procedure so far needs to be re-iterated as described above, until the list of top pressures is consolidated (i.e., confirmed as such); each case study must be studied for what it is).

Further, the confirmed first three main types of risks (pressures) have been developed into questions to be answered (with "yes" or "no"), in this order:

1. "Implement a business-friendly and efficient governance system?"
2. "Develop a resource management system?"
3. "Carry out a human capital strategy?".

Following steps 4–5 applied to the above set of questions, the logical types of possible outcomes (ontological scenarios) are obtained (with the original scenario titles from [14]):

- Answers: "No", giving scenario type "Receding City" (Fig. 2.7);
- Answers: "Yes–No", giving scenario type "Wanting City" (Fig. 2.8);
- Answers: "Yes–Yes–No", giving scenario type "Promising City" (Fig. 2.9);
- Answers: "Yes–Yes–Yes", giving scenario type "Inspiring City" (Fig. 2.10).

A visual idea about Jassy itself as a "city of inspiration" and as an "Inspiring City" is shown in Fig. 2.11, with the observation that the two notions can be understood as fully overlapping but can also be interpreted independently, e.g., the first one

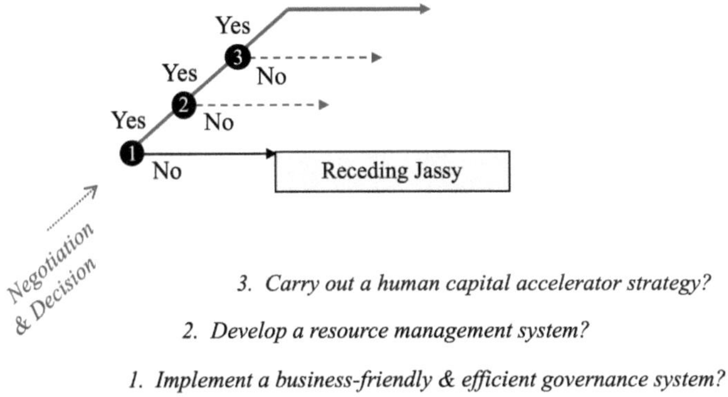

Fig. 2.7 First scenario resulting from the process of negotiation and decision along the decision tree aiming at urban sustainability [decision pathway: "No"]. *Source* redrawn after [14]

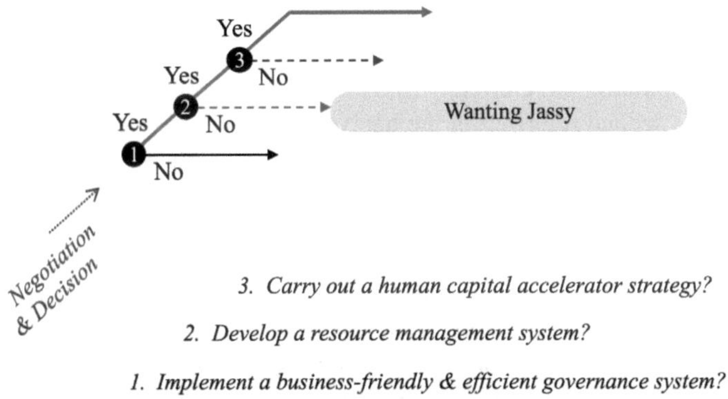

Fig. 2.8 Second scenario resulting from the process of negotiation and decision along the decision tree aiming at urban sustainability [decision pathway: "yes; no"]. *Source* redrawn after [14]

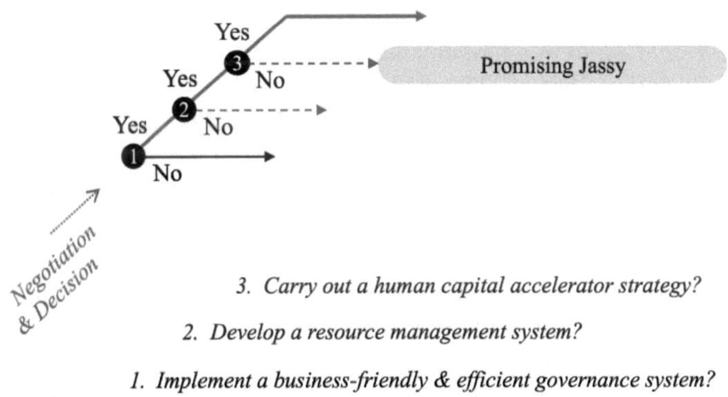

Fig. 2.9 Third scenario resulting from the process of negotiation and decision along the decision tree aiming at urban sustainability [decision pathway: "yes; yes; no"]. *Source* redrawn after [14]

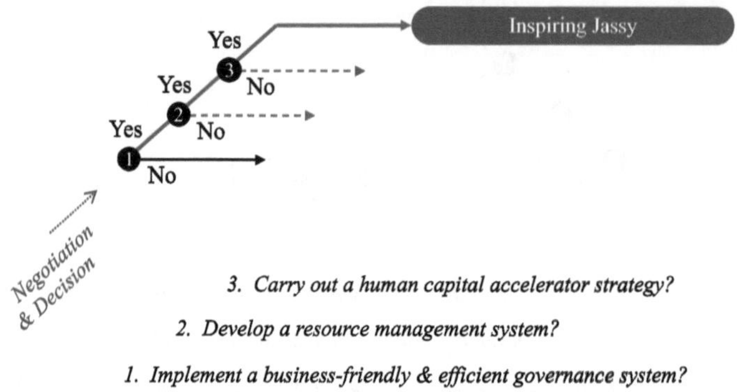

Fig. 2.10 Fourth (unstopped) scenario resulting from the process of negotiation and decision along the decision tree aiming at urban sustainability [decision pathway: "yes; yes; yes"]. This is the technical target for insuring continual progress on a city's path to sustainable development. *Source* redrawn after [14]

referring to inspiration more broadly (the city as an environment that stimulates the mind), and the second referring in the analysis to a role that a city can play for its citizens and for other cities (and their citizens) that may draw insights from this city.

Then, the full scenario set for the initial case study of Jassy but now generalized for any city is shown as such in Fig. 2.12.

At this moment, I am going to include some supplementary details (as bullets) that might want to skip if you are following strictly the core methodological discussion for the strict purposes of your own projects—but just remember that a city is never just a technical product of an abstract methodology, and that it is the real-life and experiences of the people-in-the-city that makes a city what it is.

Here is a short list of context details, selected to give some concrete background about the items displayed in Fig. 2.11. Collectively, these details tell about the city of Jassy as a truly dynamic and inspiring place and community of people:

- The *Palace of Culture* is a monumental and a main urbanistic symbol of the city, being built during 1906–1925 based on plans of I.D. Berindey in a Neogothic Flamboyant style, on the foundations of an old Princely Court (palace and dependencies) that was known in official documents since 1434. The Palace incorporated several major technological innovations at the time. Now the building hosts four museums and is surrounded by public landscaped gardens and activity areas as part of the recently finished *Palas Mall* project of urban regeneration that incorporates green areas and dedicated spaces for RDI, business, commercial, artistic and leisure activities [182]. I have visited it first as a child (school trip), then as a student in the city in the 90's and as a professional;
- The National Theater and Opera House is the oldest in country (1846), and the current building (1896, architects Fellner & Helmer), reflects the typical style of Central European opera houses of the nineteenth century and is known as one of the most beautiful in the World ([183], see also [184]). It is not the only in the city.

Fig. 2.11 Jassy as an inspiring city in the eyes of locals and visitors, combining nature and culture. *Source* original arrangement using modified images from the public domain (with [credits]) as follows. **Top**: The Palace of Culture [177]. **Middle left**: The National Opera House [178]. **Middle right**: Alexandru Ioan Cuza University [179]. **On the side**: statue of national poet Mihai Eminescu (1850–1889, who lived and worked in Jassy) in the Copou Gardens [180]. **Bottom left**: the Palas Mall complex in the central district (*Civic Center*) of the city (urban regeneration project, completed in 2012) [181]

Among similar institutions, one counts the Jassy Philharmonic, The Children's Theater, and the Tătărași Theater);

- Alexandru Ioan Cuza University (UAIC) is the country's first modern university, re-founded in 1860 by reorganization of a previous higher education institution, the Michaelian Academy and tracing its roots directly to a Princely Academy (in 1642), being also a moral and intellectual continuator of the Protestant school for governmental administrators called *Schola Latina* from Cotnari (established in 1562–1563, and taken over by a competing Catholic school). I am familiar with UAIC as a student (undergraduate, 1994–1998, BSc, and graduate, 1998–2001, MSc), and as a professional: urban and peri-urban studies in collaboration with local and international colleagues (and former teachers of mine). Some of the works cited in this book are the result of these projects and reflect the spirit of this great institution and its role in the city [185];
- In this context, it is worth highlighting that our personal and professional efforts to pursue and operationalize sustainability goals happened in a period of time (the transition from soviet-imposed command economy to a modern market economy), which was marked by economic (and more general societal) hardships, but which we also knew it was necessary on the path to the recovery of dignity all across Central Europe. In short, the beautiful images from the University and the City aren't meant to make us forget the difficulties but to remind us that a dignified future is worth fighting for. It was during the first decade of this millennium that the necessary transition to the market-based economy started to overlap with the incipient transition to sustainable development, and we, at UAIC, were fully aware of it: this is when we made the first-in-the-country coherent introductory course (Master level) to the theory and practice of sustainable development through personal efforts ([186]—a bilingual, Romanian & English book for the use of students and experts), and when the first dedicated program of life-long learning for environmental educators (oriented to a systemic and holistic view on sustainable development, not just natural sciences) was implemented through a pioneering ("pilot") project which has won (in a competition) the support of the European Commission within its *Leonardo da Vinci program* [187–194];
- Copou Park (or Gardens) is the oldest public garden in the city and in the country, officially established in 1833–1834 (on the ground of the former garden of the Post Office) that was already used informally as an outdoor meeting place and promenade by the locals since the second half of the eighteenth century, with the first documented works (albeit private but for the same purpose) dating since 1752. Located on the place of a natural forest with numerous centennial Linden trees, it now hosts a unique combination of indigenous and exotic species ([195], see also [196, 197]). The statue of the national poet is located exactly near/under such a Linden tree (which, as of 2023, is at least 467 years old, and recognized and protected as a natural monument). It has been made famous because the poet himself is known to have been a habitual visitor—seeking shade, rest and inspiration under its crown; linden tree flowers feature high in his poetry and this Linden tree itself has become a cultural symbol; a stone mosaic on the ground in front of the statue reads a few verses about linden tree flowers in poetic contemplation. It

is a usual place visited by school children in school trips, many adults guarding this childhood memory in their spirit (I am one of them). More public information (self-speaking images and some text in Romanian) is provided in [198], an online public history of linden trees in the city of Jassy; a few key moments will be summarized in the following lines;

- The Linden tree (Tilia tomentosa) behind statue of poet in Fig. 2.11 is only the 7th most remarkable tree and the second oldest linden tree in the Jassy County [199], and the planner of this public park (Gheorghe Asachi, a well-known figure of the city and of the country itself) noticed in 1833 that the location of this tree didn't fit with his plans but he decided to keep it because of its beauty and respectable age it already had at that time. During the WWII, during the occupation by the USSR, the park was used by the occupiers as a camp for the prisoners that were being deported to Siberia—during that time, the school teachers among the prisoners organized themselves to guard the tree so that the occupiers do not cut it for wood. Later, the tree benefitted some dendro-chirurgical operations and re-interventions and other interventions that finally saved it from advanced degradation. Now it is called the Linden Tree of Eminescu ("*Teiul lui Eminescu*"). I know this story from Mandache Leocov (1928–2018), one of my professors in the Master Program (in the management of natural resources) which I graduated from UAIC in 2001—a first-rank personality and public intellectual who dedicated his life to the green areas and natural patrimony of the city of Jassy (also Director of the Botanical Garden between 1973 and 1990). He had a key role in the protection and recovery (and health) of this particular monumental linden tree, in the preservation of linden trees in the city with scientific arguments against the brutish and uniformizing dictatorial rule imposed by the soviets after WWII (until the recovery of Freedom in 1989) [200], and then, after the fall of communism, in the public support for Linden trees in the city after a serious conflict between citizens and the city hall, later resolved in favor of the citizens—see below);
- The linden trees episode in the city center: over 70 linden trees have been felled in the historic city center and about the same number elsewhere in the city (during 1–15 February 2013), and the people have been revolted that the City Hall regarded this as a purely technical matter and did not initiate a public debate before such a dramatic action that affected the outlook of the city (and even some mandatory steps of the legal procedure have been skipped). In the end, the City Hall has been legally penalized and obliged to re-plant linden trees, which then happened in November 2015 after a popular consultation online in which 97% of all those expressing their opinion voted in favor of approving a plan already prepared technically by an alliance made of civil society and scientists. The pro-Linden campaign used the name "*Jassy loves the linden tree*" (in Romanian "*Iașul iubește teii*").
- The public Jassy Botanical Garden, also the oldest in the country (established in 1856, currently ca. 100 hectares) is also not far from the Copou Garden and UAIC. This is in fact a great example of ecological restoration and development of urban green areas that was extremely degraded and polluted—basically, this is

a model and inspiration for cities around the world: a successful transformation of degraded urban areas in green areas. As a personal experience from my years of undergraduate studies, I also used to go in this botanical garden (located close to our student residence) to attempt to study outdoors (at least I tried!) but usually I simply ended up doing something else (visiting the garden sectors, or doing other handwritten notes). While normally there is an entry ticket (to be paid—not huge, but something) biology students like us had a permanent free pass, which was perfectly logical and which was a great inspiration for us as students of life in all its forms. That should be a normal practice in the relation between universities and cities.

- As a key observation, the *Palas Mall* urban regeneration project started from a general need for renovation and consolidation of the Palace which required a broader systemic perspective which has led to the entire district being redeveloped along with the renovation and consolidation of the historical palace itself (and, this process also led to additional archeological research with certain underground parts of the old princely court being now made accessible to visitors).

As an observation, the medium-size city of Jassy shown in Fig. 2.11 also illustrates how a City of Inspiration tends to look like across the region of Central Europe more broadly (one naturally thinks also about other cities of different sizes, e.g., Munich in Germany, Krakow in Poland, Budapest in Hungary, and so on).

Now, beautiful city centers and urban landscapes can be found in inspiring cities around the world and across history. But this is of course not a reason to forget the less favored cities or/and city districts that require equal attention (as briefly evoked above, the city of Jassy itself went through many hard times during its long history). On the contrary, the highlight in Fig. 2.11 is meant to help us imagine (a) that every city and city district can be, and is, made beautiful by its people, and (b) that the same spirit we can see in the historic center of a city is actually present on every street corner where the people of that city live. Just like a city cannot be understood without its hinterlands, a city center cannot be separated from the rest of the city just because it looks more attractive. In fact, a recent study in the city of Jassy itself (from the Geography Department of UAIC) shows in more detail what we already knew: the bulk of the population is tied to a few high-density ("dormitory") districts that may not have the touristic glamour of the city center but which make the city be what it is given its more ancient and more recent history [201], and which require further efforts for addressing important differences (and inequity) in access to high-level urban facilities. This and other, often more technical studies (e.g., [84, 101, 190, 202–205], [138]), give the image of a city of Jassy that is growing and transforming on the long-haul, in a dynamic relation with its peri-urban areas, the broader hinterlands and the long-range communications across the country and around the world.

Equally important is that a city can be inspiring in subtler (often more technical) ways, e.g., various case studies on urban development topics and the research papers themselves can be inspiring. For example, I found a very nice and inspiring paper on using the method of Land Suitability Method for the goal of planning green areas in the great and complex city of Hanoi (Vietnam) 15 years ago [206]. This paper is also a simple and effective example of a direct international cooperation at the level of a small team of researchers, for the benefit of a city and its inhabitants. The fact that it involves international cooperation is nothing really special in itself, because this is actually a common practice in science and in the academic environment. But precisely this fact is a potential inspiration for city halls and different communities to work with each other across geographies; in the end, this is how we, cities and peoples around the world, can inspire each other.

Beyond the Jassy case study, some historical or present time events that took or take place in different cities can be inspiring, and I only need let each of us recall such moments from our own experience. For the purpose of the book, I provide a supplementary discussion and give a few ideas in Sect. 2.6 (the last section of this second Module, dedicated to Cities of Inspiration).

Back to the core method described in this book section, one also observes that the scenario set in Fig. 2.12 is a general minimal taxonomy that is derived from the scenarios generated for the city of Jassy, within which all other cases can fall. So, in any city of interest, the reader s/he can just replace the word "City" in the generic typology titles with the name of her/his city of interest. As the base study city was Jassy, the names of the scenarios in the set generated in the procedure above become, respectively: "The Receding Jassy", "The Wanting Jassy", "The Promising Jassy", and "The Inspiring Jassy". But if the target city is, for example, La Paz (in Bolivia), the scenario names in the scenario set would be: "The Receding La Paz", "The Wanting La Paz", "The Promising La Paz" and "The Inspiring La Paz. Just the same way, if the target city is Addis Ababa (in Ethiopia), then the scenario names would be, "The Receding Addis Ababa", "The Wanting Addis Ababa", "The Promising Addis Ababa", and "The Inspiring Addis Ababa".

In a broader geographic discussion, islands are demographically and functionally very similar to cities, so the logic described above also applies to small and medium

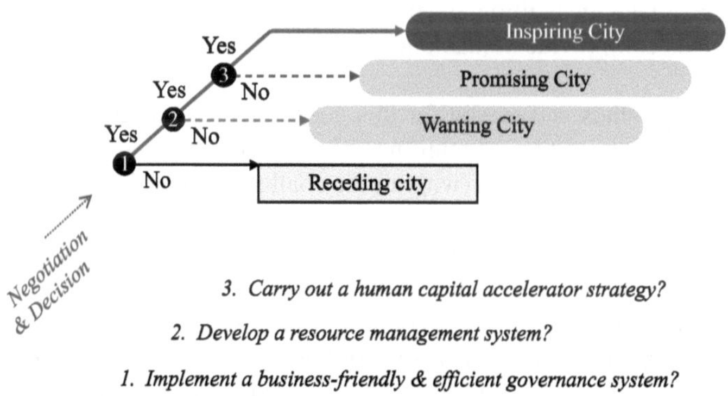

Fig. 2.12 Fourth (unstopped) scenario resulting from the process of negotiation and decision along the decision tree aiming at urban sustainability [decision pathway: "yes; yes; yes"], along with the full set of preceding scenarios. This is the technical target for insuring continual progress on a city's path to sustainable development. *Source* redrawn after [14]

size islands (or sufficiently connected archipelagoes). For instance, in Oceania, the scenario set above would become “The Receding Tuvalu”, “The Wanting Tuvalu”, “The Promising Tuvalu”, and “The Inspiring Tuvalu”. In this later case, the typical issues are related to ocean levels (sea level rises due to climate change) and distance from large population centers, but advantages can be related to modern progress, e.g., the increasing satellite-based internet access and the presence of campuses of the University of The South Pacific, both of which can connect them with each other in the Oceania region and with the whole world.

In all cases evoked here, modern science and technology allows increased participation of local communities across geographies and in the Global arenas, e.g., through Diaspora in general. Many local communities in the countries are located at higher distance from the great economic centers of the world and often depend a lot on the remittances sent from abroad (consequently, going abroad for work is sometimes even encouraged by the public authorities as a survival strategy of these communities). But skilled diaspora can have an even more profound socio-economic role in anchoring those communities in global circuits and exchanges, through internet-based joint projects and remote work (For the methodology and a discussion, see Module 3, Sect. 3.5).

The similarity between small islands and cities means that strengthening local–global connections and bidirectional fluxes of people, goods and services and value co-production emerges as a common, logical priority for remote communities, one which can transcend distances and unite people around the world. In fact, the chosen example above are not by chance: La Paz is the world’s highest capital city (average altitude of 3640 m) and the name means *Peace*, in Spanish; Addis Ababa is known by many as the *City of Eternal Spring* (because of its climate resulting from location on a high plateau at an average altitude of 2354 m near the Equator; although it is not the only city to be called that, other famous ones being Nairobi, at 1795 m altitude); Tuvalu is the world’s smallest republic, has the world’s tiniest GDP (because of the size of the population), and its name means “eight standing together” in the local (Tuvaluan) language. In addition, the past (experience), present (culture), and future of these islands can provide valuable lessons for all local communities around the world.

Finally, the grounds and reason for this generalization based on the case study of Jassy relate to the great broad relevance of the particular pressures (needs) faced by that city. In fact, the situation (with all the challenges) faced by the city of Jassy in 2007 when the public consultation (on which this post-factual analysis was made as an application of the method developed for Versailles Chantiers) happened, was representative for the general context in the entire region of Central Europe, where the painful (deep, systemic) socio-economic transition from command to market economy overlapped with the wider transition to sustainable development that had already started to manifest itself [207]. This obliged the city (both the citizens and the experts studying it, as mediated by the local authorities) to make a deep introspection, and the results logically reflect this profound transformational context: the resulted list of priorities are so fundamental that they greatly reflect systemic societal priorities in all major transitions anywhere.

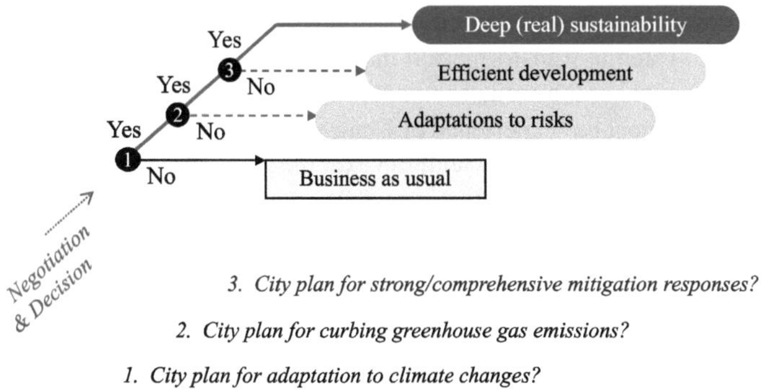

Fig. 2.13 Full set of scenarios resulting from the process of negotiation and decision along the decision tree aiming at addressing the issue of mitigation of and adaptation to climate changes (as part of the path to urban sustainability) in the city of Vancouver, Canada. *Source* redrawn and modified after [14], based on [50]

For example, the individual scenarios in the scenario set derived from the list of priorities from the city of Jassy can be reproduced by analogy through the set of scenarios imagined (initially without any ranking or Delphi methodology behind it) for the city of Vancouver, in Canada [50] (Fig. 2.13).

A supplementary discussion is provided in the next section.

2.3 Selected Illustrations and Discussions

The case shown with the city of Jassy is representative for the historical context of Central Europe (a group of countries that now constitute the eastern flank of the European Union and NATO), with three main features that are worth highlighting as relevant for our analytical purposes here:

(a) during the public consultation exercise made by the city hall in 2007, the city (and the entire region) had basically closed a difficult socio-economic transition to an open society and a market economy, and was seeking to accelerate its economic catching-up with western Europe;
(b) during the academic analysis cited in the 2013 paper (and conducted in 2011–2012), the city (and the entire region) was confronted with the aftermaths of the 2009–2010 economic crisis (so, beginning a new development cycle and the wider trend of economic convergence with western Europe);
(c) as already mentioned, the socio-economic transition in Central Europe also overlapped, especially in its later phases, and is now plenary continued by the transition to sustainable development in Europe and the whole world.

Given the depth and systemic character of this transition, this process is indeed one in which cities and society as a whole reinvented themselves for a better future and thus the experience that these cities and societies gained in that phase is directly relevant for the systemic transformation that is required in the transition to sustainability. Then, the logic of the scenarios-generation method described above is general enough to help connect the experience of Central Europe with other case studies that bring relevant experience, notably with the example from Versailles in this analysis, and with all cities around the world—especially those cities and local communities that have a memory of systemic challenges or/and or are experiencing such challenges right now.

In order to transpose the technical scenarios presented above in verbal storylines, the following paragraphs make a brief description of each of the general scenario types mentioned above in Fig. 2.12.

The "Receding City" scenario represents business-as-usual (BUS), with no progress being made/observed towards urban sustainable development. Future actions of the city authorities (reflecting the needs and will of the local community) will be essentially reactive and limited. Emigration (the so-called "brain drain" included) will continue. Potentially, a decline of the population will result in a lower rate of environmental degradation on short term but this rate will become higher on medium and longer term because of increasingly inefficient economic activities (as a shrinking, lesser attractive city also represents a shrinking market), which will lead to a situation of overuse of all local natural resources. The power of the city within its nation/country will fade; the city would be on a path of slow-but-sure decline while other cities in country (and in that part of the world) will move on and converge with the more developed regions on the continent. The prestige of the city will erode, which will contribute to holding its hinterland captive to a state of relative underdevelopment. Overall, on medium and longer term, the city in this scenario type will essentially be stuck in "a vicious circle of irrelevance and poverty".

The "Wanting City" scenario represents the general kind of situation in which urban transformation towards sustainability commenced, for example by implementing a program aiming at (and succeeding in) boosting local governance capacity (to attract high-value people and businesses), but, because subsequent questions were not answered positively (in the illustrative case study, ecosystem and resource management being the next most urgent in the global-and-local context of aggravating climate changes and other kinds of degradation of the environment), the city can expect to be confronted with increasing rates of natural disasters and pressures.

Notably, the example city of Jassy is biogeographically located in a region that is relatively dry compared to other parts of the country and will likely have to face serious episodes of drought. The other chosen example, the city of La Paz is even much more exposed to drought, and does not have the benefit of being part of a union of high-income countries like the EU, so its climatic and related economic challenges are even greater, which further means that the importance of using its specific local advantages (whatever those are as identified by the local community) to compensate for its local disadvantages is even greater. But this requires sufficient organizational/managerial capacity. So, without applying (and testing hardly) the governance capacity in a policy addressing climate and environment, the overall ability of the city to act and react will erode, and the local community will face disillusion. Other cities in other biogeographic and socio-economic contexts, say, Libreville in Gabon, Juneau in Alaska and Manilla in the Philippines, can expect different combinations of risks, but—given the intrinsic challenges described in this book in relation to urban systems—a strong governance capacity is most likely a common top requirement everywhere, and a top pressure wherever this requirement is not fulfilled to a level that can be seen by the local community as being satisfactory.

The "Promising City" scenario describes the situation in which the city develops and implements a strategy for enhancing its governance capacity, then a policy for mitigation of and adaptation to climate (and broader environmental) changes, but fails to maintain a reform-and-development momentum for the next challenge, e.g., in the chosen example, through a policy to attract human capital (a particular importance having the highly-skilled, because they help attract socio-economic development in general). In this theoretical situation, the example city of Jassy would fail to capitalize on the fact that it has the oldest university (and some of the best group of universities) in the country. As a retrospective observation 10 years after the first publication of this analysis, one can notice that the city of Jassy is booming: because of its universities, it managed to attract numerous investors and international companies, which attract further highly-skilled; all in all, the theoretical situation in this scenario did not materialize: there still is room for improvement but a virtuous circle or attractivity and development has already been triggered.

In this theoretical case (which can still happen anywhere in the world), the city of interest has some good fundamentals for progress (or can have them—to stress the conditional character of the hypothetical nature of scenarios that may or may not materialize) but lacks (/may lack) a functional relation with its universities. In other words, the conditions may be in place for the highly-productive triple-helix interaction to happen through projects covering participation from universities, businesses and the public authorities but such interactions do not happen yet (for different reasons, starting with a lack of experience in such complex projects or if one or more of the three domains of the "helix" is reluctant to engage with the others). So, this scenario corresponds to a case in which the city of interest already has (/ may have) certain strengths, thanks to its history, and a good governance capacity, but its overall success within in the worldwide network of cities is still an unfulfilled promise.

The "Inspiring City" scenario, the last in this basic series, is still only a "snap-shot" on the often-long path toward prosperity. Also, this scenario embodies the on-going development that results from a sequence of key decisions made by a city seeking sustainable development (and high resilience against shocks that can derail it from its course). In this case, the city has managed to integrate its research–education–innovation centers (i.e., universities, research institutes and other public or private organizations involved in knowledge production) within its own dynamics as a coherent urban system; this makes the city well-prepared for acting as a regional pivot of development and overall transformation towards sustainability.

How much the study city of Jassy (or any other) will manage to develop and maintain a dominant place within the international networks of cities, and what its future identity will be as a potential eco-city, depends on future prioritizations and how capable the city will be to raise to those emerging challenges. Whether or not the city will become sustainable, and how long it will keep this status (ideally, forever), only the future will tell. Sustainability, after all, is not an end-of-history, but the beginning of another era of Humanity. Technically, it is a situation of equilibrium that will need to be constantly cultivated and maintained by dynamic updates—not only in response to local risks and challenges but also (because of the dual local-and-global character of every city, as discussed in Module 1 of this book) in response to global risks and challenges of the community. In a nutshell, cities must collaborate to attain sustainability, and then they will have to collaborate even more to maintain it. To close up with a famous expression, the cities of the world have emerged from local villages and will find each other in one "global village".

Wider Scopes: From Cities to Countries, Continents, and the World

The method described above naturally extends to cover country-wide scopes: the method is the same, but the details will have to change accordingly. To give a concise example, I refer here to the Mont Fleur scenarios developed in and for South Africa at the end of Apartheid in the early 1990s [208].

In Fig. 2.14, I re-studied this well-known scenario set by adopting the backward-and-forward-look comparative perspective of historical analysis and scenario planning, i.e., based on the rationale that a historical vantage point "can help scenario planners to develop better insights into the world ahead" [209]). And, a simple inspection reveals that this scenario set (taken together with the details behind their generation) maps naturally in the logic of ontological scenario generation method described above.

As for the fate of the Mont Fleur scenarios in its own country of birth, current analysis shows that the outcome so far was less than the best situation, and even the idea of scenarios has not been translated into national policy-making. While this in fact speaks more about the nature of policy-making in general than about a national context (policy-making has serious shortcomings in all countries; see the literature review and the necessary stepwise methodology that I summarize in Module 3, Sect. 3.8 of this book), there are recent calls for recovering this intellectual tradition into mainstream South African policy-making as a way to build societal resilience, with the increased benefit of the experiences and lessons generated by other countries which used scenarios in their policy-making, many of which were actually inspired by those Mont Fleur scenarios in South Africa [210], and showing how different human communities and cultures can learn from each other.

The Mont Fleur exercise (and scenario set) is particularly relevant for our analysis on eco-city projects as a key to sustainable development, for at least three reasons. Firstly, the Mont Fleur approach and scenarios were part of the international landscape of insights and literature references when I was developing this method focused on Versailles and Jassy (10–12 years ago) [14], and thus helped inspire this Delphi-based method in the first place. In particular, my attention was drawn by the Mont Fleur scenarios because my own interest in the issues of socio-economic transitions in Central Europe (at present, the Eastern EU) from central command economy to a market economy—and many issues were alike and also more generally relevant

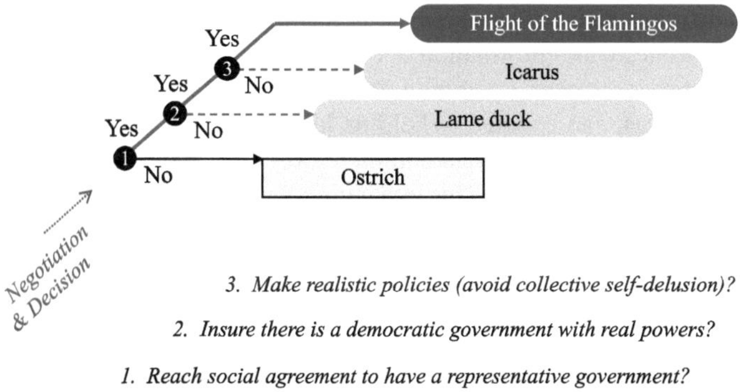

Fig. 2.14 Ontological prioritization model of both a nation-scale transition and city-scale transition out of an untenable situation: a process-oriented re-interpretation of the Mont Fleur scenarios known from South Africa (at the end of the Apartheid regime in the early 1990s), based on their explicitly identified (consensus-based) priorities (understood as necessary conditions), and the implicit ranking used in the original scenarios (the original verbal description). *Source* original reconceptualization based on the initial description of the Mont Fleur scenarios as summarized by the Mont Fleur exercise co-initiator [208]: "a non-representative alliance government (the Ostrich scenario), a democratic government incapacitated by a long transition (the Lame duck scenario), a macro-economic populist government (the Icarus scenario) and an inclusive democracy focused on socio-economic development (the Flamingos scenario)"

in the broader discussion of systemic transitions, which also includes the transition to sustainability in all cities and countries around the world. In other words, it is a matter of methodological consistency across time, contexts, and literature.

Secondly, and more importantly, it is a natural way to scale up the discussion to cover the other important scales beside cities (and which also influence how cities work), notably national (and international) scales—which is particularly important because current policies on climate and sustainability are primarily developed at national scales. In fact, the initial moderator of the Mont Fleur exercise has used that experience (and implicitly the intrinsic logic behind the structured storytelling represented by the scenarios themselves) and applied it in efforts to solve an array of other multi-stakeholder complex, societal challenges with around the world: judicial reform in Argentina, public service restructuring and local economic resilience in Canada, chronic violent conflicts in Colombia, peace negotiations in Sudan, implementation of peace accords in Guatemala, social-institutional schisms in Israel, regional development in the USA, government service delivery in Australia, and sustainable finance in the UK in the aftermath of the 2007–2008 financial crisis [211].

Thirdly, it concretely shows the versatility of the method (i.e., it can be used at different scales without having to change the method core), thus completing a series (from the local science-business innovation cluster in Versailles, to a city as a whole in Jassy; and, logically, it can be applied to higher scales, starting with local networks of cities). So, it can be used by companies, cities, nations and even continent-wide (e.g., at the scale of the European Union).

Essentially, one can observe that mapping onto this Delphi-based scenario-development strategy the information about meaningful exercises aiming at developing social consensus over a common path in cities (e.g., in Jassy) or in countries (e.g., South Africa), can concretely help us learn from the past how to deal with the future in a way that is (coherent and therefore) effective as a common methodological denominator, i.e., a basis for integrating knowledge and learning from each other across scales and around the world. This is important, because a person can do a great deal of "learning from experience" [211], but when we talk about learning as a society, personal experiences need to be supplemented and complemented with a structured methodology that can support "disciplined experimentation" [84] and advanced operational solutions for systemic transformation, and avoid being trapped in "eternal series of new beginnings" (to be clear, "initiatives" are not enough, regardless of how many they are—it's not a matter of adding up initiatives: first because those are simply too incomplete, and second because they don't just add up like a stack since they are inevitably too incompatible and there is nothing concrete to add up).

At this point, an epistemic observation is necessary: while the basic idea of the Delphi method is to develop consensus, it is not meant to achieve a situation where everyone "thinks the same way". Instead, the goal is to distill what a community of people thinks as a community, taking into full account the diversity and different weights of factors and perspectives. Therefore, the word "consensus" is here used in the sense of building a point towards which individual perspectives can agree on as being worth to be used as a goal towards which different choices can converge based on current knowledge (in the same vein as the notion of social learning), and, consequently, also as part of the process of building a basis of common denominators as a platform and enabler of co-work across disciplinary, profession and institution boundaries.

In this sense, the Delphi method, both in its general core and in its application here, is not to be confused with the ancient principle of *consensus gentium* (literally, "agreement of the people") which in the classic Greco-Roman world from two millennia ago was known (in the description given by Vergilius Ferm) as "that which is universal among men carries the weight of truth" [212]. This is especially because its validity per se is not "automatic" but depends on the correct use of induction to come as close to truth as possible [213], and this condition applies here too. One of these conditions that is particularly important in practice was summarized by [214]: "how many heads are better than one?"; "if I am to subscribe to a consensual outcome responsibly I must be personally committed to it, and this requires a form of knowledge I call 'fiduciary', in this case knowledge of the competence and trustworthiness of other participants in deliberation whose expertise may have influenced my agreement". Given its inductive core, the concept can cover interests from theology to "wisdom of the crowds" [215]. However, from here to deciding that something is knowledge there is some way to go. (For a dedicated discussion about what is knowledge, see Module 3, and especially Sect. 3.7).

One also observes that the questions in the historic-methodologic interpretation in Fig. 2.14 reflect not only the original concerns of the authors of the original Mont Fleur scenarios but also the idea that the self-organization of a society (country/

city/world) towards a socially sustainable future begins from the core requirements of respecting Human Rights (notably The Consent of the Governed principle) and closes with the necessity (necessary condition) of competent government (which also involves the Whole of Government principle). So, these scenarios are normative only in the sense that they demand us to acknowledge reality. In a sense, this is also a form of agreement on values: do we collectively decide to look at reality as it is (thus, pursue truth, as Science does) or do we make a de facto decision to delude ourselves (e.g., through a spending spree that seeks to make everybody happy and finishes by making everybody worse off—as stated in the Icarus scenario of "macro-economic populism" of the original Mont Fleur scenarios). One further observes that when things are being said (explicitly), society is in the first modus; when things are not being said (everyone avoids saying that "the king is naked"), then society is in the second modus. Basically, the "Yes–Yes–Yes" path in Fig. 2.14 reflects the path that was summarized in words back then by [208]: in the case of the Flamingos scenario, "a decisive political settlement, followed by· good government, creates conditions in which an initially slow but sustainable economic and social take-off becomes possible"; and "The key to success is the ability to combine strategies for improved social delivery with policies that create confidence in the economic sphere".

By the way, from the systemic perspective on sustainability summarized at the beginning of the book (Module 1, Sect. 1.2), this entire discussion above can be located inside the Social Sustainability Filter.

Further, the Mont Fleur scenarios are particularly relevant for the notion of "viable society", and the notion of identifying a set of necessary conditions was already used at the time of their generation in South Africa, along with the recognition that there can be more than one viable path to the desired situation, thus implicitly recognizing the existence of what I have summarized in Fig. 1.9 (Module 1, Sect. 1.3) as a space of options/space of possibilities. This is what the original Mont Fleur paper ([208]) tells in 1992: "Many different types of blueprints could potentially put South Africa on the Flight of the Flamingos. They should, however, all contain some essential elements. The necessary conditions for success in the economic sphere would be: effective negotiations; good investment environment; basic needs provision; development of experts; monetary and fiscal discipline".

Now, one point in that comparative historical and scenario analysis is to look back and see what happened within the last "long three decades" since the Mont Fleur exercise and draw practical conclusions for the next 30-plus years, knowing that the current requirement for addressing the issue of unsustainable development is actually not waiting for us to try until we succeed like children or students could do in the safe environment of a school. Because the next third of a century is substantially different than the one that started in the early 1990s, the consequences of failures will be very different and (given all tensions accumulated since then) will be less forgiving.

These being said, I believe that this process-wise interpretation and representation can help address practical questions, not only in South Africa but in any country that aims to transition to a free and open, "sky is the limit" type of society that is not only socially sustainable (as primarily aimed in the original Mont Fleur scenarios) but sustainable overall. This exact idea is reflected in the next examples: scenarios generated with the same, Delphi-based method for assimilating and streamlining the information provided in different studies and highlighting different sets of priorities from different contexts, for instance in Asian cities (Fig. 2.15).

Finally, one may observe that the same technique can absorb the priorities emerging from the systemic perspective on sustainability described in the first section of this book, namely, the model using the sustainability filters (Fig. 2.16).

Moreover, Fig. 2.17 and Table 2.7 show that it is possible to map the scenarios emerging from different case studies onto each other and see what fits and what does not fit, and I suggest (as an exercise) that you can try to convert some other case studies you know about into a set of scenarios according to the method described above, and then see what correspondences (i.e., mappings) you can detect in the big picture you obtain, and what new insights and ideas might come out of this. Obviously, this opens widely the door to comparisons between different perspectives that can be used in different projects and as reflected in the literature and experience to date.

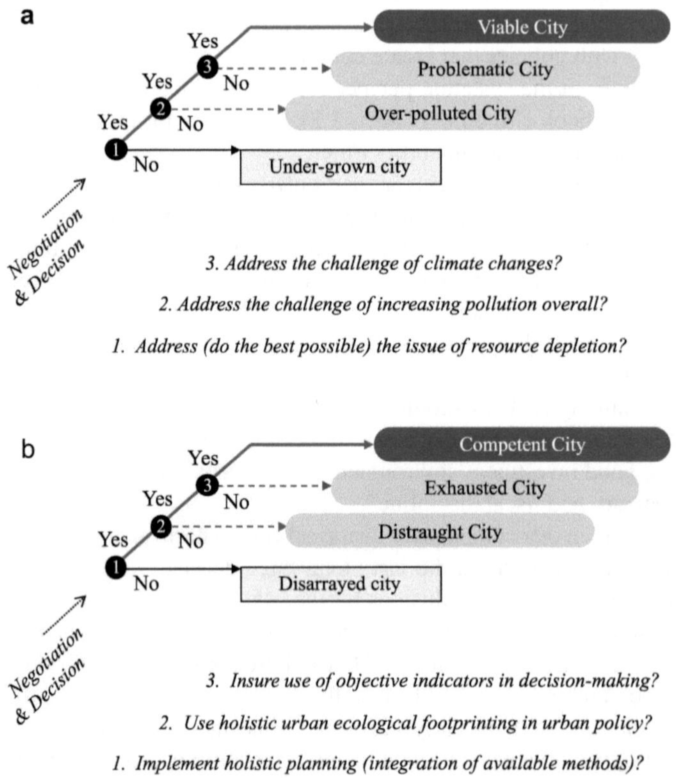

Fig. 2.15 Ontological prioritization model towards a virtual (theoretic common denominator) of Asian cities (and, by extension, of cities in all fast-urbanizing areas around the world), seen through the lens of **a** general topical/managerial priorities and, alternatively, **b** epistemic priorities. *Source* original illustration, based on nominal priorities identified in a literature review by ref. [216]

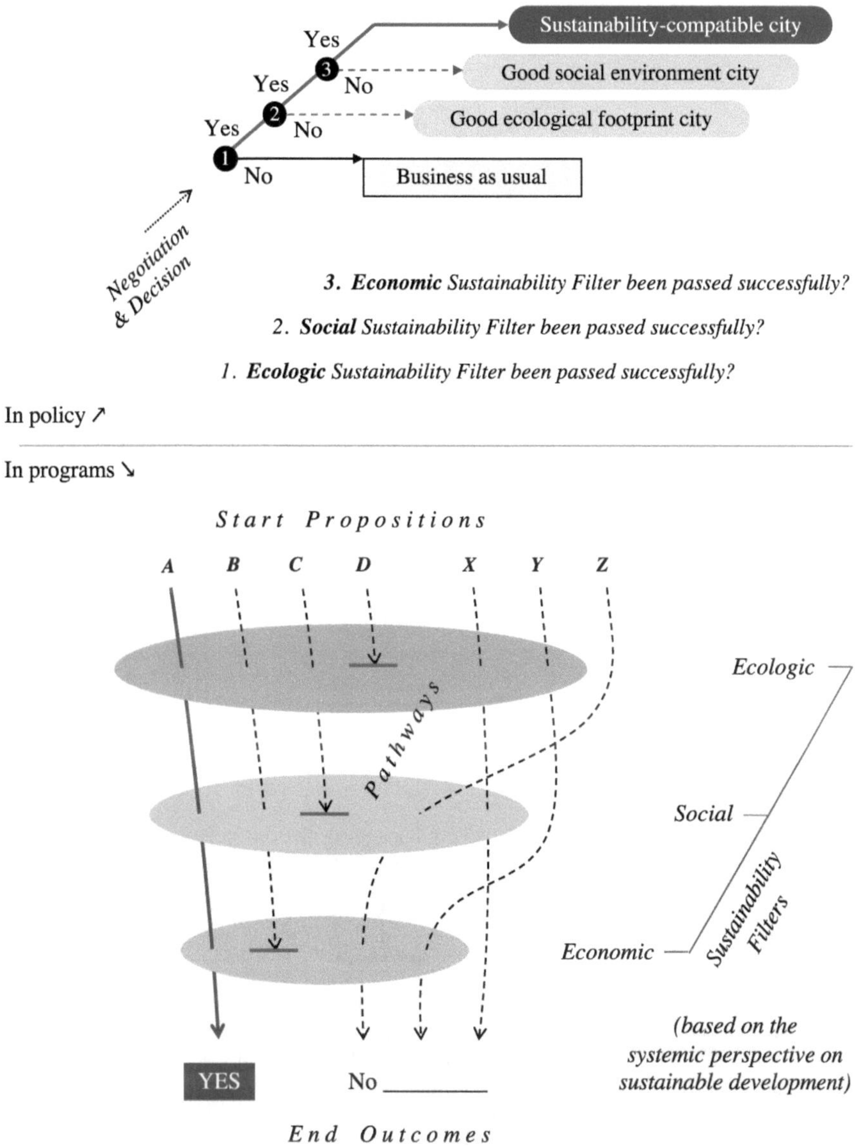

Fig. 2.16 Ontological prioritization model towards a virtual (theoretic common denominator) of solutions proposed explicitly (or headed to implicitly) in any city (or city district; or group of cities, and by extension the global network of cities). *Source* original, based on [100]. and Figs. 1.1, 1.2 and 1.3 in this book (Module 1). *Observations:* (1) Regarding the direction of the scenarios (pathways) in projects and programs (bottom half), as per Fig. 1.3: evaluation of proposed solutions must start with the Ecologic Sustainability Filter (in this order); (2) The redline stopping the arrows of scenarios {B, C, D} means that those pathways do not continue, because failing any sustainability filter renders a pathway unviable (to discard); (3) Because passing each sustainability filter is a systemic necessary condition, if an evaluation concludes that a certain path actually went around (avoided) at least one filters, the end result is the same as filter stopping (i.e., rejection), even if other filters appear to have been passed

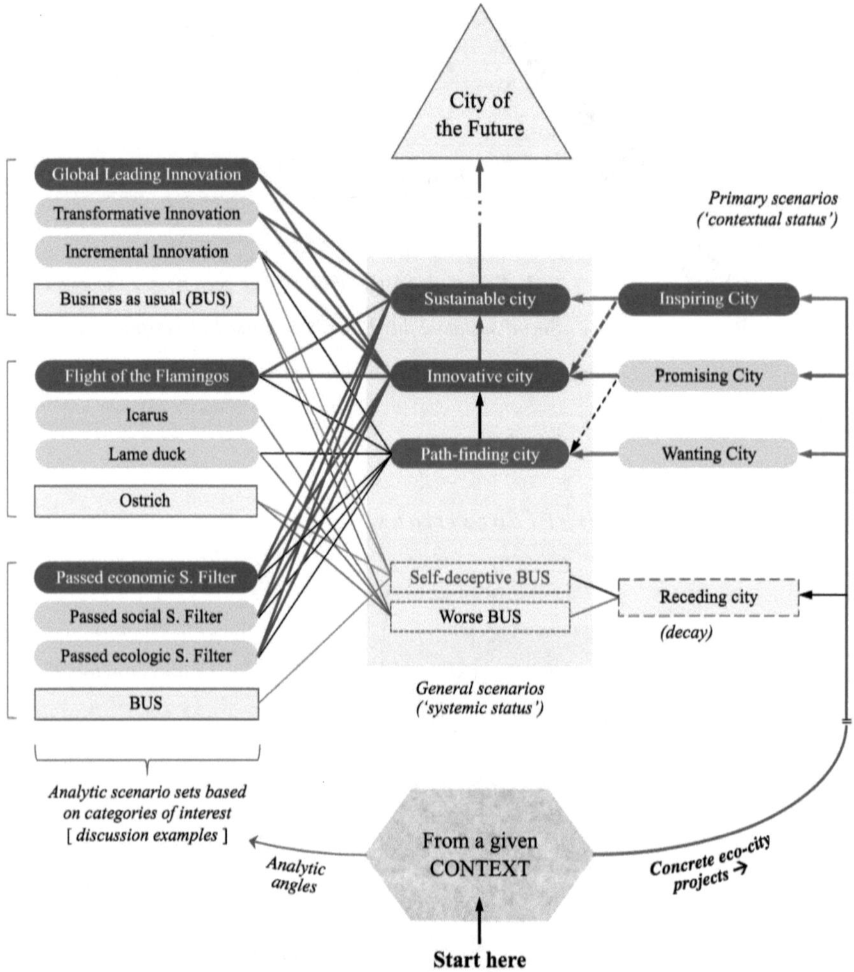

Fig. 2.17 Mappings between scenario types and analytic categories known from practice and the literature (examples). The upper right quarter of the figure containing the arrows (minus the red arrow and the red bordered boxes) represents a functionally structured space of possibilities, i.e., type of options within the general space of possible trajectories of a city as depicted in Fig. 1.9 (Module 1, Sect. 1.3) and according to the definition of an eco-city provided in Fig. 1.12 (Module 1, Sect. 1.4). The analytic scenarios set on the left bottom is derived directly from the knowledge-action model shown in Fig. 1.3 (Module 1, Sect. 1.2). *Source* original

I would also note that what strikes in the Big Picture obtained in Fig. 2.17 is that the Primary Scenarios (up-right sector of the figure) is very general and it therefore covers (maps) the evolution of any city from "trailing" to "leading" in terms of eco-city programs. But it is of course possible that somebody may find an even better (or at least complementary) set of scenarios mapping the same path. However, the "systemic status" scenarios are derived directly from the systemic perspective

Table 2.7 Mapping scenario sets (examples) on eco-city path from Fig. 1.12 which was used in Sect. 1.4 to define the eco-city

Question and answer (Yes/No)			City situation type	Human system situation type
3	Has the sustainability threshold been passed?	Y	Sustainable city	Flight of the flamingos
2	Has the innovability threshold been passed?	Y	Innovative city	Icarus
1	Has the path-finding effort begun already?	Y	Path-finding city	Lame duck
		N	Business as usual	Ostrich

on sustainability applied to urban sustainability, so they depict the core stepping milestones (the most basic necessary conditions) towards sustainability: starting the change and achieving minimal sustainability. So, perhaps the names can be improved and the detailed description will probably become better with time, but the essential idea they embody is already clear. In any case, taken together, the "contextual status" scenarios (the default Jassy set or an equivalent generated in other projects) and the "systemic status" scenarios represent a basic operational mapping of the space of possibilities for any city that wants to become a true eco-city (and ultimately, a sustainable city).

Further, using Fig. 2.17 as a platform, and based on the discussions in the book so far, one can make several other observations (and the reader is invited to continue with her/his own observations), for example:

- There is no direct connection between "Incremental Innovation" and "Sustainable City", i.e., it must pass through transformative/systemic change (it doesn't work with doing "a bit of this, a bit of that, without taking risks", and this is because cities are real complex dynamic systems, not linear additions of bits;
- The "Icarus" circumstance gets you in an even worse situation than the "Lame duck" because it is the case of moving fast in the wrong direction vs. moving nowhere or moving too slowly;
- "Sustainable City" implies that all three levels of sustainability are necessary (biophysical, social, and economic), starting and ending with economic viability, because this is the primordial concern of people.
- One can do an exercise, to see how other sets of analytic scenarios map onto the core scenarios in the general space of possible trajectories, using different sets of analytic scenarios, e.g., those in Fig. 2.15 (in fast-urbanizing Asia), or others from the literature or some produced by yourself for your particular city/-ies of interest.
- Essentially, the schematic from Fig. 2.17 provides a method for a city to find its way as an eco-city by mapping its own path (1) according to its local details, (2) based on current knowledge, and (3) using learning from (obtaining analytic insights from the studies in) other cities and contexts (at all scales, i.e., including city district and country scales). More generally, when used as just mentioned

above, this figure will help cities (all communities of people) learn from each other and thus hopefully accelerate the progress of all in a practical marketplace of perspectives, insights, ideas, methods, and solutions.

2.4 Supplementary Technical Notes

In Eco-City Projects, Decision-Making Needs to Be Both Structured and Versatile

The main point here is that, in any action within CDS, we must not lose sight of the big picture, as this is a necessary condition for orientation in dynamic situations: eyes on the goal.

The utility of building scenarios as a managerial approach lies in understanding correctly both its force and its limitations. On the one hand, they are necessary for gaining visibility into the space of options evoked in the previous part of the book (Module 1, Sect. 1.3). On the other hand, the in-built issue with scenario development is that usually—sooner or later—one or more assumptions used in their generation will turn out to be unrealistic, very often because the context, the needs, the available knowledge, the possibilities, the goals and the preferences can all change. When previous assumptions are not accepted anymore (or at least not as such) by experts and/or the decision-makers, a new set of scenarios is needed, and their generation must not have prohibitive costs.

Decisions will always have to be made, even more so in times of crisis/emergency and change. By assigning explicit formulations of choices and choice consequences, the method described above provides a tool for clear-cut and prompt decision-making. Further, the method is robust to updates of knowledge and preferences, because it can be rerun without a need to discard all previous work: one can follow what was done and simply make the necessary updates and follow through the logical consequences along the procedure. In short, the procedure allows effort minimization of updates because assumptions sit (and are applied) in an algorithm of prioritization. This insures a best tradeoff between the two simultaneous requirements in the decision-making process: to be structured (not “chaotic”) and to be versatile (not rigid).

To insure this nimble-and-effective character, combining empirical experience with abstract modeling is a logical approach. In addition, this facilitates interaction with support methods from mathematics and information sciences and technologies. In sum, decision trees have the following advantages derived from their abstract character:

- In-built feature selection, the hierarchy of priorities in the decision trees reflects the importance: issues at the top are the most informative about the target urban system. Additional priorities add lesser and lesser relevance, which justifies a cutoff line as necessary—the question is only how much more resources are you willing to commit for lesser gain;

- They reflect well how human decisions are being taken and can be easily translated in the abstract language of operational and support information tools (for example, decision trees can approximate any Boolean function in technical languages);
- They combine well accuracy and flexibility as necessary in operational modeling;
- They are easy to understand and interpret by both people and computers;
- Relatedly, they are applicable both to humans (expert judgments or public consultations) and machines (notably data mining), which brings a supplementary set of technical advantages (statistical included), which matter a lot for providing data processing and analytical support and thus assist the experts in making the best judgment possible—in a virtuous circle of work relevance and efficiency;
- More broadly, this model facilitates application in virtually any case via Lambda Calculus and the related Church-Turing thesis (see e.g., [217]): the intuitive managerial (and all other) categories used above are lambda terms and every identified situation (scenario) can be substituted with the lambda argument expression which corresponds to the string of yes/no answers, and thus automatically identify corresponding situations in other target cities. In the third module, I discuss in a bit more detail this methodology for interfacing human and machine agents in complex pioneering projects using disciplined experimentation and related approaches.

In Eco-city Projects, Decision-Making Needs a Memory

The scenario-generation method described above has yet another, related and complementary benefit: it can be used as a coherent identifier and recorder of priorities (at any moment in time) and of their respective set of consequences, thus enabling a reflexive perspective on past decisions and the lessons they hold for the present and for the future. Whenever reasons emerge for a local community to re-examine the validity of any of the assumptions used in the past, the procedure can be reiterated with the purpose to map all potential changes and consequences. Because of this complementarity, the method bodes well as a versatile tool for learning and knowledge management in general. This is particularly important when eco-city developments are, in their context, highly experimental. But this is also useful as a low-risk method for applying change in more advanced eco-city contexts.

In general, there is always "a trade-off between the need for structured thinking (allowing expeditious decision making) and the need for in-depth scrutiny of the employed assumptions" [14]. This requires that, to be effective, a scenario generation method (a) results in learning at both individual and group levels and (b) allows room for the evolution of the situation and for the evolution of the very motives driving the individual and group preferences of the stakeholders.

In this sense, one observation about the history of the development of this method might be relevant. The method above was generated in the context of the Econoving cluster at the University of Versailles (UVSQ) by combining the "decision tree" idea of the so-called "Mont Fleur scenarios" in South Africa (on the evolution of the society after the abolishment of the apartheid in South Africa) [208–211, 218–220] with the idea of passing/failing (yes/no) of the sustainability filters method that I had developed earlier in a project of urban development in the city of Jassy [100], and having in mind the vocabulary and formalisms used in relation to "decision trees" in information sciences, operational research and in organization studies (in general, and in project management in particular). The main idea from the Mont Fleur scenarios was the necessity to stimulate debate and recognize the main steps (realistic "stepping stone" situations) in the transition from an unsustainable type of social organization to a better "normal" one. This idea is very important because it goes at the heart of societal and expert discussions in societal transitions.

For instance, the original Mont Fleur set of scenarios itself has been criticized in the past for its "excessively normative character" because they depicted only three "problematic" narratives and only one positive narrative, i.e., implying a bias toward a known variation (alternative future) and away from the open character of the future, i.e., it could be understood like the replacement of a "dictated" narrative by another. However, the deep virtue of the Mont Fleur scenarios was their power to mobilize society and intellectuals on a path of transformation towards a change that the society as a whole agreed it had to happen [39]. In other words, the future is open anyways, and the important aspect is not the details of the distantly envisaged future but the capacity of a commonly agreed goal to engage a community of people on a soul-searching and self-developing path that remains coherent with the principles of a free society as enshrined in the UN Universal Declaration on Human Rights.

The method discussed in this section benefits from the straight-out style of decision-making used (by necessity) by the Mont Fleur scenarios, but it frames the decision-making process in a wider context of current knowledge and the imperative to allow for open-end, continuous betterment of a community of people from their perspective in a local-and-global context. The normative aspect of the Mont Fleur scenarios shifted in the method hereby away from the direct definition of scenarios and to the idea that it is necessary to use an updatable prioritization of the policy (in deed, societal) questions that a community of people need to decide on.

Further, the method benefits from the imaginative and heuristic thinking allowed by four-cell matrix analysis that is incorporated in the procedure to pinpoint the potential situations and effects that can be detectable at a basic, logical inspection, i.e., to avoid the situations when the conversation drifts away from reality under the influence of cognitive and/or the related behavioral biases.

How Do These Scenarios Compare with Other Approaches in the Literature on Forecasting and Scenario-Building? Brief Remarks

First of all, these scenarios are integrated as part of a coherent set, i.e., they are more than a collection of storylines, by means of a unifying ontological conception.

In a broader panorama of understanding scenario taxonomies, the here-presented set of scenarios connect the project manager to other old ("classic") typologies [107, 221]. Most importantly, because they explore the future consequences of different responses (that a constituency of actors may give) to an explicit sequence of pressures that are formulated as questions, they are forecasting (not backcasting) scenarios.

Then, these scenarios are descriptive (not normative): they articulate a set of possible events irrespective of their level of desirability. Nonetheless, the iterative process (managerial algorithm) leading to this set of scenarios can be regarded as methodologically normative; and, precisely because of their iterative character, can also be regarded as a tool for optimizing backcasting. In other words, this procedure can be used to identify the most desirable scenario(s) to reach, and then to map and study the logically-potential situations that exist as intermediate between that future goal and the present state of affairs, by means of an empirically supported, logically-derived chain of actions and decisions.

Further, these scenarios are also characterizable as essentially qualitative (i.e., not quantitative), even though there are some quantitative aspects involved in the overall procedure. This is because they are narrated, not based on numerical/statistical models. Inevitably, all the analyses preceding and supporting the ordered set of decision-oriented questions that are produced (and updated) with this procedure will involve numerous points where quantitative data matter a lot (e.g., measurements, estimations, and combined quantitatively or/and qualitatively identifiable thresholds). Therefore, the scenarios resulted from it are best understood as bearing mixed qualitative-and-quantitative features, which, to a certain extent, is similar to what is known in the literature, e.g., with climate change storylines [222–224].

Further, a key characteristic of this method and of the resultant type of scenarios is that they aim at simplifying and facilitating the decision-making process through explicit integration of, on the one hand, expert knowledge and skills, and, on the other hand, data, information and knowledge coming from public surveys. Furthermore, the fact that this scenario generation method is iterative (i.e., it allows and is efficiently enhanced by iterative runs), represents a concrete way to address efficiently the paired combination of requirements (a) for policy-action continuity and (b) for successive updates of the available expert knowledge or/and of the public priorities, e.g., as produced via public scenario workshops [225].

Then, by definition, these scenarios belong to the category "action-based scenarios" [106] because their author(s) must always have in mind that "decisions must be taken", i.e., the production of unactionable insights is not enough/would not serve a real-world purpose.

In the well-known terms of transition management theory, the journey the city of Jassy took towards re-inventing itself (and becoming a sustainable and resilient city) corresponds to the general evolution of an ensemble of techno-social innovations during and through crises, resulting in some *non-irrelevant* changes in the so-called "socio-technical regime" and "socio-technical landscape" [52, 226]; the double negation was here used to stress the profound nature of *dislocking* a city from a given "past" and engaging it in a meaningful transformation as per Fig. 1.12 and Table 1.6 (Module 1, Sect. 1.4), and, more broadly what it can do in a given context to move on and out of an unsatisfactory situation, and what the literature sometimes refers to as "dislocations" (albeit often with some different meanings—depending on what topic is being studied).

The point is that, when we talk about an urban (or other type of) system, the employment of words like "dislocation" hints (at least intuitively) to the occurrence of some profound structural and functional shift. One example (inter alia) is the dislocation (in the sense of disempowering and marginalization) of some socio-economic categories of people living in cities and societies that have driven themselves—through an over-emphasis on economic efficiency—in a vicious circle of short-term over-exploitation of existing social structures and roles to the blatant detriment of an equilibrium between exploration and exploitation: a phenomenon that indicates that the dynamics being discussed is sufficiently profound to be relevant about how the system under scrutiny works, e.g., by the massive expulsion of vulnerable people from mainstream economic circuits [227].

One important aspect here is that such dislocations can have profound consequences, as they go hand in hand with destabilizations of older institutional orders and with a disassembling of urban roles.

Another important aspect is the mentioned equilibrium that must be cultivated between exploitation and exploration, which in innovation management (both in urban projects and in organizations in general) is known as "ambidexterity" and is related to organizational resilience and a so-called "ambidextruous leadership style" (e.g., [228–237]).

This re-invention of urban selves involves an active learning process that involves experimentation and eventually leads to the development of a high capacity to manage present and future combinations of risks and opportunities. To that aim, prioritization is particularly instrumental when seeking to address (present) and prevent (future) systemic risks that may further compromise the resilience of the city (in a vicious circle of under-capacity and under-performance) and thus undermine past investments and achievements.

For example, the imagined perspectives of the Vancouver metropolitan area (in British Columbia, Canada; one of the few cities for which sufficiently reliable data was locally available decades ago, i.e., when this method was first developed), which allow the development of city scenarios [50], can be re-framed and studied logically as a decision tree connecting a set of ontological scenarios (Fig. 2.13). Like this, when the question is how a city can organize its response to climate changes, a set of four scenarios developed for Vancouver ([50], for subsequent discussions see also [238, 239]) with a different methodology (where the main concerns were the vulnerability of a city to climate change and the capacity of a city to adapt to climate changes) can be enhanced by connecting them ontologically. The first and obvious advantage of doing this reformulation is that it allows a form of prioritization of efforts under (the always) limited time and resources. Furthermore, this matters because we also know that downscaling climate change (and other environmental changes) to gain clarity at city level (1) is hard to do (as climate science works with large scales and timeframes, and must not be confused with the related distinct field of weather forecasting) and (2) attempts to scale down dynamics at local levels (with the inevitable wide ranges of uncertainties) rarely include downscaling other aspects that are relevant for sustainability transitions, like social-political and economic aspects. In other words, the method helps us make some useful distinctions and organizations of efforts. Another straightforward advantage is that its logical decision algorithm maximizes coherence and allows a good visibility on the relation between principal decisions and principal consequences in terms of the core capacity of the studied local community to cope with pressures and disturbances.

How Do These Scenarios Fit and Link with Known Mathematical Formalism? Elements for Further Modeling

As part of the more technical observations, let me note that the mathematical representations used above can, in principle, be used for converting the logic and discussions used in the description into abstract models that can support more advance works in foresight. For example, they can be used in Lambda Calculus (a formal system in mathematical logic and universal model of computation that is based on the abstraction of the function, i.e., the relations between variables, which can then be logically bound or/and substituted), and related to it, Graph Theory (for more about graphs, see also Module 3, Sect. 3.2, Box 3.1) and Category Theory (for more about it, see the technical sub-sections toward the end of Sect. 3.7, in Module 3). For example, in the Lambda Calculus, the series of decision points can be viewed as reductions, and the "No" answer signals that reductions terminated (the result being the scenarios branching out from the tree).

2.5 Broader Insights

Between Households and Geopolitics: The Complex Context in Which Cities and All Local Communities Have to Operate

Because of the double, local-and-global character of cities, eco-city projects need to consider the issue *local and global coordination.* For start, it is important to make a couple of observations.

First, although an extensive discussion of purely geopolitical aspects (i.e., beyond the rather technical mentions of the IPCC reports or the societal, cultural, and urban network dynamics) would mostly fall outside the scope of this book dedicated to eco-city projects, one can find relevant literature about the matter, starting with (I would suggest) the relations between geopolitics and the many details of the energy transitions.

In connection with our purposes here, and in the context of the armed and "hybrid" (i.e., combined armed and non-armed) conflicts in our world today, I would cite some good introductions to the matter (see, e.g., [240]), and I would highlight that the world's affairs are not only influencing the Earth's biogeophysical systems. International relations are also deeply influenced by environmental changes, notably through the distribution of resources and capacities in different countries and regions around the world (and the changes thereof), which translates into complex industrial and commercial interdependencies and vulnerabilities inside global and regional supply chains. Therefore, national and local (urban included) affairs are necessarily characterized by an interplay between distinct but overlapping agendas, e.g., as determined by energy, security and sustainability objectives (e.g., [240, 241]) (for an abstract model connecting these challenges, see also Fig. 3.15 and the related models in Module 3, Sect. 3.4).

Secondly and most importantly for the global relations between cities, nations, and world regions (and socio-economic blocks), we must never fail to remember that democracy is part of the necessary features in the social (or "socio-human") sustainability sub-focus of the systemic perspective on sustainability discussed in the first section of this book (and as per Figs. 1.1, 1.2 and 1.3 and the related discussions in this book). The old idea that Democracy remains the best way (as Winston Churchill pointed out in a speech in the House of Common in 1947: "it has been said that democracy is the worst form of Government except all those other forms that have been tried from time to time") is as actual as ever, and never ceased to be. Relatedly, assuming membership of or kinship with the community of people of a city is the highest form of civilization: famous statements like the "*Ich bin ein Berliner*" ("I am a Berliner") of John F. Kennedy on June 26th, 1963, and the "Mr. Gorbachev, tear this (Berlin) wall down" of Ronald Reagan on 12 June, 1987 [242] reflect this eternal reality of cities as communities of people, and above all, that the cities of the future are Cities of Inspiration (see also the text related to Fig. 2.23 in Sect. 2.5, dedicated to illustrating this idea).

Therefore, and thirdly, a good starting point for cooperation at any scale is represented by those instances in which cities and nations need to help each other in the aftermath of disasters. A prime example is the Sendai Framework mentioned earlier (in Module 1, Sect. 1.4): it was agreed upon during a dedicated UN conference in the Japanese city of Sendai which was at the heart of the 2011 Tohoku earthquake and tsunami (also known as Great East Japan Earthquake) that has caused the Fukushima Daiichi nuclear disaster and over 20,000 deaths, over 2400 injured and over 2500 missing. Beside the issue of immediate help and relief in the wake of majors disaster (the above example was caused by an exceptionally strong earthquake and a tsunami wave that left the coastal communities with only 8–10 min to evacuate before the water hit with a height of 6–10 m and even 40 m in some locations), a main point is that communities need to study what happened and draw the appropriate lessons for subsequent re-construction and in view of developing resilient communities (in this specific case, taking into account the common characteristics of coastal communities) (e.g., [243, 244]).

Regardless of the nature and origin of the shock underwent by a city, in the case of devastating disasters, we talk about "forced transformation" of local communities, as examples of the phases of release and then transformation of the adaptive cycle model from Table 2.3 (e.g., [245]). But in perspective, we also talk about "necessary transformation to the best of our knowledge". To continue with the example mentioned above, it is worth reminding that coastal cities in Japan (and people in the whole country) have always been keenly aware of the tsunami risk (even the word tsunami itself is a Japanese word), and that, based on known historical patterns, the public was already waiting for large earthquakes to happen any time—just not that big an Earthquake (of a magnitude of Mw 9.0–9.1) and not so much to the north of the Japanese archipelago (because higher seismic activity was known more in the south of it).

At this point, it is important to make a small technical note about earthquakes, because, even though they are not caused by human (but by natural geophysical) forces, they can affect the sustainability of human communities through their effects. Scientifically speaking, earthquakes are a prime example of a natural phenomenon that exhibits a pattern of self-organized criticality (SOC, as per Tables 1.1 and 1.5 in Module 1): the accumulation and release physical force at the contact front between tectonic plates pushing against each other is size- and time-invariant, which means that it is not possible to know in advance whether the next release event (earthquake) will be small or large (also meaning that one cannot determine when a large one will occur). And, the main type of situation-of-interest associated with SOC phenomena is that of a large sudden displacement, i.e., shock.

Further, events of such magnitude tend to have reverberating effects that can profoundly modify contexts and generate major (and often unexpected) effects at large scales. For example, a recent economic study [246] found that the successful reduction of emissions of greenhouse gases through the Tokyo Emissions Trade Scheme (ETS) has been due, in proportion of 50% to the high electricity prices caused by the coupled disasters of the Tohoku earthquake and tsunami and the Fukushima Daiichi nuclear meltdown. This raises some questions about the mechanism itself (e.g., related to its exposure to short-term effects although it is meant to serve in sustainability transitions on long-term), and also has complex consequences on all climate change policies in the country and around the world (ETSs being under scrutiny as many experts are skeptic about their overall utility).

Moreover, importantly, the Fukushima Daiichi nuclear disaster of 2011 has triggered major worries and street protests against nuclear energy in Germany, which resulted in the country deciding to do away with nuclear energy, by closing down all its nuclear power plants (an important side note being that previously, after the 1986, Chernobyl nuclear disaster, Italy held a referendum which resulted in closing down all nuclear power plants). This development had and continues to have major consequences on the economic and the general security context (and combination of advantages or/and vulnerabilities for Germany and other countries in Europe) (e.g., [247]). Even the key term "Energy Transition" (which now refers to the systemic transition to renewable energy sources and is a common expression in public conversations about climate change and sustainability in general) is a translation from the German term "Energiewende", which emerged in those conversations after 2011 [248, 249].

Eco-city Projects Must Engage in Dynamics Across Spatial and Temporal Scales: From City Districts and Multi-city Networks of Cities Moving at Different Speeds

A small place can have a great role in a large urban area or network (conurbation or megalopolis). For instance, "eco-restructuring" in the Ruhr area (Germany) started from the small town of Bottrop (see [250]), and it is a common knowledge already that the change of a city usually begins from one of its districts. An extensive discussion about the relation between (eco-)districts and (eco-)cities would not fit in this one book, but I provide a few observations here below and a brief supplementary resumé about it in Module 3 (Sect. 3.3) when discussing the importance of mapping knowledge (and cognitive maps in general) for concrete operations in urban projects. The reason for doing that association (and the main point to start with in this subsection here) is that, once a project team adopts a systemic perspective and arms itself with good methods for evidence-based work, the relations between a city and its districts (and components in general) emerge logically and the relevant elements fall into place.

A first observation to make is that the idea of eco-district is tightly linked to that of an eco-city, because: as long as a city functions as a whole, i.e., as a true city (and not by mere convention whereby several separated settlements are bundled under the same name for some subjective convenience), it is illogic (and improper) to talk about an eco-district without implicitly referring to the city itself as an eco-city.

Sure, a city may begin its own transformative journey in one specific district, for some practical reason, for example a part of the city that is less risky to begin with (if the situation is such that the emphasis must be put on risk minimization) or one that has the highest potential to drive change in other parts of the city (if the power of example or sheer functionality merits the top spot in the list of operational priorities). But the existence of an eco-district pursuing sustainability ideals implies that the city itself is pursuing those ideals, i.e., we talk about an eco-city.

At the same time, if a city is already engaged as an eco-city, each and every district of that city has the potential to be an eco-district but cannot be called a proper eco-district before that district can prove it is part of that engagement (i.e., before eco-city action involves that district). In other words, we can talk about a certain city as being an eco-city without necessarily talking about all of its districts as eco-districts, but we cannot talk about a city district as an eco-district without talking about the city (to which that district belongs) as an eco-city.

The observations above are just logical deductions from the reality that an eco-district is included in a city but not vice-versa (except when a city has only, or functions as, one district). But, the consequences are very practical: every eco-district project must also be able to convince (the experts and the stakeholders) that the city incorporating that district qualifies as an eco-city. If efforts exist towards an eco-district but the city-as-a-whole does not qualify as an eco-city, at least two logical conclusions can be drawn right away: (1) those particular efforts are under-calibrated and must be recalibrated to cover the city as a whole, and (2) if that is not possible (yet), then those efforts do not represent work "in" an eco-district (and their outcome cannot possibly be an eco-district) but "towards" an eco-district. Until this happens, we are talking about initial efforts and explorations towards eco-city (and eco-district) ideals. In other words, not every "greening" and "depollution" or other general betterment of the urban space represent a sufficient argument to qualify a city as an eco-city, and therefore reducing the discussion to "smaller targets" (perhaps as in "win small" strategies) is not a real solution, simply because a city is not the sum of its components but an objective complex dynamic system.

Put differently, a city is not a sum of buildings (nor a predictable machine), but a "living system" (actually a "system of systems") with its own dynamic and character that must be studied, understood (as much as possible) and respected. Failure to do so, can only result in naive projects which will either fail or create more harm than good (and undermine the proper development of that city on longer term), i.e., they will stray from a useful path.

Key Requirements for Eco-city: A Non-exhaustive, Start List for Future Projects

The complex process of transformation of cities towards urban sustainability involves some key requirements.

Firstly, there is the primordial need of knowledge integration for decision-making, which amounts to a "re-inventing" the relation between science and society in the twenty-first century. The best way to do this well is through experimentation, i.e., learning-by-doing and doing-by-learning (as already mentioned in Module 1).

Secondly, any trajectory involving local experimentation towards community sustainability inevitably goes through public agendas (and policies) and requires efficient decision-making, the technical purpose being to prevent unfettered accumulations of systemic risks that can converge in "perfect storms". Obviously, risk and disaster managers have always been at the forefront of sustainability issues at large and small scales; using the insight and professional experience of this domain to help generating city sustainability scenarios (e.g., by framing systemic priorities in terms of risk management) appears as both logical and practical.

Thirdly, the reality of extensive and fast urbanization means, as already noted in Module 1 of the book, that the global "battle" for a successful transition to sustainable development will be "won or lost in cities". To date, there still is no critical mass of eco-city developments around the world, and not even in rich countries. Given this reality of facts (unchanged for at least a decade since I started to work on this book, in spite of the many lessons learned in the meantime), and given an even greater (now-burning) urgency of breakthroughs in the transition to sustainability, we have a societal need to find the means to accelerate progress in terms of (1) experimentation and (2) learning from existing insights and case studies.

These scenarios suggest that an attitude where cities in countries from Central Europe that are newer in the EU (compared to older ones), would content themselves to simply "execute" priorities coming from the European Union's institutions will not help to bring those cities (Jassy included) very close to the aim and vision of being a "City of Inspiration". For that to happen, the city would have to take advantage of the unique context where the transition to the open society and the market economy already overlapped (and now also in their cultural patrimony) with the transition to sustainability. This type of experience is unique to the eastern side of the EU and can fuel developments that might not be possible in the already rich EU countries—where change is costlier and overall harder to do.

This is not to diminish in any way the importance of the first moves that have been taken in view of modernization and joining the EU project which involves continent-wide deliberation and co-operation. Being part of this is only right and the past programs that were done with the direct support of the EU have helped to build momentum and to develop options. But these actions need to be complemented with an even more powerful approach. Economic convergence, i.e., the tendency of the EU's poorer countries to grow faster than the wealthier countries is a reality in spite of the many regional and topical details [251, 252]. Now, it's less a matter of catching up and more a matter of developing a capacity to act as engines of the EU as a whole (and knowing that, at global levels, this "EU talk" can be translated, and can in principle inform various forms of multi-country cooperation on other continents). This must incorporate both the shared aspirations of a common continental culture and the local identity of every target city.

Freezing attitudes in a “catching-up” mentality (dominated by a fear to take initiative at the front) would be systemically harmful and morally inadequate. On longer term, this would make any city void of self-reliance and incapable of veritable progress in the mature cooperation-and-competition (or “coopetition”) contexts of the continental and global arenas.

Importantly, the method discussed here embodies an all-integrative approach: its power is not dependent on specific risks, but on the capacity of a local community to reflect upon itself and to organize itself.

By design (and because it emerged in the context of complex pioneering projects), the method is meant to be intrinsically versatile: it simply feeds on data, information and knowledge that is available from experimentation-oriented initiatives. It is specifically intended to help a diverse/complex constituency of people to find its way forward. In that respect, it is also a continuously evolving tool in itself, thus embodying the fundamental process of experimentation for transformation. For example, the method can be enriched in the deliberation steps with whatever complementary method the team using it wants to apply for stirring and managing the deliberation process; and, future descriptions of it will likely use all references that are available at that moment in time. By analogy, this method is like a person (or team), which becomes better with experience and learning while still remaining essentially the same.

In this sense, from a project management perspective, one can regard this method as a management and learning tool to be employed by local communities experiencing very diverse contexts and preferences.

Finally, as high-density and high-activity human settlements, cities embody organically different research and knowledge-action perspectives. As illustrated in Fig. 2.17, with one among countless other possible examples from virtually every country of the world, cities are more about socioeconomics (“internal affairs” of humanity), while city hinterlands are more about “nature out there”, and the link between the two can be easily forgotten. Competent knowledge-action models, however, understand both perspectives and the real (easily visible or not) connections between the two.

2.6 Cities of Inspiration: A Few Ideas

The Long-View: For People Every City is First and Foremost a Matter of Identity

Cities are hubs of human interaction across spaces and times. It is a fact-of-life that many of the notions used in the literature in relation to cities and sustainability are often debated, supported or contested. This reflects (a) normal updates in available information and knowledge, our professional vocabulary and communication tools being continually refined and (b) a normal diversity of perspectives.

Obviously, this semantic richness should not divert our focus from the operational aspects. On the contrary, it represents a natural motivation for us to stay open and use critical thinking, and then to be ready to update our knowledge and methods. We need both a stable core of shared views (so that we can "speak the same language" with our project partners when trying to develop new solutions) and a sufficiently dynamic environment of search and creativity (so that we avoid stasis and decay). As already mentioned in Sect. 2.1 (text and Table 2.1), managers call this ability "ambidexterity" while scientists see it as a form of "dynamic equilibrium". In general, this ability represents a main characteristic of any system that persists in time and flourishes.

For projects, this has two immediate consequences. The first is that concrete work on eco-city-related solutions is not about seeking definitive answers but acquiring enough reference points to be able to operate, and about harnessing enough human competence to generate viable solutions while remaining open to progress. Thus, the concept of eco-city is not some universal panacea for solving "by miracle" all there is bad in the world; instead, it should be understood as one of the best (or even the best) "opportunity train" to a better future, and the cities that embody this notion successfully should be regarded as practical sources of inspiration for any other city that seeks to cultivate their own dynamic equilibrium towards sustainability.

The second consequence of the semantic variability around eco-city projects is that the so-called "staying power" of a city across historical eras actually depends on the existence of a strong identity of that city. Indeed, this is the starting point of any discussion about how a future "eco-" version of a city could look like and about finding which among those versions have the highest changes to materialize.

In our discussions about insuring that we talk a common language with our project partners, identity is the very first common denominator, and the unifying force that keeps together a constituency of people. Ergo, that identity is the concrete human basis on which cities can develop viable real-world eco-city projects, and constitutes the bedrock and value core that allows and enables a project to succeed.

Some years ago, I and a few colleagues [253] have proposed in a conference talk in Ghent (a historical city in Belgium), that this "something that unites people" in a city can be adequately called a "*spiritus loci*" (i.e., the spirit of the place) and that it is a key factor in the local transition to urban sustainability. In the literature and in daily speech, this notion is often called "sense of place". In concreteness, this *Spiritus Loci* can trace its existence from very different things, from some symbolic objects that are known in the classic anthropological studies (like a tree or a stone of special meaning for the local people that know about it) or even religious or semi-religious beliefs (like the notion of *Genius loci* of the Ancient Romans or "*the way of Kami*" in the Shinto tradition of Japan) to more abstract symbolisms of modern life (like a recognizable city skyline or elements thereof, for example the Eiffel Tower or the Sydney Opera House). Furthermore, it can even take the form of internalized cultural abstractions like "*Ich bin ein Berliner*" ("I am from Berlin") or "I am a Glaswegian" (i.e., I am from Glasgow; "don't mess with me"). It can also be strongly associated with historical personalities. For example, the small-medium city of Târgu Jiu in Romania, identifies a lot with the spirit of the father of modern sculpture Constantin Brancusi, not least because some of the original works of this artist are local monuments like the *Endless Column* (or *Infinity Column*), *The Table of Silence* and *The Gate of the Kiss* [254, 255] which together constitute a major component of the global patrimony left by the artist around the world (see also [256–258]).

This local identity (in the world), when it is well-developed, be it in some abstract form or as embodied by "sacred" or/and monumental objects or/and mythical events effectively serves as an operational mental reference in the everyday life in that local community—or, as the historian of culture Mircea Eliade called it (by generalization of a traditional belief of Australian aborigines), an *Axis mundi*, i.e., axis of the world ([259], see also [260]). In simple worlds, for us humans, the world itself essentially is (and turns around) the city/community in which we live.

A tale of two neighboring cities

Importantly, this shared characteristic personality of a place tends to be stronger in old and historic cities/towns/villages: people are proud to be a part of that place, and will defend the prestige and quality of human interaction in that town. And, it tends to be weak or completely absent in newly-built cities and towns: people regard those new settlements as "just a place where they go to sleep, or to work", i.e., without any particular meaning. As it happens, I can illustrate this point with my own experience is two neighboring towns that display a clearly noticeable difference: in the historic city of Versailles, people will often salute you on the street even if they do not know you personally, and a driver would wait a pedestrian to cross the street even when the driver has the green traffic light (and would courteously invite the pedestrian with the hand like "Please! Really, it's no problem"). By contrast, in the neighboring town of Saint-Quentin-en-Yvelines (also known under the abbreviated form SQY), which is one of the "New Towns" (in French "*ville nouvelles*") that were built in the seventies and the eighties following the decision of the president of the country to "decongest" the fast-growing city of Paris by establishing a ring of new suburbs in a mostly rural countryside) … drivers would often honk impatiently while waiting at the traffic lights and behave as if they will drive over you on the spot because their time is more important than yours. The general impression people give in the "new town" is that they do not feel like they are in their town or simply feel no connection with it and its people. In fact, this is also true to a large extent, because the new town is the second largest business/economic concentration in the Paris region after the *La Défense* business district of Paris, so many of the people using its roads are actually daily commuters who only come there for the work. Now, while reckless and gentle drivers can be found anywhere, the stark difference in frequency of the two behaviors (and general atmosphere in each of the two town) essentially tells me that: what drivers would basically never do in Versailles they do feel free (even eager) to do in this other town (which they clearly do not seem to respect as a place of its own).

This different "feel" of the town comes with other practical details. The new towns have some clear advantages (notably housing is cheaper and usually more modern), but the general aspect of these towns strikes as artificial and often in line with what in architecture is known as the "brutalist" style. And, I hurry to add that this is not just about my personal impressions but it is a well-recognized fact by the people in the town (and by its local authorities, with which I did have the chance to interact in various projects and activities, as I did with local authorities in Versailles) and it is also what external visitors (usually from Paris or from other universities around the world) observed when visiting one of our university campuses located there.

This is not to say that the situation in these new cities (and especially in SQY) does not evolve for the better (actually it does!), and it is not to say that beautiful places do not exist in these new towns (they do exist!) but the difference (in terms of town identity) is very visible and the comparison itself is a great opportunity to understand how cities come to be what they are.

Further, the purpose of such an intuitive and analytical comparison is also not to suggest that there is no meaning whatsoever in such cities (there is!) but to point out that, in such new cities, the character of the place is in the making, and that this is in itself a very interesting process that students of cities can witness with their own senses (and, to which they can also participate), and that such analyses and comparisons can help us understand the profundity of a city's identity instead of taking it for granted and treating it just as a curiosity.

In the concrete case of SQY, the existence of different citizen associations is a proof that the city has a "growing soul". Occasionally, I have participated as a guest in such associative meetings of some of the locals that are united by a common desire to protect the local natural and cultural patrimony and to undertake voluntary citizen actions (like cleaning urban forests and lakes from debris) and engage in science-society interactions. I was of course happy to contribute (along with other colleagues), not the least because the invitation came from former students of mine and also because our institutional interactions with SQY was also related to the local *Agenda 21*, the famous UN's non-binding action plan with regard to sustainable development that came out of the 1992 UN's Conference on Environment and Development (the "Earth Summit") held in Rio de Janeiro, Brazil.

Also relevant for this discussion are a few details I can witness about SQY. Firstly, the 2-year exercise of scenario building that we have done in our eco-innovation cluster for Versailles Chantiers (as discussed in and around Figs. 2.4 and 2.5 and Table 2.5) have been preceded by a 2-year exercise (and case study) which we have done for the SQY as a whole, and the SQY city hall worked with us and provided guiding and hosting for a touring of the city for our students (and me), and we also had some other interactions in different projects, both local and at the EU-level. So, the concrete development of method discussed here has one of its roots in SQY and benefited from the interaction between our innovation cluster and SQY. Secondly, the establishment of the social sciences campus of our university over 30 years ago had a great contribution from SQY (starting with the land itself), as the local authorities and the people knew very well that a University campus is a great asset for a new city (and our subsequent interaction with SQY also included some of the very people who were in charge with the development of the city in the previous decades, so we had a great source of historical and operational insights). Thirdly, this campus itself is one of the beautiful places in this New City with a young and growing character.

When I first arrived on the campus, Keith Culver, the former Chairholder of the then-newly established Econoving cluster (now at University of British Columbia), took me for a walk in the campus and the urban park nearby (called *Parc de Sources de la Bièvre*, meaning the *Park of the springs of river Bièvre*, as the actual springs are there in the park, right above a string of lakes that continue down the river in a forested landscape). I admit, I was charmed by the sight in addition to being eager to start the new and interesting work projects. In addition, the original conception of the campus takes full advantage of a great perspective on the city and on the surrounding nature, with a carefully designed pattern of alleys, green spaces and buildings which another colleague and friend, Jean-Paul Vanderlinden, likes to point out proudly (also to his students at the beginning of each university year) that, when regarded from above (e.g., on zoom-ins of satellite images) looks like a great piece of art. Last but not least, the new city of SQY incorporated and urbanized the historical core of a village called Guyancourt, which now constitutes one of the much-appreciated historical components of both the university campus and the city as a whole, one that actually relates in terms of historical and present realities with the city of Versailles. As I often discussed in the past with yet another colleague and friend much more experienced than myself, Jean-Luc Vayssière, when one does work and looks closer at the academic and sustainability history and possible realistic futures, these two very different but complementary cities have a fantastic potential. Unsurprisingly, the broader picture of infrastructural and functional developments led by the national government and the public authorities of the Greater Paris region (known locally as "Ile-de-France") include both Versailles and SQY as major hubs.

However, the consolidation of the personality of a city takes time. The most important aspect of this spirit of the place is that it develops naturally and therefore needs to be cultivated. It cannot be fabricated, and it cannot be "copied" from somewhere else. To try and speed up the evolution of a city as an eco-city, all kind of problems from cities around the world can serve (if studied) and inspire much more than "ambitious plans" that are all-too-often disconnected from daily life. They can help a city advance for good.

Therefore, the natural desire we may have that eco-cities become "ubiquitous" [261] needs to be in accord with the nature of cities as complex socio-economic systems and as places that have emerged from and for human interaction. Specifically, we need to avoid falling into the trap of the "inherent inflexibility of a closed-system approach" used in the "pre-determined" city conceptions that seem to be adopted by some government-appointed planners, the result of this approach being a set of new cities that have been designed from scratch (instead of emerging from natural interactions between people). The core issue with these "designed", "by demand" cities is that they have been "produced" under the hasty presumptions of some narrow and commercially overhyped engineering perspectives (and group interests). By definition, and for all the reasons summed up in this module, such "manufactured" cities are "not sufficiently resilient to external shocks" ([262], for comparison, see also [263]) and cannot possibly be so.

This is in fact part of a broader issue-and-opportunity with all "new cities". Some planning is necessary, and history knows many cases of very prompt initial planning of cities (although those initial plans have been inevitably adjusted in the following time periods), from Alexandria in the Hellenistic Egypt (e.g., [264]), to Constantinople in the Roman Empire (see e.g., [265]), Odessa (see e.g., [266–268]) in the tsarist empire, the city of Washington DC (see e.g., [269]) in the USA, and through to present "new cities" in modern states). But one must not mistake the first plans (as an initial impulse) for the real development of a real city (which is in fact made by its people, i.e., generations after generations, during long periods of time). Indeed, the famous phrase "Rome was not built in a day" applies to every real city and complex human system, and bears profound insights for urban systems (but requires us to look beyond appearances).

This also means that "that initial impulse" might be necessary but must not be overdone: planning needs to be limited to the minimum necessary because every bit of unnecessary planning made at the beginning will translate in practice as a limitation of a city's space of options (in other words, it will "suffocate" the city later on, or will otherwise impose very high costs on the evolution of the city).

Needless to say, under-planning can be equally harmful and can have similarly negative effects, especially in fast-growing cities. What is required is finding the right balance, which further requires adequate understanding of the situation at hand.

Of course, the today's technological, institutional, and economic means can greatly improve and accelerate some processes, but they cannot change the nature of a city as a complex dynamic system. Instead of irrationally trying to "manufacture" a fully-grown city from the very beginning, local communities (and all the relevant stakeholders that want to stay on course to sustainability), have to essentially equate the notion of "eco-city" with that of a "better city", and to acknowledge the fact that the only truly viable path to the future (and the only path that is morally justifiable for future generations) is the path of sustainability—with all that this entails.

For this to happen, therefore, a city needs to understand its own soul and re-invent itself intelligently around its identity core. Importantly, this is not at all a matter of "activism", but a matter of mature introspection and competence at the level of the local community. The first of the two aspects (in-earnest collective introspection) means that, to be able to reinvent itself, a city must look in the mirror and be honest with its collective self. The second aspect (practical collective competence) means that such a city is only viable on long term if it constantly develops its overall attractiveness and its capacity to be a true unifier of people who respect each other as a true community that is tied together by a common "spirit of the place", instead of being just a spatio-temporal co-incidence of individuals.

In this sense, and to add another avenue for study, the experience I evoked earlier about driver behaviors, and the matter of local identity more broadly, can be looked at through the classic debate around the Broken Window Theory (BWT) versus the Collective Efficacy Theory (CET) [270–277]. Although it originates in criminological research, this debate brings into the conversation a series of other aspects like urban planning, urban demographics, urban security, urban health, as well as education and management in a broader sense (and the list can continue). This is why I suggest that you may want to check the references cited above as a bridge to further knowledge and experiences about "citizens and their cities". In the broad picture, one also observes that problems that would tend to be regarded as "problems" (i.e., undesirable situations) or perhaps as "uninteresting" can actually help us better understand what is happening on the streets of a city and thus can inspire us in very practical ways.

A related topic which I will only mention briefly is the wide recognition of the fact that culture itself is key in any realistic discussion about sustainability in general, knowing that the spirit of the place that one finds in real human communities (as opposed to just massive aggregations of a high number of individuals) is yet another reason why cities are central for the transition for humanity to sustainability.

Therefore, I encourage anyone who is or wants to get involved with eco-city projects to gain some familiarity with the conversations related to the topic of "cultural sustainability" (and the strongly related topic of "institutional sustainability"). In that respect, I provide below a short list of references that you might want to start with, but I also suggest that (1) you do your own searches in the available literature so as find those works that are the closest to your own kind of context and priority interests and (2) read all of those with an always-critical eye:

- Hawkes [278] even talked about culture as a "fourth pillar of sustainability". Although I regard this expression as a poor choice of words made back then in 2001, clearly reflecting the now-outdated "pillars" perspective on sustainability (as already discussed in Module 1, Sect. 1.2), the point with this reference is that culture was correctly identified as fundamental for sustainability. In the systemic perspective used in this book, culture is logically part of how social systems function, and is therefore dynamically connected to the ecological and economical systems as well.
- References [279–283], and others in the literature discuss the role of culture in the evolution of thinking about sustainability in cities of all sizes.
- Reference [284–289], and others in the literature discuss institutional sustainability for concrete management.

A thought about arts—the sound of cities and the cities in sounds.

More broadly, the profoundly historical character and identity of a city reveals itself in many ways, and this is why one needs to know well a city before attempting to help it become an eco-city. Urban systems evolve based on objective pulls and tensions and according to how people respond to challenges and opportunities, and what matters in the end is the life quality of its inhabitants and, as part of that, the place this local community has in the global concert of cities. Identity, after all, is not given by mere words; instead, words should better reflect what a local community is.

In that sense, a look at the city from different perspectives reflecting how people live in the city and see the city (or town or village) is often much more informative than officially ascribed attributes.

Some instances of visual and other arts have already been evoked or will be alluded to in the rest of the book (for more about the relation between sustainable development and arts, as well as communication and education through art in local communities, see, e.g., [290–299]).

But here I find it interesting to make a special mention about the audio aspect. For instance, one needs only look at how pop culture reflects particular cities as seen by its inhabitants (e.g., New York, Paris, Napoli, etc.) or as broader symbols for the national and international culture (e.g., Nashville and New Orleans) or cities in general (Box 2.4).

Like art in general, music is profoundly connected to cities, so that we can talk about "music and urban geography" and "music, space and place" ([300, 301]).

Box 2.4 Lyrics of the pop song Downtown, interpreted by Petula Clark, 1964 (text written by Tony Hatch) ([302], for a video, see the links therein]. Source: public domain

When you're alone, and life is making you lonely
You can always go
Downtown,

When you've got worries, all the noise and the hurry
Seems to help, I know
Downtown

Just listen to the music of the traffic in the city
Linger on the sidewalk where the neon signs are pretty
How can you lose?
The lights are much brighter there
You can forget all your troubles, forget all your cares
So go downtown, things'll be great when you're
Downtown, no finer place for sure
Downtown everything's waiting for you
Downtown, downtown

Don't hang around and let your problems surround you
There are movie shows
Downtown,

Maybe you know some little places to go to
Where they never close
Downtown,

Just listen to the rhythm of a gentle Bossa Nova
You'll be dancing with him too before the night is over
Happy again
The lights are much brighter there

You can forget all your troubles, forget all your cares
So go downtown, where all the lights are bright
Downtown, waiting for you tonight
Downtown, you're gonna be alright now
Downtown, downtown,
Downtown
Downtown

And you may find somebody kind to help and understand you
Someone who is just like you and needs a gentle hand to
Guide them along
So maybe I'll see you there
We can forget all our troubles, forget all our cares
So go downtown, things'll be great when you're
Downtown, don't wait a minute more
Downtown, everything's waiting for you
Downtown, downtown, downtown, downtown
Downtown, downtown, downtown, downtown, downtown, downtown.

Inspiring Cities: A Few Other Examples for Starting Work Conversations

At this point, allow me a technical observation about the examples that I am using. In order to insure the verity and the illustrative power of the examples in this book, I am favoring insights that are coming from those cities that I have personally lived in (at least one that I have visited and was able to know directly in some significant ways). When that experience is not available, I favored those case studies that have been reported by published in-depth analyses or otherwise bring about broadly relevant insight for as many readers as possible.

It is a sheer fact of life that, because of known historical differences due to local conditions and levels of economic fortunes, some cities and some world regions are more abundantly present in the available literature, and in the direct expertise that different authors have. This means that in the list of examples used in a book such as this, one cannot possibly do justice to all cities and countries that may provide excellent example and may hold relevant insights. For this reason, this book is only thought of as practical instrument that should help eco-city project managers and participants to start and sustain their work, and thus to assist them in using existing insights regardless of their origin.

In this sense, this discussion sits in the "*global urban: difference and complexity*" paradigm aptly described by [303]: "*The study of the city is in a phase of theoretical reflection and experimentation. There is concern to build analyses of the urban that are informed by trends and experiences of urbanization across the world, accounting for the pace, scale and form of demographic and social change as well as the overall social, economic and ecological significance of the city in global or planetary change. For example, the desire in international policy circles to 'leave no city behind' and to generate useful guidance for development in what are now some of the world's most vulnerable places is a powerful motive for ramping up urban research and for identifying key interventions that could shift cities onto a more sustainable path. There is a widespread consensus that to make sense of massive urban expansion, especially in Asia, and to illuminate lesser understood urban realities, such as those of Africa, the generation of some new concepts and approaches is necessary. Regardless of their philosophical points of origin or methodological commitments, initiatives to understand the significance of an urban world bring to the fore the tension between specificity (or difference) and universality in conceptualizing the urban and in drawing together empirical evidence, at the global scale, for innovation in policy and theory. In the field of urban studies, theory is now strongly shaped by the search for effective ways to understand the nature of the urban across a great diversity of outcomes. We argue that this concern needs to be further prioritized as policy makers seek to illuminate the drivers and outcomes of the global settlement transition*".

Thus, my purpose here is just to try and provide a general understanding and a short list of concrete methods and ideas that can help cities anywhere in the world—and I hope that this can be especially useful to those cities that benefited from lesser dedicated projects in the past. So, I encourage everybody to engage in exchanges with other people, projects, and cities, and to try to discern those general insights that can help cities help each other to evolve and improve. Hopefully, this book and the eco-city projects out there can facilitate the interaction between experts around the world for the benefit of any project and city; can help citizens find common conceptual grounds and figure out where they are in the big picture, and how to get involved in the making of their own city, town or village.

Classic and medieval cities: the emergence of strong local identities.

One prime and memorable example to begin with is the city in the Ancient Greek world, the *polis* (πόλεις or πόλις). The key detail here is that people from different cities were often speaking about each other not in terms of Athens, Miletus and Sparta (to give just a few examples), but as *The Athenians*, *The Milesians*, and *The Spartans*—so much so that, at the height of the Ancient Greek civilization, the term city ("polis") came to mean citizenry (and belonging; basically, citizenship), and a different word was reserved for the area of urban buildings: *asty* (ἄστυ). The term *polis* had actually originated in the Archaic period before the classic Greek civilization, and in time came to include the *asty* and the rural areas around it. Even more, an *asty* was not always present: for example, Sparta consisted in fact in a network of villages (what linked them together was a common spirit).

The Romans made the same distinction, respectively, using the words "*civitas*" and "*urbis*" (which is at the origin of the current word "urban"). Similarly, the word *civitas* evolved in time to mean citizenship (in a sense that was similar to the notion we use nowadays), plus the word *municipium* which was reserved to (non-sovereign) towns or cities inside the Roman world, i.e., local power centers. Of course, some other key words we use in modern times, like "citizen" and "civilization", "urbanism" and the modern day "municipium" and "municipal" come from these respective word roots.

This fundamental reference to the people being more important than its walls of a city was also present during the European Middle Ages, and carried a very strong sense of identity. To walk the talk, I can give the example of my own hometown in Romania (a country name which comes from "Romanus" which comes from ancient Rome). As it happens, this city concentrates much of the historic symbolism evoked above: its name is *Roman*, and comes from the same word root at the name of the country (it is not the capital of the country, but it was a regional capital city during the middle ages). Officially, the city of Roman is also a *municipium*, the older name in Romanian language being *Târgul Romanului* (literally, *Roman Market or Roman Fair)*, the word "târg" being a cognate of the word "torg" in Scandinavian languages, and both mean "market place". This precise meaning is reflected in the names of the city in other languages: *Romesmarkt* (in German) and *Románvásár* (in Hungarian). This triple name of the town reflects the multi-cultural environment in the whole of Central Europe, and comes specifically from the fact that the city was founded at the very beginnings of the Principality of Moldavia in the fourteenth century by a broad coalition of Romanians (at the time also known under the exonyms Wallachians and Moldo-Wallachians, although the people called their language Romanian and knew it was a Romance language, i.e., descending from Latin), Germans (which in the official documents of the time were called Saxons or/and Romans because they came from the Holy Roman Empire/HRE), and Hungarians (because the old Kingdom of Hungary which also included many Romanians and many German "Romans" was for a while part of HRE and acted as organizer of its eastward political and territorial expansion). (If you think that's complicated, you are right, but this is normal in Central Europe, as it is in many other places of the world). In the subsequent centuries, people from other ethnic backgrounds joined *Târgul Romanului*, notably Armenians, Jews and Lipovan Russians fleeing religious or other types of persecution in other countries, and all these newcomers became participants in the city's affairs based on the common identity provided by the belonging to the community of that city. For example, the town's oldest church still in use is known as "the Armenian Church", which was built in the fourteenth century by the local Armenians on a piece of estate that they bought from the local Germans.

The city's old seal recovered from seventeenth-century documents and originating from the fourteenth century bears the Latin circumscription *Sigillium Civium de Foro Romani*, i.e., the seal of the citizens from the (market town of) Roman [304]. But the most important and illustrative detail here is the fact that official expressions like "*sigillum civium*" and related derivations (see Fig. 2.18) are typical with city seals from the European Middle Ages (e.g., [12, 305–307]) because most political entities in the region regarded themselves as continuators of the classic Greco-Roman civilization.

Strictly-historical details aside, the take-home message for eco-city projects and the future of sustainable cities is that we need to understand and heed the general (and dynamic) process of development of local common identities across times and in any context. In other words, identity is the fundamental priority from which all the other start (or Priority No. Zero in the set of issues a city faces), as this reflects what a city is and how it functions. This is not to say that identity does not evolve

Seal of the Citizens of Stavenhagen, now in the federal state of Mecklenburg-Vorpommern, Germany (dated from 1513)

SIGILLUM CIVITATIS STAVENHAGEN

Citizen's Seal of the market town of Roman (/Romesmarkt) in the historical region of Moldavia, Romania (17th century, with origin in the 14th century)

SIGILLUM CIVIUM DE FORO ROMANI

Citizen's Seal from Groningen (City of Groningen, capital of the homonymous region in The Netherlands) (dated from 1313)

SIGILLUM CIVIUM GRUOENINGEN

SIGILLUM CIVIUM KILENSIUM

Citizen's Seal from Kiel, now in the federal state of Schleswig-Holstein, Germany (dated from 1365)

SIGILLUM CIVIUM DE REVALIA

Citizen's Seal of Reval (current Tallin, Estonia) (Early 20th century academic illustration; used on a document dated 1340, during Danish rule)

SIGILLUM CONSILII ET UNIVERSITATIS CIVIUM IN ROTENBURCH

Seal of the Citizens' Council and University of Rothenburg ob der Tauber (1320), in the federal state of Baden-Würtemberg, Germany

Fig. 2.18 Examples of city seals circumscribed with a Latin text including the word *civium* or *civitatis*, from the historic towns in Europe from outside the classic borders of the old Roman Empire but inside or under the political influence the Holy Roman Empire (962–1806), which had Rome as its *de jure* capital (with many other cities as de facto capitals as chosen by different emperors) and regarded itself as the continuator of the old Roman Empire, based on the cultural-religious concept of *translatio imperii* (in spite of the fact that the eastern Roman Empire still existed until the Fall of Constantinople in 1453). During the European middle ages, the word "sigillum" was often abbreviated by "si". Seals used different technologies that were predominant in different eras, e.g., wax, black smoke. *Source* original compilation and arrangement based on six images from the public domain; additional image author attribution for Kiel and Rothenburg ob der Tauber: Wolfgang Sauber, Public domain

in time—of course it does; but identity is the most salient feature and therefore the most important feature from an operational point of view. It is the "who we are (as a city)", the "personality" of that community of people, and no project or program will have a real impact if it is not coherent with that essential core. And, if there is no such core, there is no real "city of people" to talk about.

As one can see in Fig. 2.18 (and in the professional literature), the word *civium* is a common presence in official and symbolic communication as illustrated here with city seals, and this shows well the importance of a city as local community with a shared identity and interest in spite of everything else. The *civium* indicates that the city belongs to its people and has a soul by itself: an identity to be affirmed in the world, so that the city of Revalia means The Revalians, and so on. All in all, *civium* reflects the importance of cities as roots of progress and origin of the human civilization itself and represents the common local identity, which tells the others that in front of them stands a community of people (not just a crowd of individuals).

In Fig. 2.18, one also remarks a rich combination of natural and man-made elements chosen to participate in the symbolic representation of the community of people of the city and their place in the world. The lesson that history has to teach us here is that an eco-city can only be made to exist by its inhabitants, and therefore starts with their own introspection of who they are and who they want to become in the future. Interestingly, heraldic research readily points out interesting elements of identity that could help start the conversations about the relations between a community of people and natural elements and how they see themselves in the world. For example, the seven stars of the constellation *Ursa Major* depicted as white dots on the citizens' seal of the city of Roman (/Romesmarkt) in Fig. 2.18 suggest that the city is at commercial crossroads and serves as a key reference for long-range travelers. This symbolism is not by chance: it corroborates well with known details from that era. Specifically, the first written attestation of the city comes from the Novgorod Chronicle, i.e., from a distant city and land although both Novgorod and Roman were part of the long-range commercial routes between Scandinavia and the eastern Roman Empire (also known during that time as "the route from the Varangians to the Greeks"; a rare Viking helmet of seemingly royal origin is known in archaeology to have been discovered not far from the city; there are Viking stories telling about interactions with the Wallachians in the area). The city was and still at crossroads between main North–South and West–East trade routes, suggesting a huge long-term potential as a center of industrial, economic, and cultural exchanges.

Similar stories of cities playing a role in travels and exchanges are being evoked by elements used on other city seals as already visible in the compilation shown in Fig. 2.18: waters and boats on the seal of the city of Kiel denote its character as a main sea seaport; other natural elements on other seals, such as flowers and animals (e.g., lions and the Roman Aquila) stand as references to various political contexts. Unsurprisingly, military symbols are also common in the epoch but they evoke the character of the community: watchtowers denote attentiveness and preparedness, while shields refer to wisdom as ultimate protector, in a tradition that tracks back to the Ancient Greek symbolism of Athena Promachos (goddess of wisdom in battle), which was regarded as the protector of the city of Athens (and of many other cities and colonies in the Greek world) and was often shown as holding a shield.

Human frontier cities, lost and found

History is replete with inspiring examples for present understanding and deliberations. The examples in Figs. 2.19, 2.20 and 2.21 (to which we could add a long list

Fig. 2.19 City of Khara-Khoto (1032–1372) in the Gobi desert (western Inner Mongolia, China). *Source* Original compilation of photos from public domain (with credits) as follows. **Top**: the west wall [312]. **Bottom left**: Ramparts and five stupas (Buddhist hemispheric structures containing relics) from the Yuan dynasty (1271–1368) [313]. **Bottom right**: city plan (1923) by Russian explorer Pyotr K. Kozlov [314]. For more information and images, see ([315] and the link therein)

of other places) represent some open history lessons for us to understand how cities work and how they may fail to sustain themselves as communities of people.

To feed the conversation, I would only remark that there is a rather interesting detail connecting all these three examples: these were all great achievements of the human spirit (as embodied and still visible in their physical making) have at least one attribute in common: they were all special-purpose communities driven by one factor that dominated all the other aspects—one that constituted the source of both the establishment and demise of the city.

In the case of Khara-Khoto, this one factor was the commercial power conferred by serving as a main node of exchanges between different people and along the old Silk Road: an opportunity that compensated for the difficult location in the desert but one which was proven insufficient against an eventual combination of social and environmental drivers. The city survived and prospered under the Mongol rule of Genghis Han and Kublai Khan because of a convergence of favorable social

Fig. 2.20 Historic "urban experiments" that are currently UNESCO world heritage sites and treasured windows into human urbanity: Sigiriya, Sri Lanka. *Source* original compilation of photos from public domain (with credits) as follows. **Top**: Aerial view of Sigiriya rock [316]. **Middle left**: The lion gate and climbing stretch [317]. **Middle right**: Sigiriya rock from above [318]). **Bottom left**: Sigiriya gardens at the foot of the lions rock [319]. **Bottom right**: Garden complex as seen from the top of the lions rock [320]

Fig. 2.21 Historic "urban experiments" that are currently UNESCO world heritage sites and treasured windows into human urbanity: Machu Picchu, Peru (note that the name of the place by the ancient Inca was probably Huayna Picchu [321]). *Source* Original compilation of photos from public domain (with credits) as follows. **Top**: Overview of the main part of the old city ruins [146]. **Middle left**: Original terraces for agriculture and landslide prevention [322]. **Middle right**: Room of the Three Windows [323]. **Bottom left**: A rough map of Machu Picchu [324]. **Bottom right**: Temple of the Sun ("the watch tower") [325]

and economic factors, but the same city wasn't able to withstand a convergence of unfavorable social and environmental conditions and was abandoned (according to legend and available documentary sources, it is likely that this happened when it lost vital control of the surroundings, notably the access to its main water source: because of the military and civil forces of hostile political-military power (namely the Ming

Dynasty of China purposefully diverted away the course of the river Ejin, on which the city depended (see, e.g., [308, 309]).

In the cases of Sigiriya and Machu Picchu, the main driver was political: certain rulers were able to mobilize enough resources to design a place at a remote location that provided a strategic advantage but which required such a great amount of resources to sustain it that the city had to be abandoned when that stream of resources dried out. However, these two "royal court" cities survived long enough to embody and testify for a great deal of knowledge and craft of the people who made it. Unsurprising but interestingly, water was also a main feature as a main aspect in these two cities: both cities were located in (tropical) climates where water was abundant and each city is characterized by fascinating works of engineering aimed at managing water for the use of its inhabitants.

And another aspect connects all the three examples: their remote location meant that they existed at the edge of the oikumene (the inhabited areas; the "civilization" in a common speech), and their high degree of isolation was both a reflection of hard bio-geophysical conditions and a fundamental vulnerability because it diminished its options for situations when may have not be envisaged as urgent during the initial planning but which dramatically reduced the planned city's space of real options, and, in terms of the intrinsic functioning of the city, it reduced other systemic properties, notably its evolvability, potential, resilience, and ultimately, livability (for a technical enumeration with definitions, see Table 3.3 in Module 3, Sect. 3.4).

Then, the way present generations look at these ancient cities is also telling about their legacy and continued capacity to inspire.

In the case of Khara-Khoto, current tightening of relations between Russia, China, and Mongolia includes the development of new transport infrastructure (railways and roads) along some of the old routes of the ancient Silk Road that include Khara-Khoto. This helps explain the renewed interest of state authorities in reviving this old city as a reference point in international trade, which (beside other geopolitical considerations which need not be discussed here) proves that the city continues to inspire and, judging after the extensive (and therefore surely costly) excavations and work carried out at the side, this historic site is on track to becoming a major tourist attraction and perhaps more than that. From a historic-urbanistic and geographical standpoint, this is good news. In this context, one observes that, if this locality has a new future, this comes in the middle of the global challenge posed by climate changes, the likely effects of which will include even harsher drought of the area and an expansion of the Gobi Desert (a phenomenon which already touches on the area of Beijing, with sandstorms affecting the city itself [310, 311]). As this is surely known by the state authorities, one can reasonably imagine and presume that the current/intended program for Khara-Khoto includes some form of environmental engineering that would transform the city and its immediate surroundings in an oasis in the middle of the Gobi Desert. In principle, this would require a great deal of effort, basically, it could involve the development of a new, modern city focused on commerce and tourism around the ancient Khara-Khoto as historical city quarter (just the same way many cities around the world have a historic city center coexisting with modern city districts). I don't know whether this is really the case, but this is what this city and the available public information inspires me to think about. And even if this is not true (for objective reasons that the people in that area, the public authorities and the entrepreneurs on site would know), the very fact that this old city elicits such a great interest is a proof that a city can continue to inspire humanity even after its former glory was lost, and thus continue to exist in the spirit of the people around the world. Besides the archaeological site itself, the historical artifacts (notably book, paintings, and various materials that are being conserved in museums and are still studied) and the presence of all references that exist on the Internet, show that the city continues to live. Somewhat paradoxically (but in truth very logically), the fact that a city that existed 700–1000 years ago is present (at least indirectly, through current generations) in what we now call the Internet-of-Things (IoT) and the Internet-of-People (IoP) shows that a city actually has a future. How this future might look like depends on the people who want this great city of the past to live again.

Space—the final frontier

Now, if one looks around the world, there is a long list of other cities that were "lost" and sometimes rediscovered (at least their ruins) in less accessible areas in our planet's deserts, jungles or/and mountainous areas, across all climates—i.e., at the margins of our *Oikumene* (the sum of all inhabited areas of the world). All of those "frontier cities" represent windows into the complex dynamics of human settlements in their natural, social, and economic contexts.

In perspective, these lessons from "the edge of Humanity" so to speak are also quite relevant for the sustainability of future human settlements on other celestial bodies and in the outer space, notably Moon and Mars, knowing that NASA (the North American Space Agency), ESA (the European Space Agency), RSA (the Russian Space Agency), ISA (the Indian Space Agency), CSA (the Chinese Space Agency), JSA (the Japanese Space Agency), and other relevant institutions already have long histories and further plans for un-/manned explorations in space.

Of course, the stakes of sustainable extraterrestrial settlements ("colonies") is not only about learning our way in Space: it goes at the heart of what humanity is and how it functions, and finding our way in space and other celestial bodies ultimately helps us advance sustainability of human settlements on Earth. Therefore, together with some of my colleagues OVSQ (a European hub in planetary and space science and technology), I have begun to test the application of the Delphi-based method described above to the question of how to make sustainable colonies on Mars. At the moment of finalizing this book, further work is required before drawing detailed conclusions about this work thread but the central themes (/expected drivers) that have already emerged are not surprising: *Governance* (how people will manage to organize themselves to survive on Mars), *Technology* (including industry and on-site manufacturing capability, with implicit RDI capability), and *Justice* (including equity, power and access to resources), followed closely by *Health & Wellbeing* and *Mars-Earth Connections* which depends on the 26-month (730 days) cyclicity of the "lowest distance between the two planets" (which operationally translates as "minimum-energy launch windows") and which determines the logistic details.

With this "Human life on Mars" exercise in mind, any old or new human experience in harsh environments on Earth (e.g., deserts, sub-/polar regions, high mountains, and other isolated locations) look much more interesting and valuable than usually thought of. And thinking concretely about human life in the outer space is a sobering and informative exercise about what it means to be human.

Africa as cradle and future of Humanity

A topic that I estimate most important to bring to the front is the present and future of African Cities because the African Continent is so complex that it can tell by its own the story of Humanity.

This idea might sound somewhat surprising for some, so here below I will explain the rationale. (This also gives the occasion to at least touch on some aspects that cannot be fairly addressed in this book but matter for the entire discussion).

First, I think it is important to avoid the mistake of associating sustainability and eco-cities only with the intellectual tradition of the twentieth-century environmentalism in the most industrialized (and economically developed) countries. Sustainability is equally about knowledge and action, and the people on this continent were doing it long before the United Nations recognized the concept of sustainable development. The *Green Belt Movement* was established in the city of Nairobi in 1977 (and then across the continent) as energized by biologist Wangari Mutua Maathai (later also a Nobel Prize laureate for Peace) and the *National Council of Women of Kenya* who used scientific and intuitive/tacit knowledge to discuss with and mobilize tens of thousands of women who acted to restore degraded land with their own hands (by planting trees and by standing up against business-as-usual social pressure), and they did that in spite of their poverty and all the difficulties that people in the developed world were not (and still aren't) even capable of imagining. These people directly and organically addressed the relation between the degradation of ecosystems and the harsh socio-economic issues faced by people in their local environment. In fact, many of the local solutions for addressing common resource pools (alluded to in Sect. 1.2), come from hands-on knowledge-action by local communities around the world, Africa being a perfect representative thereof.

This highlight of the African continent might also appear surprising to those who look at the unsustainability challenge through the lens of the human demographics and point out that East and South Asia concentrate over half of the world population—and therefore, they believe we should look there first. Here, I need to observe that each region of the world has its particularities to be understood and addressed competently, but the action must be taken everywhere at the same time. At the essence, it would be wrong to place the weight on demographics because everyone has the right to live like anybody else, because (like with the other species and as per elementary biology) the population of *Homo sapiens* is in fact self-limiting and will eventually stabilize. The key issue is not how numerous we are but whether or not we act intelligently on this planet.

Another related demographic aspect is the very fast growth of African cities per se. At current rates of urbanization, it has been estimated that out of the total (of over 2 billion people) increase in the world's urban population until 2050, about 90% is expected to occur in Africa and Asia [326], and modeling exercises by the United Nations indicate that the world's most populous cities by 2100 will be, in order, Lagos (in Nigeria), Kinshasa (in the Democratic Republic of Congo) and Dar-es-Salam (in Tanzania), [327]. Consequently, all urbanization phases and aspects are concentrated in a short span of time (see also [328]), which multiplies and intensifies challenges and opportunities at the same time. To put it in everyday words, at least in terms of cities of the future, *we need to look where the ball will be (more than where it is right now).*

This temporal perspective also reminds us of the past, and current knowledge tells us that, in spite of some misapprehensions that might have been caused by political interests in the colonial era, sub-Saharan Africa too (not just the southern shores of the Mediterranean Sea) has a rich urban history that precedes known European influence ([329], see also [330]) and all this cultural heritage can serve as a fantastic fountain of inspiration for the future—if studied through coherent, future-oriented projects that would seek to unveil and understand this patrimony as a resource for the African cities striving to re-invent themselves as eco-cities based on all that is known and possible. In this sense, I would dare to estimate that some collaborative African-European projects (pooling financial, human, and cultural resources from the two neighboring and historically deeply connected continents) would be a logical next step for a new era of international cooperation for the cities of the future. Of course, while such programs would acknowledge that Africa and Europe are natural partners, they should remain open to other potential participants from anywhere in the world—based on professional merit and provided that the fundamental sustainability criteria described in this book are understood and respected (starting with democracy and human rights).

Surely enough, this idea of giving the priority spotlight to African cities (and more broadly to any given continent or to cities) might also appear surprising in the eyes of those who estimate that we should keep "development" and "nature" somewhat "separate", the underlying idea being that economic development and nature protection aren't really compatible, and therefore the key to everything would be the protection of the "unspoiled nature" from the least anthropized parts on any continent. To answer to this kind of worry concisely, I would highlight again that nature protection and the conservation of habitats is a necessary-but-insufficient condition for sustainable development, and finding solutions is a matter of developing knowledge-action options that recognize that reality that humans are part of nature, not separate from it. This is what the Green Belt Movement has demonstrated so humanly beautifully.

To round it up, (1) I believe that the idea of this sub-section is not very surprising for those who look at the issue from a holistic, global, and practical perspective, and (2) the central point is that, because of the great complexity of situations one finds in Africa (basically a spatio-temporal mesocosm of the whole World), the African cities are representative for the essence of urban systems and constitute a first-rank source of useful insights for eco-cities anywhere.

Then, the fact that Africa is one of the understudied regions of the world in terms of urban systems and sustainability makes it even more important for all of us on this planet to acknowledge the need for more cooperative projects between African cities themselves, and with the most resourced cities and institutions around the world, with two natural observations: (1) this is not to exclude cities from other continents but simply to stress that we have a common interest to include African cities in our international projects, and (2) universities have a central role to play in such interactions, programs and projects.

As a collateral observation, for those who are looking for ways to materialize that new social contract for science and scientists for our transition to sustainability (mentioned in the first two sections of this module, and again in the next module, Sect 3.7), here sits a great work program: finding solutions for African cities as a direct way to find solutions for us all.

Further, I provide a selection of references that can serve as entry points for exploring the literature on African cities on different topics (and as potential inputs in the use of the methods and the models provided in this book, and for future projects):

- the complex links between urbanization and socio-economic development, e.g., urban morphology [331]; urban–rural interfaces [332]; the challenging relations between urbanization, industrialization, international supply chains, and local economic growth [333, 334] and difficult issues like "urbanization without development" [335];
- the many relations between urbanization and climate change ([336], see also the many details throughout IPCC-WGIII), and the interdependencies (nexuses) between water and energy and between these two and food, and climate (e.g., [337–340]), then food systems per se [341], food insecurity [342], and food-health relation [343];
- urban security-tourism relations [344], past colonial influences and legacies on present patterns, e.g., a comparison between British and French heritage in urban planning and management [221], and city-region resilience [345].

A short, illustrative case study: Nairobi

As an example of starting point, and food for thought, here is a short list of papers about addressing sustainability in Nairobi (see also Fig. 2.22) as a lead example and as a subject of analytical comparisons with other African cities, on various topics: municipal solid waste, health and demography, electric mobility and local transport systems like the popular minibuses, flood management especially in the informal settlements, and underlying social factors ([346–358]).

In the above-evoked logic of directly addressing problems, the Kibera slum in Nairobi is a key topical case study for addressing the matter of informal settlements around the world, both because of its size (one of the largest in Africa) and because it is a test bed: different solutions are being explored and experimented with (e.g., [368, 374, 490, 488]) (also shown in Fig. 2.22, in the top-right middle). To understand the dimension of this issue, one must know that, as per 2023, out of the world population of 8.1 billion, about 55% (i.e., ca. 4.4 billion) lives in cities and the number is expected to reach 70% in 2050. One important detail is that about 1.1 billion people in cities

Fig. 2.22 Nairobi, "The green city under the sun", in the eyes of inhabitants and visitors. *Source* Original compilation of photos from public domain [credits given] as follows. Top: City skyline [359]. **Middle-top left**: the Kibera slum in the process of being upgraded to apartment blocks by the Government and UN-Habitat [360]. **Middle-top right**: entrance to the parliament of Kenya [361]. **Middle-bottom left**: University of Nairobi [362]. **Middle-bottom right**: auditorium of the Kenyatta international conference centre (KICC), built in the shape of a traditional hut [363]. **Bottom left**: walkway in the business district [364]. **Bottom right**: walkway in Central Park [365]

actually live in slums and other informal settlements, i.e., just under 25% (see details by the UN Statistics Division [366]). I think we all agree this is huge.

Indeed, Nairobi has already been and remains a city of inspiration: born as a railway station and logistic hub on the Uganda Railway that was built in the late nineteenth century to connect the African coast at the Indian Ocean with the Lake Victoria region, "The Green City in the Sun" (as its current nickname goes), has become a major socio-economic hub, and problems solved in Nairobi can inspire other cities in Africa and all around the world [367–371]. What the city of Nairobi needs (like all the other cities it evokes) is international partnership through projects.

A most inspiring feature of this city as a potential global reference for eco-cities is the strong link between nature and culture at its heart and an emerging leitmotif for eco-cities everywhere around the world. In Nairobi, this starts with the many and important open spaces that the local community cherishes (and that still require consolidated protection against "traditional" land-use changes that are frequent in fast-developing cities, reminiscent of the past two centuries when sustainability was not necessarily a conscious criterium). Here below is a short selection from a much longer list (for further details and studies, see (e.g., [372–379], and many other sources to be found)):

- The Central Park (featured in Fig. 2.22): a natural park (original forest) that has already lost in the past about 20 of its original 60 hectares. Because of its location and function for a healthy city of the future, this public natural area is essentially equivalent to the Central Park in New York and the Englische Garten in Munich;
- The Uhuru Park (a name which means *Freedom* in Swahili), adjacent to both the Central Park and the Central Business District, has a particular significance for the inhabitants because it is a public center for outdoor speeches, rallies, and various activities and services; the park was saved from being built over, by a campaign of protests led by the Nobel Peace Prize winner (2004) Wangari Mutua Maathai (a daughter of countryside, a biologist with studies in Kenya, USA, and Germany), and founding mother of the Green Belt Movement in Nairobi, Kenya, and Africa as a whole;
- The Jeevanjee Gardens is also particularly interesting and holds a special place in the heart of the Nairobians (and not only). This open garden in the Central Business District explicitly belongs to (held in trust for) the public, as it was "*donated to the magnificent people of Nairobi as a resting area*" (as usually and fittingly presented in publicly-available sources, presumably citing the original donation document) by Alibhai Mulla Jeevanjee (1856–1936) an Indian-born Kenyan entrepreneur, philanthropist and politician who had a major role in the development of Nairobi at its beginnings;
- The 7 August Memorial Park, which is located on the site of the former US Embassy, a place of quiet remembrance of a dark day (the terrorist attack of 1998 [380]);
- The City Park in the Central Business District, previously known as the Nairobi Forest Reserve [381];

- The Nairobi Arboretum, spreading on 30.4 hectares of wooded landscape with over 350 species of plants, was established in 1907 on the banks of the Kirichwa Kubwa River for the purpose of experimental introduction of new plant species. Today, it is a popular place for individual and group activities [382].
- Karura Forest, an urban forest with an area of 1 063.0 hectares and 50 km of walking and biking trails, is managed by the Kenyan Forest Service and a local community association: the Friends of Karura Forest [383].
- And, as I can testify from my own visit 15 years ago, a dense and beautiful tree cover is a usual sight on the streets of the city.

To these add some other interesting details that suggest a great vocation of Nairobi for and in the global transition to sustainable development. One is the fact that, as it happens, the headquarters of some key divisions of global institutions are located in the city, which constitutes a direct and obvious opportunity for this community of people to serve as a practical test-bed and shows case for worldwide urban and societal progress: the United Nations Environment Program (UNEP) and the United Nations Human Settlements Programme (UN-Habitat). And indeed, there are many positive outcomes of this interaction already.

However, as the eco-city paradigm is only at the beginnings of its full translation into practice, the potential benefits via eco-city projects and programs in the coming decades are immense, and the city of Nairobi can serve not only as a hub for the sustainable socio-economic developments of its hinterlands but also as a key node in national and broader international networks of cities. For that, Nairobi can inspire and get inspiration from other great hubs around the world in the complex dynamics of learning from each other.

As it was remarked in the World Cities Report 2016 by [43]

"*Urbanization has helped millions escape poverty through higher levels of productivity, employment opportunities; improved quality of life via better education and health; large-scale public investment and access to improved infrastructure and services*" and "*The transformative power of urbanization has important implications for rural areas. Cities act as magnets for rural migration; in developed countries, migration is driven by better opportunities in urban areas. However, in developing countries, rural–urban migration is more complex, in some cases driven by rural migrants seeking refuge from disasters such as famine or war*", but "*The benefits of urbanization should not be limited to large cities, but made available to small and medium towns. The adequate provision of adequate infrastructure and opportunities in small and medium cities can promote rural urbanization and contribute to achieving balanced population distribution*" (which also refers to the benefits of decentralization of the national administration and thus enhancing the functional responsibilities of local authorities). For checking out progress, see also [491].

In this sense, the fact that in 2010 the city of Nairobi joined the UNESCO *Global Network of Learning Cities* (briefly described on the official website as "an international policy-oriented network providing inspiration, know-how and best practice" [384]) suggests that this city is well aware of its potentials and is on the right way.

Another detail cuts directly to a human-level connection (and I mention it here both because the notoriety and inspirational power of the involved persons but also as a way to highlight that many other people may be less famous but are equally important in the making and the life of a city today): as a brilliant student in Kenya, the young Wangari Mutua had the chance to study in the USA through the same intergovernmental program that had led another Kenyan to that country: the father of the now-former US President Barack Obama. In 2006, i.e., even before becoming President, Barack Obama visited Nairobi as a US Senator and met with Wangari Maathai (2 years after her Nobel Award). Also, interesting though unsurprising, she became herself inspired by the traditional Japanese concept (of Buddhist origin) of "Mottainai", i.e., conscious avoidance of waste in our daily life, and promoted it at the UN and around the world. At present times, the University of Nairobi (where Wangari Maathai was a professor and which hosts her remains at the eponymous Institute of Sustainability) also embodies and reflects this tradition and vocation of Nairobi to be a meeting place of people and ideas, as reflected for example in the names of buildings in its campuses: the "Mahatma Ghandi Wing", the "American Wing", and so on.

At this point, let us remember that, in the process of evolution of cities as eco-cities, certain types of institutions have a crucial role to play and are currently underrated: the obvious one is the University, and this needs no further explanation. But another, albeit related, category of institutions is represented by the Botanical Gardens, the Arboretums, the Museums of Natural Sciences and the related institutions (which are often departments of Universities or of the Government) that are in charge with forests, natural reserves and other forms of protection and management of the natural environment. This book cannot fairly address this extensive topic, but it is a great theme for exploration by project teams and all those interested. The motto "*Look deep into nature and you will understand everything better*" (a saying usually attributed to Albert Einstein) is a perfect summary of the benefits for eco-innovation and eco-city projects.

Other cities around the world

Zooming out and in other locations around the world, different cities stand out as inspiring for at least one aspect that matters. Technically, the overall global diversity of insights multiplies exponentially the chances of each city to learn and become knowledgeable about urban systems in general, and then to articulate its own list of best priorities (and, consequently, potential pathways/scenarios for the future): not only in terms of addressing urgent local needs and struggles but also in connection with the operational opportunities that a city can have in national, continental, and global networks of cities. In short, many challenges that may look insurmountable in the local dynamics and constraints (at a given moment in time and in the life of a city) can actually be addressed successfully by harnessing complementarities and by engaging in various partnerships with other cities (or rural communities) in that region or from around the world.

For brevity reasons, I will only cite some well-known examples but, of course, one can easily identify numerous others. Thus, Curitiba (in Brazil) and Bogotá (in Colombia) have emerged in recent decades as great examples of sustainability-oriented policies and community-based innovations in South America, and in fact the entire continent is emerging as a laboratory for urban solutions, in the context in which cities spread across very different sets of local conditions but can very easily learn from each other because of a shared linguistic (mostly Spanish and Portuguese) and wider cultural base (combining Indigenous, European, and other heritage elements from Africa and Asia-Oceania) (see e.g., [385–389]).

Similarly, Surabaya and other Indonesian cities are examples of community-based innovation laboratories in the Asia–Pacific region, a famous example being the *Kampung Improvement Programme*, which addresses the ensemble of social, economic and environmental conditions in housing, transport and public spaces, initiated and promoted by another local champion with a strong empathy for people of all categories and at the same time a strong international orientation, Johan Silas ([390–398]). In this case, an important highlight is that Kampungs are villages (that are fast urbanized because are located near a fast-growing city) that, because their inhabitants have a strong sense of place (local identity), may appear as problems for the large urban developers (because these locals tend to resist those projects about which they are not being consulted) but are in fact a solution-in-disguise for large and growing cities, because they constitute the germs of that local identity which a city (with its city districts) needs for insuring long-term viability of new urban developments (for more discussions about Kampungs, see [399–402, 489]). The fact that people in Kampungs have a very collaborative and deliberative spirit is an inspiration for addressing the issue of chaotic urban growth (and informal settlements) anywhere in the world and especially in the so-called Global South, because this is where the fast urbanization mostly happens. However, these insights can be corroborated with those coming from the history of urbanization in the industrialized countries too, where numerous peri-urban villages have been urbanized and retain a distinguishable local identity of those city districts as part of the human heritage of the city (see also [403]).

At this point, it is worth highlighting that the sense of place manifests itself not only in the early growing phases of a city, like in the case of Kampungs. It is there also in the socio-economic contexts of mature phases of cities as well. Making an imaginary jump from Oceania to the other side of the world, I would illustrate this point with the phenomenon that emerged recently in some European cities that have to deal with mass tourism. Namely, citizens in Venice and Barcelona have grown fairly hostile to tourism in their city because this is taking too much the form of mass tourism, and this is "too much": a nuisance and a threat to their local heritage, so much so that the locals started to reject the argument of the economic benefit of tourism for their city. Basically, the message that one can distill in the attitude of the locals is something like "*This is our city and we do not accept that you come here and destroy it*".

Yet, in more remote and less economically well-off communities, e.g., at Machu Picchu, the situation is very different: the local people are making constant pressure on the government to keep the famous Inca site open to as many tourists as possible: they need the short-term income provided by tourism. However, one still needs to look beyond the surface to understand the place. First of all, those locals do not live in the old city so the degradation of the ruins (due to a combination of geological conditions, landslides, and tourism (see e.g., [404, 405])) does not affect them directly and it is unlikely that it will deter tourists from coming in. Secondly, they (the local community) are actually the live descendants and representatives of the ancient Inca civilization: so they need to make a real living, and, in those high mountains covered with tropical forest, nothing compares to a constant flux of international tourists.

These two examples (and others already discussed, as tourism is an important topic in most of those cities) hold a fundamental lesson to remember: local identity is what provides the balance and gives the correct direction in a local community.

At this point, one also remarks that tourism and visibility of a city go hand in hand: visible cities tend to attract tourists and a good touristic brand can also increase the chances of a city to attract investments and to sustain socio-economic development more broadly. In this sense, an interesting case of urban experimentation and related lessons are provided by the tourist town of Cancún, located at the tip of the Yucatan peninsula, in Mexico.

Beside the information available from technical reports in the literature, I can use a bit of live experience of my own to attest about how things feel on the ground, and to make some interesting connections in this discussion about eco-cities. I have briefly visited Cancún over a decade ago, not for leisure but because I attended a science-and-technology conference (the biennial *World's Congress on Biosensors*) that took place there, and I talked (in front of a mostly international audience) about the interaction between technology and sustainability from an RDI viewpoint: "*Biosensors, eco-innovation and markets*", i.e., the role of biosensor technology in the context of sustainable development goals in cities and in general [406]. Tellingly, the conference took place there because this world congress on biosensors takes place every two years by rotation on different continents, and, that time it was the turn of the Americas—and one of the grand tourist venues in Cancún offered enough space for so many people (several hundreds) combined with a great vista for everyone to remember. The touristic fame and attraction was clearly a big factor in the decision to hold this important global meeting there. But what does this mean for the local community? Now, the event was a big success, and many people met again or met for the first time with this occasion—a great gain for RDI and the world. Myself too, following my scheduled talk before those hundreds of people, I have got another invitation and gave a follow-up talk on eco-innovation (this time on transitioning markets) at another technology conference on the other side of the Atlantic Ocean [100] in Bordeaux (France), itself a great example of an inspiring city (see, e.g., [407, 408]). So, for me (and in general the participants at this event), all was basically excellent, as we did what we were meant to do. However, our international meeting had very little to do with the local community, which was just marginal or even invisible to us in our very specialized technical conversations: it was just the basic attractiveness of Cancún—a quintessential tropical "paradise" at the Caribbean Sea, with fantastic blue shades waters and long white-sand beaches right outside the conference venue (so, just for us to quickly admire in the breaks between work sessions and during the evenings). To this adds a great culture and the world patrimony represented by Mayan sites scattered throughout the nearby jungle. Some vestiges of a Mayan settlement are even located in the city itself, namely the *El Rey* ("The King") archaeological site, which I didn't miss to visit by foot because I have the habit of studying the cities I visit and I like long walks for observing both nature and human systems.

All and all, this exquisite tourist town has all the appearances of a great success. But, the story behind it is more complex and requires a deeper regard for the eco-city interest of this book. The tourist district (in Spanish, "*Zona Hotelera*", i.e., the Hotels Area) of Cancún has a life of itself, with socio-economic conditions that are very different from the local town itself; but also, the city districts outside the tourist area have significant differences between themselves. In short, Cancún tells a story that goes at the heart of the local-and-global character of a city, with all the gains and tensions that come with it. This story basically started in the 1970s as a planned city in a project aimed at fostering local development based on tourism, led by the government who provided the concrete planning-and-development of the basic infrastructure and the funding of the first nine hotels. Once the ball started to roll, this former village with a few fishermen "lost" in an almost uninhabited area had such a great success that it started to attract huge numbers of people (looking for jobs) from all across the Yucatan peninsula and the rest of Mexico, and continues to do so. The city has now about 1 million inhabitants and has established itself as the main tourist attraction in Mexico, at least on the east coast (of course, Acapulco is a counterpart attraction on the west coast, and Mexico City in the center of the country). The result of this great success and spectacular exceeding of expectations is that the local authorities are now struggling to keep up with the provision of urban facilities (knowing that the tourist area is given priority as the main economic engine for the city) in the fast-growing districts. Just a few years before my visit, Cancún was cited as an example of extreme inequality where the locals do not have sufficient access to clean drinking water while "everything needed" was reserved for foreign tourists [409].

Because of its origins in a governmental initiative, Cancún has also been called a "manufactured tourist city" [263]. In any case, the city has become a great laboratory and research platform for urban development, especially (but not only) in relation with tourism, with relevance for the entire Mexico, Latin America, and the whole world (for more discussions, see also, e.g., [410, 411]).

Along with many other cities around the world, Cancún shows that, for the most people (at least outside the already developed countries), the emphasis in the notion "sustainable development" is on "development" and this fact must be respected (for reasons already discussed in this book since Module 1, going from moral aspects to sheer logic and the operational requirements in concrete projects).

To cite only one lead aspect, the city of Cancún keeps growing and has three main areas with informal settlements but these three are very different in terms of their capacity to cope and have quite different prospects (both in qualitative and quantitative terms) for proper incorporation in urban planning and services. In sum, "the differences lie in the purchasing power and educational level of the population in each city district, which allows them to cope with these deficiencies" but also the history and the technical details in different districts [412].

In perspective, studies exist and will surely multiply about the local-and-global character of this city, and here are two illustrative examples. In a study of the attitudes of Cancún residents towards sustainable tourism in their city (documentary research, random field survey on 429 respondents and ANOVA statistical analysis) [413] found that "*though Cancún residents proved to have little knowledge of the meaning of sustainability in general, when associating the concept to tourism their attitude was straightforward. They saw tourism as a favorable activity in economic and socio-cultural terms but less favorable in environmental terms. A more detailed analysis showed significant variations among survey participants according to their birthplace, educational level, and occupation, especially when the latter was related to tourism*".

In another study on classroom-and-vocational education in the hotel industry in Cancún (the "duality" consisting in a close cooperation between schools/universities and companies/employers), using the results from ten face-to-face expert and semi-structured interviews, [414] found that there are opportunities for collaborative learning, but that (in order for such cooperation to be successful) "*certain principles are needed*" and these principles need to be properly looked for and identified because they "*are related to the specificity of a region or a particular industry*".

Given all these and many other details that the interested professionals can find, the city of Cancún has the potential to be a city of inspiration in the very concrete terms of solution development for sustainable tourism but also more broadly, as a springboard for holistic urban development, and, in perspective, for overcoming the dependence of a city on a single major industry by using their particular strengths at a moment in time as a basis for further socio-economic diversification for the mutual benefit of all stakeholders, starting with the local community.

Then, and zooming out and in again at the scale of the world, studies that compare practices in different cities (or any other type of human settlements) in different countries are also particularly valuable, and this applies to any topic. For example, [415] studied "stormwater runoff management practices and governance under climate change and urbanization" in Bangkok (in Thailand), Hanoi (in Vietnam) and Tokyo (in Japan). Reports by some international organizations make periodic reviews that compare cities on different topics (for a start, see, e.g., [43]). In other projects, e.g., [416] studied the mitigation of urban heat island in Kuala Lumpur (Malaysia), Singapore and Hong Kong via urban vegetation, and [417] compared local perceptions and practices in coastal communities faced with the consequences of climate changes in very different local climates and socio-economic conditions, which I cite here because it is an excellent illustration (and because I know this work indirectly through some of my colleagues): Uummannaq, Greenland; Tiksi, Sakha Republic (Yakutia), Russia; Wainwright, Alaska (USA); Cocagne-Grande-Digue, New Brunswick (Canada); Bay of Brest, Brittany (France); Mbour (Senegal); and the Vypin Island-Chellanam Peninsula, Kerala State (India).

Then, to develop understanding along the in-depth dimension that complements such "horizontal" comparisons, one can search and find papers from anywhere and especially those fast-growing cities (and communities) that have inspired detailed studies because of their impressive size and dynamics. For example, one finds that climate changes impact both peri-urban farming communities and the informal settlements in and around the city of Lagos, Nigeria [418, 419]. One also finds, to give another example from a different climate, that cities in Siberia and other polar and subpolar regions have a serious problem with the permafrost (i.e., permanently frozen layers of the ground), as its thawing due to climate changes causes numerous problems for the cities in the region, starting with damaged and collapsing buildings and infrastructure (see, e.g., [420–424], and the most recent IPCC reports).

But, this horizontal/vertical discussion is more complex than its name might suggest: in fact, systematic and thorough "horizontal" comparisons can also help us gain in-depth knowledge because it illuminates some fundamental systemic dynamics that are underlying human systems. This realism is more important than technical vertical/horizontal distinctions because this is not about making purely academic distinctions but about improving the life of those communities in concreteness and without unneeded "armchair" formalisms. After all, the purpose of eco-city projects is not (and must not be) "to show that we are doing something" and it is not to create more problems that we solve (see, e.g., [43, 60, 61, 357]).

Other challenges result from the dynamics of urbanization itself. A very high rate of urban growth (combined with insufficient extant capacity) can make that urban planning is hardly possible at all and may result in situations where high-quality projects become disjunct from the life of most people, as it happens, e.g., with the districts of Badia and Eko Atlantic City (EAC) in Lagos [425–431]. This phenomenon has been called "splintered urbanism" [328] or "enclave urbanism" [432]. Such situations might be inevitable as a managerial compromise on short term (since, without the establishment of some sort of "protected space", high-quality projects might not be possible at all) but this also means that the city must find further ways to close the gap thus created; otherwise this kind of gap will become permanent and a source of other problems on the long-term. Here is an illustrative snapshot text fragment from the latest IPCC-WGII report [60]: "*In Lagos, Nigeria, efforts to achieve climate resilience and sustainability through future city practices risk perpetuating the enclosure and commodification of land* [433]. *The exclusionary outcomes of some adaptation interventions can therefore further heighten the risk to communities that are socioeconomically more vulnerable. See* Sect. 6.3 *for further discussion of equity and justice considerations in local climate adaptation*".

As a side but relevant observation (and to close a loop with some citations made earlier in the book), an expression related to "splintered urbanism" mentioned above is "splintering urbanism": this may refer to a different topic (an abusive understanding of nature as "infrastructure in cities" resulting in the destruction of nature while pretending we "return to nature" (see [2]), in a logic that is more in line with the notion of "Frankenstein urbanism" evoked in the premises of this book (Sect. 2.1). Nonetheless, both expressions speak about the same essence: artificial introduction of divisions and incoherences in the name of a misapprehended progress.

Ultimately, such examples are also illustrations of urban experimentation and are part of the global dynamics of cities inspiring each other to try and see what works and how to find ways forward. And, they indicate the need for new ideas to be tested in very difficult contexts in which usual patterns of urban development may not fit. This will take thorough analysis and experimentation in dedicated projects and programs. But as a general reflection that I can do in this section of the book, (perhaps as a food for thought for interested project teams), based on current knowledge, it looks to me that, in high-density urban areas in West Africa (and in similar regions anywhere else on the continent, for example around Lake Victoria—just to cite another area with high population densities), the development of new and future- and sustainability-oriented urban projects (that would also generate socio-economic development for the countries themselves) might need to switch focus away from trying to create space in current high-density city centers and towards generating new urban nuclei in rural areas with lowest population density (and therefore experiencing lesser land-use pressures) that might be found in between current urban agglomerations, preferably in the least fertile areas (so that good agricultural land is not destroyed by new urbanization). This could help eco-city (actually any urban) projects in several ways: (1) it would avoid creating new problems for the existing poor—it would let them live without making their life even worse, so to speak; (2) it would help decongest existing city centers that absorbs all in-migration fluxes, which will facilitate their management; (3) it would lower the cost (and overall complications, of course pending on contextual details) for developers and all actors and stakeholders; (4) it would create new types of opportunities, e.g., by involving those rural communities (that would become urbanized and modernized) in the process and thus creating some positive precedents for democratic and fruitful participation in new urban developments; and (5) it would expand practical infrastructure across the national territory (which will further help spread out opportunities) and will generate experiences and expertise that can be deployed in similar and ever-better projects (which will help strengthen administrative/managerial capacity of all involved actors, and will be a factor favoring foreign investments in those cities and countries). In perspective, if many new small-and-modern urban centers appear along the Atlantic coast of West Africa (with a strong African identity but also oriented towards international and intercontinental cooperation), the region will have a very powerful network of cities and towns all along the coast (and the northward hinterland). Given this and all the competence and potential created in the process of generating this constellation of small-and-beautiful urban settlements, along with the benefit of a young and booming population, the entire region will become a magnet for investors, and a major or even the greatest engine of sustainable economic development (with each word deliberately underlined individually) in the South Atlantic, in a dynamic of international exchanges between Africa and South America (which will lift both continents), and further along the entire Atlantic for shared prosperity, north and south together. In concreteness, this simple vision is perfectly achievable.

Relatedly, we have to acknowledge that we know much more about large cities, and much less about smaller communities. But this only increases the practical value of the information we can obtain from all highly visible cities and projects. For a long time, small cities have benefitted less attention (at least compared to large cities and urban agglomerations) and they require their own research agenda. This is how [434] described the situation some years ago and argued in favor of a dedicated effort: "small cities have been ignored by urban theorists who, in seeking to conceptualize broad urban agendas and depict generalizable models (for example relating to epochal urbanism, the structure and nature of the urban hierarchy, global cities and global city-regions), have tended to obscure as much as they illuminate. Given that study of 'the city' has been vital to broader advances in the social sciences, this neglect of smaller urban centres has profound consequences for urban studies. (…) this situation needs to be rectified. (…) conceptual and empirical research into small cities can contribute to broader arguments that have championed the necessity of developing sophisticated and nuanced comparative approaches to understanding the complexity of cities and urban life".

While the fact that large and global cities attract more attention is understandable, smaller communities are at least as important for our understanding of urban life and trajectories. In any case, this constitutes a key reason for smaller communities to be connected and collaborative in the regional and global networks via concrete projects (See also Module 3, Sect. 3.4, Table 3.3—Connectedness). After all, who cares where the useful insights come from? Important is that they come and that useful light be shed on the problems at hand so that solutions are being developed.

In the next (and last) part of this section, I use some more well-known examples from the history of humanity (and my own experience) but the fundamental point in all these discussions is rather straightforward: in the end, what matters is not what history has made of you but what you do with what history has made of you.

A Balancing Act

In the strive to find an optimum balance between planning and evolution of urban systems, and as the multi-disciplinary concept of landscape garden (and the holistic field of landscape architecture more broadly) tells us, we need to work with nature, not against it. In this sense, an urban landscape which I personally find most inspiring and which combines nature and culture at the very heart of the city is the well-known National Mall in Washington DC, USA (Fig. 2.23), which, it may also be noted, is listed in the National Register of Historic Places and the District of Columbia Inventory of Historic Sites.

I have seen it in 2007, with the occasion of my participation in a double conference with a long but fitting name, where I had the privilege to present some of the results from our projects in Jassy (Iași) about sustainable vs. unsustainable scenarios in the systemic perspective on sustainability (as shown earlier in Sect. 2.2, Fig. 2.16). That double event was *Biological Sciences for the 21st Century: Meeting the Challenges of Sustainable Development in an Era of Global Change* (10–12 May) combined with the *29th International Union of Biological Sciences General Assembly & Scientific Symposium* (9–13 May). My technical communication was a poster titled '*Transition to sustainability through learning-by-doing capacity building: the pilot project Centre of Expertise for Sustainable Exploitation of Ecosystems in Iasi, Romania*' [435]. And, in fact, I had flown to Washington DC directly from Leipzig and Berlin (Germany), where I had discussed the same work details in a talk [436] that I had in a conference called '*Sustainable Neighbourhood—from Lisbon to Leipzig through Research (L2L)—"A conference on research for sustainable development in Europe*" (08–10 May); and note the tight schedule.

Fig. 2.23 An example of promoting nature and culture in a landscaped city center: the extended National Mall in Washington DC, from the Lincoln Memorial (in the foreground; west) to the Washington Monument (reflected in the Reflecting Pool) and on to the US Capitol building (in the background; east). The western part includes the West Potomac Park (not all of it is visible here). *Source* Public Domain, credit [437]

Needless to say, I used all my extra time to visit the city up close: I walked along the extended National Mall from the Capitol all the way to the Lincoln Monument and the Arlington Memorial Bridge over the Potomac River, in-between visiting the Smithsonian Castle, the WWII Memorial, strolling on the alleys lined with American Elm Trees, and visiting or just admiring some of the other numerous great institutions in the neighborhood, notably the *United States Botanic Garden* (100 Maryland Ave. SW), the *National Museum of Natural History* (Constitution Ave. at 10th St. NW), and along the *Pennsylvania Avenue* (on the north side). The latter is (as per the original city plan drawn by Peter Charles L'Enfant) the corresponding equivalent of *Avenue de Sceaux* in the famous urban trident of Versailles conceived by architect André Le Nôtre, and which inspired his own plan for Washington DC. In this plan, the National Mall corresponds to the middle avenue of the trident (which in Versailles is Avenue de Paris, the city's main street, also surrounded by historical places).

In Fig. 2.24, I make a comparative historical perspective between Washington DC and another city that is inspiring for me, namely Berlin: great cities as great meeting places.

In their historical context, the images in Fig. 2.24 also show some other interesting details, of which I will only cite a few as an exercise (and example) of exploration of the temporal dimension of a city. In the top left quarter, one sees the Main Navy and Munitions buildings (the white roofs behind the tree row) which were raised by the US Government during WWII and demolished in 1970. Much of that area later became the Constitution Gardens (named so in 1976) which one can see in the top

The National Mall filled with people at Martin Luther King's speech 'I have a dream' (1963)

The National Mall as a quiet promenade lined with American elm trees (2010, facing east)

The Berlin Wall (painted on the western side) separated the city between 1961 and 1989

The fall of Berlin Wall on 9 November 1989. Germans officially reunited on 3 October 1990

Fig. 2.24 The city as a meeting place, an engine of progress through affirmation and recovery of human values. Instances from Washington DC (USA) and Berlin (Germany). *Source* original arrangement of modified images from the Public Domain, with credits as follows: [438] (top left), [439] (top right), [440] (bottom left), [441] (bottom right)

right image (the large green area with trees to the left from the walkway), and in Fig. 2.23 (to the north of the Reflecting Pool, i.e., on the left). Then, in these images, one also notices some subtle details indicative of an open vs. closed society: the open water surface of the Reflecting Pool in Washington CD, around which people are gathered, vs. a (rubble) filled-in water course Luisenstadt Canal in the dictatorial East Germany (in German, *Luisenstaedtischer Kanal*; visible as a darker ground in the space between the walls, and people being kept away from it.

Most tellingly, the bottom-right image shows the fall of the Berlin Wall, with people climbing it to overcome a historic division of the city in front of the city symbol itself: the Brandenburg Gate. Anyone who has visited the place in the young twenty-first century (and I speak from lived experience), notices a beautiful place that marks the entry in the historic city center of Berlin represented by the *Unter den Linden* boulevard (literally meaning *Under the Linden Trees*) and its many and impressive cultural landmarks (one of which is Humboldt University) (in the photo from Fig. 2.24 bottom right, the wall is seen from the east side, behind the Brandenburg Gate, at the west end of *Unter den Linden*). It is a perfect example of how unnatural such a separation is (and how unnatural and how socially and humanly unsustainable the political regime that imposed it was), and it shows that a true city that is conscious about itself does not permit something like this to persist for too long a historical time.

In this context, it befits to remind ourselves the speech that US President Ronald Reagan had in 1987 (12 June, in West Berlin, in front of the then still-standing Berlin Wall near the Brandenburg Gate), when he called for the leader of URSS to "tear down this wall", and continued with these words that epitomize the spirit of the people of a city, a spirit that defines human civilization itself: "(…) *In these four decades, as I have said, you Berliners have built a great city. You've done so in spite of threats, the Soviet attempts to impose the east mark, the Blockade. Today the city thrives in spite of the challenges implicit in the very presence of this wall. What keeps you here? Certainly, there is a great deal to be said for your fortitude, for your defiant courage. But, I believe there is something deeper, something that involves Berlin's whole look and feel and way of life—not mere sentiment (no one could live long in Berlin without being completely disabused of illusions), something instead that has seen the difficulty of life in Berlin but chose to accept them, that continues to build this good and proud city in contrast to a surrounding totalitarian presence that refuses to release human energies or aspirations, something that speaks with the powerful voice of affirmation that says "Yes" to the city, yes to the future, yes to Freedom. In a word, I would submit that what keeps you in Berlin is Love—love, both profound and abiding. Perhaps this gets to the root of the matter, to the most fundamental distinction of all between East and West: the totalitarian world produces backwardness because it does such violence to the spirit, thwarting the human impulse to create, to enjoy, to worship. The totalitarian world finds even symbols of love and of worship an affront. (…) There, in (east) Berlin, like the city itself, symbols of love, symbols of worship cannot be supressed. As I looked out a moment ago from the Reichstag, that embodiment of German unity, I noticed words crudely painted on the Wall, perhaps by a young Berliner, quote: 'This wall will fall. Beliefs become reality'. Yes, across Europe this wall will fall for it cannot withstand faith, it cannot withstand the truth. The wall cannot withstand Freedom. And I would like, before I close, to say one word: I have read and I have been questioned since I've been here about certain demonstrations about my coming (…) and I would like to say just one thing and to those who demonstrate so: I wonder if they have ever asked themselves, that if they should have the government they apparently seek, no one would be able to do what they're doing, again*" (the author's transcript of a historic video by the Ronald Reagan Foundation [242]).

By the way, again, personal experiences are always tightly connected to a city and how much, and especially in which way, they inspire. I visited Washington DC in the same year when (as mentioned in the Module 1, Sect. 1.2 in relation to Fig . 1.5), William C. Clark published his well-known paper "Sustainability Science: A Room of Its Own" in *Proceedings of the National Academy of Sciences* of the

USA [442]. And he was one of the key figures in the above-mentioned conferences in Washington DC, and I remember well his joking comments about how hard it is for people to attend the works of a scientific event in such a beautiful metropolis (I cite from memory, so any wording mistake belongs to me): “So, it’s springtime in Washington and you want me to stay inside and listen to some scientific models … are you crazy?”. To me, with the benefit of the hindsight and current knowledge, this convivial remark in a break between two sessions of scientific communications and debates, continues to resonate in my mind: the historical city centers of cities with a strong identity is about ‘Nature & Culture’, and this is what inspires people to do what they do; ultimately, this is what gives a city its future as an eco-city.

To go on with the personal-professional story, as it happened in a fortunate series of events in that eventful year of 2007, 2 months later I have visited the city of Versailles (for the first time) with the occasion of another international conference, a workshop called *University in Society—UNISO 2007* (9–12 July) where I had a talk [443] on the international circulation of talent and its role in local development via internet-based co-working (“Brain Networking”). The event was hosted by the University of Versailles (UVSQ), an institution where I returned three years later to work in the newly established Econoving cluster mentioned earlier in this book. In between the two conferences, my multi-annual personal-professional journey included some other events: professional talks at the *1st German-Romanian Conference on Research for Sustainability* (03 June, Bucharest) [104], *University in Society—UNISO 2008* (16–20 July, in Jassy) [444], *ISEE 2008: Applying Ecological Economics for Social and Environmental Sustainability (10th biennial conference of the International Society for Ecological Economics*, 7–11 August, in Nairobi) [101, 205], and some other technical contributions in Glasgow (UK), Berlin (Germany), Prague (Czechia) and Timișoara (Romania) [207, 443–445, 486, 487]. During all this time I was based actually in Strasbourg (France): finding ways via Brain Networking.

Such are the travels and human interaction details that shaped my own experience. But this is just a glimpse about one person. More importantly, there are many tens of thousands of similar personal-and-professional journeys, and all this experience and the associated technical materials constitute a tremendous global resource for cities to use in concrete eco-city projects. The fact that the experience of different people is predominantly linked to one or another city (and/or part of the world) is ultimately irrelevant: this is life. What really matters is that experts around the world naturally talk to each other and they can contribute concretely in actual projects in cities located around the world. What counts is that “*Our Common Future*” on Earth depends on us getting our act together for real, in life and work.

In Fig. 2.25, I tried to give a visual sense to this simple reflection by joining together several contexts, from a person to a city to the planet and to the outer space.

Ancient knowledge about the world: A short comparison

From an epistemic perspective, current knowledge can be better understood if we keep an eye on how our current knowledge came to be. In this sense, I find very interesting to put next to each other two ancient (/pre-modern) representations of the

Fig. 2.25 Study scales, from individuals to cities to planet and the outer space (see technical details in Module 3, Sect. 3.2). **Top**: Paris seen from the Belleville Park (looking southwest-ward) located on the homonymous hill in the city's northeast). *Source* the author (2013). **Middle left**: View of Versailles from Avenue de Saint-cloud looking towards the Chateaux. **Middle right**: Yours truly near a natural size replica of the Mars rover Perseverance exhibited in 2017 in the Premises of the Observatory of Versailles (OVSQ). Source of the photos: the author. **Bottom left**: The "Pale Blue Dot" photograph of planet Earth taken in 2020 from about 6 billion km away, by Voyager 1 space probe; the apparent size of the Earth is less than a pixel. Source [446, 447], public domain. **Bottom right**: All humanity in one picture except the photographer, showing Earth and the Lunar Module returning to the command Module after aselenization during the mission Apollo 11 (1969). *Source* public domain (credit: Michael Collins/NASA)

place of humans in the world: the classic worldview from the city of Alexandria in the Hellenistic Egypt and the city of Khara-Khoto shown earlier in Fig. 2.18.

As a technical observation, the two models of the cosmos in Fig. 2.26 are examples of cognitive maps (details in Module 3, Sect. 3.3) and the temporal (historical) dimension is important because it helps us understand the process of development of human cognition about the world, and highlights them as results of the sense-making efforts by people in different spatio-temporal contexts. Specifically, from our perspective in the present, these representations are only Conceptual Maps and not yet Knowledge Maps (because they cannot be justified based on our current knowledge), but, from their perspective at the moment of their creation, these two models can be considered Knowledge Maps (of their time) because they integrated the best-knowledge available at the time. Indeed, both models can serve to explain mathematically the then-known patterns in the movements of the celestial bodies, with the difference that the geocentric model does so using some more complicated calculations than those necessary for a heliocentric model.

Further the same kind of cognitive effort of people to establish a relation between themselves and the world is identifiable in all cultures and the city itself is the embodiment of that effort. Beside the details already discussed in the examples given so far, another particularly interesting example that can serve as a source of insights about cities and their profound nature is the pre-Columbian city of Cahokia, currently a UNESCO site in Illinois, USA. Beside the basic details (also summarized in a short case study discussion I make in Module 3, Sect. 3.4, Table 3.3—at "Urban Potential"), these human settlements in the western hemisphere also reflected what we already know from ancient cities in the eastern hemisphere: cities are themselves a mirror of the cosmos and of how people saw themselves in the world based on their current knowledge [448–451], and, ultimately, an *Axis mundi* [452].

Ultimately, the point is that people from every epoch and location use the best knowledge available to them to make sense of the world and to try (1) to improve their lives and (2) to try and anticipate.

Making an in-depth historical discussion would fall outside the scope of this book. Therefore, I will only quickly enumerate below a few general ideas as food for thought around the purpose of this book and the importance of knowledge, science and learning as a never-ending endeavor.

Among these quick details, it is important to note that Greek and Roman natural philosophy of Classical Antiquity have built on the earlier contributions in mathematics, astronomy and medicine made by Ancient Egypt, Mesopotamia, and (via long-range political-military, commercial, and other human interactions) other ancient civilizational centers, especially in current day India and China (also knowing that Alexander the Great and his general Ptolemy himself, had reached as far east as northwest India, and that historic proofs exist of commercial and human exchanges between Ancient China and the Roman Empire). For example, from the history of mathematics we know that "zero" was invented in India. Important to also highlight is that Ancient Greeks (and the subsequent civilizations) already knew that the Earth was round (approximately spherical, and knew its approximate circumference as

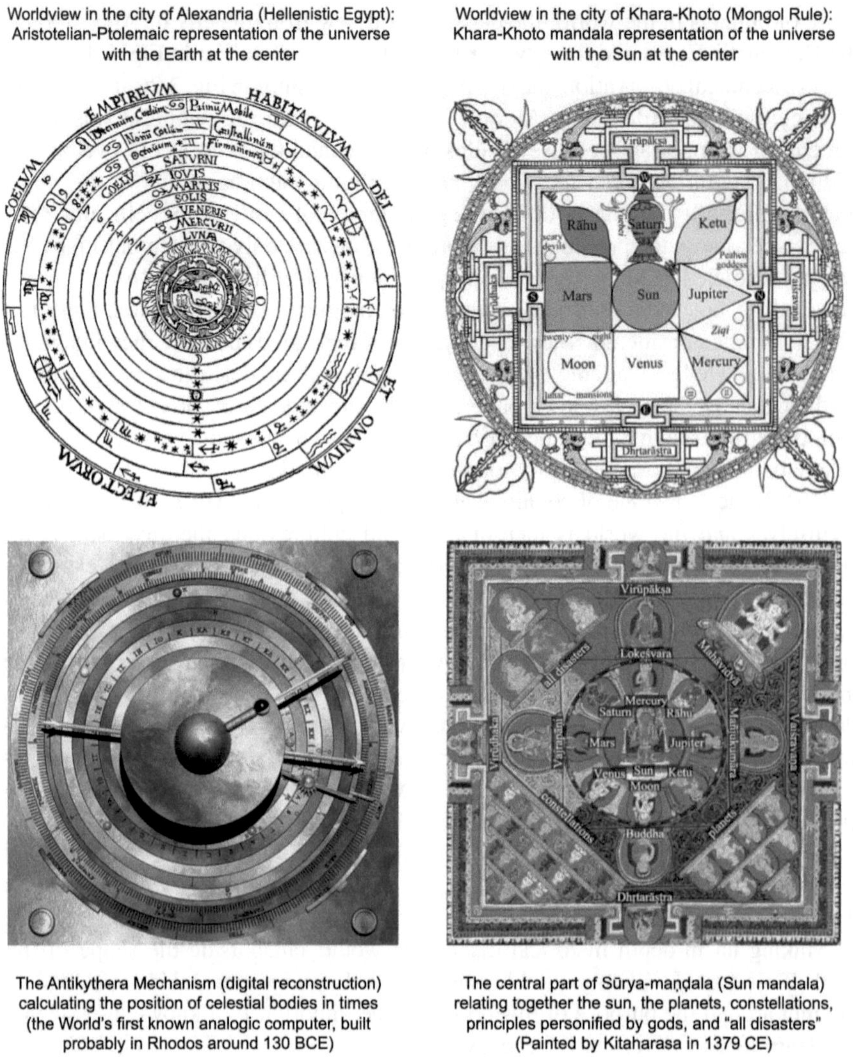

Fig. 2.26 Pre-modern models of the cosmos, integrating existing knowledge and beliefs by sense-making efforts in their respective contexts: the Geocentric Model from the ancient city of Alexandria (331 BCE to 415 CE) and the Heliocentric Model from the city of Khara-Khoto (1032–1372). *Source* original synthesis, based on four modified images from different sources from the public domain as follows. **Top left**: from Ptolemy's work "Almagest" (Cipozy (reproduction), Peter Apian (original), [453]), edited by Theon of Alexandria and (his daughter) Hypathia of Alexandria [454, 455], based on the Greek natural philosophy of classical antiquity and involving layers of spherical shells, and able to correctly support simple calculations based on empirical astronomical observations [456]. **Top right**: Khara-Khoto [308]; **Bottom left**: [457], an artifact which was discovered in 1901 in a shipwreck near the Antikythera island in southern Greece and continues to be studied; for details see also [458]. **Bottom right**: Ngor mandala (ca. 1429–1456., pigment on cotton, Ngor Monastery in Tibet), Kimbell Art Museum, cited by [308, 459]

calculated by Antisthenes based on the shade made by sun in two different cities of Egypt: Alexandria and Aswan).

Importantly, the contemporary ideas about past knowledge (or lack thereof) are actually myths. One such myth is the idea that before Christopher Columbus (and his discovery of America in 1492) people in Europe/Old World believed the Earth was flat. They didn't, as all learned people (usually monks) knew what the ancient Greeks knew.

Another myth is that there is a fundamental conflict between science and religion (a thesis that was refuted a long-time ago). In fact, the relation between Science and Religion is certainly not easy but these two aspects of Humanity actually refer to different things that are ultimately complementary: the human aspiration to know how out world works and the equally human aspiration to understand the nature of existence. In fact, earlier progresses in science have been made by monks, and many if not most modern universities can trace their origins in some schools that had been established by or in relation to monasteries and other religious or/and ethical organizations. Further, what may appear to be a conflict of the fundamentals tend to be in fact more political conflicts (e.g., Hypathia of Alexandria, Galileo Galilei and Charles Darwin)—basically, the same way different "controversies" appear because of some specific (often commercial) interests.

History keeps showing its lessons … (Do we learn?)

From a recent scholarly text which I find rather fitting for this discussion [460], let me include a short fragment from a book chapter called "Natural and legendary origins of local memory", which cites Plato's dialogue *Phaedrus*, as it mentions the exact location in the ancient city of Athens between the Olympion temple and the city gate across the hill where, in later times, the Panathenaic Stadium was built (about half a kilometer south-east from the Acropolis; for a visual, artistic representation, see, e.g., [461]): "*In a famous passage, Plato pictures Socrates and Phaedrus conversing about love as they stroll among the trees and shrines of the Ilissos river at the place where it ran close to the city walls of Athens. Phaedrus has spent the morning in a house near the Olympeion, the Temple of Olympian Zeus, and is on his way to take a refreshing walk along country roads when he meets Socrates, who suggests they walk along the river. Phaedrus leaps at the idea: 'How lucky, then, that I am barefoot today—you, of course, are always so. The easiest thing to do is to walk right in the stream; this way, we'll also get our feet wet, which is very pleasant, especially at this hour and season'. Along the way to the shady plane tree where they will read and discuss the speech on love Phaedrus has just been listening to, they pass a sequence of primordial shrines—to Boreas the north wind, on whom Athens' pleasant climate so depends, and to the river god Achelous and the nymphs. Phaedrus asks whether Socrates believes the local legends, to which Socrates replies that his time is better spent trying to know himself than debating the veracity of myths*".

Fast forward to modern times, one notices that different lessons have been learned in different places, and not necessarily the way we might have expected.

First, modern Athens is a large city region where the pressure of built areas is so great that most rivers in the city have already been culverted, i.e., covered and converted into underground rainwater/sewage canals. Basically, the rivers inside the city of Athens are hidden under concrete, and only 10% of the total lengths of rivers in the geological basin of Athens remain uncovered—while, at the same time, serious technical and economic issues exist related to underground erosion under culverts causing ground subsidence and risk to citizens (of which problem the same Ilissos river in which Socrates wet his feet is a central example) [462].

Of course, we know the right technical approach: nature-based solutions, opening the rivers and creating "blue and green corridors" and so on. And such propositions have already been made and are being discussed in the public space of the city of Athens (and, unsurprisingly, first proposed for "opening up" is exactly that portion of the Illissos river that is mentioned by Plato in *Phaedrus*) (see e.g., [463]). But how do we address the (socio-economic) root cause? Based on all that has been said and cited in this book so far, solutions will require a holistic approach that looks at the city of Athens as a system. This is of course not new, and I expect that urban planners know that fully well. The challenge is in *How to do it without creating more harm than good*. And this means a series of large and complex projects.

This problem represents an opportunity to begin a process of transformation of the city that is more in equilibrium with nature itself, from geological processes to ecosystem goods and services known today in the literature on "blue a green infrastructure", i.e., nature-based solutions. In this sense, this issue in Athens is a great example of problem as opportunity—with due diligence so that the city avoids what some called "replacing one mistake with another". One fact to be considered is that the great extent of urbanization of the Athens area, including (but not limited to soil imperviousness) also makes that the debit of the Ilissos river is greatly reduced at the present (compared to the Ancient era), which further means that opening up the river in the city center may not be enough, and a broader re-thinking of the entire urban and peri-urban landscape is probably necessary. For further information, see [464, 465].

Second, an interesting phenomenon that evokes at present times the ancient stroll of Socrates and his interlocutors is the existence of the peaceful walkways, typically along rivers or otherwise close to nature (blue-green areas), and in the vicinity of learning or cultural institutions and sites. Of these, the following three are perhaps the most famous and I particularly want to cite them because they come from three different continents.

The first is the Philosopher's Walk (in Japanese 哲学の道, Tetsugaku-no-michi, literarily Path of Philosophy), a pedestrian path along a cherry-tree-lined historic canal from Lake Biwa in Kyoto, the ancient capital city of Japan. This walkway is named so because it was used by twentieth-century philosophers (especially philosophy of science and mathematics) and professors from the University of Kyoto, starting with Kitaro Nishida (1870–1945). In the context of this book, it is worth pointing out that that epistemology and moral philosophy of the latter K. Nishida aimed at bringing East and West together in a creative tension that resolves not in dissolution of tension but in a complementarity of perspectives that helps define individuals, and communities. (This is akin to other intellectual threads around the world converging to the idea that a person can harmoniously be a citizen of his town and country while also being a "world citizen". This idea has later been called "rooted cosmopolitanism" and "cosmopolitan patriotism" (e.g., [466–468]).

This philosophers path passes a series of temples and shrines (which also includes the UNESCO World Heritage Site of Jinkaku-Ji (in Japanese, 銀閣寺, i.e., Silver Pavilion) and is very popular with locals and tourists, especially during Hanami (in Japanese 花見, literally "flower viewing"), a nation-wide tradition of taking time to enjoy the seasonal beauty of flowers while they last [469–473].

On another continent, there is another Philosopher's Walk, a scenic footpath that exists in the Saint George Campus of the University of Toronto. This other path in the middle of nature runs along Taddle Creek, a ravine landscape, which was once a natural waterway and is now flowing underground as it was culverted (i.e., covered to be made subterranean) during the Industrial Age. Nowadays, this is used by those who want to enjoy what has been called "forest bathing", a type of active emersion in meditating while reconnecting with nature [474–476], in the context of the well-known benefits that sheer walking has for our mental and general health (from emotional equilibrium and mind sharpness to a well-functioning cardiovascular system and through to a better immune system and overall metabolism), or, as the Romans used to say (and as we emulate nowadays and as we need to do in the future), *Mens sana in corpore sano*, i.e., "healthy mind in healthy body".

And on yet another continent, a Philosophers' Walk (in German, *Philosophenweg*, literally Philosophers' Path) goes along the right bank of Neckar river, in the historic district of Heidelberg. This is also called so because it was used by philosophers and university professors who would walk and talk along the scenic pathway which offers scenic views of both nature and the old town, which includes the Heidelberg Castle and is not far from the University of Heidelberg, right across the Old Bridge connecting the two river banks (farther up the mountain, this path also leads to the ruins and historical places going back to Celtic tribes). In context, it is worth knowing that the city of Heidelberg itself is a hub of science and arts, and was also designated "City of Literature" by the UNESCO Creative Cities Network, and that, in time, it was the home of many historic figures from around Germany, Europe and the whole world, e.g., Alexandru Ioan Cuza (1820–1873), the prince who kickstarted the political process of unification of modern Romania; U.S. Army general George S. Patton (1885–1945); American architect Daniel H. Burnham (1846–1912) who designed the iconic Flatiron Building in New York City; British Indian poet and philosopher Muhammad Iqbal (1877–1938); national hero of the Philippines José Rizal (1861–1896); German philosopher Hannah Arendt (1906–1975) known for her deep explorations about our humanity; and, inter alia, two German and universal philosophers, Hans-Georg Gadamer (1900–2002) and Jürgen Habermas (born in 1929; also a sociologist) who have open a famous debate over whether we can transcend history and culture towards an objective critique of society. Most relevantly for this book and the Delphi-based method for generating eco-city scenarios (introduced at the beginning of this module), is Habermas' concept of Communicative Action, i.e., cooperative action undertaken by a set of individuals based upon mutual deliberation and argumentation [152, 175, 477]; and Gadamer's concept of philosophical hermeneutics, a pursuit of the nature of human understanding (what happens to us when we understand), as notably embodied in his *opus magnus* "Truth and Method" ([478], see also [479] for a discussion connecting Gadamer's philosophy with current sustainability issues), and its follow-up book "Who am I and who are you" where hermeneutics is "a philosophical effort to account for understanding as an ontological—the ontological—process of man" [480].

Essentially, these philosophical walks are short inspiring physical journeys and constitute just a sub-category of a broader category of "taking some fresh air" in a park, garden or other public landscapes and lanes that combine nature and history (and culture in general). Examples in this sense abound in all historical districts (and water fronts) of cities around the world.

Inspiration is only a beginning

In this book section, I used just a few examples, chosen based on my own experience with and in different cities, but this is merely an invitation for each of you, the readers, to do the same, as what matters for each of us is our own experience and what matters for us as a community is sharing our understanding about the world and learning from others about the world. To use again the words of Carl Sagan (1934–1996), a scientist who loved Humanity and inspired many others (generations), the journey for each of us begins in our mind, and "*we make our world significant by the courage of our questions and by the depth of our answers*" [481, p. 193], and we might recall the memorable introduction to the TV series Cosmos (1980; written by Carl Sagan, Ann Druyan, Steven Soter in 1978–1979): "*The surface of the Earth is the shore of the cosmic ocean. On this shore, we have learned most of what we know. Recently, we've waded a little way out... maybe ankle-deep, and the water seems inviting. Some part of our being knows this is where we came from. We long to return. And we can. Because the cosmos is also within us. We're made of star-stuff. We are a way for the cosmos to know itself. The journey for each of us begins here*" (in the film, when saying that last word, Carl Sagan points his finger to his head).

But an inspired vision is only a beginning, and the summary made in the quotation above is more profound than it may appear.

For our concern here, Cities of Inspiration are those cities into which their own people and friends have put an immense amount of work. Inspiration only means something when it is a constant process of search of the possible in its operational details (indeed, this is the meaning of the word "possibilities") so as to find what works and what doesn't, and then to identify what shall be the best options forward.

It was certainly not by chance that the famous old Library of Alexandria included about half a million works in the main building and at least the same number in its annex facilities. And, the fact that this treasure of Humanity was burned in dogmatic wars (when philosopher Hypathia herself, was humiliated and killed by a mob made of ideological "activists" and fanatics) should act as a stark warning that the lack of pondering and tolerance for alternative views is at least as consequential as inspiration itself, albeit in the opposite way: it can destroy the work of entire generations of brilliant minds, industrious hearts, and skilled hands.

At present times, to take seriously the challenge of urban and overall sustainability introduced in Module 1 (and illustrated in Sect. 1.4, Fig. 1.13 with the model charting possible paths of Humanity), one question needs to be answered pragmatically in eco-city project: *Actually, how much do we learn from history*?

A famous quote by writer and philosopher Aldous Huxley suggests the beginning of an answer: "that men do not learn very much from the lessons of history is the most important of the lessons that history has to teach" [482]. However, based on knowledge to date, we can only reckon that history (taken alone) is not enough: we need evidence-based decision-making, and history is but one among other necessary-but-insufficient resources. To learn from history or from present, it is not enough to "listen to it". As a popular saying that I heard around me as a child, "Many can see, few actually understand" (in Romanian, "*Mulţi văd, puțini cunosc*").

We need to actively use purposeful and perseverant Research, Development and Innovation (RDI) in problem-solving-oriented projects. Whenever we did that in the past, we have actually proved that we are in fact very good at learning, history included—with a caveat: generations change and each generation needs to learn its lessons.

Even if, in complex process of transition to sustainability, we cannot possibly (and we need not) "know everything from the beginning", we have to "know what we are doing". This means that cities need to prioritize on learning and building a real capacity to act.

In this respect too, as always, we must avoid the avoidable mistakes. Most notably, this active transition to sustainability is <u>not</u> a matter of "ambition" and "visionary" "leadership" (to cite only a few popular "memes") but indeed a matter of legitimate aspirations and professional adequacy maintained through intelligent work because, as the popular wisdom tells us, "the Devil hides in details", and "the road to Hell is paved with good intentions". Good policies and management require a good understanding of different concepts that are being proposed and handled in both academic and public agoras. We know that politics is often described as "art of the possible" but this is not (and cannot be) so if science is ignored or dismissed and replaced with ideologies or/and pure voluntarism. The results of such approaches can only be catastrophic. To the point, history tells us that social experiments attempting to do that (like Michurinism and Lysenkoism in the soviet-dominated parts of the world: "we cannot wait for Nature's good graces—to take them from her is our goal" ([483, see also [484, 485]) have failed, and current knowledge corroborates these insights to tell us that (and why) such attempts will always fail if attempted again.

Relatedly, some worrying trends requite caution at present times. One example is the propensity of many to restrict their attention and focus to the "1% inspiration" part and to ignore the "99% perspiration" part. Another and related example is the preference (of many, perhaps a majority) for acting based on clichés like "we must show that we are doing something" (even when the objectively best course of action is doing less or nothing) [1]. Both of these trends betray a naive and dangerous way of conducting personal and social affairs by certain people: they seem to just ignore the reality (/fact of life) that we exist and live in complex dynamic systems.

What we need on the difficult path towards urban and society sustainability is not "ambition" (we already have that, and it can also turn destructive). Instead, truly and productively inspiring are those cities and people who face reality and demonstrate grace under duress.

The next module of this book is focused on project management.

References

1. Ciumasu IM (2023b) Mapping Industry 4.0 onto eco-city transitions: a knowledge-action matrix. In: Da Silva Portela CF (ed) Sustainable, Innovative and Intelligent Societies and Cities. Springer, Cham, CH. https://doi.org/10.1007/978-3-031-30514-6_13
2. Gabrys J (2022) Programming nature as infrastructure in the smart forest city. J Urban Technol 29(1):13–19. https://doi.org/10.1080/10630732.2021.2004067
3. Cugurullo F (2018) Exposing smart cities and eco-cities: Frankenstein Urbanism and the sustainability challenges of the experimental city. Environ Plann A: Econ Space 50(1):73–92. https://doi.org/10.1177/0308518X17738
4. Maitlis S, Christianson M (2014) Sensemaking in organizations: taking stock and moving forward. Acad Manag Ann 8(1):57–125
5. Sandberg J, Tsoukas H (2020) Sensemaking reconsidered: towards a broader understanding through phenomenology. Organ Theory 1(1):2631787719879937. https://doi.org/10.1177/2631787719879937
6. Weick KE (1995) Sensemaking in organizations. Sage, Thousand Oaks, Calif. https://doi.org/10.1002/hrdq.3920090211
7. Weick KE (2020) Sensemaking, organizing, and surpassing: a handoff. J Manage Stud 57(7):1420–1431. https://doi.org/10.1111/joms.12617
8. Westley F, Carpenter SR, Brock WA, Holling CS, Gunderson LH (2002) Why systems of people and nature are not just social and ecological systems. In: Gunderson LH, Holling CS (eds) Panarchy: understanding transformations in systems of humans and nature. Washington, Island Press, pp 103–119. http://hdl.handle.net/10919/65569
9. Wieland A (2021) Dancing the supply chain: toward transformative supply chain management. J Supply Chain Manag 57(1):58–73. https://doi.org/10.1111/jscm.12248
10. Darby JL, Fugate BS, Murray JB (2019) Interpretive research: a complementary approach to seeking knowledge in supply chain management. Int J Logistics Manage 30:395–413. https://doi.org/10.1108/IJLM-07-2018-0187
11. Granovetter M (1985) Economic action and social structure: the problem of embeddedness. Am J Sociol 91:481–510. https://doi.org/10.1086/228311
12. Dincă AC (2022) Istorie–percepţie istorică–istoriografie. Despre originea şi etnia primilor coloniști occidentali ai Transilvaniei medievale. Anuarul Institutului de Istorie George Bariţiu Ser HISTORICA 61(61):15–33. https://www.ceeol.com/search/article-detail?id=1091208
13. Ortman SG, Lobo J, Smith ME (2020) Cities: complexity, theory and history. PLoS ONE 15(12):e0243621. https://doi.org/10.1371/journal.pone.0243621
14. Ciumasu IM (2013a) Dynamic decision trees for building resilience into future eco-cities. Technol Forecast Soc Chang 80(9):1804–1814. https://doi.org/10.1016/j.techfore.2012.12.010
15. Friedman G (2023) Forecasting Russia. Geopolitical futures, July 5, 2023. https://geopoliticalfutures.com/forecasting-russia/
16. Birkinshaw JM, Cohen J (2013) Make time for the work that matters. Harvard Bus Rev 91(9):115–118. https://hbr.org/2013/09/make-time-for-the-work-that-matters
17. Simon D (ed) (2016) Rethinking sustainable cities. Accessible, green and fair. Policy Press, University of Bristol, Bristol, UK. ISBN 978-1-4473-3284-8

18. Clar C, Steurer R (2018) Why popular support tools on climate change adaptation have difficulties in reaching local policy-makers: qualitative insights from the UK and Germany. Environ Policy Gov 28:172–182. https://doi.org/10.1002/eet.1802
19. Lubchenco J (1998) Entering the century of the environment: a new social contract for science. Science 279:491–497. https://doi.org/10.1126/science.279.5350.491
20. Lubchenco J, Rapley C (2020) Our moment of truth: the social contract realized?. Environ Res Lett 15(11):article 110201. https://doi.org/10.1088/1748-9326/abba9c
21. Esslin M (1960) The theatre of the absurd. Tulane Drama Rev 4(4):3–15. https://doi.org/10.2307/1124873
22. Bennett MY (2011) Reassessing the theatre of the absurd: Camus, Beckett, Ionesco, Genet, and pinter. Palgrave Macmillan, New York. https://doi.org/10.1057/9780230118829
23. Vassilopoulou K (2008) Possible worlds in the theatre of the absurd. In: Gabrielatos C, Slessor R, Unger JW (eds) Lancaster university postgraduate conference on linguistics and language teaching, vol 1, pp 120–138. https://faculty.uobasrah.edu.iq/uploads/teaching/1642439873.pdf
24. Bouckaert G (2020) From public administration in utopia to utopia in public administration. In: Bouckaert G, Jann W (eds) European perspectives for public administration: the way forward. Leuven University Press, Leuven, Belgium pp 71–84. https://doi.org/10.11116/9789461663078
25. Heald D (2003) Fiscal transparency: concepts, measurement and UK practice. Public Adm 81(4):723–759. https://doi.org/10.1111/j.0033-3298.2003.00369.x
26. Jurkiewicz CL, Mujkic E (2021) Deconstructing NPM: public service values in a global public administration. The Palgrave handbook of the public servant, pp 225–239. https://doi.org/10.1007/978-3-030-29980-4_80
27. Bossy MA (2001) Artists, writers, and musicians: an encyclopedia of people who changed the world. Oryx Press, Westport, Connecticut. ISBN 978-1-57356-154-9. https://www.bloomsbury.com/us/artists-writers-and-musicians-9781573561549/
28. Nasir MA (2012) Logic of the absurd: reading Kafka in a Kafkaesque world. Adm Theor Prax 34(1):40–59. https://doi.org/10.2753/ATP1084-1806340103
29. Clegg S, Pina e Cunha M, Munro I, Rego A, de Sousa MO (2016) Kafkaesque power and bureaucracy. J Polit Power 9(2):157–181. http://www.tandfonline.com/10.1080/2158379X.2016.1191161
30. Jones DR (2017) Opening up the pandora's box of sustainability league tables of universities: a Kafkaesque perspective. Stud High Educ 42(3):480–503. https://doi.org/10.1080/03075079.2015.1052737
31. McCabe D (2014) Light in the darkness? Managers in the back office of a Kafkaesque bank. Organ Stud 35(2):255–278. https://doi.org/10.1177/0170840613511928
32. Nisar MA, Masood A (2020) Dealing with disgust: street-level bureaucrats as agents of Kafkaesque bureaucracy. Organization 27(6):882–899. https://doi.org/10.1177/1350508419883382
33. Grubb M, Wieners C, Yang P (2021) Modeling myths: on DICE and dynamic realism in integrated assessment models of climate change mitigation. Wiley Interdiscip Rev: Clim Change 12(3):article e698. https://doi.org/10.1002/wcc.698
34. Bocken NMP (2015) Sustainable venture capital—catalyst for sustainable start-up success? J Cleaner Prod 108(part A):647–658. http://www.sciencedirect.com/science/article/pii/S0959652615006460
35. Voß JB, Bornemann B (2011) The politics of reflexive governance: challenges for designing adaptive management and transition management. Ecol Soc 16(2):article 9. https://www.jstor.org/stable/26268901
36. Beaverstock JV, Smith RG, Taylor PJ (1999) A roster of world cities. Cities 16(6):445–458. https://doi.org/10.1016/S0264-2751(99)00042-6
37. Beaverstock JV, Smith RG, Taylor PJ (2000) World-city network: a new metageography? Ann Am Assoc Geogr 90(1):123–134. https://doi.org/10.1111/0004-5608.00188

38. Sassen S (2002) Locating cities on global circuits. In: Global networks, linked cities. https://doi.org/10.4324/9781315538808
39. Evans G (2009) Creative cities, creative spaces and urban policy. Urban Stud 46(5–6):1003–1040. https://doi.org/10.1177/0042098009103853
40. Tsolakis N, Anthopoulos L (2015) Eco-cities: an integrated system dynamics framework and a concise research taxonomy. Sustain Cities Soc 17:1–14. https://doi.org/10.1016/j.scs.2015.03.002
41. Blake SE (2015) New York city's water supply: a case for a sustainable growth machine (Doctoral dissertation, Washington State University, Pullman, US-WA). https://rex.libraries.wsu.edu/view/pdfCoverPage?instCode=01ALLIANCE_WSU&filePid=13349961700001842&download=true
42. Molotch H (1976) The city as a growth machine: toward a political economy of place. Am J Sociol 82(2):309–332 (1976). https://doi.org/10.1086/226311
43. UN-Habitat (2016) Urbanization and development. Emerging futures (World cities report 2016). The United Nations Human Settlements Programme (UN-Habitat), Nairobi. ISBN 9789211327083. https://unhabitat.org/world-cities-report-2016
44. McDonald RI (2008) Global urbanization: can ecologists identify a sustainable way forward? Front Ecol Environ 6:99–104. https://doi.org/10.1890/070038
45. Breuste J, Qureshi S (2011) Urban sustainability, urban ecology and the society for urban ecology (SURE). Urban Ecosyst 14:313–317. https://doi.org/10.1007/s11252-011-0186-3
46. Ciumasu IM, Culver K (2011a) Eco-cities in a globalized future—from constantinopolis to cosmopolis? In: The 9th International Conference of the European Society of Ecological Economics (ESEE), 14–17 June, Istanbul (2011). https://www.academia.edu/26227428/Eco_cities_in_a_globalized_future_from_Constantinopolis_to_Cosmopolis_
47. Rees W, Wackernagel M (2008) Urban ecological footprints: why cities cannot be sustainable—and why they are a key to sustainability. In: Marzluff JM, Shulenberger E, Endlicher W, Alberti M, Bradley G, Ryan C, Simon U, ZumBrunnen C (eds) Urban ecology: an international perspective on the interaction between humans and nature. Springer, Boston, MA, pp 537–555. https://doi.org/10.1007/978-0-387-73412-5_35
48. Wackernagel M, Kitzes J, Moran D, Goldfinger S, Thomas M (2006) The ecological footprint of cities and regions: comparing resource availability with resource demand. Environ Urban 18(1):103–112. https://doi.org/10.1177/0956247806063978
49. Portney KE (2003) Taking sustainable cities seriously: economic development, the environment, and quality of life in American cities. MIT Press, Cambridge, Massachusetts
50. Shaw A, Sheppard S, Burch S, Flanders D, Wiek A, Carmichael J, Robinson J, Cohen S (2009) Making local futures tangible—synthesizing, downscaling, and visualizing climate change scenarios for participatory capacity building. Glob Environ Chang 19(4):447–463. https://doi.org/10.1016/j.gloenvcha.2009.04.002
51. Weber KM, Kubeczko K, Kaufmann A, Grunewald B (2009) Trade-offs between policy impacts of future-oriented analysis: experiences from the innovation policy foresight and strategy process of the City of Vienna. Technol Anal Strateg Manag 21(8):953–969. https://doi.org/10.1080/09537320903262314
52. Geels FW, Schot J (2007) Typology of sociotechnical transition pathways. Res Policy 36(3):399–417. https://doi.org/10.1016/j.respol.2007.01.003
53. Kemp R, Rotmans J (2009) Transitioning policy: co-production of a new strategic framework for energy innovation policy in the Netherlands. Policy Sci 42(4):303–322. https://doi.org/10.1007/s11077-009-9105-3
54. Kemp R, Loorbach D, Rotmans J (2007) Transition management as a model for managing processes of co-evolution towards sustainable development. Int J Sustain Dev World Ecol 14(1):78–91. https://doi.org/10.1080/13504500709469709
55. Williams K (2010) Sustainable cities: research and practice challenges. Int J Urban Sustain Dev 1(1–2):128–132. https://doi.org/10.1080/19463131003654863
56. Wheeler SM, Beathley T (2009) The sustainable urban development reader, 2nd edn. Routledge, London

57. Świąder M, Lin D, Szewrański S, Kazak JK, Iha K, van Hoof J, Belčáková I, Altiok S (2020) The application of ecological footprint and biocapacity for environmental carrying capacity assessment: a new approach for European cities. Environ Sci Policy 105:56–74. https://doi.org/10.1016/j.envsci.2019.12.010
58. Flyvbjerg B (2014) What you should know about megaprojects and why: an overview. Proj Manag J 45(2):6–19. https://doi.org/10.1002/pmj.21409
59. IPCC-WGI (2022) The physical science basis. Working Group I contribution to the sixth assessment report. The International Panel for Climate Change. https://www.ipcc.ch/report/sixth-assessment-report-working-group-i/
60. IPCC-WGII (2022) Impacts, adaptation and vulnerability. Working Group II contribution to the sixth assessment report. The International Panel for Climate Change. https://www.ipcc.ch/report/sixth-assessment-report-working-group-ii/
61. IPCC-WGIII (2022) Mitigation of climate change. Working Group III contribution to the sixth assessment report. The international panel for climate change. https://www.ipcc.ch/report/sixth-assessment-report-working-group-3/
62. Botzen WW, Deschenes O, Sanders M (2019) The economic impacts of natural disasters: a review of models and empirical studies. Rev Environ Econ Policy 13(2):167–341. https://doi.org/10.1093/reep/rez004
63. Coronese M, Lamperti F, Keller K, Chiaromonte F, Roventini A (2019) Evidence for sharp increase in the economic damages of extreme natural disasters. Proc Natl Acad Sci 116(43):21450–21455. https://doi.org/10.1073/pnas.1907826116
64. Formetta G, Feyen L (2019) Empirical evidence of declining global vulnerability to climate-related hazards. Global Environ Change 57:article 101920. https://doi.org/10.1016/j.gloenvcha.2019.05.004
65. Hallegatte S, Vogt-Schilb A, Rozenberg J, Bangalore M, Beaudet C (2020) From poverty to disaster and back: a review of the literature. Econ Disasters Clim Change 4:223–247. https://doi.org/10.1007/s41885-020-00060-5
66. Hemmati M, Ellingwood BR, Mahmoud HN (2020) The role of urban growth in resilience of communities under flood risk. Earth's Future 8:article e2019EF001382. https://doi.org/10.1029/2019EF001382
67. Jones RL, Guha-Sapir D, Tubeuf S (2022) Human and economic impacts of natural disasters: can we trust the global data?. Sci Data 9(1):article 572. https://doi.org/10.1038/s41597-022-01667-x
68. UN-ISDR (2009) Risk and poverty in a changing climate: invest today for a safer tomorrow, United Nations international strategy for natural disaster reduction global assessment report on disaster risk reduction, 2009, p 207. https://www.undrr.org/publication/global-assessment-report-disaster-risk-reduction-2009
69. UN-ISDR (2011) Revealing risk, redefining development, United Nations international strategy for natural disaster reduction global assessment report on disaster risk reduction, 2011, p 178. https://www.undrr.org/publication/global-assessment-report-disaster-risk-reduction-2011
70. UN-GAR (2022) Global assessment report on disaster risk reduction (GAR2022): our world at risk. Transforming governance for a resilient future. united nations office for disaster risk reduction. https://www.undrr.org/gar/gar2022-our-world-risk-gar#container-downloads
71. UN-GAR-SD (2023) Global assessment report on disaster risk reduction (GAR) Special report: mapping resilience for the sustainable development goals. https://www.undrr.org/gar/gar2023-special-report
72. ESC (2022) European state of climate 2022. Copernicus climate change service (C3S), implemented by the European centre for medium-range weather forecasts (ECMWF) on behalf of the European commission (EC). https://climate.copernicus.eu/esotc/2022
73. ESWD (2023) European severe weather database. https://eswd.eu/cgi-bin/eswd.cgi
74. Morss RE, Wilhelmi OV, Meehl GA, Dilling L (2011) Improving societal outcomes of extremeweather in a changing climate: an integrated perspective. Annu Rev Environ Resour 36:1–25. https://doi.org/10.1146/annurev-environ-060809-100145

75. Solecki W, Leichenko R, O'Brien K (2011) Climate change adaptation strategies and disaster risk reduction in cities: connections, contentions, and synergies. Curr Opin Environ Sustain 3:135–141. https://doi.org/10.1016/j.cosust.2011.03.001
76. Fünfgeld H (2010) Institutional challenges to climate risk management in cities. Curr Opin Environ Sustain 2(3):156–160. https://doi.org/10.1016/j.cosust.2010.07.001
77. Landauer M, Juhola S, Klein J (2019) The role of scale in integrating climate change adaptation and mitigation in cities. J Environ Planning Manage 62:741–765. https://doi.org/10.1080/09640568.2018.1430022
78. Aguiar APD, Collste D, Harmackova ZV, Pereira L, Selomane O, Galafassi D, Van Vuuren D, Van Der Leeuw S (2020) Co-designing global target-seeking scenarios: a cross-scale participatory process for capturing multiple perspectives on pathways to sustainability. Glob Environ Chang 65:102198. https://doi.org/10.1016/j.gloenvcha.2020.102198
79. Herberg J, Schmitz S, Stasiak D, Schmieg G (2021) Boundary speak in sustainability studies: computational reading of a transversal field. Sci Public Policy 48(3):398–411. https://doi.org/10.1093/scipol/scab006
80. Leydesdorff L, Etzkowitz H (1996) Emergence of a Triple Helix of university—industry—government relations. Sci Public Policy 23(5):279–286. https://doi.org/10.1093/spp/23.5.279
81. Rowan NJ, Casey O (2021) Empower eco multiactor HUB: a triple helix academia-industry-authority approach to creating and sharing potentially disruptive tools for addressing novel and emerging new green deal opportunities under a United Nations sustainable development goals framework. Curr Opin Environ Sci Health 21:100254. https://doi.org/10.1016/j.coesh.2021.100254
82. Klein JT (2014) Discourses of transdisciplinarity: looking back to the future. Futures 63:68–74. https://doi.org/10.1016/j.futures.2014.08.008
83. Ciumasu IM (2013b) We need large-scale experimentation platforms for cross-sector innovation in Europe, Science|Business, 16 Jan 2013. https://sciencebusiness.net/news/76003/We-need-large-scale-experimentation-platforms-for-cross-sector-innovation-in-Europe
84. Ciumasu IM (2023a) Smart solutions for sustainability: RDI for urban and societal transitions requires cross-sectoral experimentation platforms. In: Da Silva Portela CF (ed) Sustainable, innovative and intelligent societies and cities. Springer, Cham, CH. https://doi.org/10.1007/978-3-031-30514-6_10
85. Chesbrough H, Lettl C, Ritter T (2018) Value creation and value capture in open innovation. J Prod Innov Manag 35(6):930–938. https://doi.org/10.1111/jpim.12471
86. Galende-Sánchez E, Sorman AH (2021) From consultation toward co-production in science and policy: a critical systematic review of participatory climate and energy initiatives. Energy Res Soc Sci 73:article 101907. https://doi.org/10.1016/j.erss.2020.101907
87. Biggs D, Biggs RO, Dakos V, Scholes RJ, Schoon M (2011) Are we entering an era of concatenated global crises? Ecol Soc 16(2):27. https://www.jstor.org/stable/26268899
88. Beddington J (2009) Food, energy, water and the climate: a perfect storm of global events? In: Conference presentation given to the sustainable development UK annual conference, QEII Conference Centre, London, 19 March (2009). http://www.bis.gov.uk/assets/goscience/docs/p/perfect-storm-paper.pdf. archived; Accessed 20 March 2020
89. Culver K, Guilloteau R, Hue C (2011) Hard nodes in soft surroundings: a dream of islands strategy for urban sustainability. Development 54(3):336–342. https://doi.org/10.1057/dev.2011.58
90. Dearing J, Bullock S, Costanza R, Dawson T, Edwards M, Poppy G, Smith G (2012) Navigating the perfect storm: research strategies for socialecological systems in a rapidly evolving world. Environ Manage 49(4):767–775. https://doi.org/10.1007/s00267-012-9833-6
91. Falconer RA, Norton MR (2012) Global water security: engineering the future. In: Fernando HJS, Klaić ZB, McCulley JL (eds) National security and human health implications of climate change. NATO Science for Peace and Security Series C: Environmental Security. Springer Netherlands pp 261–269. https://doi.org/10.1007/978-94-007-2430-3_22
92. Fermeglia M, Lughi V, Pavan AM (2020) How to avoid the perfect storm: the role of energy and photovoltaics. MRS Energy Sustain 7:e34. https://doi.org/10.1557/mre.2020.36

93. Schlör H, Venghaus S, Fischer W, Märker C, Hake JF (2018) Deliberations about a perfect storm–the meaning of justice for food energy water-nexus (FEW-Nexus). J Environ Manage 220:16–29. https://doi.org/10.1016/j.jenvman.2018.04.097
94. Zhou H, Wang J, Wan J, Jia H (2010) Resilience to natural hazards: a geographic perspective. Nat Hazards 53(1):21–41. https://doi.org/10.1007/s11069-009-9407-y
95. Whitehead M (2012) The sustainable city: an obituary? On the future form and prospects of sustainable urbanism. In: Flint J, Raco M (eds) The future of sustainable cities: critical reflections. Policy Press, Bristol, UK, pp 29–46. ISBN 9781847426666
96. Raskin PR (2005) Global scenarios: background review for the Millennium Ecosystem Assessment. Ecosystems 8:133–142. https://doi.org/10.1007/s10021-004-0074-2
97. Ahern J (2011) From fail-safe to safe-to-fail: sustainability and resilience in the new urban world. Landsc Urban Plan 100(4):341–343. https://doi.org/10.1016/j.landurbplan.2011.02.021
98. Moldoveanu MC, Martin RL (2008) The future of the MBA, Oxford University Press, New York. ISBN: 978-0-19-534014-3
99. Scholz RW, Spoerri A, Lang DJ (2009) Problem structuring in transitions: the case of Swiss waste management. Futures 41(3):171–181. https://doi.org/10.1016/j.futures.2008.09.013
100. Ciumasu IM (2012b) Technological transitions on shifting markets: the emergence of eco-innovation. Invited plenary talk at Micro Nano Systems—Micro Nano Fluidics (Journées Nationales du GDR Micro Nano Systèmes—Micro Nano Fluidique), 02–03 July. University of Bordeaux
101. Ciumasu IM, Lupu A, Costica N, Netedu A, Stratu A, Miftode V, Stefan N (2008c) Integrating social perception of nature with economic development expectations in a peri-urban area in Iasi, Romania. Talk and paper. In: ISEE 2008: Applying Ecological Economics for Social and Environmental Sustainability (10th Biennial Conference of the International Society for Ecological Economics), 7–11 August, Nairobi-UNEP (2008)
102. Muñoz-Erickson TA, Miller CA, Miller TR (2017) How cities think: knowledge co-production for urban sustainability and resilience. Forests 8(6):203. https://doi.org/10.3390/f8060203
103. Frick V (2023) Understanding the democratic promise of the city. Philosophy and Social Criticism, published online on 7 January. https://doi.org/10.1177/01914537221150456
104. Ciumasu IM (2018a) A coordination lattice model for building urban resilience. In: Proceedings of the 15th Information Systems for Crisis Response and Management (ISCRAM) conference, Rochester, NY, USA, 23–25 May, 419–427. http://idl.iscram.org/files/ioanmciumasu/2018/2119_IoanM.Ciumasu2018.pdf
105. Nguyen TAV, Tucek D, Pham NT (2023) Indicators for TQM 4.0 model: Delphi method and analytic hierarchy process (AHP) analysis. Total Q Manage Bus Excellence 34(1–2):220–234. https://doi.org/10.1080/14783363.2022.2039062
106. Marchais-Roubelat A, Roubelat F (2008) Designing action-based scenarios. Futures 40:25–33. https://doi.org/10.1016/j.futures.2007.06.008
107. Rotmans J, van Asselt M, Anastasi C, Greeuw S, Mellors J, Peters S, Rothman D, Rijkens N (2000) Visions for a sustainable Europe. Futures 32:809–831. https://doi.org/10.1016/S0016-3287(00)00033-1
108. Boon E, Wright SJ, Biesbroek R, Goosen H, Ludwig F (2022) Successful climate services for adaptation: what we know, don't know and need to know. Clim Serv 27:article 100314. https://doi.org/10.1016/j.cliser.2022.100314
109. Kenworthy JR (2006) The eco-city: ten key transport and planning dimensions for sustainable city development. Environ Urban 18(1):67–85. https://doi.org/10.1177/0956247806063947
110. Ratcliffe J, Krawczyk E (2011) Imagineering city futures: the use of prospective through scenarios in urban planning. Futures 43:642–653. https://doi.org/10.1016/j.futures.2011.05.005
111. Tourki Y, Keisler J, Linkov I (2013) Scenario analysis: a review of methods and applications for engineering and environmental systems. Environ Syst Decisions 33:3–20. https://doi.org/10.1109/MC.2020.2993623

112. Bradfield R, Wright G, Burt G, Cairns G, Van Der Heijden K (2005) The origins and evolution of scenario techniques in long range business planning. Futures 37(8):795–812. https://doi.org/10.1016/j.futures.2005.01.003
113. Cairns G, Wright G, Fairbrother P (2016) Promoting articulated action from diverse stakeholders in response to public policy scenarios: a case analysis of the use of 'scenario improvisation' method. Technol Forecast Soc Chang 103:97–108. https://doi.org/10.1016/j.techfore.2015.10.009
114. Lehr T, Lorenz U, Willert M, Rohrbeck R (2017) Scenario-based strategizing: advancing the applicability in strategists teams. Technol Forecast Soc Chang 124:214–224. https://doi.org/10.1016/j.techfore.2017.06.026
115. Lang T, Ramírez R (2017) Building new social capital with scenario planning. Technol Forecast Soc Chang 124:51–65. https://doi.org/10.1016/j.techfore.2017.06.011
116. Favato G, Vecchiato R (2017) Embedding real options in scenario planning: a new methodological approach. Technol Forecast Soc Chang 124:135–149. https://doi.org/10.1016/j.techfore.2016.05.016
117. Burt G, Mackay DJ, van der Heijden K, Verheijdt C (2017) Openness disposition: readiness characteristics that influence participant benefits from scenario planning as strategic conversation. Technol Forecast Soc Chang 124:16–25. https://doi.org/10.1016/j.techfore.2016.11.024
118. Kunc M, O'Brien FA (2017) Exploring the development of a methodology for scenario use: combining scenario and resource mapping approaches. Technol Forecast Soc Chang 124:150–159. https://doi.org/10.1016/j.techfore.2017.03.018
119. Ravasz E, Somera AL, Mongru DA, Oltvai ZN, Barabási AL (2002) Hierarchical organization of modularity in metabolic networks. Science 297:1551–1555. https://doi.org/10.1126/science.1073374
120. Miller K, Kumar MP, Packer B, Goodman D, Koller D (2012) Max-margin min-entropy models. In: Proceedings of the 15th International Conference on Artificial Intelligence and Statistics (AISTATS), PMLR, 22, 779–787. http://proceedings.mlr.press/v22/miller12.html
121. Peel L, Larremore DB, Clauset A (2017) The ground truth about metadata and community detection in networks. Sci Adv 3(5):article e1602548. https://doi.org/10.1126/sciadv.1602548
122. Caire JM, Maurel-Techene S, Letellier T, Heiske M, Warren S, Schabaille A, Destruhaut F (2022) Canadian occupational performance measure: benefits and limitations highlighted using the Delphi method and Principal Component Analysis. Occup Ther Int. Article 9963030 (published: 02 March 2022) https://doi.org/10.1155/2022/9963030
123. Kaufmann PR (2016) Integrating factor analysis and the Delphi method in scenario development: a case study of Dalmatia, Croatia. Appl Geogr 71:56–68. https://doi.org/10.1016/j.apgeog.2016.04.007
124. Östberg J, Sandberg K, Wiström B (2021) Rating of parameters used to assess tree vitality by urban foresters and ecologists in Sweden, using the Delphi method. Urban Forest Urban Greening 62:article 127134. https://doi.org/10.1016/j.ufug.2021.127134
125. Cohen-Addad V, Kanade V, Mallmann-Trenn F, Mathieu C (2019) Hierarchical clustering: objective functions and algorithms. J ACM 66(4):article 26. https://doi.org/10.1145/3321386
126. Meilă M (2007) Comparing clusterings—an information based distance. J Multivar Anal 98:873–895. https://doi.org/10.1016/j.jmva.2006.11.013
127. Anastasiadou K, Gavanas N, Pyrgidis C, Pitsiava-Latinopoulou M (2021) Identifying and prioritizing sustainable urban mobility barriers through a modified Delphi-AHP approach. Sustainability 13:10386. https://doi.org/10.3390/su131810386
128. Khorramshahgol R, Moustakis VS (1988) Delphic hierarchy process (DHP): a methodology for priority setting derived from the Delphi method and Analytical Hierarchy Process. Eur J Oper Res 37(3):347–354. https://doi.org/10.1016/0377-2217(88)90197-X
129. Saaty TL (1980) The Analytical Hierarchy Process: planning, priority setting, resource allocation. McGraw Hill, New York, NY, USA
130. Vidal LA, Marle F, Bocquet JC (2011) Using a Delphi process and the analytic hierarchy process (AHP) to evaluate the complexity of projects. Expert Syst Appl 38(5):5388–5405. https://doi.org/10.1016/j.eswa.2010.10.016

131. Linstone HA, Turoff M (eds) (1975) The Delphi method. Reading, MA, Addison-Wesley pp 3–12. https://citeseerx.ist.psu.edu/document?repid=rep1&type=pdf&doi=863472a67f5bdc67e4782306efd883fca23e3a3d
132. Dalkey N, Helmer O (1963) An experimental application of the Delphi method to the use of experts. Manage Sci 9(3):458–467. https://doi.org/10.1287/mnsc.9.3.458
133. Orizont 2020 (2007) Strategia de dezvoltare economico—socială durabilă a Municipiului Iaşi. http://www.primaria-iasi.ro/uploads/ORIZONT2020.pdf2007. Accessed 23 March 2011
134. Muñoz-Erickson TA (2014) Co-production of knowledge–action systems in urban sustainable governance: the KASA approach. Environ Sci Policy 37:182–191. https://doi.org/10.1016/j.envsci.2013.09.014
135. Muñoz-Erickson TA, Cutts BB, Larson EK, Darby KJ, Neff M, Wutich A, Bolin B (2010) Spanning boundaries in an Arizona watershed partnership: information networks as tools for entrenchment or ties for collaboration? Ecol Soc 15(3):article 22. http://www.ecologyandsociety.org/vol15/iss3/art22/10.1890/100035
136. Hannibal B, Meerow S, Woodruff SC, Roy M, Gilbertson PG, Matos M (2022) Who collaborates on urban resilience? An analysis of flood resilience planning networks in four coastal cities. J Urban Aff (online first). https://doi.org/10.1080/07352166.2022.2137033
137. Krätke S (2010) Regional knowledge networks: a network analysis approach to the interlinking of knowledge resources. Eur Urban Reg Stud 17(1):83–97. https://doi.org/10.1177/0969776409350794
138. Neamtu M, Ciumasu IM, Costica N, Costica M, Bobu M, Nicoara MN, Catrinescu C, Becker van Slooten K, De Alencastro LF (2009) Chemical, biological, and ecotoxicological assessment of pesticides and persistent organic pollutants in the Bahlui river, Romania. Environ Sci Pollut Res 16:S76–S85. https://doi.org/10.1007/s11356-009-0101-0
139. Castells-Quintana D, Dienesch E, Krause M (2021) Air pollution in an ur-ban world: a global view on density, cities and emissions. Ecol Econ 189:article 107153. https://doi.org/10.1016/j.ecolecon.2021.107153
140. Ciumasu IM, Costica N (2010) Impacts of air pollution on the ecosystem and human health. Health and environmental impacts. In: Gurjar BR, Molina LT, Ojha CSP (eds) Air pollution. Health and environmental impacts. CRC Press (Taylor & Francis), Boca Raton pp 447–492. eBook ISBN 9780429075599. https://doi.org/10.1201/EBK1439809624
141. Rizzati MCP, Florenzio N, Guastella G, Pareglio S (2023) Kuznets and the cities: urban level EKC evidence from Europe. Ecol Ind 148:110143. https://doi.org/10.1016/j.ecolind.2023.110143
142. Sinha A, Bhattacharya J (2016) Environmental Kuznets curve estimation for NO_2 emission: a case of Indian cities. Ecol Ind 67:1–11. https://doi.org/10.1016/j.ecolind.2016.02.025
143. Sinha A, Bhattacharya J (2017) Estimation of environmental Kuznets curve for SO_2 emission: a case of Indian cities. Ecol Ind 72:881–894. https://doi.org/10.1016/j.ecolind.2016.09.018
144. Hjørland B (2016) Knowledge organization (KO). KO Knowl Organ 43(6):475–484. https://www.isko.org/cyclo/knowledge_organization.htm
145. Hjørland B (2021) Information retrieval and knowledge organization: a perspective from the philosophy of science. Information 12(3) :article 135. https://doi.org/10.3390/info12030135
146. Martin S-A (2009) Machu Picchu, sunset panorama. Public domain, CC BY-SA 3.0. https://commons.wikimedia.org/w/index.php?curid=8124183
147. Israilidis J, Odusanya K, Mazhar MU (2021) Exploring knowledge management perspectives in smart city research: a review and future research agenda. Int J Inf Manag 56:article 101989. https://doi.org/10.1016/j.ijinfomgt.2019.07.015
148. Supriana CE (2021) Designing knowledge sharing systems to support integrated eco-city planning and management. IOP Conf Ser: Earth Environ Sci 737(1), paper 012006. IOP Publishing. https://doi.org/10.1088/1755-1315/737/1/012006
149. Perveen S, Kamruzzaman M, Yigitcanlar T (2017) Developing policy scenarios for sustainable urban growth management: a Delphi approach. Sustainability 9(10):1787. https://doi.org/10.3390/su9101787

150. Ameen RFM, Mourshed M (2019) Urban sustainability assessment framework development: the ranking and weighting of sustainability indicators using analytic hierarchy process. Sustain Cities Soc 44:356–366. https://doi.org/10.1016/j.scs.2018.10.020
151. Bagheri M, Zaiton Ibrahim Z, Mansor S, Manaf LA, Akhir MF, Talaat WIAWB, Pour A (2021) Land-use suitability assessment using Delphi and analytical hierarchy process (D-AHP) hybrid model for coastal city management: Kuala Terengganu, Peninsular Malaysia. ISPRS Int J Geo Inf 10(9):621. https://doi.org/10.3390/ijgi10090621
152. Habermas J (1981) Theorie des kommunikativen Handelns. Suhrkamp Verlag. ISBN 978-3-518-28775-0. https://www.suhrkamp.de/buch/juergen-habermas-theorie-des-kommunikativen-handelns-t-9783518287750
153. Bächtiger A, Dryzek JS, Mansbridge JJ, Warren M (eds) The Oxford handbook of deliberative democracy, 2018. ISBN 978-0-19-180969-9. https://www.worldcat.org/oclc/1057358164
154. Dryzek JS, Bächtiger A, Chambers S, Cohen J, Druckman JN, Felicetti A, Fishkin JS, Farrell DM, Fung A, Gutmann A, Landemore H, Mansbridge J, Marien S, Neblo MA, Niemeyer S, Setälä M, Slothuus R, Suiter J, Thompson D, Warren ME (2019) The crisis of democracy and the science of deliberation. Science 363(6432):1144–1146. https://doi.org/10.1126/science.aaw2694
155. Ercan SA (2014) Deliberative democracy. In: Phillips D (ed) Encyclopedia of educational theory and philosophy, pp 214–216. SAGE Publications, Thousand Oaks, CA. https://doi.org/10.4135/9781483346229
156. Landemore H (2017) Deliberative democracy as open, not (just) representative democracy. Dædalus 146(3):51–63. https://doi.org/10.1162/DAED_a_00446
157. Pickering J, Bäckstrand K, Schlosberg D (2020) Between environmental and ecological democracy: theory and practice at the democracy-environment nexus. J Environ Planning Policy Manage 22(1):1–15. https://doi.org/10.1080/1523908X.2020.1703276
158. Romsdahl R, Blue G, Kirilenko A (2018) Action on climate change requires deliberative framing at local governance level. Clim Change 149(3):277–287. https://doi.org/10.1007/s10584-018-2240-0
159. Ciumasu IM (2013c) Is there a future for eco-cities? Prospective roles of Versailles chantiers within the greater paris: a hub of multi-modal communication. Technical report of the econoving integration seminar (2011–2013). ECONOVING chair for generating eco-innovation, REEDS center, University of Versailles Saint-Quentin-en-Yvelines. https://doi.org/10.13140/RG.2.1.3866.8240
160. City of Versailles (2022) Versailles Chantiers (in French). https://www.versailles.fr/actualites/toute-lactualite/detail-dune-actualite/actualites/versailles-chantiers-un-nouveau-quartier-en-2019/. Accessed 31 July 2022
161. City of Versailles (2023) Quartier de Gally, un éco-quartier innovant (in French). https://www.versailles.fr/223/decouvrir-versailles/les-grands-projets/quartier-de-gally-un-eco-quartier-innovant.htm. Accessed 29 Oct 2023
162. SMBG (2013) Classement SMBG des Meilleurs masters, MS & MBA 2013, section Management of Innovation. SMBG p 61. https://www.meilleurs-masters.com/uploads/presse/Classement_2013.pdf. Accessed 29 Oct 2023
163. Belkind L (2021) From impasse to improvisation: grand paris express as a negotiation agent in a fragmented metropolis. Built Environ 47(1):75–95. https://doi.org/10.2148/benv.47.1.75
164. Gallez C (2014) Contrats de développement territorial in the Grand Paris project: towards negotiated networked development?. Town Plan Rev 85(2):273–287. https://www.jstor.org/stable/24579238
165. Traoré I, Gavan V, Riffonneau Y, L'Henoret B, Drouet E (2013) Development of a generic and scalable Modelica-based model of a typical French railway station building. In: Proceedings of the 13th Conference of International Building Performance Simulation Association, 26–28 August, Chambéry, France, pp 25–28. https://www.aivc.org/resource/development-generic-and-scalable-modelica-based-model-typical-french-railway-station
166. Kuznetsova E, Ruiz C, Li YF, Zio E (2013) Reliable microgrid energy management under environmental uncertainty and mechanical failures: an agent-based modeling and robust

optimization approach. In: Safety, Reliability and Risk Analysis: Beyond the Horizon (ESREL 2013), pp 2873–2882
167. Kuznetsova E, Li YF, Ruiz C, Zio E (2014) An integrated framework of Agent-Based Modelling and Robust Optimization for microgrid energy management. Appl Energy 129:70–88. https://doi.org/10.1016/j.apenergy.2014.04.024
168. Ahat M, Ben Amor S, Bui M (2013) Agent Based Modeling of ecodistricts with smart grid. In: Nguyen NT, Do TV, Thi HAL (eds) Advanced computational methods for knowledge engineering. Springer, Heidelberg pp 307–318. https://doi.org/10.1007/978-3-319-00293-4_23
169. Hermans Y, Le Cun B, Bui A (2012) Individual decisions & schedule planner in a vehi-cle-to-grid context. In: 2012 IEEE International Electric Vehicle Conference, Greenville, US, pp 1–6. https://doi.org/10.1109/IEVC.2012.6183275
170. Ebrahimi S, Bridgelall R (2020) A fuzzy Delphi analytic hierarchy model to rank factors influencing public transit mode choice: a case study. Res Transp Bus Manag 39:100496. https://doi.org/10.1016/j.rtbm.2020.100496
171. Hsueh SL, Sun Y, Gao M, Hu X, Meen TH (2022) Delphi and analytical hierarchy process fuzzy model for auxiliary decision-making for cross-field learning in landscape design. Sens Mater 34(5):1707–1719. https://doi.org/10.18494/SAM3817
172. Lee SW, Xue K (2021) An integrated importance-performance analysis and modified analytic hierarchy process approach to sustainable city assessment. Environ Sci Pollut Res 28:63346–63358. https://doi.org/10.1007/s11356-021-15235-0
173. Vaidya OS, Kumar S (2006) Analytic hierarchy process: an overview of applications. Eur J Oper Res 169(1):1–29. https://doi.org/10.1016/j.ejor.2004.04.028
174. Wey WM, Peng TC (2021) Study on building a smart sustainable city assessment framework using Big Data and Analytic Hierarchical Process. J Urban Plann Dev 147(3):04021031. https://doi.org/10.1061/(ASCE)UP.1943-5444.0000704
175. Greason M (2018) Connecting findings to meaningful change: the benefits of using qualitative Delphi in empirical ethics and policy research in long-term care. Int J Qual Methods 17(1):1609406918803271. https://doi.org/10.1177/1609406918803271
176. Matemane R, Moloi T, Adelowotan M (2022) Appraising executive compensation ESG-based indicators using Analytic Hierarchical Process and Delphi techniques. J Risk Fin Manage 15(10):469. https://doi.org/10.3390/jrfm15100469
177. Sebastiaan ter Burg (2019) Europeana EU—exposing online the European cultural heritage: the impact of cultural heritage on the digital transformation of the society. Public domain via CC BY 2.0/CC BY 4.0. https://commons.wikimedia.org/w/index.php?curid=78355135
178. Manole Alexandru (2020) Vasile Alecsandri national theater in Iași, Iași county. Public domain under CC BY-SA 4.0. https://fr.wikipedia.org/wiki/Fichier:Teatrul_National_„Vasile_Alecsandri“_(7).jpg
179. Argenna (2010) University Alexandru Ioan Cuza, building a, detail central side. Public domain, under CC BY-SA 3.0. https://commons.wikimedia.org/wiki/File:Iași_,_%22Alexandru_Ioan_Cuza%22_University_,_detail_1.JPG
180. Alesia17 (2008) Bustul lui Eminescu din parcul Copou, Iaşi. public domain, via CC BY 3.0. https://commons.wikimedia.org/w/index.php?curid=4465423
181. Argenna (2016) Iasi, public garden palas. By argenna—own work, CC BY-SA 4.0. https://commons.wikimedia.org/w/index.php?curid=51430605
182. Palace of Culture (2023) Palace of culture, Jassy, Romania. (Palatul Culturii, Iasi, Romania). Official webpage: https://palatulculturii.ro. Accessed 28 Oct 2023
183. TNI (2023) Teatrul National Iasi (National theater of Jassy, official webpage, in Romanian). https://teatrulnationaliasi.ro/despre/. Accessed 28 Oct 2023
184. Romania Dacia (2015) Iasi theatre–teatrul Vasile Alecsandri. https://romaniadacia.wordpress.com/2015/02/16/iasi-theatre-teatrul-vasile-alecsandri/
185. UAIC (2023) Universitatea Alexandru Ioan Cuza/Alexandru Ioan Cuza University. Official webpage: https://www.uaic.ro/en/about-uaic/university/timeline/. Accessed 28 Oct 2023

186. Ciumasu IM, Stefan N (2008) Introduction to the theory and practice of sustainable development—introducere in teoria si practica dezvoltarii durabile. Bilingual: English and Romanian. Alexandru Ioan Cuza University Press, Iasi, Romania. ISBN 978-973-703-393-2
187. Ciumasu IM, Costica N, Costica M (2006a) Needs of methodological training of trainers in the field of environmental education. Talk and book module. University within society—UNISO 2006. In: Zaharia SE, Ivan MS (eds) UNISO 2006. University and economy: partners in education and training, ISBN (11) 978-973-596-337-8, pp 227–232. https://biblio-hepl.be/pmb//opac_css/index.php?lvl=notice_display&id=70479
188. Ciumasu IM, Costica N, Costica M, Stefan N (2006b) Academic assistance for investors in sustainable development: the approach of the centre of expertise for sustainable exploitation of ecosystems—CEXDUREC at UAIC, Iasi, Romania. Talk and book module. University within society—UNISO 2006, 18–23 July, Cluj-Napoca, Romania. In: Zaharia SE, Ivan MS (eds) UNISO 2006. University and economy: partners in education and training. ISBN (11) 978-973-596-337-8, pp 331–340. https://biblio-hepl.be/pmb//opac_css/index.php?lvl=notice_display&id=70479
189. Ciumasu IM, Costica N, Gavrilovici O (2010) The age of the global universities: focus on education for sustainable development. Talk and book module, University in society—UNISO 2010, 7–11 July, West University of Timisoara, Timisoara, Romania
190. Ciumasu IM, Costica N, Popa G (2008a) Research for sustainability: focus on Alexandru Ioan Cuza university. Talk in the 1st German-Romanian Conference on Research for Sustainability, 03 June, Bucharest, Romania. https://www.deruge.org/wp-content/uploads/2018/12/DRH_2008-02.pdf
191. Costica N, Ciumasu IM, Costica M (2006) Partnership for human resources development in Romania and the European Union: training the trainers in environmental education. Proceedings of the 2nd European Fair on Education for Sustainable Development "Promoting Education for Sustainable Development in Europe", September 13–15, Hamburg
192. Costica N, Stratu A, Costica M (2012) The key role of higher education institutions in implementing education for sustainable development (ESD). In: Conservation of plant diversity, pp 524–529. https://ibn.idsi.md/sites/default/files/imag_file/524-529.pdf
193. Costică N, Tănăsescu FV (2012) Equal opportunities to education: aspects on ecological education of children with special needs. In: Proceedings of conservation of plant diversity, Chișinău, Moldova, 16–19 Mai 2012, pp 516–524. https://ibn.idsi.md/vizualizare_articol/108882
194. Manolescu I, Borza M, Talmaciu M, Costica N, Ciumasu IM (2010) Extracurricular, nonformal undergraduate education for sustainability: the ECO-FUN project. In: International Society for Ecological Economics—ISEE 2010: advancing sustainability in a time of crisis, 22–25 August, Oldenburg & Bremen, Germany
195. Grigoraș N (1973) Complexul de monumente de la Copou–Iași. Cercetări Istorice IV (Serie Nouă, Muzeul de Istorie a Moldovei, Iași), pp 177–190
196. Lupu I (2014) Parcul Copou: Studiu Dendrologic (cu note consultative privitoare la proiectul edilitar de Reabilitare și Restaurare a Grădinii Copou din Municipiul Iași). Septembrie-Noiembrie 2013. (Studiu de specialitate disponibil în Arhivele Asociaţiei Dendro-Ornamentale „Anastasie Fătu" Iași și în Arhivele Primăriei Iași—Proiectul municipal „Reabilitare și Restaurare a Grădinii Copou din Municipiul Iași", Anexa „Studiu Dendrologic"), 2014, pp 1–16
197. Mititiuc M, Oprea A (2008) Parcuri şi grădini din Iaşi. Revista Botanică 1(1):192–196. https://ibn.idsi.md/sites/default/files/imag_file/19.Parcuri%20si%20gradini%20din%20Iasi.pdf
198. Istoria teilor în Iași (2023) Istoria teilor în Iași. Wikipedia (in Romanian). https://ro.wikipedia.org/wiki/Istoria_teilor_în_Iași. Accessed 31 Oct 2023
199. Păduraru C (2014) Teiul lui Eminescu, datat cu precizie. Care este vârsta exactă a arborelui considerat un miracol de supravieţuire (26–04–2014). Adevărul. https://adevarul.ro/stiri-locale/iasi/teiul-lui-eminescu-datat-cu-precizie-care-este-1531625.html. Accessed 28 Oct 2023
200. Toniuc A, Ifrim C (2018) Professor Dr. engineer Mandache Leocov (1928–2018) (in memoriam). Analele Stiintifice ale Universitatii Al. I. Cuza din Iasi 64:83–85. http://cercetare.bio.uaic.ro/publicatii/anale_vegetala/issue/2018/09-2018.pdf

201. Roşu L, Oişte A-M (2014) Urban landscape patterns of Iaşi city. Analysing city relations between urban land use, topography and inhabitants pressure upon urban morphology. Bull UASVM Agric 71(1):96–104
202. Ciumasu IM, Buzdugan L, Stefan N, Culver K (2012a) Ecosystems: planning. In: Jorgensen SE (ed) Encyclopedia of environmental management, Four volume set. CRC Press, pp 632–642. eISBN 9781351235860. https://www.routledgehandbooks.com/doi/10.1081/E-EEM-120047256
203. Ciumasu IM, Costica M, Costica N, Neamtu M, Dirtu AC, De Alencastro LP, Buzdugan L, Andriesa R, Iconomu L, Stratu A, Popovici OA, Secu CV, Paveliuc-Olariu C, Dunca S, Stefan M, Lupu A, Stingaciu-Basu A, Netedu A, Dimitriu RI, Gavrilovici O, Talmaciu M, Borza M (2012b) Complex risks from old urban waste landfills—a sustainability perspective from Iasi, Romania. J Hazard, Toxic Radioactive Waste 16(2):158–168. https://doi.org/10.1061/(ASCE)HZ.2153-5515.0000090
204. Ciumasu IM, Costica N, Stefan N (2008d) Urban development pressures and land-use conflicts in a Peri-urban area, North of Iasi City, Romania. In: 12th European roundtable on sustainable consumption and production—ERSCP2008, and 5th BMBF-forum for sustainability, 23–25 September, Berlin, Germany. https://nachhaltigwirtschaften.at/de/fdz/veranstaltungen/2008/20080925-european-roundtable-on-sustainable-consumption-and-production-erscp2008.php
205. Ciumasu IM, Manolescu I, Costica N, Talmaciu M, Costica M, Nita V, Stefan N (2008b) Sustainable exploitation of ecosystems using scenarios of sustainable versus unsustainable economic development: the ciric river basin, Iasi, Romania. In: ISEE 2008: applying ecological economics for social and environmental sustainability (10th Biennial Conference of the International Society for Ecological Economics), 7–11 August, Nairobi-UNEP (2008)
206. Uy PD, Nakagoshi N (2008) Application of land suitability analysis and landscape ecology to urban greenspace planning in Hanoi, Vietnam. Urban For Urban Greening 7(1):25–40. https://doi.org/10.1016/j.ufug.2007.09.002
207. Ciumasu IM (2009) From the transition to market economy to the transition to sustainability in Central Europe—a civic scientist's perspective. Talk and paper. European Society for Ecological Economics—ESEE 2009: transformation, innovation and adaptation for sustainability, 29 June–02 July, Ljubljana, Slovenia. https://www.researchgate.net/profile/Ioan-Manuel-Ciumasu/publication/242404444_FROM_THE_TRANSITION_TO_MARKET_ECONOMY_TO_THE_TRANSITION_TO_SUSTAINABILITY_IN_CENTRAL_EUROPE_-_A_CIVIC_SCIENTIST'S_PERSPECTIVE/links/5463839f0cf2c0c6aec4c99d/FROM-THE-TRANSITION-TO-MARKET-ECONOMY-TO-THE-TRANSITION-TO-SUSTAINABILITY-IN-CENTRAL-EUROPE-A-CIVIC-SCIENTISTS-PERSPECTIVE.pdf
208. Le Roux P (1992) Flight of the flamingoes: South Africa 1992–2002. Indic South Africa 9(4):109–119. https://journals.co.za/doi/pdf/10.10520/AJA0259188X_1427
209. Schoemaker PJ (2020) How historical analysis can enrich scenario planning. Futures Foresight Sci 2(3–4):e35. https://doi.org/10.1002/ffo2.35
210. Nkomo ML (2020) Finding Mont Fleur via Singapore: what role for scenario planning in South African public policy?. J Public Adm 55(3):270–282. https://www.researchgate.net/profile/Desiree-Manicom/publication/370472552_Public_Participation_in_Public_Policy_Making_the_Case_of_the_Lesotho_National_Decentralisation_Policy_Khiba_Manuebe_and_Desiree_Manicom/links/645233ae809a53502147dd5b/
211. Kahane A (2010) Learning from experience: the Mont Fleur scenario exercise. REOS website, 13. https://globaltfokus.dk/images/Kapacitetsudvikling/Arrangementer_kap_forum/Lessons_from_Mont_Fleur.pdf
212. Ferm V (1962) Consensus gentium. In: Runes DD (ed) Dictionary of philosophy. Totowa, Littlefield, Adams p 64. ISBN 978-1162957395
213. Meierding L (1998) The consensus gentium argument. Faith Philos 15(3):271–297. https://doi.org/10.5840/faithphil199815325
214. Caws P (1991) Committees and consensus: how many heads are better than one? J Med Philos: Forum Bioeth Philos Med 16(4):375–391. https://doi.org/10.1093/jmp/16.4.375

215. Warman J, De Brasi L (2023) Religious belief and the wisdom of Crowds. Sophia 62(1):17–31. https://doi.org/10.1007/s11841-022-00938-9
216. Dong H, Fujita T, Geng Y, Dong L, Ohnishi S, Sun L, Dou Y, Fujii M (2016) A review on eco-city evaluation methods and highlights for integration. Ecol Ind 60:1184–1191. https://doi.org/10.1016/j.ecolind.2015.08.044
217. Boker U, Dershowitz N (2022) What is the Church-turing thesis? In: Ferreira F, Kahle R, Sommaruga G (eds) Axiomatic thinking II. Springer, Cham pp 199–234. https://doi.org/10.1007/978-3-030-77799-9_9
218. Derbyshire J (2020) History and scenario planning: a commentary on Schoemaker 2020. Futures Foresight Sci 2(3–4):article e35. https://doi.org/10.1002/ffo2.47
219. Gordon AV (2020) Limits and longevity: a model for scenarios that influence the future. Technol Forecast Soc Chang 151:119851. https://doi.org/10.1016/j.techfore.2019.119851
220. Van Notten PW, Rotmans J, Van Asselt MB, Rothman DS (2003) An updated scenario typology. Futures 35(5):423–443. https://doi.org/10.1016/S0016-3287(02)00090-3
221. Baruah NG, Henderson JV, Peng C (2021) Colonial legacies: shaping African cities. J Econ Geogr 21(1):29–65. https://doi.org/10.1093/jeg/lbaa026
222. Faber A, Idenburg AM, Wilting HC (2007) Exploring techno-economic scenarios in an input–output model. Futures 39(1):16–37. https://doi.org/10.1016/j.futures.2006.03.011
223. Martens P, Rotmans J (2005) Transitions in a globalizing world. Futures 37:1133–1144. https://doi.org/10.1016/j.futures.2005.02.010
224. Rotmans J, Loorbach D (2008) Transition management: reflexive governance of societal complexity through searching, learning and experimenting. In: Van den Bergh JCJM, Bruinsma FR (eds) Managing the transition to renewable energy, theory and practice from local, regional and macro perspectives. Edward Elgar publishing Ltd, Cheltenham, UK
225. Street P (1997) Scenario workshops: a participatory approach to sustainable urban living. Futures 29(2):139–158. https://doi.org/10.1016/S0016-3287(96)00073-0
226. Schneidewind U, Augenstein K (2012) Analyzing a transition to a sustainability-oriented science system in Germany. Environ Innov Soc Trans 3:16–28. https://doi.org/10.1016/j.eist.2012.04.004
227. Sassen S (2014) Expulsions. Brutality and complexity in the global economy. Harvard University Press, Cambridge MA
228. Andriopoulos C, Lewis MW (2009) Exploitation-exploration tensions and organizational ambidexterity: managing paradoxes of innovation. Organ Sci 20:696–717. https://doi.org/10.1287/orsc.1080.0406
229. Bresciani S, Ferraris A, Del Giudice M (2018) The management of organizational ambidexterity through alliances in a new context of analysis: internet of things (IoT) smart city projects. Technol Forecast Soc Chang 136:331–338. https://doi.org/10.1016/j.techfore.2017.03.002
230. Colberg T (2022) Strategic resilience: a systematic review of leading literature. J Bus Manag 20:article 1. https://doi.org/10.32025/JBM22004
231. Gregory RW, Keil M, Muntermann J, Mahring M (2015) Paradoxes and the nature of ambidexterity in IT transformation programs. Inf Syst Res 26(1):57–80. https://doi.org/10.1287/isre.2014.0554
232. Lee K, Woo H, Joshi K (2017) Pro-innovation culture, ambidexterity and new product development performance: polynomial regression and response surface analysis. Eur Manag J 35:249–260. https://doi.org/10.1016/j.emj.2016.05.002
233. Lubatkin MH, Simsek Z, Ling Y, Veiga JF (2006) Ambidexterity and performance in small-to-medium-sized firms: the pivotal role of top management team behavioral integration. J Manag 32:646–672. https://doi.org/10.1177/0149206306290712
234. O'Reilly CA, Tushman ML (2008) Ambidexterity as a dynamic capability: resolving the innovator's dilemma. Res Organ Behav 28:185–206. https://doi.org/10.1016/j.riob.2008.06.002
235. Pertusa-Ortega E, Molina-Azorin J (2018) A joint analysis of determinants and performance consequences of ambidexterity. BRQ Bus Res Q 21:84–98. https://doi.org/10.1016/j.brq.2018.03.001

236. Zhang JA, Edgar F, Geare A, O'Kane C (2016) The interactive effects of entrepreneurial orientation and capability-based HRM on firm performance: the mediating role of innovation ambidexterity. Ind Mark Manage 59:131–143. https://doi.org/10.1016/j.indmarman.2016.02.018
237. Zimmermann A, Hill SA, Birkinshaw J, Jaeckel M (2019) Complements or Substitutes? A microfoundations perspective on the interplay between drivers of ambidexterity in SMEs. Long Range Plann 53(6):article 101927. https://doi.org/10.1016/j.lrp.2019.101927
238. Dale A, Robinson J, King L, Burch S, Newell R, Shaw A, Jost F (2020) Meeting the climate change challenge: local government climate action in British Columbia, Canada. Climate Policy 20(7):866–880. https://doi.org/10.1080/14693062.2019.1651244
239. Shaw A, Burch S, Kristensen F, Robinson J, Dale A (2014) Accelerating the sustainability transition: exploring synergies between adaptation and mitigation in British Columbian communities. Glob Environ Chang 25:41–51. https://doi.org/10.1016/j.gloenvcha.2014.01.002
240. Hafner M, Tagliapetra S (eds) (2020) The geopolitics of the global energy transition. Springer, Cham, Switzerland. https://doi.org/10.1007/978-3-030-39066-2
241. Bressand A (2012) The changed geopolitics of energy and climate and the challenge for Europe. a geopolitical and European perspective on the triple agenda of competition, energy security and sustainability. In: Clingendael international energy programme (CIEP), CIEP paper 04. https://www.clingendaelenergy.com/inc/upload/files/The_changed_geopolitics_of_energy_and_climate_bressand.pdf
242. Reagan R (1987) Berlin wall speech—President Reagan's address at the Brandenburg gate—6/12/87. Ronald Reagan Foundation. Video available at: https://www.youtube.com/watch?v=5MDFX-dNtsM
243. Koshimura S, Hayashi S, Gokon H (2014) The impact of the 2011 Tohoku earthquake tsunami disaster and implications to the reconstruction. Soils Found 54(4):560–572. https://doi.org/10.1016/j.sandf.2014.06.002
244. Sugawara D (2021) Numerical modeling of Tsunami: advances and future challenges after the 2011 Tohoku earthquake and tsunami. Earth-Sci Rev 214:article 103498. https://doi.org/10.1016/j.earscirev.2020.103498
245. Desmarais E (2023) Natural disasters and adaptive cycles: a case study of forced transformation from the 2011 Tohoku earthquake and tsunami. 岡山県立大学保健福祉学部紀要 29:123–129. https://doi.org/10.15009/00002447
246. Arimura TH, Abe T (2021) The impact of the Tokyo emissions trading scheme on office buildings: what factor contributed to the emission reduction? Environ Econ Policy Stud 23:517–533. https://doi.org/10.1007/s10018-020-00271-w
247. Henderson J, Mitrova T (2020) Implications of the global energy transition on Russia. In: Hafner M, Tagliapetra S (eds) The geopolitics of the global energy transition. Springer, Cham, Switzerland pp 93–114. Lecture Notes in Energy 73, https://doi.org/10.1007/978-3-030-39066-2_5
248. OSCE (2013) Energy concept for an environmentally sound, reliable and affordable energy supply. Germany's Federal Ministry of Economics and Technology, and the Federal Ministry for the Environment, Nature Conservation and Nuclear Safety, 24 April 2013. Organization for security and co-operation in Europe. https://www.osce.org/eea/101047
249. Trüby J, Schiffer HW (2018) A review of the German energy transition: taking stock, looking ahead, and drawing conclusions for the Middle East and North Africa. Energy Transitions 2(1–2):1–14. https://doi.org/10.1007/s41825-018-0010-2
250. Schepelmann P, Kemp R, Schneidewind U (2016) The eco-restructuring of the Ruhr district as an example of a managed transition. In: Brauch HG, Oswald Spring U, Grin J, Scheffran J (eds) Handbook on sustainable transitions and sustainable peace. Hexagon series on human and environmental security and peace, vol 10. Springer, Switzerland pp 593–612. https://doi.org/10.1007/978-3-319-43884-9
251. Cartone A, Postiglione P, Hewings GJ (2021) Does economic convergence hold? A spatial quantile analysis on European regions. Econ Model 95:408–417. https://doi.org/10.1016/j.econmod.2020.03.008

252. Nagy SG, Šiljak D (2022) Is the European Union still a convergence machine? Acta Oeconomica 72(1):47–63. https://doi.org/10.1556/032.2022.00003
253. Laermans W, Costica M, Costica N, Koos-Morar TA, Ntanou K, Ciumasu IM (2015) People and urban parks: searching the new spiritus loci. In: Nature and Urban Wellbeing. Nature-based Solutions to Societal Challenges, 18–20 May, Ghent, Belgium. http://www.alter-net.info/outputs/conf-2015
254. Newton R (2006) Reclaiming sacred space. Brancusi's endless column ensemble: Targu Jiu, Romania. In: Beck E, (ed) ICON Magazine, summer 2006. London, World Monuments Fund pp 32–39. https://www.wmf.org/publication/reclaiming-sacred-space. PDF directly at: https://www.wmf.org/sites/default/files/article/pdfs/pg_32-39_brancusi.pdf
255. WMF (2019) Brancusi's endless column ensemble. World monuments fund. https://www.wmf.org/project/brancusi%92s-endless-column-ensemble
256. Masteller RN (1997) Using Brancusi: three writers, three magazines, three versions of modernism. Am Art 11(1):47–67. https://doi.org/10.1086/424278
257. Miholca A (2015) Brancusi's involvement with African art in New York. Crit Interv 9(3):179–190. https://doi.org/10.1080/19301944.2016.1156978
258. Quin J (2017) WB Yeats and the sculpture of Brancusi. Int Yeats Stud 2(1):article 4. https://tigerprints.clemson.edu/iys/vol2/iss1/4/
259. Eliade M (1957) Das heilige und das profane. Rowohlts Deutsche Enzyklopädie Rowohlt Taschenbuchverlag, Hamburg
260. Berry BJL (2008) Urbanization. In: Marzluff JM, Shulenberger E, Endlicher W, Alberti M, Bradley G, Ryan C, Simon U, ZumBrunnen C (eds) Urban ecology: an international perspective on the interaction between humans and nature. Springer, Boston, MA pp 25–48. https://doi.org/10.1007/978-0-387-73412-5_35
261. Joss S, Cowley R, Tomozeiu D (2013) Towards the ubiquitous eco-city: an analysis of the internationalisation of eco-city policy and practice. Urban Res Pract 6(1):54–74. https://doi.org/10.1080/17535069.2012.762216
262. Mullins PD (2017) The ubiquitous-eco-city of Songdo: an urban systems perspective on South Korea's green city approach. Urban Plann 2(2):4–12. https://doi.org/10.17645/up.v2i2.933
263. Azcárate MC (2020) Fueling ecological neglect in a manufactured tourist city: planning, disaster mapping, and environmental art in Cancun, Mexico. In: Mostafanezhad M, Norum R (eds) Anthropocene ecologies. Routledge pp 82–100. eBook ISBN 9781003000099. https://doi.org/10.1080/09669582.2018.1478839
264. Howe T (2014) Founding Alexandria: Alexander the great and the politics of memory. In: Acta classica: proceedings of the classical association of South Africa vol 2014, no sup-5. Classical Association of South Africa (CASA) pp 72–91. https://hdl.handle.net/10520/EJC165412
265. Baghos M (2021) The founding of constantinople: an interdisciplinary approach. Ancient West East 20:121–161. https://doi.org/10.2143/AWE.20.0.3289532
266. Richardson T (2005a) Walking streets, talking history: the making of Odessa. Ethnology 44(1):13–33. https://doi.org/10.2307/3773957
267. Richardson T (2005b) The place(s) of Moldovanka in the Making of Odessa. Anthropol East Eur Rev 23(2):72–89. https://scholarworks.iu.edu/journals/index.php/aeer/article/view/1138. Also available at: https://scholars.wlu.ca/anth_faculty/9
268. Schlör J (2011) Odessity: in search of transnational Odessa (or Odessa the best city in the world: all about Odessa and a great many jokes). Quest: Issues in Contemporary Jewish History, 2:124–148. https://doi.org/10.48248/issn.2037-741X/730. Also available at: https://www.quest-cdecjournal.it/wp-content/uploads/file/Q02/F_5_SCHLOER.pdf
269. Harris CM (1999) Washington's gamble, l'enfant's dream: politics, design, and the founding of the national capital. William Mary Q 56(3):527–564. https://doi.org/10.2307/2674560
270. Bruinsma GJ, Pauwels LJ, Weerman FM, Bernasco W (2013) Social disorganization, social capital, collective efficacy and the spatial distribution of crime and offenders: an empirical test of six neighbourhood models for a Dutch city. Br J Criminol 53(5):942–963. https://doi.org/10.1093/bjc/azt030

271. Gau JM, Pratt TC (2010) Revisiting broken windows theory: examining the sources of the discriminant validity of perceived disorder and crime. J Crim Just 38(4):758–766. https://doi.org/10.1016/j.jcrimjus.2010.05.002
272. Gerell M, Kronkvist K (2017) Violent crime, collective efficacy and city-centre effects in Malmö. Br J Criminol 57(5):1185–1207. https://doi.org/10.1093/bjc/azw074
273. Hipp JR (2016) Collective efficacy: how is it conceptualized, how is it measured, and does it really matter for understanding perceived neighborhood crime and disorder? J Crim Just 46:32–44. https://doi.org/10.1016/j.jcrimjus.2016.02.016
274. Lanfear CC, Matsueda RL, Beach LR (2020) Broken windows, informal social control, and crime: assessing causality in empirical studies. Annu Rev Criminol 3:97–120. https://doi.org/10.1146/annurev-criminol-011419-041541
275. O'Brien DT, Farrell C, Welsh BC (2019) Broken (windows) theory: a meta-analysis of the evidence for the pathways from neighborhood disorder to resident health outcomes and behaviors. Soc Sci Med 228:272–292. https://doi.org/10.1016/j.socscimed.2018.11.015
276. Oteng-Ababio M, Owusu AY, Owusu G, Wrigley-Asante C (2017) Geographies of crime and collective efficacy in urban Ghana. Territory, Politics, Gov 5(4):459–477. https://doi.org/10.1080/21622671.2016.1159602
277. Weisburd D, White C, Wooditch A (2020) Does collective efficacy matter at the micro geographic level?: Findings from a study of street segments. Br J Criminol 60(4):873–891. https://doi.org/10.1093/bjc/azaa007
278. Hawkes J (2001) The fourth pillar of sustainability: culture's essential role in public planning. Common Ground Publishing Pt Ltd. in association with The Cultural Development Network, Victoria, Australia, 2001. https://apo.org.au/node/253826
279. Anheier HK, Hoelscher M (2015) Cultural sustainability in small and medium-sized cities: what are the issues? In: Culture and sustainability in European cities. Routledge, London, UK pp 37–51. https://www.taylorfrancis.com/chapters/edit/10.4324/9781315772028-3/cultural-sustainability-small-medium-sized-cities-helmut-anheier-michael-hoelscher
280. Hristova S (2015) We, European cities and towns: the role of culture for the evolving European model of urban sustainability. In: Culture and sustainability in European cities. Routledge: London, UK, pp 62–74
281. Milfont TL, Osborne D, Yogeeswaran K, Sibley CG (2020) The role of national identity in collective pro-environmental action. J Environ Psychol 72:article 101522. https://doi.org/10.1016/j.jenvp.2020.101522
282. Tam KP, Milfont TL (2020) Towards cross-cultural environmental psychology: a state-of-the-art review and recommendations. J Environ Psychol 71:101474. https://doi.org/10.1016/j.jenvp.2020.101474
283. Wiktor-Mach D (2020) What role for culture in the age of sustainable development? UNESCO's advocacy in the 2030 Agenda negotiations. Int J Cult Policy 26(3):312–327. https://doi.org/10.1080/10286632.2018.1534841
284. Agrawal A (2002) Common resources and institutional sustainability. In: Ostrom E, Dietz T, Dolsak N, Stern PC, Stonich S, Weber EU (eds) Drama commons. National Academy Press, Washington, DC, USA, pp 41–85. https://psycnet.apa.org/record/2004-15858-002
285. Chatzichristos G, Nagopoulos N (2020) Social entrepreneurship and institutional sustainability: insights from an embedded social enterprise. Voluntas 31:484–493. https://doi.org/10.1007/s11266-019-00188-3
286. Massiri SD, Nugroho B, Kartodihardjo H, Soekmadi R (2019) Institutional sustainability of a community conservation agreement in Lore Lindu national park. For Soc 3(1):64–76. https://doi.org/10.24259/fs.v3i1.5204
287. Pfahl S (2005) Institutional sustainability. Int J Sustain Dev 8:80–96. https://doi.org/10.1504/IJSD.2005.007376
288. Spangenberg JH, Pfahl S, Deller K (2002) Towards indicators for institutional sustainability: lessons from an analysis of Agenda 21. Ecol Ind 2(1–2):61–77. https://doi.org/10.1016/S1470-160X(02)00050-X

289. Warneryd M, Håkansson M, Karltorp K (2020) Unpacking the complexity of community microgrids: a review of institutions' roles for development of microgrids. Renew Sustain Energy Rev 121:article 109690. https://doi.org/10.1016/j.rser.2019.109690
290. Azmat F, Ferdous A, Rentschler R, Winston E (2018) Arts-based initiatives in museums: creating value for sustainable development. J Bus Res 85:386–395. https://doi.org/10.1016/j.jbusres.2017.10.016
291. Baztan J, Vanderlinden JP, Jaffrès L, Jorgensen B, Zhu Z (2020) Facing climate injustices: community trust-building for climate services through arts and sciences narrative co-production. Clim Risk Manage 30:article 100253
292. Duxbury N, Hosagrahar J, Pascual J (2016) Why must culture be at the heart of sustainable urban development?. Agenda 21 for culture. United Cities and Local Governments (UCLG). www.agenda21culture.net. PDF available at: https://estudogeral.uc.pt/bitstream/10316/44016/1/Why%20must%20culture%20be%20at%20the%20heart%20of%20sustainable%20urban%20development.pdf
293. Eernstman N, Wals AE (2013) Locative meaning-making: an arts-based approach to learning for sustainable development. Sustainability 5(4):1645–1660. https://doi.org/10.3390/su5041645
294. Kirchberg V, Kagan S (2013) The roles of artists in the emergence of creative sustainable cities: theoretical clues and empirical illustrations. City Cult Soc 4(3):137–152. https://doi.org/10.1016/j.ccs.2013.04.001
295. Molderez I, Ceulemans K (2018) The power of art to foster systems thinking, one of the key competencies of education for sustainable development. J Clean Prod 186:758–770. https://doi.org/10.1016/j.jclepro.2018.03.120
296. Qu M, Cheer JM (2022) Community art festivals and sustainable rural revitalisation. In: Smith A, Mair J (eds) Events and sustainability. Routledge pp 18–37. https://doi.org/10.4324/9781003314295
297. Quinn B (2006) Problematising festival tourism: arts festivals and sustainable development in Ireland. J Sustain Tour 14(3):288–306. https://doi.org/10.1080/09669580608669060
298. Ratiu DE (2013) Creative cities and/or sustainable cities: discourses and practices. City Cult Soc 4(3):125–135. https://doi.org/10.1016/j.ccs.2013.04.002
299. Shrivastava P, Ivanaj V, Ivanaj S (2012) Sustainable development and the arts. Int J Technol Manage 60(1–2):23–43. https://doi.org/10.1504/IJTM.2012.049104
300. Krims A (2007) Music and urban geography. Routledge, New York. https://doi.org/10.4324/9780203942376
301. Krims A (2012) Music, space, and place: the geography of music. In: The cultural study of music. Routledge pp 140–148. eBook ISBN 9780203149454. https://www.taylorfrancis.com/chapters/edit/10.4324/9780203149454-15/music-space-place-adam-krims
302. Hatch T (1964) Downtown (song interpreted by Petula Clark). Videos at these two links: https://www.youtube.com/watch?v=9jMuRfoSDrg and https://www.youtube.com/watch?v=Wv5nd-3BRr4. Accessed 30 Aug 2023
303. Parnell S, Robinson J (2017) The global urban: difference and complexity in urban studies and the science of cities. In: Handbook of social science. London, Routledge, pp 13–31. ISBN 9781032022567
304. Gorovei ŞS (2020) Din nou despre vechimea unor oraşe din Moldova medievală. Câteva însemnări. Carpica XLIX:152–159. https://www.ceeol.com/search/article-detail?id=1107704
305. Brunner T (2018) Le martyre de saint Georges sur la roue. Le sceau médiéval de Molsheim mis en perspective. Revue d'Alsace 144:367–420. https://doi.org/10.4000/alsace.3526
306. Deigendesch R (2014) Adler oder hirschhorn. Zur Geschichte des Wappens der Reichsstadt Reutlingen. Reutlinger Geschichtsblätter (Neue Folge), vol 53, pp 43–64. https://publikationen.uni-tuebingen.de/xmlui/bitstream/handle/10900/82920/RGB_Ausgabe_2014_Online.pdf
307. Kluge MF (2014) II geburt: adaption und assimilation (1234–1304). In: Die Macht des Gedächtnisses: Entstehung und Wandel kommunaler Schriftkultur im spätmittelalterlichen Augsburg. Brill pp 29–137. https://doi.org/10.1163/9789004266766_003

308. Liao Y (2023) Astral cult in the encompassing mind: rethinking the stellar magic circle from Khara-Khoto. Acta Orientalia Acad Scientiarum Hung 76(1):47–91. https://doi.org/10.1556/062.2022.00124; Top-right image from Fig. 40. made available in the public domain via CC BY-NC 4.0 deed. https://www.researchgate.net/figure/Diagram-of-the-Khara-khoto-mandala-Four-deities-inside-the-four-inner-gates-E-Buddha_fig1_369328397. Accessed 22 Oct 2023
309. Webster D (2009) Alashan plateau—China's unknown Gobi. National Geographic, December 02. Available online at: https://www.nationalgeographic.com/science/article/alashan-plateau-article. Accessed 6 Oct 2023
310. Han J, Dai H, Gu Z (2021) Sandstorms and desertification in Mongolia, an example of future climate events: a review. Environ Chem Lett 19:4063–4073. https://doi.org/10.1007/s10311-021-01285-w
311. Miao L, Moore JC, Zeng F, Lei J, Ding J, He B, Cui X (2015) Footprint of research in desertification management in China. Land Degrad Dev 26(5):450–457. https://doi.org/10.1002/ldr.2399
312. Roberts R (2008) West wall of Khara-Khoto seen from the ground. Public domain via CC BY-SA 1.0, provided from the international dunhuang project. https://commons.wikimedia.org/wiki/File:Kharakhoto_BLP465_PHOTO1187_1_1.jpg
313. BabelStone (2016) View of northwest corner of Khara-Hhoto from the west. Public domain via CC BY-SA 3.0. https://commons.wikimedia.org/wiki/File:Kharakhoto_2016_northwest_corner_stupas_from_outside_B.jpg
314. Pyotr Kozlov (1923) Pyotr Kozlov (Life time: October 3, 1863–September 26, 1935)—original publication: Mongolia and Amdo and the dead city of Khara-Khoto. Immediate source: the book itself, public domain. https://commons.wikimedia.org/w/index.php?curid=39678840. Also available at: https://en.wikipedia.org/wiki/Khara-Khoto#/media/File:Khara_Khoto_Kozlov_Map.jpg
315. Patowary K (2022) The ruins of Khara-Khoto in the middle of the gobi desert. The archaeologist, 11 September 2022. https://www.thearchaeologist.org/blog/the-ruins-of-khara-khoto-in-the-middle-of-the-gobi-desert. Accessed 31 Oct 2023
316. Binuka poojan (2020) Aerial view of Sigiriya rock, Central Province, Sri Lanka. Public domain via CC-BY-SA 4.0. https://commons.wikimedia.org/w/index.php?curid=90567505
317. Cherubino (2013) The lion gate and climbing stretch. Public domain via CC BY-SA 3.0. https://commons.wikimedia.org/w/index.php?curid=30088456
318. Dronepicr (2016) Sigiriya rock from above. Public domain via CC BY 2.0. https://commons.wikimedia.org/w/index.php?curid=52016328
319. Shoka (2006) A pool in the garden complex at the foot of Sigiriya rock, Sri Lanka. Public domain via CC BY-SA 3.0. https://commons.wikimedia.org/w/index.php?curid=16296182
320. Chamal N (2008) Garden complex as seen from the top of the Lions Rock. Public domain via CC BY-SA 3.0. https://commons.wikimedia.org/w/index.php?curid=4499684
321. Gonzales DA, Bauer BS (2022) The ancient Inca town named Huayna Picchu. Ñawpa Pacha 42(1):17–31. https://doi.org/10.1080/00776297.2021.1949833
322. RAF-YYC (2009) Terraces at Machu Pichu, Peru. Public domain via CC BY-SA 2.0. https://commons.wikimedia.org/w/index.php?curid=30584279
323. WMrapids (2021) Room of the three windows at Machu Picchu, Peru. Public domain. https://commons.wikimedia.org/w/index.php?curid=112735155
324. Hobe/Holger Behr (2007) A rough map of Machu Picchu. Public domain. https://commons.wikimedia.org/w/index.php?curid=1568662
325. Guzmán F (2006) Temple of the sun (the watch tower) at Machu Picchu, Peru. Public domain. https://commons.wikimedia.org/w/index.php?curid=1264337
326. Bassett R, Young PJ, Blair GS, Samreen F, Simm W (2020) The megacity Lagos and three decades of urban heat island growth. J Appl Meteorol Climatol 59(12):2041–2055. https://doi.org/10.1175/JAMC-D-20-0059.1
327. Hoornweg D, Pope K (2017) Population predictions for the world's largest cities in the 21st century. Environ Urban 29(1):195–216. https://doi.org/10.1177/0956247816663557

328. Agwor A, de Lourdes Melo Zurita M, Munro PG (2023) Underground urbanism in Africa: splintered subterranean space in Lagos, Nigeria. Urban studies. First published online on May 31, 2023. https://doi.org/10.1177/00420980231174996
329. Monroe JC (2018) Elephants for want of towns: archaeological perspectives on West African cities and their hinterlands. J Archaeol Res 26:387–446. https://doi.org/10.1007/s10814-017-9114-2
330. Wengrow D (2015) Cities before the state in early Eurasia. Goody lecture 2015, Max Planck Institute for Social Anthropology. https://discovery.ucl.ac.uk/id/eprint/1539997/1/2015_Goody_Lecture_RO151126_1_1.pdf
331. Antos SE, Lall SV, Lozano-Gracia N (2016) The morphology of African cities. World Bank Policy Research Working Paper (7911). https://ssrn.com/abstract=2883394
332. Simon D, McGregor D, Nsiah-Gyabaah K (2004) The changing urban-rural interface of African cities: definitional issues and an application to Kumasi, Ghana. Environ Urbanization 16(2):235–248. https://doi.org/10.1177/095624780401600214
333. Ahimah-Agyakwah S, Nketiah-Amponsah E, Agyire-Tettey F (2022) Urbanisation and economic growth causal nexus: evidence from panel data analyses of selected positively urbanizing countries in Sub-Saharan Africa. Afr Rev Econ Finance 14(2):16–36. https://hdl.handle.net/10520/ejc-aref_v14_n2_a2
334. Goodfellow T, Huang Z (2022) Manufacturing urbanism: improvising the urban–industrial nexus through Chinese economic zones in Africa. Urban Studies 59(7):1459–1480. https://doi.org/10.1177/00420980211007800
335. Boadi K, Kuitunen M, Raheem K, Hanninen K (2005) Urbanisation without development: environmental and health implications in African cities. Environ Dev Sustain 7:465–500. https://doi.org/10.1007/s10668-004-5410-3
336. Tiba S (2019) A non-linear assessment of the urbanization and climate change nexus: the African context. Environ Sci Pollut Res 26:32311–32321. https://doi.org/10.1007/s11356-019-06475-2
337. Bellezoni RA, Meng F, He P, Seto KC (2021) Understanding and conceptualizing how urban green and blue infrastructure affects the food, water, and energy nexus: a synthesis of the literature. J Cleaner Prod 289:article 125825. https://doi.org/10.1016/j.jclepro.2021.125825
338. Chirisa I, Bandauko E (2015) African cities and the water-food-climate-energy nexus: an agenda for sustainability and resilience at a local level. Urban Forum 26:391–404. https://doi.org/10.1007/s12132-015-9256-6
339. Macharia P, Kreuzinger N, Kitaka N (2020) Applying the water-energy nexus for water supply—a diagnostic review on energy use for water provision in Africa. Water 12(9):2560. https://doi.org/10.3390/w12092560
340. Naidoo D, Nhamo L, Mpandeli S, Sobratee N, Senzanje A, Liphadzi S, Slotow R, Jacobson M, Modi AT, Mabhaudhi T (2021) Operationalising the water-energy-food nexus through the theory of change. Renew Sustain Energy Rev 149:111416. https://doi.org/10.1016/j.rser.2021.111416
341. Karg H, Drechsel P, Akoto-Danso EK, Glaser R, Nyarko G, Buerkert A (2016) Foodsheds and city region food systems in two West African cities. Sustainability 8(12):article 1175. https://doi.org/10.3390/su8121175
342. Mazenda A, Molepo N, Mushayanyama T, Ngarava S (2022) The invisible crisis: the determinants of local food insecurity in Gauteng municipalities, South Africa. Br Food J 124(13):274–289. https://doi.org/10.1108/BFJ-11-2021-1234
343. Mutono N, Wright J, Mutembei H, Muema J, Thomas M, Mutunga M, Thumbi SM (2020) The nexus between improved water supply and water-borne diseases in urban areas in Africa: a scoping review protocol. AAS Open Res 3. https://doi.org/10.12688/aasopenres.13063.2
344. Musavengane R, Siakwah P, Leonard L (2020) The nexus between tourism and urban risk: towards inclusive, safe, resilient and sustainable outdoor tourism in African cities. J Outdoor Recreation Tourism 29:article 100254. https://doi.org/10.1016/j.jort.2019.100254
345. Nhamo L, Rwizi L, Mpandeli S, Botai J, Magidi J, Tazvinga H, Sobratee N, Liphadzi S, Naidoo D, Modi AT, Slotow R (2021) Urban nexus and transformative pathways towards

a resilient Gauteng city-region, South Africa. Cities 116:article 103266. https://doi.org/10.1016/j.cities.2021.103266
346. Abuje SJ (2021) Influence of urban form on climate vulnerability in the city county of Nairobi. Doctoral dissertation, Jomo Kenyatta University of Agriculture and Technology (JKUAT), college of engineering and technology (COETEC). http://ir.jkuat.ac.ke/handle/123456789/5679
347. Beguy D, Elung'ata P, Mberu B, Oduor C, Wamukoya M, Nganyi B, Ezeh A (2015) Health and demographic surveillance system profile: the Nairobi urban health and demographic surveillance system (NUHDSS). Int J Epidemiol 44(2):462–471. https://doi.org/10.1093/ije/dyu251
348. Corburn J, Hildebrand C (2015) Slum sanitation and the social determinants of women's health in Nairobi, Kenya. J Environ Public Health, Article 209505. https://doi.org/10.1155/2015/209505
349. Kimote J (2020) A study of the adoption and maintenance of green roofs as part of the urban green spaces for the city of Nairobi. Doctoral dissertation, University of Nairobi. http://erepository.uonbi.ac.ke/handle/11295/154392
350. Klopp JM, Cavoli C (2019) Mapping minibuses in Maputo and Nairobi: engaging paratransit in transportation planning in African cities. Transp Rev 39(5):657–676. https://doi.org/10.1080/01441647.2019.1598513
351. Makunda C (2019) The unsustainability of urban habitat transformation: a case study of Kileleshwa in Nairobi, Kenya. Africa Habitat Rev J 13(1). http://journals.uonbi.ac.ke/index.php/afr/index.
352. Mudhune SA, Okiro EA, Noor AM, Zurovac D, Juma E, Ochola SA, Snow RW (2011) The clinical burden of malaria in Nairobi: a historical review and contemporary audit. Malar J 10:1–10. https://doi.org/10.1186/1475-2875-10-138
353. Mwaniki BW, Gakuya DW, Mwaura AM, Muthama NJ (2019) Addressing urban challenges in Nairobi City, Kenya: the scope for collaborative planning approaches in achieving sustainable development goals. Urban Dev, Plann Environ Res J 3(2):1–17. http://academicresearchinsight.com/udpaerj/3_2_1_17.pdf
354. Njoroge BNK, Kimani M, Ndunge D (2014) Review of municipal solid waste management: a case study of Nairobi, Kenya. Res Inventy: Int J Eng Sci 4(2):16–20. Issn(e): 2278–4721, Issn(p):2319–6483. http://erepository.uonbi.ac.ke/bitstream/handle/11295/92590/Njoroge_Review%20of%20municipal%20solid%20waste%20management:%20a%20case%20study%20of%20Nairobi,%20Kenya.pdf?sequence=1
355. Ogutu FA, Kimata DM, Kweyu RM (2021) Partnerships for sustainable cities as options for improving solid waste management in Nairobi city. Waste Manage Res 39(1):25–31. https://doi.org/10.1177/0734242X20967735
356. Ongoma V, Muange P, Shilenje Z (2016) Potential effects of urbanization on urban thermal comfort, a case study of Nairobi city, Kenya: a review. Geogr Pannonica 20(1):19–31. https://doi.org/10.18421/GP20.01-03
357. Sovacool BK, Daniels C, AbdulRafiu A (2022) Transitioning to electrified, automated and shared mobility in an African context: a comparative review of Johannesburg, Kigali, Lagos and Nairobi. J Transp Geogr 98:103256. https://doi.org/10.1016/j.jtrangeo.2021.103256
358. Tom RO, George KO, Joanes AO, Haron A (2022) Review of flood modelling and models in developing cities and informal settlements: a case of Nairobi city. J Hydrol: Reg Stud 43:101188. https://doi.org/10.1016/j.ejrh.2022.101188
359. Afromusing (2008) Nairobi skyline. Public domain via CC BY 2.0, https://commons.wikimedia.org/w/index.php?curid=29600773
360. Genvessel (2007) Kibera, Nairobi May 2007. Public domain, under CC BY 2.0. https://commons.wikimedia.org/w/index.php?curid=5739874
361. PK (2023) Entrance to the parliament of Kenya. (Public domain). http://www.parliament.go.ke/the-senate/history. Accessed 14 Oct 2023
362. Kenyaverification (2006). Public domain, under CC BY-SA 4.0. https://commons.wikimedia.org/w/index.php?curid=7832264

363. Jorge L (2009) Auditorium of the Kenyatta international conference centre (KICC). Public domain, under CC BY-SA 2.0. https://commons.wikimedia.org/w/index.php?curid=15942651
364. Bobokine (2008) Image from the business district of Nairobi. Public domain under CC BY-SA 3.0. https://commons.wikimedia.org/w/index.php?curid=8335981
365. Onchester454 (2016) Walkway in Central Park (of Nairobi). CC BY-SA 4.0. https://commons.wikimedia.org/w/index.php?curid=48610702
366. UN Statistics Division SDG11 (2023) Sustainable development goals, SDG11—sustainable cities & communities. https://unstats.un.org/sdgs/report/2023/Goal-11/#:~:text=Approximately%201.1%20billion%20people%20currently,public%20spaces%20persist%20in%20cities. Accessed 21 Oct 2023
367. Gallaher CM, Kerr JM, Njenga M, Karanja NK, WinklerPrins AM (2013) Urban agriculture, social capital, and food security in the Kibera slums of Nairobi, Kenya. Agric Hum Values 30:389–404. https://doi.org/10.1007/s10460-013-9425-y
368. Mitra S, Mulligan J, Schilling J, Harper J, Vivekananda J, Krause L (2017) Developing risk or resilience? Effects of slum upgrading on the social contract and social cohesion in Kibera, Nairobi. Environ Urbanization 29(1):103–122. https://doi.org/10.1177/0956247816689
369. Mukeku J (2018) Urban slum morphology and socio-economic analogies: a case study of Kibera Slum, Nairobi, Kenya. Urbanisation 3(1):17–32. https://doi.org/10.1177/2455747118790581
370. Mutisya E, Yarime M (2011) Understanding the grassroots dynamics of slums in Nairobi: the dilemma of Kibera informal settlements. Int Trans J Eng, Manage, Appl Sci Technol 2(2):197–213. https://urbanleaders.org/655LandRights/05LocalResponses/Nairobi/197-213.pdf
371. Ono H, Kidokoro T (2020) Understanding the development patterns of informal settlements in Nairobi. Jpn Architectural Rev 3(3):384–393. https://doi.org/10.1002/2475-8876.12161
372. Makworo M, Mireri C (2011) Public open spaces in Nairobi city, Kenya, under threat. J Environ Plann Manage 54(8):1107–1123. https://doi.org/10.1080/09640568.2010.549631
373. M'Ikiugu MM, Kinoshita I, Tashiro Y (2012) Urban green space analysis and identification of its potential expansion areas. Procedia Soc Behav Sci 35:449–458. https://doi.org/10.1016/j.sbspro.2012.02.110
374. Mulligan J, Bukachi V, Clause JC, Jewell R, Kirimi F, Odbert C (2020) Hybrid infrastructures, hybrid governance: new evidence from Nairobi (Kenya) on green-blue-grey infrastructure in informal settlements. Anthropocene 29:article 100227. https://doi.org/10.1016/j.ancene.2019.100227
375. Nyambane DO, Njoroge JB, Watako AO (2016) Assessment of tree species distribution and diversity in the major urban green spaces of Nairobi city, Kenya. J Hortic For 8(2):12–23 (article 04B1C6958737). https://doi.org/10.5897/JHF2016.0439
376. Okech EA, Nyadera IN (2022) Urban green spaces in the wake of COVID-19 pandemic: reflections from Nairobi, Kenya. GeoJournal 87(6):4931–4945. https://doi.org/10.1007/s10708-021-10540-0
377. Rajé F, Tight M, Pope FD (2018) Traffic pollution: a search for solutions for a city like Nairobi. Cities 82:100–107. https://doi.org/10.1016/j.cities.2018.05.008
378. Simwanda M, Ranagalage M, Estoque RC, Murayama Y (2019) Spatial analysis of surface urban heat islands in four rapidly growing African cities. Remote Sens 11(14):article 1645. https://doi.org/10.3390/rs11141645
379. Tibaijuka A (2007) Nairobi and its environment (Module 5). In: Kenya Atlas. Nairobi, UN-Habitat, United Nations Environment Programme (UNEP) pp 145–161. https://na.unep.net/atlas/kenya/downloads/chapters/Kenya_Screen_Chapter5-End.pdf
380. August 7 MP (2023) August 7 Memorial Park, Nairobi. https://memorialparkkenya.org. Accessed 10 Oct 2023
381. FoCP (2023) Friends of City Park. https://naturekenya.org/about/friends-of-city-park/. Accessed 10 Oct 2023

382. Nairobi Arboretum (2023) Nairobi arboretum, official page. https://nairobiarboretum.org/index.php/nairobi-arboretum/. Accessed 10 Oct 2023
383. FoKF (2023) Friends of Karura Forest. http://www.karurafriends.org/. Accessed 10 Oct 2023
384. UNESCO-GNLC (2016) The UNESCO Global Network of Learning Cities. Official website, https://www.uil.unesco.org/en/learning-cities. Accessed 15 Oct 2023
385. Cervero R (2016) Transport infrastructure and the environment: sustainable mobility and urbanism. In: Etingoff K (ed) Urban development for the 21st century: managing resources and creating infrastructure. CRC Press (Taylor & Francis), Boca Raton, FL, pp 155–182. ISBN 13: 9781771882583
386. Hidalgo D, Huizenga C (2013) Implementation of sustainable urban transport in Latin America. Res Transp Econ 40(1):66–77. https://doi.org/10.1016/j.retrec.2012.06.034
387. Larbi M, Kellett J, Palazzo E (2022) Urban sustainability transitions in the global south: a case study of Curitiba and Accra. Urban Forum 33(2):223–244. https://doi.org/10.1007/s12132-021-09438-4
388. Montero S (2020) Leveraging Bogotá: sustainable development, global philanthropy and the rise of urban solutionism. Urban Stud 57(11):2263–2281. https://doi.org/10.1177/0042098018798555
389. Suzuki H, Cervero R, Iuchi K (2013) Transforming cities with transit: transit and land-use integration for sustainable urban development. World Bank Publications. https://doi.org/10.1596/978-0-8213-9745-9
390. Colombijn F (2016) I am a singer: a conversation with Johan Silas, architect and urban planner in Surabaya, Indonesia. Indonesia 102:7–30. https://doi.org/10.5728/indonesia.102.0007
391. Ernawati R, Santosa HR, Setijanti P (2013) Facing urban vulnerability through kampung development, case study of kampungs in Surabaya, Indonesia. Development 1(1):1–6. https://doi.org/10.11648/j.hss.20130101.11
392. Funo S, Yamamoto N, Silas J (2002) Typology of kampung houses and their transformation process a study on urban tissues of an Indonesian city. J Asian Archit Build Eng 1(2):2_193–200. https://doi.org/10.3130/jaabe.1.2_193
393. Kustiwan I, Ukrin I, Aulia A (2015) Identification of the creative capacity of Kampong's community towards sustainable Kampong (case studies: Cicadas and Pasundan Kampong, Bandung): a preliminary study. Procedia Soc Behav Sci 184:144–151. https://doi.org/10.1016/j.sbspro.2015.05.074
394. Shirleyana, Hawken S, Sunindijo RY (2018) City of Kampung: risk and resilience in the urban communities of Surabaya, Indonesia. Int J Build Pathol Adapt 36(5):543–568. https://doi.org/10.1108/IJBPA-02-2018-0025
395. Silas J (1989) Marginal settlements in Surabaya, Indonesia: problem or potential? Environ Urban 1(2):60–70. https://doi.org/10.1177/095624788900100207
396. Silas J (1992) Government-community partnerships in Kampung improvement programmes in Surabaya. Environ Urban 4(2):33–41. https://doi.org/10.1177/095624789200400204
397. Silas J (2019) Surabaya. The Wiley Blackwell Encyclopedia of Urban and Regional Studies, pp 1–4. https://doi.org/10.1002/9781118568446.eurs0333
398. Wijayanti DR, Suryani S (2015) Waste bank as community-based environmental governance: a lesson learned from Surabaya. Procedia Soc Behav Sci 184:171–179. https://doi.org/10.1016/j.sbspro.2015.05.077
399. Budianta M (2019) Smart kampung: doing cultural studies in the global south. Commun Crit/Cult Stud 16(3):241–256. https://doi.org/10.1080/14791420.2019.1650194
400. Ghazali S (2012) Sense of place and the politics of insider-ness in villages undergoing transition: the case of City Kampung on Penang Island. In: Cleavage, connection and conflict in rural, urban and contemporary Asia. Dordrecht: Springer Netherlands pp 117–142. http://ndl.ethernet.edu.et/bitstream/123456789/14768/1/211.pdf#page=121
401. Purnamasari LS, Yudana G, Rini EF (2017) Spatial transformation of Surakarta's peripheral rural villages under in-situ urbanization phenomenon: the case of Gentan village. Geoplanning: J Geomatics Plann 4(1):83–96. https://doi.org/10.14710/geoplanning.4.1.83-96

402. Zhu J, Simarmata HA (2015) Formal land rights versus informal land rights: governance for sustainable urbanization in the Jakarta metropolitan region, Indonesia. Land Use Policy 43:63–73. https://doi.org/10.1016/j.landusepol.2014.10.016
403. Ramadhani AN, Faqih M, Hayati A (2018) Inhabitant's sense of place in the context of tourism Kampung. J Archit Environ 17(2):151–168. http://iptek.its.ac.id/index.php/joae/article/view/3894
404. Canuti P, Margottini C, Mucho R, Casagli N, Delmonaco G, Ferretti A, Lollino G, Puglisi C, Tarchi D (2005) Preliminary remarks on monitoring, geomorphological evolution and slope stability of Inca Citadel of Machu Picchu (C101-1). In: Sassa K, Fukuoka H, Wang F, Wang G (eds) Landslides: risk analysis and sustainable disaster management. Springer, Berlin, Heidelberg pp 39–47. ISBN 978-3-540-28664-6. https://doi.org/10.1007/3-540-28680-2_3
405. Zan L, Lusiani M (2011) Managing change and master plans: Machu Picchu between conservation and exploitation. Archaeologies 7:329–371. https://doi.org/10.1007/s11759-011-9167-7
406. Ciumasu IM (2012a) Biosensors, eco-innovation and markets. Talk at Biosensors 2012, 15–18 May 2012, Cancun, Mexico. https://www.elsevier.com/events/conferences/world-congress-on-biosensors/about/previous-events/biosensors-2012. Program available here: https://www.elsevier.com/__data/assets/pdf_file/0006/234681/2012_oral_programme.pdf
407. Appendino F (2017) Balancing heritage conservation and sustainable development–the case of bordeaux. IOP Conf Ser: Mater Sci Eng 245(6):062002. IOP Publishing. https://doi.org/10.1088/1757-899X/245/6/062002
408. Martone A, Sepe M (2012) Creativity, urban regeneration and sustainability/the Bordeaux case study. J Urban Regeneration Renewal 5(2):164–183. https://www.ingentaconnect.com/content/hsp/jurr/2012/00000005/00000002/art00009
409. Aguilar M, De Fuentes AG (2007) Barriers to achieving the water and sanitation-related millennium development goals in Cancun, Mexico at the beginning of the twenty-first Century. Environ Urban 19(1):243–260. https://doi.org/10.1177/0956247807076918
410. Camargo BA, Chávez ML, del Carmen Ginocchio M (2020) Tourism in Latin America: an overview and new experiences in city tourism. In: Morrison AM, Coca-Stefaniak JA (eds) Routledge handbook of tourism cities. eBook ISBN 9780429244605, pp 446–461
411. Dieleman H, Martínez-Rodríguez MC (2019) Potentials and challenges for a circular economy in Mexico. In: Franco-García ML, Carpio-Aguilar J, Bressers H (eds) Towards zero waste. Greening of industry networks studies, vol 6. Springer, Cham pp 9–24. https://doi.org/10.1007/978-3-319-92931-6_2
412. Cador CEM, Alayola PA (2022) Irregular settlements in tourist cities and their relationship with social exclusion. J Strateg Innov Sustain 17(3):31–50. https://doi.org/10.33423/jsis.v17i3.5640
413. Balbuena Vázquez A, López López Á (2020) Residents attitudes towards sustainable tourism in Cancun. J Tourism Hospitality Manage 8(1):99–109. https://doi.org/10.15640/jthm.v8n1a11
414. Vogelsang B, Röhrer N, Fuchs M, Pilz M (2021) Cooperation between learning venues and its limits: the hotel industry in Cancún (Mexico). Int J Res Vocat Educ Training 8(4):67–89. https://doi.org/10.13152/IJRVET.8.4.4
415. Saraswat C, Kumar P, Mishra BK (2016) Assessment of stormwater runoff management practices and governance under climate change and urbanization: an analysis of Bangkok, Hanoi and Tokyo. Environ Sci Policy 64:101–117. https://doi.org/10.1016/j.envsci.2016.06.018
416. Aflaki A, Mirnezhad M, Ghaffarianhoseini A, Ghaffarianhoseini A, Omrany H, Wang ZH, Akbari H (2017) Urban heat island mitigation strategies: a state-of-the-art review on Kuala Lumpur, Singapore and Hong Kong. Cities 62:131–145. https://doi.org/10.1016/j.cities.2016.09.003
417. Berman M, Baztan J, Kofinas G, Vanderlinden JP, Chouinard O, Huctin JM, Kane A, Mazé C, Nikulkina I, Thomson K (2020) Adaptation to climate change in coastal communities: findings from seven sites on four continents. Clim Change 159:1–16. https://doi.org/10.1007/s10584-019-02571-x

418. Ajibade I, McBean G (2014) Climate extremes and housing rights: a political ecology of impacts, early warning and adaptation constraints in Lagos slum communities. Geoforum 55:76–86. https://doi.org/10.1016/j.geoforum.2014.05.005
419. Lawanson T, Orelaja O, Simire M (2014) Effects of climate change on a peri-urban farming community in Lagos, Nigeria. Afr J Sci, Technol, Innov Dev 6(6):511–517. https://doi.org/10.1080/20421338.2014.976990
420. Badina SV (2021) Estimation of the value of buildings and structures in the context of permafrost degradation: the case of the Russian Arctic. Polar Sci 29:article 100730. https://doi.org/10.1016/j.polar.2021.100730
421. Hjort J, Streletskiy D, Doré G, Wu Q, Bjella K, Luoto M (2022) Impacts of permafrost degradation on infrastructure. Nat Rev Earth Environ 3(1):24–38. https://doi.org/10.1038/s43017-021-00247-8
422. Revich BA, Eliseev DO, Shaposhnikov DA (2022) Risks for public health and social infrastructure in Russian arctic under climate change and permafrost degradation. Atmosphere 13(4):article 532. https://doi.org/10.3390/atmos13040532
423. Shiklomanov NI, Streletskiy DA, Swales TB, Kokorev VA (2017) Climate change and stability of urban infrastructure in Russian permafrost regions: prognostic assessment based on GCM climate projections. Geogr Rev 107(1):125–142. https://doi.org/10.1080/1088937X.2017.1329237
424. Vincent WF, Lemay M, Allard M (2017) Arctic permafrost landscapes in transition: towards an integrated Earth system approach. Arct Sci 3(2):39–64. https://doi.org/10.1139/as-2016-0027
425. Adekola PO, Allen AA, Tinuola FR (2017) Socio-economic and health implications of urban renewal on internally displaced persons in Ogun state, southwestern Nigeria. J Intern Displacement 7(1):16–30. https://core.ac.uk/download/pdf/79125309.pdf
426. Adekola PO, Azuh D, Adeloye D, Amoo E (2019) Urban renewal in Nigeria: a slash and burn approach? Environ Dev Sustain 21:2165–2182. https://doi.org/10.1007/s10668-018-0130-2
427. Ezema IC, Opoko PA, Oluwatayo AA (2016) Urban regeneration through state-led, new-build gentrification in Lagos inner city, Nigeria. Int J Appl Environ Sci 11(1):135–146. https://www.ripublication.com/ijaes16/ijaesv11n1_10.pdf
428. Jimoh HO, Omole FK, Omosulu SB (2013) An examination of urban renewal exercise of Badia East of Lagos State, Nigeria. Int J Educ Res 1(5):1–14. https://www.researchgate.net/profile/kayode-Omole/publication/252322622_An_examination_of_urban_renewal_exercise_of_Badia_East_of_Lagos_State_Nigeria/links/02e7e51f2fd817a69b000000/An-examination-of-urban-renewal-exercise-of-Badia-East-of-Lagos-State-Nigeria.pdf
429. Morka FC (2007) A place to live: a case study of the Ijora-Badia community in Lagos, Nigeria. Unpublished case study prepared for global report on human settlements. https://staging.unhabitat.org/downloads/docs/GRHS.2007.CaseStudy.Tenure.Nigeria.pdf
430. Olaniyi WA (2021) The proposed Eko Atlantic city project, Victoria Island, Lagos: preliminary impact assessment of land reclamation on the aquatic lives and climate change. Niger J Anim Prod 48(4):194–200. https://doi.org/10.51791/njap.v48i4.3014
431. Popoola A, Olatunde MO, Magidimisha HH, Amuda A, Adeleye BM, Chipungu L (2020) Urban forced eviction: experiences in the metropolitan city of Lagos, Nigeria. Indonesian J Geogr 52(1):112–127. https://doi.org/10.22146/ijg.40555
432. Titilayo JA (2023) Enclave urbanism and infrastructure outcomes: the Eko Atlantic city and urban sustainability issues in Lagos, Nigeria. Technische Universität Darmstadt, Darmstadt, Germany. https://tuprints.ulb.tu-darmstadt.de/24195/
433. Ajibade I (2017) Can a future city enhance urban resilience and sustainability? A political ecology analysis of Eko Atlantic city, Nigeria. Int J Disaster Risk Reduction 26:85–92
434. Bell D, Jayne M (2009) Small cities? Towards a research agenda. Int J Urban Reg Res 33(3):683–699. https://doi.org/10.1111/j.1468-2427.2009.00886.x
435. Ciumasu IM (2007b) Transition to sustainability through learning-by-doing capacity building: the pilot project centre of expertise for sustainable exploitation of ecosystems in Iasi, Romania.

In: Biological Sciences for the 21st Century: Meeting the Challenges of Sustainable Development in an Era of Global Change, 10–12 May, and the 29th International Union of Biological Sciences General Assembly and Scientific Symposium, 9–13 May, Washington DC, USA
436. Ciumasu IM (2007a) Triggering sustainable development: challenging local authorities with problem-solving oriented research. In: Sustainable Neighbourhood—from Lisbon to Leipzig through Research (L2L)—A Conference on Research for Sustainable Development in Europe, 08–10 May, 2007, Leipzig, Germany, section 1.1 European partnership for sustainability, 2007. http://www.fona.de/en/5015. Accessed 31 Jan 2008. Bundesministerium für Bildung und Forschung (German federal ministry for education and research). https://cordis.europa.eu/event/id/110767-sustainable-neighbourhood-from-lisbon-to-leipzig-through-research-121/. Accessed 18 Sept 2022
437. Blume M (2010) National Mall (photo taken on 25 March 2010, 15:45:06). Public domain, attribution: Matti Blume, under CC BY-SA 4.0 via Wikimedia Commons. File https://upload.wikimedia.org/wikipedia/commons/d/d9/National_Mall%2C_Washington%2C_D.C._%2820100325-DSC01310%29.jpg. Page https://commons.wikimedia.org/wiki/File:National_Mall,_Washington,_D.C._(20100325-DSC01310).jpg
438. US-IA (1963) View of crowd at 1963 March on Washington. US Information Agency, Press and Publications Service. US National Archives at College Park: NWDNS-306-SSM-4D(80)10 by USIA (NARA), Public Domain. https://commons.wikimedia.org/w/index.php?curid=2724744
439. APK (2010) Walking path, facing east on the National Mall in Washington, D.C., as viewed near the 1300 block of Jefferson Drive SW. Source: public domain (attribution: APK, 14 April 2010). File link: https://upload.wikimedia.org/wikipedia/commons/2/26/National_Mall_-_walking_path.JPG
440. Noir T (1986) Berlin Wall at Bethaniendamm in Berlin-Kreuzberg. Public Domain, under CC BY-SA 3.0. https://commons.wikimedia.org/w/index.php?curid=1402275
441. Lear 21 (1989) The fall of the Berlin Wall. At English Wikipedia, Public Domain by CC BY-SA 3.0. https://commons.wikimedia.org/w/index.php?curid=3692038
442. Clark WC (2007) Sustainability science: a room of its own. Proc Natl Acad Sci 104(6):1737–1738. https://doi.org/10.1073/pnas.0611291104
443. Ciumasu IM (2007c) International circulation of talent—a solution for the development of Romania. Talk & book module. University in society—UNISO 2007, 9–12 July. University of Versailles Saint-Quentin-en-Yvelines, Versailles, France. In: Zaharia SE, Ivan MS (eds) UNISO 2007. Lifelong learning and qualifications in higher education, pp 207–217. https://citeseerx.ist.psu.edu/document?repid=rep1&type=pdf&doi=2fd5e4f57e599ebf184acb04827a285eab325154
444. Ciumasu IM (2008) Brain networking—how to do it? Talk and book module. University in society—UNISO 2008, 16–20 July. Alexandru Ioan Cuza university of Iasi, Romania. In: Zaharia SE, Ivan MS (eds) UNISO 2008. Competence-based higher education: challenges and solutions, pp 197–203
445. Ciumasu IM (2010) Turning brain drain into brain networking. Sci Public Policy 37(2):135–146. https://doi.org/10.3152/030234210X489572
446. NASA-JPL (1996) PIA00452: solar system portrait—Earth as pale blue dot. Photojurnal, PIA00452. Jet Propulsion Laboratory (JPL), National Aeronautics and Space Administration (NASA). https://photojournal.jpl.nasa.gov/catalog/PIA00452. Accessed 30 March 2023
447. NASA-JPL (2020) PIA23645: pale blue dot revisited. Photojurnal, PIA23645. Jet propulsion laboratory (JPL), national aeronautics and space administration (NASA). https://photojournal.jpl.nasa.gov/catalog/PIA23645. Accessed 30 March 2023
448. Hele K (2004) Cahokia: mirror of the cosmos (book review, summary). Urban Hist Rev 32(2):65–66. https://doi.org/10.7202/1015723ar
449. Kelly JE, Brown JA (2014) Cahokia: the processes and principles of the creation of an early Mississippian city. In: Creekmore III, AT, Fisher KD (eds) Making ancient cities: space and place in early urban societies. Cambridge University Press pp 292–336

450. Pauketat TR, Susan MA, Alleen MB, Jeffery DK, Benson EM (2023) Cahokia as urban anomaly. J Urban Archaeol 7:253-274. https://doi.org/10.1484/J.JUA.5.133458
451. Stauffer JG, Grooms SB, Hu LW, Mersmann J, Kidder TR, Henry ER (2023) Reimagining the development of downtown Cahokia using remote sensing visualizations from the western edge of the Grand Plaza. Land 12(2):article 342. https://doi.org/10.3390/land12020342
452. Romain WF (2017) Monks Mound as an Axis Mundi for the Cahokian world. Ill Archaeol 29:27–52
453. Cipozy (2015) Ptolemaic system—digital reproduction.png. Cipozy (reproduction), Peter Apian (original). Public domain via CC BY 4.0. https://commons.wikimedia.org/wiki/File:Ptolemaic_System_-_Digital_Reproduction.png
454. Cameron A (1990) Isidore of Miletus and Hypatia: on the editing of mathematical texts. Greek, Roman, Byzantine Stud 31(1):103–127. https://www.torrossa.com/en/resources/an/5189879
455. Deakin MA (1994) Hypatia and her mathematics. Am Math Mon 101(3):234–243. https://doi.org/10.1080/00029890.1994.11996935
456. Pedersen O (1993) Early physics and astronomy: a historical introduction. Cambridge University Press. ISBN 978-0-521-40899-8
457. Freeth T, Higgon D, Dacanalis A, MacDonald L, Georgakopoulou M, Wojcik A (2021) A model of the Cosmos in the ancient Greek Antikythera Mechanism. Sci Rep 11:article 5821. https://doi.org/10.1038/s41598-021-84310-w
458. Efstathiou K, Efstathiou M, Basiakoulis A, Kokkinos N (2021) The Antikythera Mechanism: the prove of the accuracy of the astronomical calculations based on it. Heritage 4:3848–3878. https://doi.org/10.3390/heritage4040211
459. KAM (2023) Ngor mandala, Kimbell art Museum, cited by [Liao, 2023, Fig. 2 therein, acc. no. AP 2000.01, original image from KAM: https://kimbellart.org/collection/ap-200001]. The original image was visited at https://kimbellart.org/collection/ap-200001
460. Fowden EK (2022) Rituals of memory at the Olympieion precinct of Athens. In: Martinez Jiménez J, Soulsby SO (eds) Remembering and forgetting the ancient city. Oxbow Books, Oxford & Philadelphia pp 297–326. (location details of the phragment cited in the main text: 229a–230d, translated by Nehamas & Woodruff. Ferrari, listening to the Cicadas, 1–12; Chiotis, Springs). Digital ISBN 978-1-78925-816-5. https://library.oapen.org/bitstream/handle/20.500.12657/54112/external_content.pdf?sequence=1#page=320
461. National Geographic (2023) Greek city-states. National Geographic Education. https://education.nationalgeographic.org/resource/greek-city-states/. Accessed 15 July 2023
462. Krassakis P, Kazana S, Chen F, Koukouzas N, Parcharidis I, Lekkas E (2021) Detecting subsidence spatial risk distribution of ground deformation induced by urban hidden streams. Geocarto Int 36(6):622–639. https://doi.org/10.1080/10106049.2019.1622601
463. BBC (2019) Athens to open up ancient river. News from Elsewhere, 27 Feb 2019. https://www.bbc.com/news/blogs-news-from-elsewhere-47385140
464. Chiotis ED (2021) Springs, sanctuaries and aqueducts in the llissos valley, Attica, and the enchanting scenery in Plato's dialogue Phaedrus. In: Bouffier S, Belvedere O, Vassalo S (eds) Gérer l'eau en Méditerranée au premier millénaire avant J.-C., pp 25–37. https://books.openedition.org/pup/39138?lang=en
465. GCT (2019). Athens to uncover historic river that flows through the heart of the Greek Capital. https://greekcitytimes.com/2019/03/11/athens-to-uncover-historic-river-that-flows-through-the-heart-of-the-greek-capital/
466. Appiah KA (1997) Cosmopolitan patriots. Crit Inq 23(3):617–639. https://doi.org/10.1086/448846
467. Eze C (2014) Rethinking African culture and identity: the Afropolitan model. J Afr Cult Stud 26(2):234–247. https://doi.org/10.1080/13696815.2014.894474
468. Mukherjee M (2020) Tagore's rooted-cosmopolitanism and international mindedness against institutional sustainability. Asia Pac J Educ 40(1):49–60. https://doi.org/10.1080/02188791.2020.1725430
469. Clancy J (2008) Exploring Kyoto: on foot in the ancient capital. Stone Bridge Press. ISBN 978-1-933330-64-8

470. Richmond S, Dodd J, Branscombe S, Goss R, Snow J (2011) The rough guide to Japan. Penguin. ISBN 978-1-84836-615-2
471. Rowthorn C (2012) Kyoto. Lonely Planet Publications Pty Ltd. ISBN 978-1-74179-401-4
472. Santini T, Taji T (2019) Natural urban heritage and preservation policies: the case of Kyoto's waterways. Environ Sci Sustain Dev 4(2):95–106. https://doi.org/10.21625/essd.v4i2.559
473. Shevtsova GV, Parkhomchuk MS (2022) Lake Biwa Canal at Kyoto Japan: sustainable development and revitalization. IOP Conf Ser: Earth Environ Sci 1049(1):012078. https://doi.org/10.1088/1755-1315/1049/1/012078
474. Cardoso J (2020) Forest bathing for students: current barriers and future opportunities. University of Toronto. https://hdl.handle.net/1807/99010
475. Honderich T (2002) Philosopher: a kind of life. Routledge p 62. ISBN 0-415-27631-4. https://philpapers.org/rec/HONPAK
476. Sylvester E (2016) Taking a stroll back in time through University of Toronto's Philosopher's Walk. Torontoist. https://torontoist543.rssing.com/chan-63635420/all_p1.html. Accessed 17 Oct 2023
477. Tett P, Mee L (2015) Scenarios explored with Delphi. In: Turner RK, Schaafsma M (eds) Coastal zones ecosystem services: from science to values and decision making. Stud Ecol Econ 9(1389–6954):127–144. Springer. https://doi.org/10.1007/978-3-319-17214-9_7
478. Gadamer HG (1960) Truth and method (Weinsheimer J, Marshall DG, Trans.) New York. NY, Continuum
479. Szaj P (2021) Hermeneutics at the time of the Anthropocene: the case of Hans-Georg Gadamer. Environ Values 30(2):235–254. https://doi.org/10.3197/096327120X15752810324039
480. Gadamer HG (1997) Gadamer on Celan: Who am I and who are you? and other essays. SUNY Press. https://sunypress.edu/Books/G/Gadamer-on-Celan2
481. Sagan CE (1980) Cosmos. New York: Random House. LCC QB44.2.S235. ISBN 0394502949. See also https://en.wikiquote.org/wiki/Carl_Sagan
482. Huxley A (Aldous) (1959) Collected essays. Harper and Bros Publishers, New York
483. Soyfer VN (2001) The consequences of political dictatorship for Russian science. Nat Rev Genet 2:723–729. https://doi.org/10.1038/35088598
484. Borinskaya SA, Ermolaev AI, Kolchinsky EI (2019) Lysenkoism against genetics: the meeting of the Lenin all-union academy of agricultural sciences of August 1948, its background, causes, and aftermath. Genetics 212(1):1–12. https://doi.org/10.1534/genetics.118.301413
485. Gordin MD (2022) Lysenkoism. Encyclopedia of the history of Science (February 2022). Carnegie Mellon University. https://ethos.lps.library.cmu.edu/article/id/560/
486. Ciumasu IM, Bos D, Bakker J (2007c) The effect of cattle grazing upon plant diversity in salt marshes: the importance of the sample size. In: The British Ecological Society's Annual Meeting 2007, University of Glasgow, 10–12 September, Glasgow, Scotland
487. Ciumasu IM, Stefan N, Lacatusu C (2009) Conflicting economic and conservation interests in the sugau gorges nature reserve in the Romanian Carpathian Mountains. Talk. In: The 2nd European Congress of Conservation Biology. Conservation biology and beyond: from science to practice, 01–05 Sept, Prague, Czechia. https://www.fzp.czu.cz/cs/r-6897-veda-a-vyzkum/r-7292-konference/r-9126-archiv-konferenci/r-9127-eccb-2009
488. Georganos S, Abascal A, Kuffer M, Wang J, Owusu M, Wolff E, Vanhuysse S (2021) Is it all the same? Mapping and characterizing deprived urban areas using Worldview-3 superspectral imagery. A case study in Nairobi, Kenya. Remote Sens 13(24):article 4986. https://doi.org/10.3390/rs13244986
489. Jones P (2017) Formalizing the informal: understanding the position of informal settlements and slums in sustainable urbanization policies and strategies in Bandung, Indonesia. Sustainability 9(8):article 1436. https://doi.org/10.3390/su9081436
490. Mulligan J, Bukachi V, Gregoriou R, Venn N, Ker-Reid D, Travers A, Benard J, Olang L (2019) Participatory flood modelling for negotiation and planning in urban informal settlements. Proc Inst Civil Eng—Eng Sustain 172:1–14. https://doi.org/10.1680/jensu.17.00020
491. UN-NUA (2017) New Urban Agenda. United Nations. ISBN 9789211327311. https://habitat3.org/wp-content/uploads/NUA-English-With-Index-1.pdf

Module 3
Coordination Across Expert Areas

3.1 Premises

Obviously, communities want to develop a *favorable reference story* about themselves, one that (a) maximizes the positive attributes and potentials and minimizes the less desirable attributes and (b) provides a benchmark (i.e., a set of reference values) against which progress generated by their efforts (and others) can be measured. Then, self-improvements may result in a virtuous circle of desire to improve and actual improvement (while failure to do so would produce a vicious circle of reluctance to change and accumulation of problems).

Progress therefore is done by actively resolving problems, especially the hard ones. But successful generation of solutions requires the existence (in relevant constituencies of people) of a set of shared knowledge and models, i.e., a basis of common denominators which is good enough for (1) generating and (2) integrating/making use of all available information. In other words, it is perfectly possible that the required information for solving a certain problem (i) is there but it is not made use of (because there is a lack of knowledge/models to do so) or/and (ii) it is not there and the people needing it do not even know it (or they do know it but the fact of knowing it doesn't cause a change in the overall outcome). In light of the discussions in the previous two Modules, one observes that this idea is also related to the interactive character of learning itself and the phenomenon of social learning—with multiple implications.

So, to be successful in expert interactions and generating relevant list of priorities and scenarios, we need to overcome the grand challenges of (A) knowledge fragmentation between knowledge domains and (B) the fact that knowledge is distributed and mobilizing it requires potent methods that also have sufficiently low (i.e., affordable) transaction costs and opportunity costs, in addition to a sufficiently high convergence of interests to motivate relevant (albeit beautifully different) actors and stakeholders to pool together resources and competences in joint projects—and then to manage the projects towards insuring sufficient results for those actors and stakeholders (or

I. M. Ciumasu, *Eco-cities*, Green Energy and Technology,
https://doi.org/10.1007/978-3-319-14702-4_3

rather enough of them, since we know from practice that there is always a natural turnover; partners come and go) to insure that the core project team is motivated to continue working together to address shared challenges.

Relatedly, we know from Cultural Theory that there exist conceptual splits and gaps between personal worldviews, e.g., positivist versus constructivist. There is no need to discuss these aspects in detail, but we should know that this diversity and its analysis has merits and can provide insights for problem-solving, and I encourage methodological open-mindness, and attention to details (see also [134, 311, 367, 415, 549]).

Then, the personal and cultural value-load of decisions is also a matter demanding attention, but I will only mention some real-world challenges that require pragmatic management. Notably, one can find situations in which "Machiavellian" methods are being used, like the deliberate organization of expert or/and public consultations in formats which allow true experts (or involved citizens, or other legitimate actors) to be "outvoted or outfoxed" [28, 257, 260]. These can be successful in the pursuit of some narrow, short-term political or bureaucratic goals but will undermine the capacity of society as a whole to close the gap between what we (really) know and what (really) we do collectively. Put differently, if the legitimate authorities (starting with the elected actors) and the expertise-holding persons and institutions allow projects to be undermined by illegitimate power games, as discussed in Module 1 (Sect. 1.3), unsolved problems will accumulate and severe systemic crises and collapse are the inevitable outcome. So, the power of a democratic or otherwise culture can make possible or, respectively, impossible the conduction of true eco-city projects. A supportive policy and culture favor community building, and the projects and programs in this culture show a certain "ésprit de corps" which develops in time from the awareness of the people sharing a common goal [192, 262].

As you can see, there is quite a handful of major challenges that can be enumerated in one breath. This means that conceptual models from different knowledge areas need to be connected so as to insure coherence of our understanding of reality from different angles: in the words of Melanie Mitchell (referring to the central value of analogy when trying to make sense of reality based on different, disparate insights) [479, 481] similarities and differences tell us what to expect, and model-based analogies help us figure out what happens and what is likely.

In this sense, and in light of the discussions about resilience, crises are revelators of realities and natural occasions to improve the status of a city of interest. But we already know that dealing with crises is not possible without (a) inter-organizational learning and (b) effective citizen involvement. And, in the absence of a sufficient basis of common denominators, different requirements (necessary but insufficient conditions for success) result in bargaining and sub-optimal decisions [496, 703]. This is fully understandable (and, to a great extent even inevitable). Still, complex problems like urban resilience (vs. vulnerability) and urban (un-)sustainability) can only be addressed effectively as part of a holistic endeavor. There is some scope for breaking some problems into smaller problems (to be addressed one by one), but, most of the times, disciplinary experts can't "just do one thing": they have to "jump in the fray", find ways to work together (even in messy ways if this is what it takes at times), so as to learn from each other and to learn (as quick as possible) (1) how to make expert decisions as a group, and (2) how to engage in "chaotic" collaboration, ultimately (3) how to be good at social learning—a term already mentioned in the book so far (see also [171], as an observation, in the literature, one should carefully check the meaning of terms).

In short, systemic value and a high and sustainable quality of life is generated in human communities through a history of complex projects, which requires a good problem definition to start with. This, in turn, is based on agreements on values and knowledge and is a basis (necessary condition) for further agreements on work goals and methods [121].

Figure 3.1 summarizes this idea in terms of how well-prepared are projects as per how well structured is the problem they address and how well coordinated is the project team in that effort.

Figure 3.2 follows up with the very same logic but in terms of the feasibility and adequacy of a project.

This greatly underrated activity of developing common denominators starts with project conception and preparation.

However, because we talk about complex dynamic systems (CDS) and all the consequences discussed so far in the book, and because it takes dedicated projects that must successfully bridge across academic disciplines and between science and society (e.g., [90, 338, 459, 569, 632]), developing a starting set of common denominators is inevitably (1) a long-term activity and (2) one that never stops.

a

Primary degrees of problem structuredness in pioneering, trans-disciplinary projects		AGREEMENT ON KNOWLEDGE	
		not agreed	**AGREED**
AGREEMENT ON VALUES	**not agreed**	(1) Unstructured problems	(3) Partly structured problems, with disagreements on values
	AGREED	(2) Partly structured problems with disagreements about knowledge	**(4) Well-structured problems**

b

Primary degrees of coordination of endeavors to solve well-structured problems		AGREEMENT ON TECHNOLOGY	
		not agreed	**AGREED**
AGREEMENT ON GOALS	**not agreed**	(4.1) Unprepared endeavor (where endeavors = problem solving attempts)	(4.3) Partly structured endeavor, with disagreements about goals
	AGREED	(4.2) Partly prepared endeavor with disagreements about technology	**(4.4) Well-coordinated endeavor**

Fig. 3.1 Organization of transdisciplinary work: coordination. Technology here is used in the largest sense, including "applied knowledge": techniques, skills, methods, and (all non-automated and automated) processes that are used in the efforts towards accomplishing the project goals; it comprises both hardware (machines) and software (executable protocols). *Source* original, adapted after [121]

a

Primary degrees of feasibility in pioneering, trans-disciplinary projects		SUITABILITY TO TEAM	
		not SUITABLE	**ADEQUATE**
SUITABILITY TO INDIVIDUAL	**not suitable**	(1) Un feasible project	(3) Partly suitable project, with unsuitabilities for the individual(s)
	ADEQUATE	(2) Partly suitable project, with unsuitabilities for the group	**(4) Feasible project**

b

Primary degrees of adequacy in pioneering, trans-disciplinary projects		ADEQUACY TO SUSTAINABILITY	
		not adequate	**ADEQUATE**
ADEQUACY TO CHANGE	**not adequate**	(4.1) Inadequate project (or series of projects, i.e., a program)	(4.3) Partly adequate project, with inadequacies to the change
	ADEQUATE	(4.2) Partly adequate project, with inadequacies to sustainability	**(4.4) Adequate project**

Fig. 3.2 Organization of transdisciplinary work: adequacy. The "individual" is understood as an agent in agent-based models. The "team" is understood as a cluster or network and can be either a set of agents (with common goals) and/or a unitary collective agent. "Change" and "sustainability" are understood as respectively defined in the innovability and sustainability thresholds (Fig. 1.12, Module 1). In a program (4.1), project portfolio selection is based on the non-violation of objective (logical and thermodynamical/physical) constraints and then on the order of priorities identified in the process of project management. *Source* original, adapted after [121]

Getting very diverse partners on the same page involves a great deal of uncertainty, substantial investment, and hard-to-do experimentation, as well as constant learning, conflict management, and—more often that is usually thought—institutional building (either from scratch or via the Schumpeterian "creative destruction"). This kind of work (projects or series of projects, i.e., programs) includes a broad range of fields of expertise, socio-economic actors, terminologies, and context-dependencies. Such projects tend to be triggered by convergences of interests that happen to be matching public demands (or other kinds of circumstances that are perceived by a given set of actors as opportunities), but initial coordination efforts are nearly always constrained by competing demands faced by the persons taking the initiative and are therefore limited to a minimal alignment of the formal requirements of the funders/resource providers. This means that many essential structural and functional mismatches are postponed to later stages ("One problem at a time, please!").

In Sect. 3.7 of this Module, I describe a generalized model of the RDI logic behind TRLs, called Solution Readiness Level (SRL). But before anything else, one needs to find ways to build those necessary common denominators in spite of the inertia of knowledge fragmentation and business-as-usual (BUS).

To overcome BUS inertia (and related project misconceptions and setbacks) a group of actors and stakeholders (from citizens to scientists to businesses to public authorities) need to recognize that (and understand why) they must avoid having a poorly-structured problem (PSP). By definition, projects that are built around PSPs have latent disagreements on both values (materialized as general rules of operation) and on knowledge (theories, models, methods, and terminology). In the absence of such agreements, there is no realistic basis for real coordination on goals and means to reach those goals. This situation is actually usual in pioneering projects, and even more so in large and transformational projects that are connecting science and society on multiple levels [338, 459, 473, 477, 632].

From a more technically formal point of view (specifically, in terms of Game Theory in mathematics), a constituency of actors lacking that necessary minimal basis of common denominators, is just another case of a group of agents that are locked in a situation of Prisoner's dilemma. In this type of situation, because it wants to avoid being duped, it is actually logical for each rational agent to make non-cooperative choices: choosing to defect (i.e., not cooperate) will yield lower (but sure) individual payoffs, although this means the lowest payoff for the group-as-a-whole.

Finally, it is important to highlight that "getting the people on the same page" is not at all a matter of "regimenting" everyone so that they march in lock-step in the same direction as indicated by a "commander". That would be just the opposite (and equally damaging) extreme to having people talk past each other like in the mythical "Tower of Babel". The productive equilibrium is reached in a project when there is a minimally functional basis of common denominators as a necessary platform for meaningful conversations, but not much more than that—because the value of the conversation comes exactly from having different perspectives exist freely so that different types of agents "spread out" and explore at maximum the space of possible options (for understanding a problem and generating new solutions) before coming together to make sense of their collective understanding and capacities in a coherent big picture that ultimately leads to a best-possible work result.

3.2 Orientation Across Study Scales and Domains: The CIARPO/Systems Table

Observations About Scale

For finding the way in the context, a project team must have a start reference, so that their action is not blind, progress is measurable and resources are not wasted. Any "big picture" is first and foremost about scale. Figure 3.3 proposes a visual model based on the systemic perspective on sustainability, and Figs. 3.4 and 3.5 show a general model for locating different types of expertise and inputs from individual experts, entire domains of knowledge and practice or/and organizations like academic and applied transdisciplinary institutions.

In the CIARPO model below one makes two further analytical observations that matter for a correct understanding of the big picture, and at the same time, stand as practical insights made possible by the big picture regard. First, one notices that the scales in the {Cell; Individual; Assembly; Region; World; Outer Space} series are

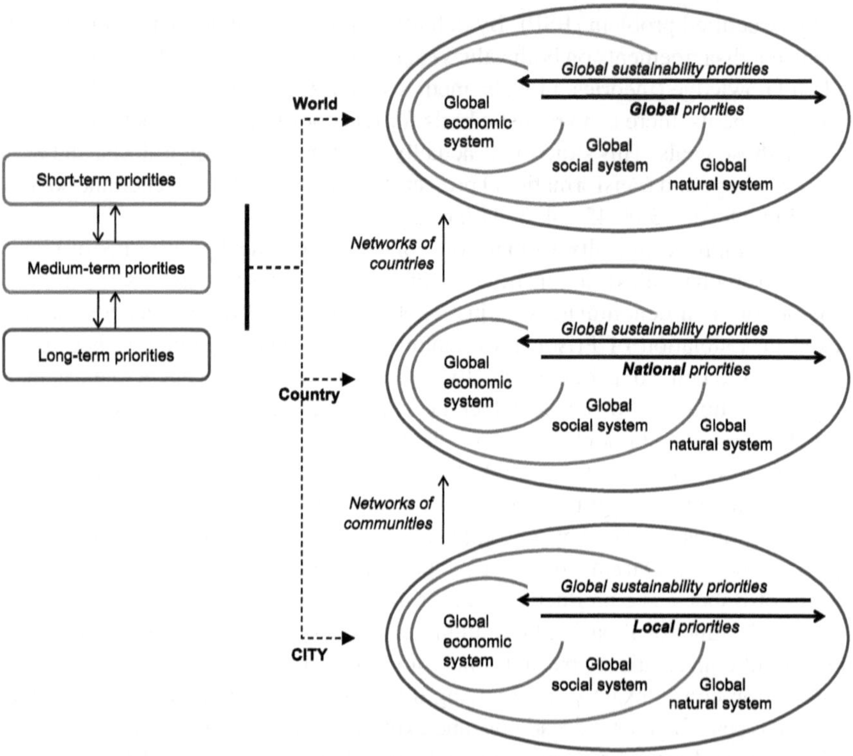

Fig. 3.3 Declination of the systemic perspective (from Fig. 1.1: the nested inclusion relation between economic, social, and natural systems) at different scales. Observation: "global sustainability" is a partial pleonasm (sustainability can only be defined at a planetary scale) but is used here because it provides an analytical highlight of the relations between different scales. *Source* original

nested, which means that the larger ones include the smaller ones (also considering "outer space" as just a label for "space" without excluding Earth from it unless one chooses to do so for some specific reasons). This means that all categories (ordered pairs) resulting from the Cartesian Product in Fig 3.4 are also nested along both scales and diagonally along the shortest distances from bottom left (smallest) to top right (largest)—namely, from the "economic aspects at the level of the individual" (e.g., economic preferences and choices and all the reasons behind it) to the "natural reality of space" (astrophysics, origin of living organisms, life in space, and so on). Idem, the "economic individual" is nested in the "social individual", which is nested in the "biological individual", in the sense (to make a reminding deduction from the first Module of the book) that a "biological individual" can exist without the "social individual" (here the term "social" being used in the sense of social aspects, not in the mathematical, informatic or biological sense) but not vice-versa; and the "social individual" can exist without the "economic individual" but not vice-versa.

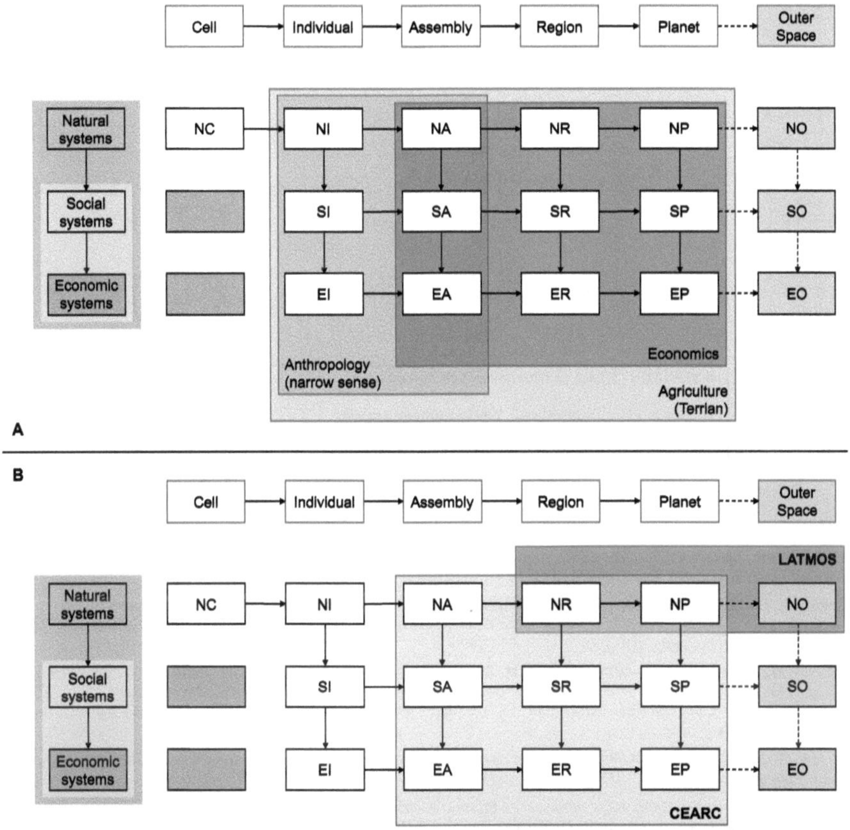

Fig. 3.4 The *CIARPO/systems* scale model of knowledge (expertise) domains as identified through the Cartesian Product between the series of nested systems from the systemic perspective on sustainability and the series of main scales known from research and practice: the cell (C), the human individual (I), the assembly (A) of humans, an immediate example being a local community (for instance, a village, town or city), the Region (R) of the world that is identifiable based on some common attributes (e.g., Europe or Sub-Saharan Africa), the planet (P) itself taken as a whole, and the outer space (OS), which is included because humanity has extended its presence outside Earth. Shown on the top (A), are three examples of mapping (non-exclusively) academic fields onto the model. Grey boxes indicate empty cells. Shown at the bottom (B), are two examples of mapping the scales/systems covered by two academic institutions: the CEARC and LATMOS within the Observatory of Versailles Saint Quentin-en-Yvelines (OVSQ), University of Versailles (OVSQ) and University of Paris-Saclay (UPS). Labelled cases indicate systemic-and-scale domains in which different experts may find themselves (and their inputs to a given project) or not. Grey boxes indicate empty cells. *Source* original by the author (credit: I thank Jean-Paul Vanderlinden and Jean-Luc Vayssière from the CEARC/OVSQ—University of Paris-Saclay, for valuable feedbacks on this model during late 2018)

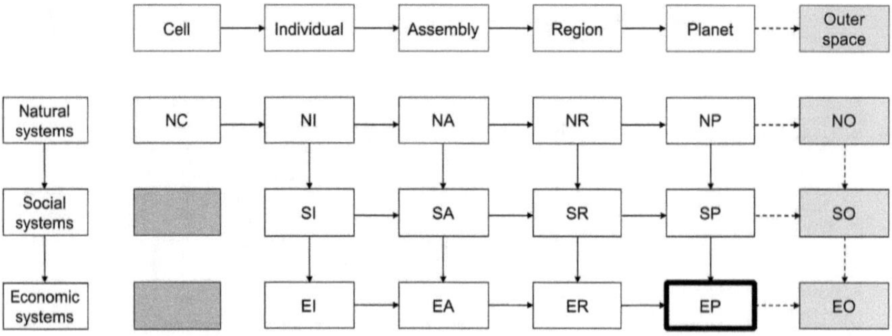

KEYWORDS (& lists of broader topics and issues) EP

Global economy sustainability; Whole-Earth economics;
Global economic variables; Global macroeconomic models;
Global supply chain management; Global economic crisis;
Global economic system;

References

Georgescu-Roegen, N. (1971). *The Entropy Law and the Economic Process*. Harvard University Press. https://doi.org/10.4159/harvard.9780674281653.

Murray, A., Skene, K., Haynes, K. (2017). The circular economy: an interdisciplinary exploration of the concept and application in a global context. *Journal of Business Ethics*, *140*(3), 369–380. https://link.springer.com/article/10.1007/s10551-015-2693-2.

Pascual, U., Palomo, I., Adams, W.M., Chan, K.M., Daw, T.M., Garmendia, E., Gómez-Baggethun, E., De Groot, R.S., Mace, G.M., Martín-López, B., Phelps, J. (2017). Off-stage ecosystem service burdens: a blind spot for global sustainability. *Environmental Research Letters*, *12*(7), 075001. https://iopscience.iop.org/article/10.1088/1748-9326/aa7392/meta.

Binkley, S. (2003). The seers of Menlo Park: The discourse of heroic consumption in the 'Whole Earth Catalog'. *Journal of Consumer Culture*, *3*(3), 283–313. https://doi.org/10.1177%2F14695405030033001.

Brown, P.G., Garver, G. (2009). *Right relationship: Building a whole earth economy*. Berrett-Koehler Publishers. https://www.bkconnection.com/static/Right_Relationship_EXCERPT.pdf.

Ratti, R.A., Vespignani, J.L. (2016). Oil prices and global factor macroeconomic variables. *Energy Economics*, *59*, 198–212. https://www.sciencedirect.com/science/article/pii/S0140988316301372.

Berger, T., Grabert, S., Kempa, B. (2016). Global and Country-Specific Output Growth Uncertainty and Macroeconomic Performance. *Oxford Bulletin of Economics & Statistics*, *78*(5), 694–716. https://onlinelibrary.wiley.com/doi/abs/10.1111/obes.12118.

Dafermos, Y., Nikolaidi, M., Galanis, G. (2017). A stock-flow-fund ecological macroeconomic model. *Ecological Economics*, *131*, 191–207. https://www.sciencedirect.com/science/article/pii/S0921800916301343.

Koberg, E., Longoni, A. (2019). A systematic review of sustainable supply chain management in global supply chains. *Journal of Cleaner Production*, *207*, 1084–1098. https://www.sciencedirect.com/science/article/pii/S0959652618330476.

Papatheodorou, A., Rosselló, J., Xiao, H. (2010). Global economic crisis and tourism: Consequences and perspectives. *Journal of Travel Research*, *49*(1), 39–45. https://journals.sagepub.com/doi/abs/10.1177/0047287509355327.

Benatar, S.R., Gill, S., Bakker, I. (2011). Global health and the global economic crisis. *American Journal of Public Health*, *101*(4), 646–653. https://ajph.aphapublications.org/doi/abs/10.2105/AJPH.2009.188458.

Wigley, T.M., Richels, R., Edmonds, J.A. (1996). Economic and environmental choices in the stabilization of atmospheric CO2 concentrations. *Nature*, *379*(6562), 240. https://www.nature.com/articles/379240a0.

Fig. 3.5 Illustration of content filling of a cell in the *CIARPO/Systems* framework model based on selected keywords: [59, 63, 73, 94, 186, 267, 370, 501, 547, 550, 584, 754]. One can of course do her/his own lists as relevant for specific interests and projects. *Source* original

In the same way, the scale of the Individual (i.e., everything related to an individual) is nested in the scale of the Assembly (e.g., a local community/city), itself nested in the scale of a Region of the planet (a regional network of all human communities that share the set of common contextual attributes that are characteristic to that region of the planet); and the Region itself is nested in the World (planet Earth), which is itself part of the Solar system (some of which Humans have already reached physically by modern technology and space travel).

This observation alone is enough to realize the immense complexity we are dealing with when are making evaluations and decisions such as those involved in the Delphi-based method and scenario-generation, in eco-innovation and RDI in general, in policy-making, and in the management of complex projects in general.

The second observation regards the fundamental mathematic aspects of the model itself: one can do a simple graph analysis of the CIARPO/Systems model (for the definition of graphs and graph analysis, see Box 3.1).

Box 3.1 Detail about Graphs and Graph Analysis

Essentially, graphs are dots (or other geometrical shapes, if needed for communication; which represent abstracted objects) connected by lines (or arrows, when directed) denoting the relation(s) between those objects. When lines/arrows are used, the graph is called "undirected"/"directed".

The word "graph" here should not be confused with "graph" as vernacular notion referring to any diagram/technical visual displays in general. For the latter, in this book I only use standard numbered figures (like "Fig. 1.1", etc.).

In Graph Theory, a dot is called "vertex" (plural "vertices"), and a line is called "edge" (or "transformation").

Importantly, graphs can also include sets of dots that may not be connected; when connections exist, the graph is called connected (but there can be situations when some dots remain without connection with the rest of the dots; there are also "fully connected graphs"). Similarly, there are "strongly connected graph" (with many direct connections between the vertices), "complete graphs" (when each vertex is directly connected with every other vertex in the graph), and so on, up to very complex patterns and geometrical motifs.

With this system, virtually anything can be represented, and a large number of patterns can be identified, the benefit of this method being the possibility to study intrinsic, i.e., context-independent) effects of networks—knowing that all systems can be represented as networks. In applied research (and RDI), complex and multiply-interconnected "connected graphs" are usually being used, and they are oftentimes called "networks" instead of graphs (please note however that all networks are graphs but not all graphs are networks).

Graphs describing real-world objects and realities can be very simple or very complex and their structure carries useful information about the phenomenon/problem/topic being represented, so they can be used to carry out in-depth study of the underlying dynamics and properties of natural and human systems. This study method is called Graph Analysis or Network Analysis. Graphs Analysis is therefore a standard method for the study of complex systems, i.e., complex phenomena that emerge from sets of simple rules.

For the purpose of sustainability studies in general, and as illustrated in Module 1, Sect. 1.2, Fig. 1.5, current bibliometric analyses (clouds of topics detected by network analyses) can be (if studied) highly instructive or at least technically informative for on-going efforts and projects (e.g., [107, 695, 375])—and this is why it is particularly important to find and ask good questions ([153, 154]; see also [127]). In Sect. 3.6 of this Module, I discuss the matter of placing questions in context.

In this book, graphs have been first mentioned briefly in Module 1, Sect. 1. 2, Table 1.1, at the "Network character" of complex systems [and I paste here the text again, for convenience: "Operational representability as networks: sets of objects (and relations between them) coded as mathematical graphs (dots/ vertices connected by links/edges). This allows the detailed study of individual and collective properties and interactions between systems components."] and again in Table 1.3 at "Nested networks" ("mathematical graphs: sets of vertices/dots (agents) and edges (links)"), and in some other places throughout the book. But I have reserved the details about graphs for this more technical Module, in order to try and keep the main text easier to read.

A special category of graphs that can be very useful in practice is represented by hypergraphs: in this case, a connection unites more than 2 vertices and is not represented by a line but by a surface. Other motifs and terminology exist but need not be discussed here.

For more on the Graph Theory (hypergraphs included), see, e.g., [537, 596, 630, 668]. Here, suffice it to say that graphs and graph analysis are a fundamental basis for all research and management, and are a necessary tool in all work related to eco-cities, both directly and indirectly. To give some examples, graphs can be explicitly mentioned in city—and sustainability-related engineering [340, 374, 491, 492, 596, 710]. Relatedly, graph analyses have a particularly set of interesting application in geography (and are therefore even more important for cities), because it can easily represent spatial effects, e.g., the gravity model of trade and migration or/and cultural influence between two cities.

The fundamental importance of graphs is also reflected in notions like "Big Graph Systems" and statements like "The future is big graphs" (e.g., [625]); while reflecting certain narrow perspectives of some specialty field (and sometimes in very peculiar, field-specific or commercially driven language that may confuse people outside that field of professional interest), such examples collectively reflect the underlying ubiquity of the graphs and graph analysis and invite at better communication across different fields.

In this very book, there are a good deal of models and displays which are essentially graphs (even if not explicitly mentioned, for brevity reasons). In this book module, one can see such examples, starting with the Cognitive Maps in the next Section.

Most importantly, the decision trees connecting the scenarios discussed in Module 2 are examples of graphs (in that case, both the yes/no decision points and the End Outcomes (the situations described by scenarios) are vertices; and the arrows connecting them are directed edges (which concretely make those scenario sets be understood as directed graphs). In this sense, the dendrogram in Fig. 2.1 is a directed graph archetype.

In this Module, some advanced examples of graphs are embodied by the DIKAR_process model in Sect. 3.4 and Knowledge-Action Matrix in Sect. 3.10 (two cases of lattice graphs) and the related models and discussions (the KPN model included).

At this point, and related to these models, it is also important to notice a last technical detail: graphs can be represented also as tables of correspondences or their equivalent mathematical matrices, showing all the functional relations existing between the vertices or/and between the vertices and the edges of a given graph (usually "adjacency matrices" are being used—see a direct example in the main text). In the case of lattice graphs, these tables/ matrices are defined by the lattice coordinates.

An adjacency matrix indicates whether or not two vertices are adjacent, i.e., connected: the element entry is 1 when a connection exists and 0 where there is none. An adjacency matrix can be used to store relevant information, by replacing the non-zero elements with another value, e.g., edge weights (in which case the 1 is replaced with a number that indicates a weight). An incidence matrix shows the relationships between vertices and edges: the entry is 1 if a relation exists and 0 if it does not exist). Other variations can be used, depending on the chosen purpose.

Thus, an adjacency matrix can be used in graph analyses to store the Centrality value (i.e., the number of direct connections) of each vertex of the graph (also called a "node" in a network, i.e., each domain represented as a box) with the neighboring domains as an indicator of relative importance:

$$C_{CP} = \begin{pmatrix} NC & NI & NA & NR & NP \\ \emptyset & SI & \mathbf{SA} & \mathbf{SR} & SP \\ \emptyset & EI & EA & ER & EP \end{pmatrix} = \begin{pmatrix} 1 & 3 & 3 & 3 & 2 \\ 0 & 3 & \mathbf{4} & \mathbf{4} & 3 \\ 0 & 2 & 3 & 3 & 2 \end{pmatrix}$$

$$C_{CO} = \begin{pmatrix} NC & NI & NA & NR & NP & NO \\ \emptyset & SI & \mathbf{SA} & \mathbf{SR} & \mathbf{SP} & SO \\ \emptyset & EI & EA & ER & EP & EO \end{pmatrix} = \begin{pmatrix} 1 & 3 & 3 & 3 & 3 & 2 \\ 0 & 3 & \mathbf{4} & \mathbf{4} & \mathbf{4} & 3 \\ 0 & 2 & 3 & 3 & 3 & 2 \end{pmatrix}$$

where C_{CP} and C_{CO}, respectively, denote centrality *from Cell to Planet* and *from Cell to Outer space*, Ø indicates empty cases, and numbers indicate the number of basic direct connections.

In this, one notices right away that, in the mathematical graph represented by the *CIARPO/systems* model, when *Outer space* (O) is not included, the vertices with the highest number of connections (i.e., possessing the highest Centrality) are SA (*Social systems* x *Assembly*) and SR (*Social systems* x *Region*), i.e., cities and local networks of cities.

When *Outer space* is included, then a third one joins this set of central vertices, namely, SP (*Social systems* x *Planet*). In any of the two situations, SA is the lowest of the central scales, i.e., the most accessible/least hard to manage. This makes the (systemic) Assembly, i.e., the city, the mathematically obvious unit for systemic change of humanity—which idea is in line with the observation made already in Module 1 (Sect. 1.3) that a city is the best tradeoff between relevance and feasibility, and therefore cities are a natural test bed for policies towards sustainability (and arguably the best one), and thus eco-cities are obvious agents for playing a key role in developing adequate solutions (technologies, products and services, complex smart systems, policies) for nations and entire regions, by close collaboration with the national level.

One also notices that all these three vertices include the level of *Social systems*, which basically means—to make a long story short—that social systems are at the heart of it: it all depends on how we organize and interact between cities, world regions, and across the whole world. In other words, a simple graph analysis (that can be done intuitively by anyone by just counting connections and comparing "boxes") reveals from a mathematical angle what experience already indicates, namely, that cities are central to the entire study and action about and towards sustainable development, and it also highlights again the double local-and-global character of cities, i.e., the fact that a city is embedded in wider networks of cities as already discussed in Module 2 (see again Fig. 1.8) of this book.

This has important consequences for concrete issues. For example, it is in this context that we need to understand the Big Puzzle in the relation between the topics comprised by *Climate Action* (the UN's Sustainable Development Goal SDG13) and the other SDGs shown in Fig. 1.10 (Module 1, Sect. 1.3) in the context of the different levels and thematic filters related to the systemic perspective on sustainability and the specific responsibilities of individuals, cities/communities and countries. And of course, this multi-scale logic applies to the ensemble of environmental changes, beginning with ecosystem goods and services and the demand and supply thereof (for a start discussion, see, e.g., [642]).

Relatedly, another example comes from the basic discussion on the importance of scale for the understanding of the relations between the two main sub-topics of climate changes: mitigation and adaptation (Fig. 3.6). By definition, climate change is a planetary phenomenon, but impacts of (and causal contributions to) it occur at all scales. This provides an "automatic" reminder about two realities already discussed in this book: cities are only one (albeit a key) scale of action, and therefore the actions of the city need to be coordinated with other cities at national and global levels, and

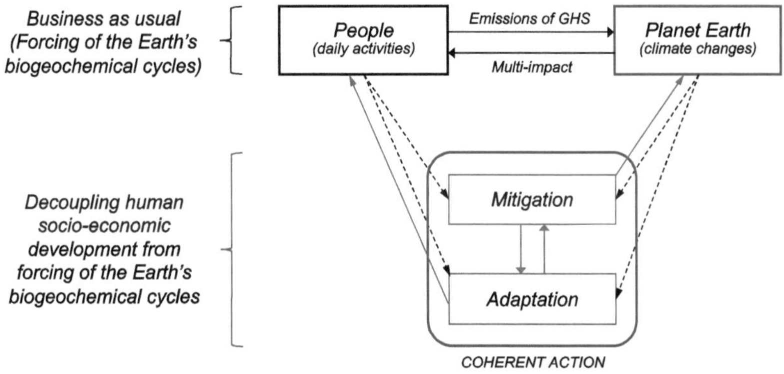

Fig. 3.6 Simplified model of the relation between mitigation of and adaptation to climate changes, and the two levels corresponding to business-as-usual (top) and engaging in decoupling from the current forcing of the planets' biogeochemical cycles through coherent action that would break the vicious circle of GHS emissions and multi-impacts of climate change that harden the life of people. *Source* modified after [161]

cities represent the best tradeoff between the relevance for human systems and the feasibility of problem-solving projects.

3.3 Articulating Expertise for Addressing New/hard Problems: Cognitive Maps

In complex projects involving large networks and clusters of experts, the distributed character of knowledge relates not only to knowledge organization (KO) [315, 317] and knowledge management (KM) ([156, 454]; see also [335, 673]) in general, but also specifically (a) to the relation between problem structuring in the space between knowledge and action and (b) to the matter of knowledge representation (as related to knowledge-action models).

Table 3.1 clarifies the relation (and thus hopefully dispelling potential confusions) between three popular notions that oftentimes are being used interchangeably although they are not equivalent: mind maps, concept maps, and knowledge maps.

In the literature, some other terms may be used to referring to any of these notions. Figure 3.7 illustrates two examples of the "causal loops" model of representation of the Big Picture of "what is known to happen" in a studied matter. The same idea applies when we want to describe a methodology, for example, (inter alia) the key issue of problem structuring for/in/and Multi Criteria Decision Analysis (MCDA) [58] or "issue mapping for problem structuring" [183], "problem structuring in transitions" [159, 160, 311, 633, 637], and related processes (e.g., [58, 313, 476]). Some authors simply talk about "cognitive mapping" [236] which actually includes

Table 3.1 Basic types of cognitive maps: definitions

Notion	Short definition (main idea)
Mind maps	Visual representation of someone's (or a group's) ideas in the form of boxes and arrows, no matter how chaotic or organized those ideas are. They do not decide whether the represented ideas are good or bad, smart or not. They are meant to simply show the thoughts about a given topic In the technical language provided by fundamental mathematics and used across science and engineering, such representations are called "graphs" (as per Box 3.1 in Sect. 3.2), a term which is not to be confused with the word "graph" that is used daily speech which refers to every technical used in science reporting). However, all forms of representing correspondences between categories are equivalent to graphs (and can provide alternative representations of graphs, e.g., usual tables and mathematical matrices
Concept maps	Like Mind Maps, but, in addition, they must be coherent and logical: they must make sense. In other words, they can still be "crazy" in the sense of being rather unconvincing and disconnected from available evidence from reality, but at least they must be internally coherent (there should be no internal contradictions)
Knowledge Maps	Like Concept Maps, but, in addition, everything shown in there must be anchored in (and justified by) current knowledge as reflected in science. In principle this condition can be extended to include other knowledge systems beyond modern science, for example, the knowledge embedded in language or the empirical indigenous knowledge systems (which is usually also embedded in language, behaviors, and culture); but, in this book I only use knowledge from science and related project experience

all those in Table 3.1, but one must carefully check the definition used or implied by each author.

Figure 3.8 shows a key example from an operation research paper: a Knowledge Map on urban parking areas, used as a means to fulfill "the need to aggregate different viewpoints according to their different levels of importance" in the context of a collaborative decision process taking place in a city, and aiming at distilling possible alternatives options and their respective value according to the different criteria that reflect local needs or stated preferences (and combinations thereof).

The practical interest of knowledge maps (and cognitive maps more broadly, especially because knowledge maps are the result of a process and effort starting with mind maps) is that, when having such a common representation under their eyes, project team members can use it to locate, compare and discuss their respective insights and propositions in a coherent way (i.e., avoiding to talk "past each other") and to explicitly identify relevant correlations. Very often, different insights emerge directly (both logically and intuitively) from such a representation, and are recognized as such by all those interested.

One slightly technical observation is also important: the quality of the knowledge map is of the essence, and this means that both the knowledge justifying the details in this representation and the technical execution (at least basic design principles) are necessary. To put if differently, poorly executed (unrefined) cognitive representations will look more "complicated" than they are, while actually hiding the relevant intrinsic complexities that the model would reveal if made easy to read. (As an exercise, you may want to compare the representation I show in Figs. 3.7 and 3.8 with the originals in the cited studies. Of course, what matters first is that the research homework behind each knowledge map is being done; but, for communication purposes in multi-disciplinary and multi-profession projects, clarity can make the difference between fruitful exchanges and exhausting misunderstandings).

For the purposes of this book, another practical observation is that the development of clear cognitive models by mapping (i.e., drawing the one-to-one correspondences between categories) is in fact an example of "zooming-in" the

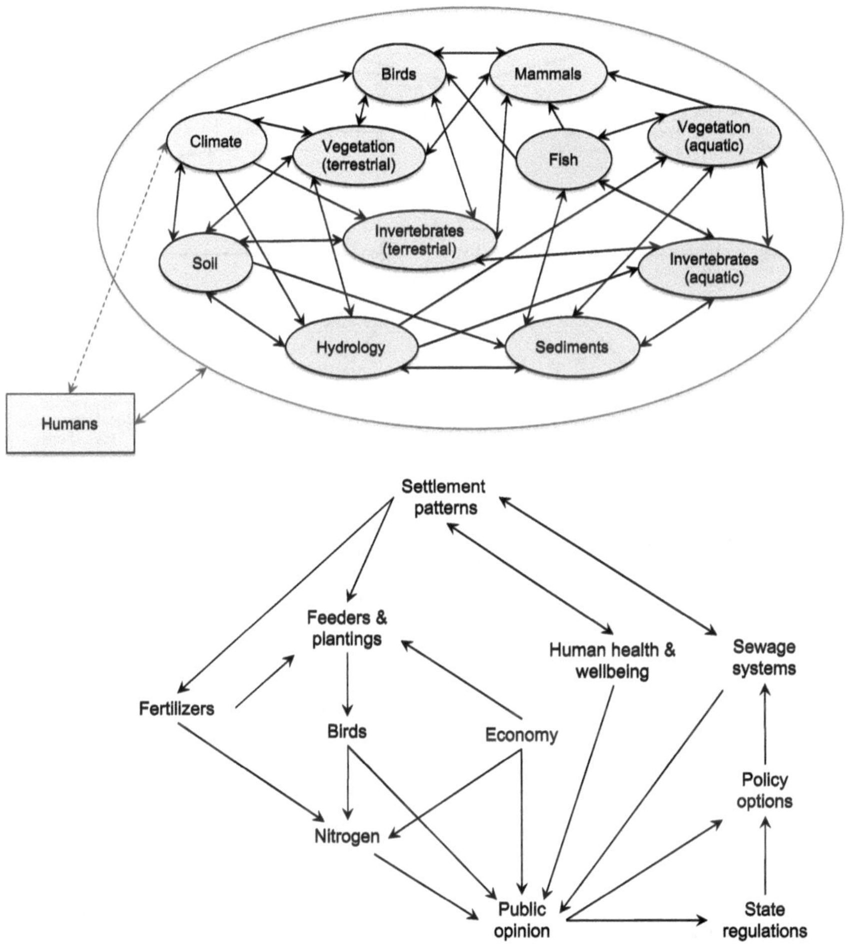

Fig. 3.7 Illustrative examples of cognitive mapping: general causal loops representing the relation between humans and the biosphere (top) and the issue of bird feeders in urban ecology (bottom). Because these represent current knowledge (at the time of their generation), these two models are examples of simple knowledge maps. *Source* redrawn and modified after [457]

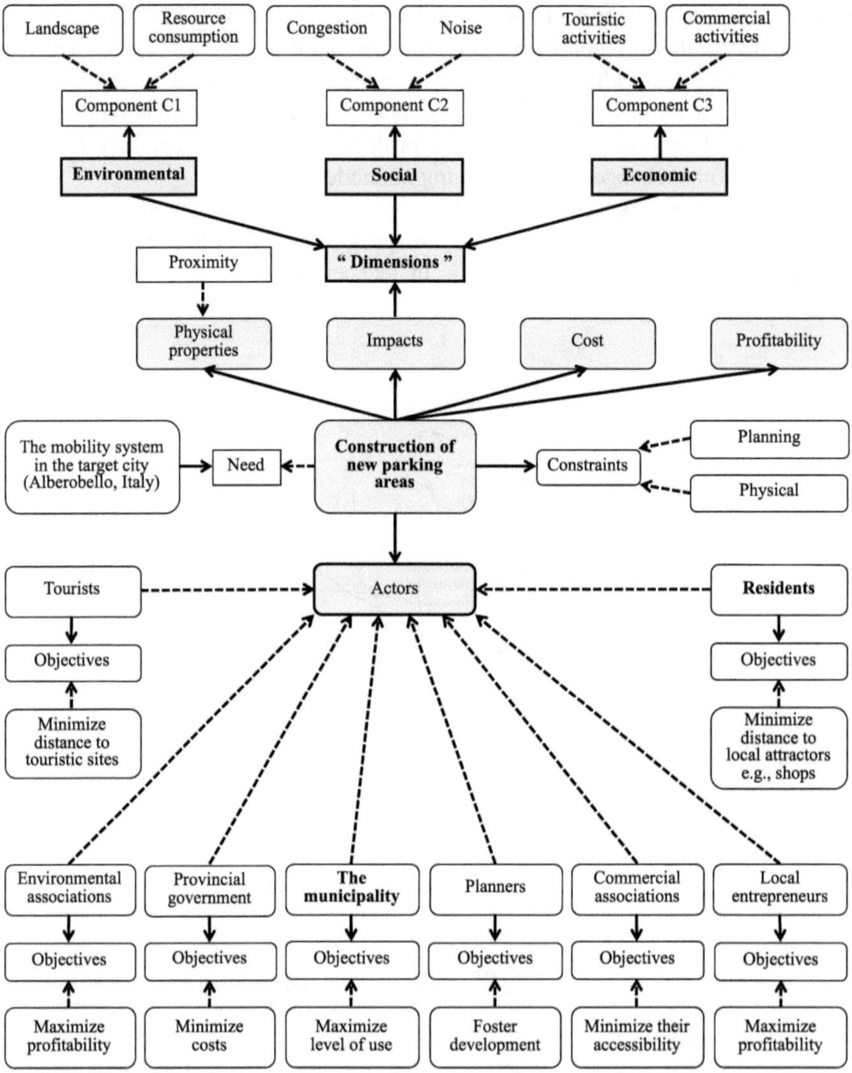

Fig. 3.8 Knowledge map for representing a study problem using the multi-attribute value theory: the management of parking areas in the city of Alberobello, Italy. *Source* Redrawn after (the already simplified version of) [236]. As per the original publication, a continuous line arrow indicates an "is a" relationship, while a dotted line arrow denotes "has a" relationship. Technical observation: when referring to environmental, social, and economic considerations, the cited author uses the term "dimensions" (and no other specification). In addition, in the original figure of the cited case study, the author uses this order from left to right: "environmental, economic, social". This means that this empirical-driven, bottom-up representation is one of the many serious efforts that can still gain in value by integrating the systemic perspective upon sustainability. Without it, such works remain undervalued

study issue as described in *Step 2* of the Delphi-based method described previously in Module 2 (Sect. 2.2): the 4-cell visual matrices described in there (Fig. 2.3) constitute one way of drawing a knowledge map. In the same way, the explanatory dendrogram shown in Fig. 2.1 (Sect. 2.2) and the illustrative network from Fig. 2. 2 in there are themselves knowledge maps as defined here.

Related to the content of the knowledge map from Fig. 3.8, one can observe some key connections with the systemic perspective on sustainability summarized and used in this book. Thus, none of the six examples of issues in the three "components" (C1; C2; C3—near the top of this figure, and in the details discussed in the study cited for it), is in fact related to only one of the three "dimensions" mentioned by the cited author (environmental; social; economic): in fact, each is related to all three (just in different ways), but the author had to show only the connections that were deemed to be a priority (she mentions that in her study)—for the sake of operational clarity. Unsurprisingly, the cited author (1) correctly identified the relation as "has in" (not as "is a"—as per her own coding of categories of links), and (2) mentioned that this is a simplified representation. She (inevitably) had to simplify the entire picture to make a point and to help clarify, because this is what a "big picture" on an issue of interest really is: the ensemble of priority aspects with the priority connections. I did the same with the diagrams and models that I show in this entire Module and throughout the book, the prime examples being the decision trees connecting different situations in the scenario sets described in Sect. 2.2 and the whole-city models shown in the next section (and in which the issue of "parking areas" would fit as a sub-topic of a more general topic, like "urban transport" or "land uses"/"landscape", or others, and shown or not in different mappings depending on the concrete priority concerns in each study). In fact, the example in Fig. 3.8 stands out as an example of connection between categories "landscape" and "transport" in the urban sustainability nexus that I describe in the next section of this Module.

Another observation occasioned by this great example is that it illustrates very well the nature of the expert decision: the decision to tell/show or not a specific detail must be understood together with the justifications behind it—otherwise, one may draw erroneous conclusions. In other words, knowledge maps (and cognitive maps in general) are purpose-driven while remaining truthful. Such practical representations of what we know are intrinsically "alive" (i.e., they are to be updated continuously): when new questions arise, or when new knowledge is acquired, the set of priority details being used in communication changes to reflect the new state of knowledge.

Furthermore, the example above serves to make a point about the content itself. Namely, the knowledge map and the discussion in the cited study illustrate well the fundamental (sustainability-oriented) knowledge-action model that I introduced at the beginning of the book Fig. 1.3. First, one notices that the three main concerns of sustainable development (environment, society, economy) are correctly identified by the cited author. Then, if we look at the objectives of all actors, we see that the more immediate they are (by function or actors), the more they overlap with economic aspects, which shows clearly what motivates the decision to address a usual problem in a city: first economics, then social aspects, then environmental. Ergo, as, described in the basic operational model in Fig. 1.3, this is where the study and the structuring of the problem must begin (before ending, of course, with the development of long-term solutions in the logical order described in Fig. 1.3).

In terms of broader utility for project management, in addition to serving as a common denominator for members and a project team, such cognitive maps/models can facilitate the communication between teams, projects, and cities. They can serve as cognitive support (and evidence base) in the transfer of lessons between case studies; and for understanding what adaptations might need to be operated in the model deployed in a different context, where the differences between contexts are, and what they mean in the whole picture on that problem.

For instance, and to close the loop with the case studies mentioned in the previous Module, one issue faced at present by the city of Versailles is precisely of the type evoked above, namely, the complexity of the relation between parking areas and the city-as-a-whole (see Fig. 3.9), and that projects concerned with this issue in Versailles could benefit from the insights produced by the study cited at Fig. 3.8 (and I don't need to go into how, because the teams in charge with those projects will know the specifics and will be able to do what needs to be done); and of course, this applies to any city around the world.

About the issue of parking areas in Versailles, I would only mention some other details that give the measure of the importance of this specific issue represented by topical example of parking areas and reflect the current tensions between different and sometimes conflicting views, e.g., a city for cars versus a city for people (and what can be a smart and sustainable solution).

Namely, the train station and district Versailles Chantiers are directly connected with Chateau de Versailles through a main street, Avenue de Sceaux, which is one (the southernmost) of the three streets forming the famous "trident" of radiating streets that start from the Place d'Armes (in front of the Chateaux) where this large parking is located (582 parking places), and which is now subject to some changes that are planned by the local and regional authorities and that are meant to free the front of the chateau from the massive presence of cars, possibly through subterranean parking and redistribution of land uses, with the overall spatial and architectural reorganization works having started in June 2023 and meant to continue into the next years (details in [246, 408]). Then, important in this case is also the fact that Avenue de Sceaux itself has a large parking area in the middle of it (620 places) and that another large parking (405) places is located in the newly developed district of Versailles Chantiers (just near and underground the train station and bus stations). In total, the city has two surface and five underground parking areas with almost 3200 places in total, plus a total of almost 20,000 (twenty thousand) parking places along the roads throughout the city, with different regimes that may vary in time (for example, becoming taxed during the month of August, i.e., the "holidays month"). Also, some other redevelopments are in course on or near Avenue de Sceaux, notably related to the fact that this main street passes nearby another (though smaller) train station coming from Paris (Versailles Rive Gauche, which is the one typically used by tourists heading to the Chateau de Versailles), and nearby the oldest district of the city: Saint Louis.

Fig. 3.9 Author's snapshots from Versailles, a city known as an open-air museum (original photos, 2019). **Top-left**: public photographic exhibition (in occurrence, on modern architecture) in the public park *Domaine de Madame Elisabeth* (a landscape garden which belonged to a sister of King Louis XVI). **Top-right**: Saint Louis Cathedral (one of the main churches associated with the Kings of France). **Middle**: *Place d'Armes*, in front of the Gate to the Chateau de Versailles, showing the (still) heavy presence of cars and motorized traffic in this historic city center. **Bottom**: public photographic exhibition (in occurrence, on remarkable natural and semi-natural landscapes around Versailles and Paris) in the city center near the building and public gardens of the City Hall

Finally, discussing the entire issue of transportation in Versailles would fall outside the scope of this book, but I would only mention that, after the re-development of Versailles Chantiers train station and district as a pole of multi-modal communication, the entire structure of bus lines also changed in the city, notably improving the North–South connectivity inside the city—which highlights the importance that a key city district for a city-as-a-whole, and the importance of re-developing a city district only by having in mind the city-as-a-whole that is a complex dynamic system and not just "a sum of its parts". Then, connectivity was also improved between Versailles and the other cities and towns in the western suburbs of Paris (notable through the recently launched *Tram 13* line [245] and with the city of Paris and the Paris region as a whole. In other words, many cognitive maps can be developed to represent these complexities, and ultimately, they need to be coherent with each other and useful for projects, ultimately serving the purpose of knowledge integration for problem-solving.

All in all, what matters with knowledge maps and the topics they address is to correctly identify the most important categories and the relations between them in spite of, but correctly reflecting, the complexity of the real-world problem they represent, so as to insure both efficiency and flexibility of works. Once this is done, a project team (or individual actor or stakeholder) has a good default basis and a good start for further works and improvements in each specific context and through a process of learning about and from other projects and cities regardless of their specifics.

The models that I show throughout the book are essentially knowledge maps, aiming at capturing evidence and making it available for practical uses in projects. They have been developed from empirical realities up, using direct knowledge about cities and topics, project experience, available knowledge, and the lengthy work of finding and verifying different insights, and integrating them together based on the principles of scientific modeling and having in mind the priorities of an eco-city project manager, starting (beside the classical requirements) with the transaction cost (resources, practicality, and computational feasibility) and the opportunity cost (what would be the most resource-efficient course of action; managerial optimality).

3.4 Platforming Evidence-Based Projects: The Eco-city Reference Model (ECRM)

The Issue and Primary Imperative of Building Integrative Competence and a Supportive Institutional Context

For people who think that they have sufficient experience to be a good professional in their field but are not necessarily experienced in complex multi-disciplinary and integrative projects about complex problems requiring holistic work, it is very easy to underestimate the challenges posed by such projects and to overestimate their ability and the likelihood of project success. Based on my and others' experience in such projects and related decades-long work in method development and in different contexts, I can testify (if needed, but this should be common knowledge) that whoever feels (or looks) unreservedly optimistic about such projects does not understand the challenge at hand, and probably extrapolates from her or his experience in some "traditional" projects. What matters however, is not how optimistic or pessimistic someone is but the capacity to deal with the higher challenge by learning her/his way beyond her/his previous experience, knowing that a diversity of such experiences within such new, complex projects is an asset to begin with.

Then, to ensure that this collective experience is put to the right use (instead of resulting in silo-thinking and turf-fights), a number of conditions need to be met.

First off, to make sure that the best, evidence-based pathways are being followed in practice, it is important to remind ourselves that such a project requires a good understanding of natural systems, i.e., the project owners and managers must make sure that those team members who have this type of experience use it to clarify the basic aspects from which we start.

To give a quick but first-rank example, this is why the first report coming out in any round of reports published by the IPCC is the one focused on the fundamentals (the Physical Science Basis) of global climate changes [330], which is then followed by a report focused on the immediate consequences and actions that are needed to deal with the phenomenon on short and medium-terms (Impacts, Adaptation, and Vulnerability) [331], and afterwards comes the report focused on the deepest and the hardest aspect of it (Mitigation), because this is a long-term issue as per current knowledge on climatic dynamics at planetary scale, although the effects of climate change are already visible and are urgent [332]. In other words, we cannot afford to forget that urgency requires not hasty action but the fastest-possible good understanding as the only viable basis for as-fast-as-possible competent action. (As a quick reminder, the details in the big puzzle shown in Fig. 1.10 earlier on, also making direct reference to cities, come from WGII, 2022, but a related discussion is also in WGIII).

In short, the fact that a coherent understanding of how a city works must start from the "ground zero" represented by the laws of physics and a functional big picture understanding of urban metabolism (as a basis for life quality of its people), means that eco-city project teams must include natural scientists right from the beginning (specifically starting with biologists, and more broadly life scientists, and earth sciences, with the support of all the other natural scientists).

Relatedly, a project must also start from "ground zero" of basic tenets in every field and all participants need (a) to exercise patience and (b) to be ready to corroborate those field-specific aspects towards a collective understanding that can be shared by all project participants (not only those from the same field that are already familiar with them). This doesn't mean that everyone needs to become expert in everything, it only means they need to make sure there is a sufficient basis of common understanding. Otherwise, the project will morph into a dialogue of the deaf and will fail to attain its objectives. Thus, potential observations (occasionally by some project member in-project conversations) such as (impatient) "we already know that" would be beside the point. There are many things that we "already know", and such projects are not about rushing to make scientific discoveries (although those are also welcome) but about integrating the pieces that are not yet integrated. The point is to bring those "knowns" on the table to see how they relate: to actually solve the puzzle from the pile of small pieces that keeps growing (and thus, in perspective, to put ourselves in a position to close the gap between what we know and what we do; there is always some gap, the challenge is to reduce it as humanely possible).

However, even that narrow competence and willing to learn is not enough: required are those natural scientists that fall within the general requirement of competence in inter-, cross- and transdisciplinary, problem-solving oriented projects. In other words, it is not helpful to include natural scientists that only have some narrow fields of expertise or/and are not capable of working with experts from other fields. Of course, such skills can be learned, and most experts need to learn and improve those skills; this is why individual project participants need to be given the time and opportunity to test and develop their capacity to work together (more about that, in the topical sections of this Module). At the same time, social scientists and other types of experts are necessary, for symmetric reasons: the final objective of eco-cities is sustainable high quality of life for its people.

The point here is not about turf fighting or which academic field or profession would be "better" (all knowledge fields are complementary, and each has its distinctive role that must be respected for what it is); the point is that things must be done in a practical order (and coherently all the way) as indicated by knowledge and objective necessity.

More generally, the physical reality in which cities exist dictates that competence in natural science is a first-rank (zero-priority) requirement to be met by an eco-city project team. Current experience (my own included) and knowledge from projects tells us that if that expertise is not there, then it is better to not even start an eco-city project because the project will be flawed from the beginning. Unfortunately, some well-known shortcomings like "academic marketization" act as aggravating factors because they generate low-quality literature (notably a flow of buzzwords, as noted in the section about resilience; and scientifically unfounded claims) which is a very harmful phenomenon hurting the academic environment, and, through projects of societal interest, is hurting society as a whole. This is some bad news, but the good news about it is that science and the academic environment is able to self-diagnose and self-heal. At diagnosis level, systematic studies into this issue show that, by its multi-disciplinary nature, the field of sustainability studies is particularly exposed to such problems (e.g., [49, 310]), but the issue of buzzwords harming our collective capacity to address real problems occurs in both natural and social systems as a whole and is known as such since decades (e.g., [110, 513, 524]). A related disease is that of tokenism, i.e., showing that you're "doing something about it" but in fact it's only a show that masks the reality of business-as-usual.

This is why, eco-city project teams and experts need to "do their homework", build correct and shared understanding, and make sure they agree on the core values that are underlying the project—not least because it is important to have that reciprocal understanding of what we agree and not agree on as a constituency of people with a common goal (like an eco-city, or any other important target). It is better to have a contested exchange as a method for "getting to 'no'" (e.g., [121]) than to pretend that the differences between people and professional cultures do not exist and to suppose that we have enough common ground when in fact (a) nobody checked and (b) that common ground is absent or insufficient for the intended work.

Relatedly, this also means we must not succumb to the temptation of getting over technical details without proper checks. Of course, this is very hard to do in complex projects with high stakes, as it implies huge work volumes—but science in particular and true professionalism in general has never been easy.

The fact that interdisciplinary science (at least) is being increasingly undermined and its role is being supplanted with publishing for the sake of publishing is not only a major problem to be aware of (and combat to defend project quality). It also points to other institutional issues which eco-city project teams cannot make abstraction of, as this is part of their professional context and one of the main factors weighing in the overall quality (and, ultimately, success or failure) of eco-city projects.

In the limited space of this book, I can only point to this important aspect for eco-city projects, and it would not be possible to describe the entire panoply of problems waiting to be addressed by us/our institutions. Instead, I use a few examples to draw attention to the fact that the context of an eco-city project is as important as its content, as the two are inevitably interrelated and competently-run projects must face them and deal with them well enough: the first basic goal of a project manager (and, of the project team) is to create and protect the space in which that project can exist and function. If this fails, the business-as-usual will run over it as if the project never existed.

One such issue comes from academia: the evaluation of the academics by bureaucratic indicators like numbers of publications and numbers of citations instead of the quality of the content of those publications (leading to an inflation of low-quality publications citing each other to thus multiply their citation numbers). Specifically, in the literature which is meant to represent interdisciplinary efforts as contributions to sustainability studies, it is quite common to find scientifically unfounded claims and numerous fundamental mistakes. Often, these are made by social scientists that venture and make suppositions and uneducated guesses about natural systems. But the vice-versa is also true, natural scientists often make hasty suppositions about human systems, usually oversimplifications that do not survive a critical scrutiny or/and naiveties that are due to a superficial understanding of the domains with which they are not familiar but about which they think they can speak competently nevertheless.

In short, the chronic problem of knowledge fragmentation between professions and between disciplines is not just a theoretical gloss: it is a real and toxic phenomenon that needs to be addressed with practical epistemic and managerial solutions; left unchecked it will stifle any project, no matter how well-intended and resourced that project is.

Another issue comes from business but is related to the one above: most academic publications are run by private businesses for which the priority is not science per se but increasing business volumes by augmenting the overall volume (and price) of publications regardless of their academic quality. And this is only one example of situations in which science and business need to work together in complementarity (this being a condition for a successful modern society) but fail and create major disequilibria. The fundamental difference (between science and business) that needs to be understood and managed properly is that science is about exploration while business is about exploitation (of what is known), and that proper collaboration between the two means societal ambidexterity: the best possible tradeoff (i.e., resultant; equilibrium) between the two forces.

And a third issue I want to bring forward comes from governmental institutions. Namely, the fact that the policies of national and local/city public authorities do not have yet in place proper and consolidated systems for insuring that national/local policies are based on evidence inevitably creates vicious circles (and "races to the bottom") that destroy quality standards in both the private and public sectors. For example, it diminishes the societal importance of science and of academic research in general, and rewards low-quality "contributions" from across the triple-helix of participant types that are required for generating solutions for sustainability—and this happens exactly at a time when high-quality contributions are needed the most because they are critical for a fast-enough transition out of the current unsustainability crisis.

Finally (in this section), one easily observes that the issue signaled above is one of the shortcomings caused by the current state of play in the academic environment, and is a symptom of the systemic problem represented by strong fragmentation and weak communication between fields of expertise. Of course, this means not that we must wait for science and the academic environment in general "to solve its problems". Quite on the contrary: (1) we do not have time to wait, and (2) the problem of knowledge fragmentation can only be solved by concrete problem-solving oriented projects in which participants from different fields "jump in" and make honest efforts (i.e., hard real work) to step out of their "comfort zones", and learn from each other.

This means that true work mistakes are normal (as one popular saying goes, "only s/he who doesn't work makes no mistake"), some of them being useful in different ways: they can be interesting mistakes that open new view angles upon the problem at hand or/and are enlightening about the processes involved in the studied problem or/and in the work processes of the projects themselves (for example, a mistake that happens very often can signal a pattern or/and be a sign of some phenomenon about the studied problem (or about the project itself) that has passed unobserved and requires further investigation. In other words, there are "mistakes of work" and there are unnecessary and avoidable mistakes that indicate low work standards or/ and opportunism and unjustified claims of professionalism.

The bottom line is that the right approach is not to fear work mistakes (or, for that matter, the hard questions) but to understand them as constituent part of the necessary process of learning as a person and as a team. The health of a project is well served when the highest-possible quality is sought about the human resources involved in a project, and the greatest-possible space of action they have in their work. Once this is assured, the more mistakes they do, the more we know (1) that they are advancing deep in unexplored territory, i.e., they are doing their work and push the frontier of knowledge about the problem of interest, and (2) that they are going to use those mistakes to maximize progress in the desired direction.

Building the Big Picture: The Extended Urban Metabolism Model

In a nutshell, Fig. 3.10 sums-up what we know about a city as a complex dynamic system aiming at sustainability from a simplest operational input–output perspective that can be taken by an eco-city project manager, while Fig. 3.11 does the same thing from the perspective of fundamental physics, i.e., thermodynamics, and Table 3.2 explains the terms.

The literature offers supplementary descriptions that go along the thermodynamic perspective in Fig. 3.11. For example, "cities, like natural ecosystems, are self-organizing far-from-equilibrium dissipative structures because they grow and survive by continuously degrading and dissipating available energy and matter from the biosphere and the sun" ([576], see also [591]). Along this process, "low-entropy materials enter, undergo a series of changes in their energy and entropy state and, after a time lag, the residual high-entropy materials or wastes are dissipated or returned to the environment" [96]. For more, see (e.g., [123, 612]).

This thermodynamic perspective in Fig. 3.11 directly connects urban metabolism modeling to the notion of Low-Entropy City (LEC) introduced in Module 1 (Sect. 1. 3, Table 1.4), i.e., a city that enhances its internal complexity/order (thus reducing internal entropy/disorder), while minimizing waste, i.e., entropy/disorder exported to the Earth System (see also [80, 554]).

Whoever wants to study the details will quickly see that, inside the overall urban metabolism, different metabolic sub-cycles can be identified in principle (derived from current research and available knowledge), with different degrees of understanding. For example, and connecting back with the discussion around Fig. 1.9 (on the space of the physically-possible), e.g., Pelorosso et al. [554] suggested that urban metabolic models include as sub-components a number of urban metabolic subcycles which reflect the participating "socio-ecological and technical systems" such as these (non-exhaustive list): energy, water, transportation, food, buildings, industry, commerce, health care, security, education, governance, and general services. All

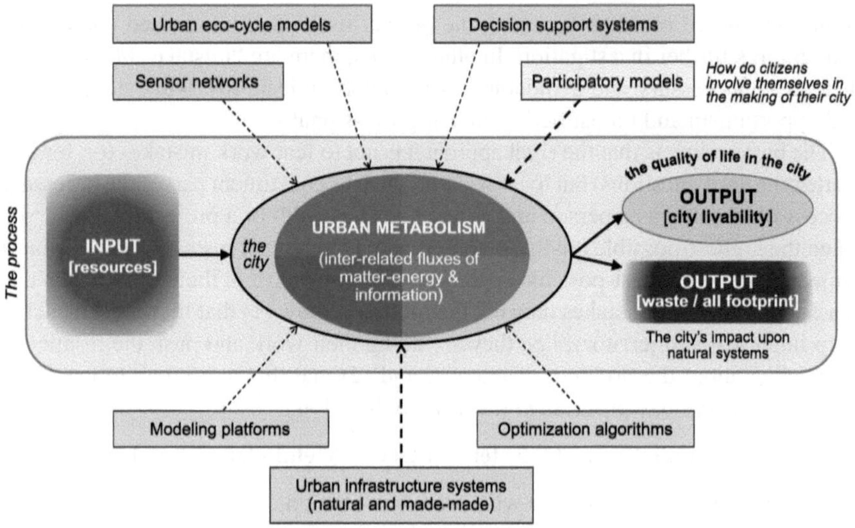

Fig. 3.10 A management-oriented extended urban metabolism model of the core dynamics from resource inputs to result outputs. The model starts from the notion of urban metabolism placed in the context of the systemic perspective on sustainability, under the requirement of a holistic (multi-disciplinary and integrated) management in eco-city projects. Terms are explained in Table 3.2. *Source* original, based on the notion of urban metabolism (input–output analyses in urban systems) (previous versions and components have been communicated by [152, 376, 400, 514])

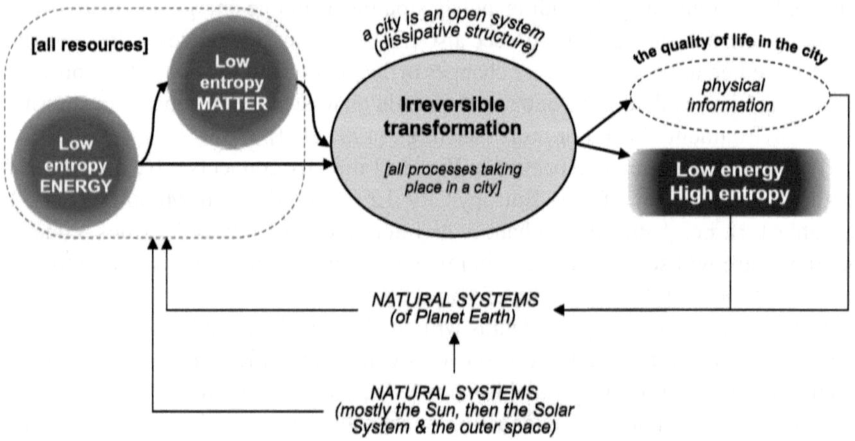

Fig. 3.11 The urban metabolism model from Fig. 3.10 from the perspective of fundamental physics (thermodynamics), knowing that matter is a form of energy, and the notions of entropy and information are closely related via probabilities/statistics. *Source* original, as inspired by the cited model itself and other extant thermodynamic models of the city in the context of sustainability goals (for comparisons, see, e.g., [554, 612]). For a supplementary general discussion about entropy and energy flows in complex dynamic systems from stars to human societies, see [123]

these metabolic subsystems (corresponding to different functions that are being insured inside the urban system) have their own more detailed sub-components but, crucially, they also interact with each other. Thus, a city can be a low-entropy city (LEC) or high-entropy city (HEC), depending on how good the city is at using the sources of low-entropy functions represented by nature-based solutions (NBSs), because, by definition, these are low-entropy.

As a prime example of NBSs, the paper cited above i.e., Pelorosso et al. [554] refers to urban green infrastructure (UGI), i.e., natural landscapes and elements. But I must observe that all nature-based solutions are to be included whenever possible, e.g., natural materials that can be obtained with low ecological footprints (low levels of required total energy, matter, and pollution) instead of man-made materials that have higher ecological footprints; or nature-inspired structures that are (as per current knowledge) demonstrably more efficient to obtain/produce and use. And, in general, NBS needs to go hand in hand with systemic thinking in the development of solutions for sustainable development (see also [452]).

In simple terms, the goal of the focus on low-entropy is that a city consumes less resources (less low-entropy) and generates a lesser ecological footprint for the same or better results in terms of life quality. Logically, using as much as possible nature-based low-entropy solutions is the obvious way forward.

For the purpose of scenario-generation and finding the way forward, these models highlight why scenarios for eco-city development programs and projects must be part of an understanding of a holistic, realistic understanding of the process within which a city functions. The red text in Fig. 3.10 is meant to promptly remind everyone that a city's inhabitants must be part of the process—this requirement being a primordial context and detail for any work aiming at generating eco-city scenarios.

The Urban Sustainability Nexus Model

Contained in the "belly" (central feature) of the urban metabolism model but not shown in Fig. 3.10 for simplicity reason is a sub-model called the Urban Sustainability Nexus (USN), which is described separately in Figs. 3.12 and 3.13, and is a visual synthesis of the most important factors (as study topics) and of their interdependencies as identified based on experience, literature (method details in Box 3.2).

Table 3.2 Notions involved in the urban metabolism model in Figs. 3.10 and 3.11

Concept	Explanation of the idea, with examples from real life
Urban metabolism model	An input/output model of functioning of a city-as-a-whole imagined by analogy with a human (/any living) body, describing the internal fluxes of energy, matter and information [359, 764]. In concrete terms on the ground, urban metabolism links with ecological and human aspects, notably via the notion of "ecosystem services", and with "smart" and "sustainable" goals (e.g., 37, 46, 71, 92, 96, 202, 280, 325, 333, 434, 487, 494, 558, 615, 667])
Waste	All waste (including the usual urban solid waste and all pollution)
Urban footprint model	An accounting method for estimating the amounts of ecosystem goods and services needed by a city ([359, 764]; see also [5, 170, 247, 263, 487, 564, 592, 593, 675, 700, 730])

(continued)

Table 3.2 (continued)

Concept	Explanation of the idea, with examples from real life
Urban eco-cycle models	Experimental models aiming at local developments, usually at city district level, which try to mold the provision of urban functions and services after the natural cycles of matter known from ecosystems (finding inspiration in how nature works, if well done, is actually just another form of pursuing "nature-based solutions") Their purpose is fundamentally exploratory although their public presentation and discussions often appear normative (being advertised with expressions aiming at suggesting that this or that district anticipates how cities will look like in the future). Of course, we are still a long way from knowing how cities will look like in the future (in concrete functioning), because we are at the beginning of the process of transformation as eco-cities. The key aspect and benefit of these experimental districts is learning and finding ways forward. They may be already inhabitable and inhabited, but these are just some examples of how urban districts can look like in the transition to urban sustainability (how urban districts will look like a long time in the future remains to be seen; but of course a great diversity exists now and is to be expected in the long run as well) A (if not the) lead example is Hammarby Sjöstad city district in Stokholm, Sweden, but other examples projects and programs are known from other countries, sometimes trying to emulate each other but each trying to integrate very different local conditions. See, e.g., [464, 725, 755] It is however important to warn that expressions like "sustainable city district" that are sometimes used to describe those projects are scientifically and logical misleading, and this constitutes a serious problem. For example, some people might understand that there is such a thing as a "sustainable district", implying that a district can be sustainable even if the rest of the city is not sustainable yet. That idea is obviously wrong: a city district is not just a territory sliced from the city territory and treatable like a piece of cake; a city district is a functional part of the city as a complex dynamic system, ergo a city district can only be sustainable if the entire city is sustainable (and, knowing of course—as already discussed in this book—that a city per se cannot be sustainable in isolation from its hinterland and the other cities with which it organically interacts). Pretending otherwise induces in the public consciousness the mistaken idea that urban sustainability can be "constructed" "piece by piece" like with some easy linear planning; in fact this kind of thinking and wording betrays a superficial understanding of the matter by those communicators (especially if they are not themselves experts on the matter and only do "public relations"), and promotes the same kind of superficial thinking that created the unsustainability crisis in the first place So, care must be taken to understand the issues at hand and to use the appropriate words; it is batter to use some technical vocabulary (and provide the explanation nearby, inviting the public to learn something new) than to describe a project in incorrect or/and untruthful or/and misleading ways Similarly, such projects are sometimes tied to another misleading suggestion: that there is a "circular economy" which can "save us" (i.e., a "silver bullet" solution). For details about this issue, see the discussion I made in this Module, Sect. 3.10 about "circular economy" and other "buzzwords", and the cautionary discussion made in the last IPCC report [332], Chap. 5, the sub-section dedicated to circular economy, and the references therein. Care must be taken that everybody understands that this is about new systemic solutions, and these require development (as already known from RDI and as everybody actually experiences in their personal lives with different personal and professional projects). It is advisable not to underestimate or underrepresent the difficulty of the process of developing systemic solutions. The entire third Module of this book is dedicated to issues related to project management and Sect. 3.8 is dedicated to stepwise development of systemic solutions through disciplined experimentation: this is in fact what such experimental districts are all about, whether they are officially described as such or not For further reading about eco-cycle models (with a critical spirit being required from readers) see, e.g., [61, 185, 254, 373, 602, 725] Also, the interconnections between different topics and sub-topics in urban sustainability is discussed in the next sub-section below, namely, the Urban Sustainability Nexus (USN) (Fig. 3.13) and its application in the Eco-City Reference Model (ECRM) (Fig. 3.14). Urban eco-cycle model thinking was in fact part of the development of USN and ECRM. Thus, in a talk my team from INTRAS had a decade ago [152, 514] at a conference in Rotterdam (a large international event bringing together science, government and businesses), we asked a direct question: "*Is it possible to develop a model of sustainable urban water cycle?*". Using the USN as a big picture (tool for knowledge integration), our answer was basically "Maybe", but only partly so and even that takes a lot of dedicated holistic development which realistically involves social and natural sciences to the bottom plus everything we can use in a given context, i.e., clearly not just some "engineering tricks" sellable with PR ("public relations") rhetorical supplements In general, the discussion about eco-cycle models has its uses and limitations, and must be placed inside that of urban metabolisms, which is itself located inside the reality of complex dynamic systems (CDS) and coupled human and natural systems (CHANS) which was summarized already in the introduction and the entire Module 1 of this book

(continued)

Table 3.2 (continued)

Concept	Explanation of the idea, with examples from real life
Decision support systems (DSS)	An information system that supports decision-making in organizations. Usually this is computerized, but the essence is not in having the computers per se but in the automation protocols that are used to generate, display, and use available information for a given type of problem and project (Relatedly, see Sects. 3.6 and 3.7 in this Module, discussing the relation between data, information, and knowledge as applied to projects) It also includes the notion of "digital twin" of a city, i.e., detailed computerized model of an entire city: at least in terms of certain aspects of a city that can be done (buildings and road and all land uses, location of services and recurrent issues, etc.), and is therefore also closely linked to the notion of "smart city" This is a very dynamic domain, so project teams need to do their own search of relevant DSS ideas and tools. Here are only some suggestions (examples from the literature, as applied to cities) to open the discussion: [67, 70, 88, 247, 324, 351, 528, 566, 740]
Participatory models	Any form of citizens' involvement in democratic and evidence-based conversations with public authorities in place to serve their interest (and within the society as a whole). These need to be consistent with good governance, starting with the Consent of the Governed (CoG) principle and the Whole of Government (WoG) approach to public administration, and as related to transparency and accountability
Modeling platforms	Any infrastructure that supports advanced computerized modeling, thus helping cities "look in the mirror" and re-invent themselves for the better
Optimization algorithms	Algorithms are human or/and machine action procedures (professional action protocols, often as computer sofware). Optimization seeks maximization or minimization of an attribute of interest, e.g., maximal resource use efficiency or/and minimal ecological footprint
Urban infrastructure systems	This is the ensemble of natural infrastructure (basically consisting in the bio-geophysical settings of a given city) and human-made infrastructure (usually referred to as the built environment, but it may also include modifications of the natural environment)

In effect, the USN is yet another example of knowledge map under the terminology introduced in Sect. 3.3. It resulted about a decade ago from my methodological explorations related to the necessity to handle the seemingly overwhelmingly complex amount of disconnected insights coming from different research projects reports as a "knowledge puzzle" [152, 155].

Box 3.2 How the Urban Sustainability Nexus (USN) model was obtained [152, 514].

The basic approach was to add together all the insights and experiences (from different projects on urban sustainability) and to solve the pile as an "eco-city puzzle" (see also [155]*). A key observation that kickstarted further investigation was that the literature abounded in expressions using the word 'nexus' to describe situations in which 2-n topics were so interdependent that they could not be properly understood if studied separately. For example, in studies concerned with systemic sustainability, an 'energy-water nexus' mean that energy can only be understood adequately if studied as a part of its complex set of relations with water, and vice-versa. To anyone studying the relevant literature, such nexuses are not a surprise (see, e.g.,* [75, 136, 239, 344, 352, 423, 455]*). But, the problem was that the literature provided multiple arguments about other nexuses too, often depending on the context: a water-landscape-food nexus, a climate-energy-transport nexus, a water-pollution-health nexus,* *and so forth—leading to the basic idea that, in urban sustainability studies, we are confronted with a total set of interdependencies, one we need to understand better.*

To that aim, I used my experience to make a starting "short list" of main topics (but long-enough to make sure no important topic was left out), which I then refined and consolidated by an extensive review of the literature. Then, based on another round of literature review combined with basic bibliometrics, a rank of three classes of topics emerged, from the most important to the least, and three categories emerged:

1. *a central club of four interdependent topics (energy, water, landscape, transport) showing four one-to-one nexuses: energy-water, water-landscape, landscape-transport, transport-energy, i.e., a closed circle equivalent to a 4-nexus. Then, cross-relations could also be identified or inferred between opposite pairs (water-and-transport and landscape-and-energy), but these "opposite topics" are interdependent anyways through*

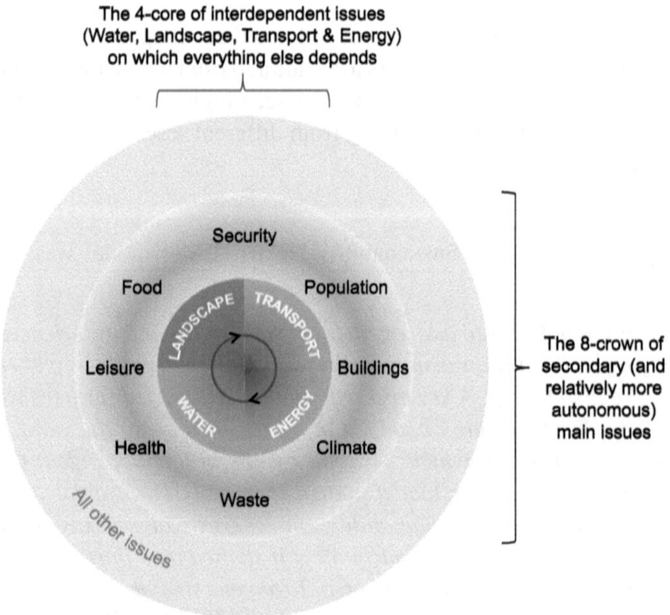

Fig. 3.12 Simplified representation of the notion of Urban Sustainability Nexus (USN), i.e., complex and variable interdependencies between principal topics of interest, based on literature reviews, bibliometrics, and experience in multi-disciplinary and integrated projects on urban sustainability. The 4-core together with the 8-crown of issues make the 12-bulk of central issues in urban sustainability on which all the other issues depend. *Source* Modified after [152, 376, 514]

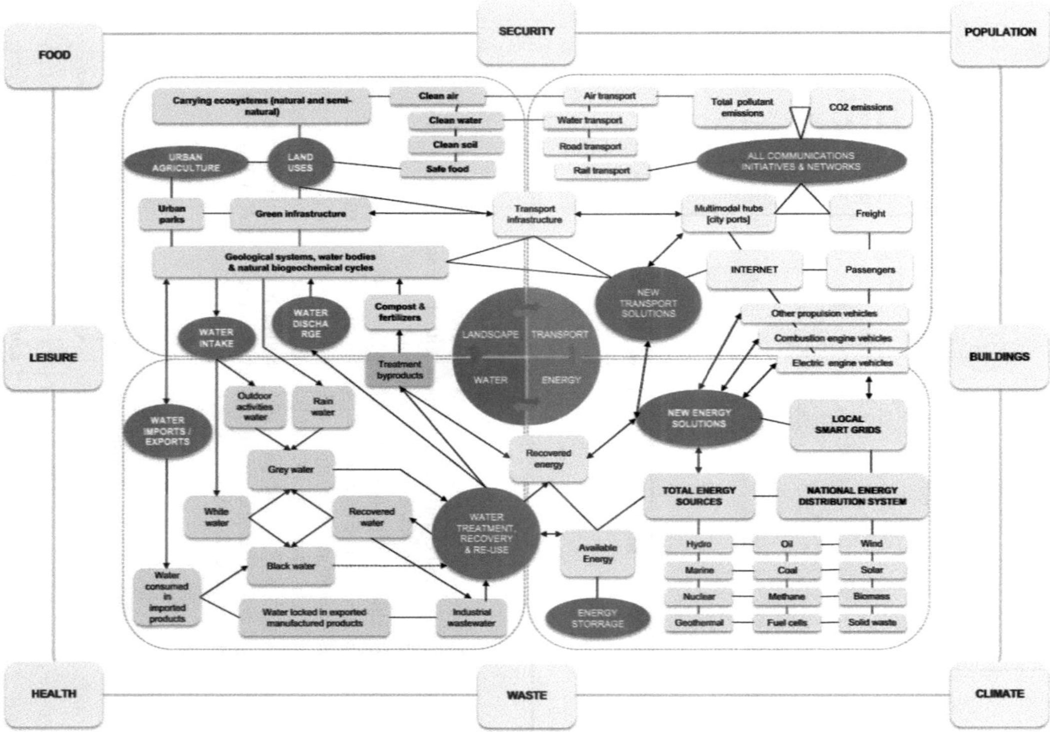

Fig. 3.13 A version of the USN, with a next level of detail shown for the core topics (water, landscape, transportation, and energy). It is a knowledge map, i.e., a representation of the relations between topical sub-categories based on current knowledge. Not shown (for visual simplicity): each category has three levels, corresponding to the sustainability filters derived from the systemic perspective (Figs. 1.1–1.3): biophysical, social, and economic. *Source* Modified after [152, 514]

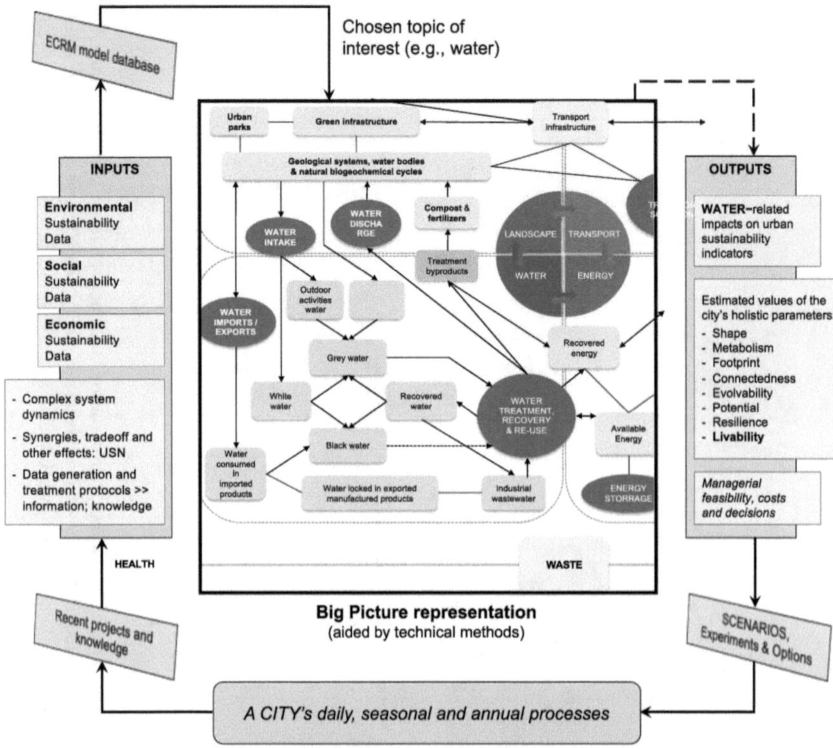

Fig. 3.14 A holistic eco-city reference model (ECRM) developed based on the dynamic process from resource inputs to result outputs. Illustration is made with the topic of interest "water" (a start question being, e.g., *How is water impacted by and impacting a city that seeks to be an eco-city?*). In the output box, "estimated values" refers to quantitative aspects of those parameters (numbers) and they result from combining what can be measured with what can only be estimated due to inherent systemic complexities and the objective limitations in knowledge and resources. When generalized for application in any human settlement, the ECRM is a Sustainable Community Reference Model (SCRM). USN is defined in Figs. 3.12 and 3.13, and the visualization illustration in the central sector shows a segment of it. The holistic parameters are defined in Table 3.3. *Source* original, based on [152, 514] and the models shown in Figs. 3.10, 3.11, 3.12 and 3.13

their two common neighbors. Technically (based on graph theory in fundamental and applied mathematics—see Box 3.1, *Sect.* 3.3*), this full set of interdependencies means that the 4-cluster is a 'complete graph', and this has certain important analytical and practical consequences for modelers. And, because it is in effect a functional unit, this 4-cluster can also be represented as a hypergraph, i.e., a single connection (represented mathematically as a surface instead of a line) that covers all four (by comparison, in graphs, every connection links only two vertices and is represented as a line), again, with important consequences;*

2. *a set of eight topics that were necessary for any city and were related between them more loosely while individually (and collectively) dependent on the 4-core nexus—in other words, a 'cloud' of eight satellite topics gravitating around the 4-core, together with which they form an extended 12-core of topics)*;
3. *and all the other topics; regardless of the connections between them these are dependent on the extended 12-core of topics (so a proper study of any of those begins with clarifying the relations with the 12-nexus).*

In practice, this means that any project or program or policy that, for whatever reason, is focused strictly on one topic of choice, for example energy, the knowledge that must ground any energy-oriented policy and action actually include the need to understand the core 4-nexus of topics as a whole (cognitive and operation unit)—and, of course, the fact that this is not easy (or that we are not used to working like that) is not a valid reason to ignore it.

Epistemically, the USN is a structured cluster model of the relations that exist between the most important topics in urban sustainability (and thus have to be considered as part of urban metabolism) as reflected by current knowledge and is therefore updatable as knowledge progresses. Its reason to exist is to serve in kick-starting analysis and, more broadly, to assist with orientation and consultations in all multi-disciplinary project teams.

The main feature of the USN, at the center of the model, is a core of four central topics (energy, water, landscape, and transportation) that are interdependent, i.e., none of them can be properly understood (and addressed) in isolation from the other three. Put differently, even if the ultimate purpose of an expert team and institution is to address one specific issue, for example, all transportation in a city, the knowledge-gathering phase must include all those four as a functional unit (any attempt to restrict investigation to the topic "transport" will result in insufficient/inadequate understanding of that topic of interest, which will then translate into an inadequate transport-policy (or program or project) in the city—if sustainability is pursuit in that city. In short, the recognition of this set of interdependencies is one key difference between traditional "business-as-usual" policies and management in a city and the concrete, systems-oriented approach that is necessary if that city aims to become a true eco-city. So, the frank choice for local communities is between the hard but right way (of objective evidence that things are intimately interrelated) versus the easy but wrong way (of persisting in traditional fragmentation). To highlight and explicate the main details of these interdependencies, Fig. 3.13 shows an expanded "zoom-in" version (the next level of detail) in the central core of four interdependent topics depicted in Fig. 3.12.

Around this hard, central core gravitates a ring of eight topics that are also necessary for the basic understanding of the city but are not part of the central 4-core of interdependencies. Each topic of the 8-ring is strongly connected (and in some specific ways) to the 4-core and they also have numerous links of various strengths with the other topics in this "satellite ring" but not as compact as in the 4-core. This means that it is easier to obtain a good understanding of a topic in the 8-ring by studying it separately from the other 7 (even if this will leave some knowledge gaps, the importance of which needs to be analyzed on a case-by-case-basis), but not in separation from the 4-core, as this will automatically create critical gaps that cannot (realistically speaking) be allowed by any responsible manager.

Lastly, this 4 + 8 core constitutes the grand nexus on which all the other topics in urban sustainability depend. That is, for any topic one might possibly want/have to study, it is judicious to start by looking at the relations between that topic of interest and the twelve explicit topics identified in the USN.

Overall, the USN model can be used as a default map of topics based on which one can start to ask project questions and operate further modifications and updates in each concrete context as needed. Technically, this map of topics is a "knowledge map" and represents an evidence-based template for organizing knowledge in complex projects with participants coming from different disciplines and professions, and operating with different field-specific mental representations, embedded in different institutional standards and cultures (details in Sect. 3.3 on cognitive maps as tools for project management). For project team members, the USN can serve as a start common reference—which they can agree to further develop with specific minutiae and actualize with their specific context details, but which nevertheless is available to serve as a common mental representation of the "cognitive landscape" in urban sustainability.

Naturally, when confronted with the challenge of sorting out what is going on in a city as a system, different authors might find some other ways and might map the topics differently, especially when the operational objectives are different or/and when they start from a different initial question. Still, such differences are most likely associated with methodological complementarities and it is only a matter of time until the puzzle is solved again and our methods and knowledge get updated. What is important to highlight here is that "brute force" computing methods that use keywords in publications to automatically generate "clouds" of topics are of limited utility and are only one type of input that can inform any effort of building a "Big Picture" model about a matter of interest: overall, they show which keywords are the most frequent but they bring no (or only a marginal) understanding about the content itself (what is going on inside and between the issues of interest); what we need is real knowledge integration.

Relatedly, the USN can serve as a reference for comparisons and as a visual checklist. In other words, a project can start from very different questions inside different topics, but in the end, the core interdependencies cannot be avoided; having a cognitive map at hand can help with maintaining awareness of the inherent complexities that characterize urban systems at all times, and can help project teams "scan" what other important aspects might have been missing in previous or planned works.

And finally for this sub-section, there is yet another, next level of complexity that is not shown in Figs. 3.12 and 3.13 but consists in the fact (which we know and must keep in mind) that each topic has to be regarded through the operational lenses of the systemic perspective on sustainability with its three systemic levels (and general cognitive filters): natural/physical, social, and economic. This further means that the links between topics have not only a width dimension as shown in the figure but also a depth dimension given by those three levels. Thus, based on one and the same model, one can generate different model avatars characterized by a specialized focus on the natural/physical or social or economic aspects of each topic (instead of the general definitions shown in Figs. 3.12 and 3.13)—or mixed avatars characterized by a combination of filters of interest. Additionally, the user of this model may want to understand how the three levels interact with each other in the case of each topic and sub-topic (bubble/box in the USN) taken separately; and then the user can compare those topics to detect relevant differences, limits, and potentialities that each topic may have from a systemic perspective.

The Eco-city Reference Model

The quality of life that is referred to in Figs. 3.10 and 3.11 is the ultimate characteristic of a city as a complex dynamic system. Figure 3.14 shows an operational model towards the goal of sustainable high quality of life as an attribute of urban systems. For the purpose of coherent modeling, I labeled this attribute Urban Livability and defined it in Table 3.3 as the ultimate state in a series of "holistic parameters" of an eco-city, i.e., system-level attributes (which I established using the same general approach I used for building the USN model: experience- and logic-based sense-making of all insights available from projects and the literature, using both bibliometric and concrete topical information). Importantly, the identity of a city, that essential feature of any city, is given by the spectrum of details that come along with these holistic parameters for any given city, from its origin to its current status.

As a supplement to the description given in Table 3.3, given the importance of large-scale ecological processes and how a city interacts with them, here are the two examples promised in the table at the discussion on urban ecology (at Shape).

Firstly, the intense interactions between cities around the globe result in some global homogenization effects—at least within similar biogeographical areas, because of intense material exchanges and cultural similarities between different cities around the world, e.g., a preference for similar ornamental tree species inside cities regardless of the origin of those species. The result is that the cities that have high degree of integration (influence into and from) the global network of cities are (becoming) very similar with each other not only in terms of socio-economic and technological aspects but also in biological/ecological aspects.

Secondly, local biodiversity is higher at the margin of a city (than in surrounding ecosystems) because of "ecotone" effects (i.e., increased habitat diversity results in increased overall biodiversity of plants, birds and so on), which is further related to complex effects of disturbances as known from ecology and biogeography (and the management of natural resources, and the related types of land uses): in ecosystems, intermediate levels (intensity and frequency) of disturbance favors greater biodiversity (because it creates some space for the less dominant species, i.e., it stimulates functional competition between species), while excessively low or excessively high disturbance correlate with low biodiversity levels (because the dominant species suppress and eliminates dominated species in the competition for space, light/energy,

Table 3.3 The holistic parameters of eco-cities (HPECs): a concise conceptual key

A city's holistic parameters[a]	Symbol and definition	Short description, with suggested start references for further study in concrete projects
Shape	$S_u = S_0 + a$	The set of fundamental geophysical and biogeographic details starting with location (latitude, longitude, altitude), the climate (Köppen-Geiger climate classification [52, 552]), ecological factors, and up to urban geometry and morphology (and the changes thereof) Apparent urban forms like territorial dimensions, skyline, sprawl, or shrinking are only a *part of* urban shape, *not equal to* it (but can have impacts on the performance of a city, including the economics [207, 601]) Less intuitive (more technical) aspects can be at least as important. For example, excessive soil sealing (i.e., covering with impervious materials like concrete or asphalt) [362, 557, 603, 628, 692] distorts the natural biogeochemical flows: beside depriving the city from green areas (key ecosystem services), it aggravates the Urban Heat Island (UHI) and Urban Pollution Island (UPI) effects [704], *inter alia* because rain water quickly exits the city (feeding floods downstream) instead of helping local ecosystems, which further aggravates the thermal and chemical stress on the human body and the remaining ecosystems [on soil-sealing-related floods and how to minimize it, see e.g., [261, 362, 579, 603, 627, 628] For a city of interest, both generalist and peculiar physical details matter, and so in-depth study of the literature is required in every real project [to start, see, e.g., [42, 43, 97, 218, 289, 355, 388, 489, 503, 580, 617, 647, 679, 733, 736, 745]. Just to give an example, morphological parameters of a city influence urban ventilation and therefore the intensity and spatio-temporal patterns of UHI at city and city district levels [542] As urban ecology reflects CDSs, there are complex tradeoffs between planning aspirations and ecological processes and benefits, as reflected in difficult notions bridging science and management, e.g., "ecosystem planning" [159, 166]. See also (e.g., [12, 13, 22, 34, 45, 84, 199, 108, 217, 287, 326, 354, 355, 366, 390, 401, 426, 462, 510, 551, 564, 565, 587, 594, 626, 687, 703, 757, 712, 721, 736] and the Intergovernmental Platform for Biodiversity and Ecosystem Services (IPBES) [329] And, cities have both positive and negative impacts and are impacted by large-scale regional, continental, and planetary processes, e.g., ecological homogenization or/and heterogenization, and insights from landscape ecology [146, 165, 166, 435, 456, 710]; see two examples in the main text right before this table Another topic that is worth mentioning is the health effects of human interaction with urban green spaces, e.g., on immunity (e.g., [343, 363, 614, 691, 752])
Metabolism	$M_u = S_u + b$	This is directly related to the urban metabolism models already discussed (see Table 3.2) and all the knowledge to date (for further study, see, e.g., [57, 105, 117, 187, 188, 359, 764 The difference is that, in those previous discussions, urban metabolism is understood as simply what it is from an objective research perspective. Instead, *Urban Metabolism* as HPEC is meant to also include the concern of "how good or bad (favorable or not) the metabolism of a city is" with respect to aspirations, as metabolism must support advancement towards those, and can therefore be subject to interventions to improve different aspects of it (in terms of all technical aspects that matter for sustainability and for the ultimate goal of sustainable life quality) One observes that *Urban Shape* and *Urban Metabolism* are tightly intertwined and strongly influence each other. For example, the UHI and UPI effects mentioned above at *Shape* also anticipate *Metabolism*. However, shape is the more basic, and metabolism results from shape plus other features As an observation, many research and innovation insights or ideas about metabolism can be drawn from the literature dedicated to some well-known topics like the UHI (to start, see, e.g., [2, 26, 39, 47, 230, 303, 305, 349, 560, 645, 656, 704, 705, 716])

(continued)

Table 3.3 (continued)

A city's holistic parameters[a]	Symbol and definition	Short description, with suggested start references for further study in concrete projects
Footprint	$F_u = M_u + c$	This is an accounting method (and area-based indicator of the sustainability of human systems) that estimates the amount of land (productive ecosystems) that is required to produce the material and energetic resources that are needed by, and to process the wastes generated by, a city (or rather metropolitan area, i.e., metropolis). In other words, it is a way to estimate the ecosystem carrying capacity that is appropriated by a city and a human system in general [589, 590] The important detail to retain is that, by definition, a city is a node in a network of material fluxes, and therefore cannot be sustainable within the bounds of the territory it occupies. Ergo, *Urban Footprint* must be understood in relation to all (total) impacts it has on the local ecosystems from (within its perimeter and) its hinterlands and, through its exchanges within the regional and global networks, with the Earth's biosphere. For example, when this method was established in the mid-90s, it was estimated that the city of Vancouver (Canada) required a land that was 180 times its own land surface. This is "why the cities cannot be sustainable—and why they are a key to sustainability" and have been recognized as such since the early days of the main concept of ecological footprint [564, 592, 593] Currently, Urban Footprint Models exist that can be used to get an idea about the situation of a target city from this point of view [65, 406, 460, 734] Obviously, the way a city is organized and how it functions (shape and metabolism) have an impact on its footprint, and important topics matter for this HP, e.g., urban green and blue infrastructure (see also, e.g., [197, 722]), and nature-based solutions and thinking more broadly in terms of the place of a city in the world
Connected-ness	$C_u = F_u + d$	This is about the degree to which a city is connected to the regional and global networks of cities, including through the global fluxes of goods, services, and people Logistically, very important is the distance between sinks and sources, and this metabolic need can be fulfilled much better when the city has a great connectedness in the global network of cities [233] These being said, a two-decade old summary by [86] remains valid today and will probably remain so: "A central feature distinguishing global cities from nonglobal ones is transnational connectedness. But another important consideration in urban globalization is the disproportionately high presence of upper middle class (UMC) whose membership includes institutional professionals at the forefront of postmodern awareness and international experience" Illustrative in the discussion are also notions like "global city" and "everyday digital connectedness in the global city" [410], and "cosmopolitan city" or "Cosmopolis" [64, 164, 482], knowing that smaller cities and rural communities, also matter in the global concert (see, e.g., [30, 54, 175, 428, 720]) Equally important are the related demographic details. For example, the fact that younger generations tend to be more connected online (while at the same time being more active in live socializing in the city itself) speaks about *Urban Connectedness* and its place in the future cities in general An expression summing up the complexity involved in *Urban Connectedness* is "urban geography: world cities, or a world of cities" [600] In this sense, professional networks are key for the success of all projects in cities. In this sense, Sect. 3.5 (in this Module) is dedicated to the notion of Brain Networking as applied to potential eco-city projects. Also important are all the details discussed in this book related to network dynamics in complex systems. Another key detail is the importance of different institutions, notable universities, and large private internationals (see, e.g., [30, 424]) which, by their nature, connect different human communities Other notions are also relevant here and can be studied and used in practice (that need not be detailed here): twinning, participation in regional and global alliances of cities, "the Global Village", "network economy", the Internet-of-People (IoP) and the Internet-of-Things (IoT), and so forth Finally, this HP can be both easy to understand intuitively and challenging technically, precisely because of the multitude of types of interactions. Basically, global connectedness can be thought in terms: people-to-people, businesses-to-businesses, institution-to-institution, and so on. It is only when we recognize that people are part of local communities with their own dynamics and identities we realize that cities communicate with each other through numerous autonomous agents, and this can be studied mathematically (e.g., using different forms of centrality in graphs representing networks of cities; and, because a city can be itself represented as a network of individual and institutional actors, which themselves can be understood as networks, we can talk about networks of networks)

(continued)

Table 3.3 (continued)

A city's holistic parameters[a]	Symbol and definition	Short description, with suggested start references for further study in concrete projects
Evolvability	$E_u = C_u + e$	Connectedness of a city also favors its evolution through the interaction with other cities. Thus, *Urban Evolvability* is the capacity to evolve and adapt to new contexts and find solutions in a space of options that itself changes in time. It implies dynamism and flexibility/versatility as afforded by all the other properties of a city Of course, the matter of the evolution and dynamics of cities is an immense topic which cannot be properly covered in this table. Here, I only include the main idea as required for our HPEC list (and the related models and operations) and give a few hints for further study One will notice that *Urban Evolvability* is another (and more general) way of speaking about the capacity of a city to pass the Innovability Threshold mentioned in Fig. 1.12 and Table 1.6 (in Module 1, Sect. 1.4, at the definition of an eco-city). *Urban Evolvability* results from the interaction between innovability and all the other features of the urban system One notable manifestation of *Urban Evolvability* is visible in how well it adapts to climate changes or/and other pressures, and the relations between adaptation and mitigation, i.e., the management of the complex relations between the two (as per Fig. 3.6). Conversely, the capacity of a city to be successful in its climate change mitigation and adaptation policy (CCMAP) depends on the evolvability of that urban system. Further, this means that CCMAPs must start from urban evolvability and its relations with the other HPECs Evolvability implies adaptability, but not in the superficial understanding of "wait and see" or "muddling through" but by a true systemic evolution. Lack of the latter, it is only a matter of time until the city will face some fatal bottlenecks Because it is directly related to the properties of urban systems as complex dynamic systems (including scaling laws, fractal patterns, self-organized criticality, and other non-linear dynamics as discussed in Module 1, starting with Table 1.1), evolvability manifests itself at different scales: from the scale of the city-as-a-whole to the *scale of specific* components. This is why, in the literature, evolvability is discussed at different levels, and especially at the level of different components of the urban system (as per different interests of each project). However, when studying evolvability at the level of urban system components, we must always keep in mind the evolvability of the urban system as-a-whole—the capacity of the city to change in time and stay fit as a city In the following lines, I provide a selection of hints one can find in the literature (for further study of different aspects of evolvability): "The city evolves to maintain itself and its evolvability" [750]; "The evolution of urban sprawl: Evidence of spatial heterogeneity and increasing land fragmentation" [334]; "Urban spatial organization, multifractals, and evolutionary patterns in large cities" [680]; "Evolvable traffic signal control for intersection congestion alleviation" [141]; In infrastructure planning and the field of systems engineering more broadly, one talks about "evolvability analysis" and "evolvability analysis framework" that includes "transition paths" and "stakeholder diversity" [678]. In this sense, the analyst needs to actively look for anything that can impede the evolvability of the studied city, and point it out (as an obstacle to be removed). See also, [300] who asks "Are we intentionally limiting urban planning and intelligence?" One can find numerous cultural analogies speaking about evolvability or the loss of it by negligence of inadequate (even absurd) actions The absurdist play "The Chairs" by Eugen Ionesco illustrates how people tend to surround themselves with "items over items" until they find themselves cut from the world and eventually cannot move anymore. In the realm of public policy, this reminds of the problem and already chronic challenge of policy accumulation [4] (see also the discussion on policy development in Sect. 3.8) As positive examples stand all those cities that grew during ancient civilizations and have survived until present times by re-invented themselves during different eras (to give an example, see the ancient Decapolis region: a local network of cities in Southern Levant [74]

(continued)

Table 3.3 (continued)

A city's holistic parameters[a]	Symbol and definition	Short description, with suggested start references for further study in concrete projects
Potential	$P_u = E_u + f$	Evolvability of a city constitutes the basis on which another holistic parameter builds, namely, the *Urban Potential*, i.e., the sum-total of what a given city is capable of doing and achieving at a given moment of its existence Because of the systemic and dynamic character of any city, different details of *Urban Potential* might or might not be anticipated (or even anticipable) and need to be discovered by study and practice whenever possible. And, obviously, it can be cultivated and enhanced (or diminished) regardless of how much of it is known In practice, beside the obvious remarks one can make about the connections between *Urban Potential* and scenarios for the future (as discussed in this book and the wider literature and in projects), *Urban Potential* manifests itself through real-life phenomena. In this sense, a rather popular topic is that of "city attractiveness" as seen through the lenses of different particular interests, from international tourists to international investors, often combined with some specific socio-demographic categories, e.g., attractiveness of a city for (prospective) students and the highly skilled (in general or with respect to specific sectors that might be of priority in city-level or institution-level strategies). In management, a related topic (and one of particular interest for local authorities and local communities) is the "brand" of a city (and its cultivation); in this, everything about a city matters, and particularly relevant are the discussions (also made in this book) about identity and the space of options Obviously, first a city needs to have these two (an identity and a space of options), and then the city needs to protect and develop them The known historic cases of "lost cities" (like Khara-Khoto, Sigiriya and Machu Picchu mentioned in Module 2, Sect. 2.6) are concrete examples of cities which, at a certain moment in their existence, have encountered a combination of exhausted (or severely diminished) potential and unforgiving circumstances, i.e., bottlenecks. The main lesson here is that, to have a future, cities must avoid structural and functional bottlenecks; when this is not possible, the local community must prioritize on either eliminating that bottleneck (if possible) or on compensating for it in some other ways History provides many other thought-provoking insights, and right from the beginnings of urban types of human settlements. For example, the archaeological evidence about the pre-urban (very large rural) settlements in the *Cucuteni-Trypilia* culture (mentioned in Module 1, Sect. 1.3), one of the three regions of the world where the wheel and the potter's wheels are known to have emerged for the first time around 4000–3500 BCE) suggests that every 75–80 years (or so) all buildings were deliberately burned to the ground and rebuilt in a way which cannot be explained as accidents (in the archeological evidence, the fires are precisely localized, all houses have the same fate periodically, and each house was built on top of the previously burned one; this pattern repeats tens of times). The detailed reasoning behind that practice is still debated but the most likely explanation that fits all the available evidence to date is that they were *symbolically burned at the end of their life-time* before a new house was built in relation to the respective worldview in that culture (notably animistic, according to which objects and elements of nature have a soul; similarly, in the medicine known from the Ancient Egypt, diseases were explained by "bad spirits" penetrating the body; related examples are the myth of the Phoenix bird "reborn" from its own ashes, the myth of the "eternal return" known from Australian aborigines, and the Buddhist concept of Samsara, i.e., cyclical change (for these and other examples, see; e.g., [214]). But besides anthropological details, knowing the strong connections of prehistoric communities with their natural environment, the scale and the highly structured and recurrent character of this phenomenon suggest a practical combination of ecological and social factors (see, e.g., [29, 120, 200, 483, 669, 698]) Another example is provided by the pre-Columbian cities in the Mississippian Culture (currently, southeastern USA), best known by the city of Cahokia, which existed between 1050 and 1350 CE (the Cahokia Mound is now a UNESCO site near Collinsville, Illinois). The site was abandoned after 1350 (before the arrival of the Europeans in the early century century), i.e., most likely due to the degradation of environmental conditions because of over-hunting, deforestation [758], pollution [571] and the local effects of climate changes during that time (which current knowledge suggests was related to the Medieval Warming Period (MWP) also known from Europe, Iceland, and Greenland), such as increased flooding and droughts [62, 305, 550, 572, 749] (deforestation followed by intensive agriculture, supporting a population of 15,000–20,000 concentrated in the Greater Cahokia, also meant of course that the protection that was provided by those forests against floods and drought was removed) None of the two civilizations mentioned above left written records but the fact that archaeological evidence shows two complex and sophisticated societies means that understanding how people lived in those civilizations is technically very informative about who we, humans, are as a species and how we live in relation with nature For further ideas and discussions about city attractiveness and its links with other notions, notably identity and the so-called "place branding", see, e.g., [24, 83, 99, 172, 196, 378, 419, 522, 575, 584, 609, 643, 652, 659, 660, 741]

(continued)

Table 3.3 (continued)

A city's holistic parameters[a]	Symbol and definition	Short description, with suggested start references for further study in concrete projects
Resilience	$R_u = P_u + g$	In light of the discussions already made in this book about resilience, the holistic parameter *Urban Resilience* can now be defined as how good is a city at guarding all the possibilities in its real space of option, and not only to preserve them but also to bounce forward after a significant challenge Obviously, the discussion can also be carried in terms of the relation between resilience and vulnerability (see also Fig. 3.26 at the end of Sect. 3.7 in this Module). As a technical note, I would also mention that, because of its systemic, city-level focus, resilience as defined here (and related to vulnerability as also discussed in Module 1, Sect. 1.3), is also coherent with the notion of "inherent urban vulnerability" [611] Further, given the fact that resilience has already been discussed extensively in this book (and in the literature), there is no need for abundant details in this table. And, as some plugs into the literature, I would also suggest a few further readings: Romero-Lankao et al. [610] discusses "urban sustainability and resilience: from theory to practice"; Blanke and Walmsley [74] provides a discussion from an archaeological perspective on urban resilience, focused on several Ancient Roman cities of region of Decapolis (a local network of cities) from late antiquity and early Middle Ages (this citation is also mentioned in Module 1, Sect. 1.3 in the discussion about resilience); Meerow and Newell [465, 466, 468, 469] make a series of discussions on: "resilience and complexity", "urban resilience for whom, what, when, where, and why?", and "multi-functional green infrastructure: growing resilience"
Livability	$L_u = R_u + h$	***The grand-total result of all attributes that can be used to describe a city.*** It represents the ultimate holistic parameter, as it includes all the others: what a city provides (as a whole) for the well-being of its citizens, with an important difference of perspective to bear in mind: Livability of a city is about cities (as experienced by people), while the term well-being refers to people as individual persons within their overall life environment As all systems are dynamic and change in time, the lived experience is then a basis for new questions and reflections leading to the identification of new problems and renewed efforts resulting in the expansion of knowledge and in the development of new solutions (for further technical details on this iterative work cycle of lived experience and problem-solving oriented study, see Sect. 3.6)

Listed names only include the distinctive descriptor label; the full names include "urban": Urban Shape, etc.

water and minerals, or, respectively, because the harsh disturbances overwhelms all or most local species) [31, 149, 165, 166, 265, 277, 365, 435, 451, 456, 577, 620, 671].

Together, these two points can tell us something about the potentially optimal shape of a city. For example, and without going into much technical details, because the anthropic impact (heat island, pollution, soil sealing, and immediate ecological footprint) is cumulative, it tends to be higher at the central part of the city: this is where nature suffers the heaviest "elimination" or/and impact from the built environment, and this effect amplifies with the overall size of the city. This means that one obvious way to minimize urban ecological footprint and increase quality of life is to preserve or create large natural areas in the center of the city, and even more so in large cities [149]. Some of the best examples of this reasoning (or at least intuition; without discounting the benefit of favorable circumstances) in urban planning are the Central Park in New York [302, 674, 682] and the English Garden (in German *Englische Garten*) in Munich [266, 346, 608]. Relatedly, I would also briefly mention that the literature on landscape architecture provides lots of technical insights about gardens in cities, and one needs only do a proper literature search to find useful details (for a start, see, e.g., [404, 511, 578]).

Then, the last IPCC-WGIII report (focused on climate change mitigation) [332] sums-up and illustrates these complex interactions, e.g., in their sections C.6.2, C.6.3, and C.6.4 (Box 3.3; paragraphs shown in full for the sake of brevity and because they illustrate well the complexity of the matter as discussed so far in the book)—and therefore the importance of having, for concrete projects, at least some operational models that are based on a Big Picture understanding of cities—and keeping in mind that climate change is only one of the United Nation's 17 sustainable development goals (SDGs).

Box 3.3 Selected text extracts from the last IPCC report on climate changes [IPCC-WGIII], illustrating the role of cities [332]. (Each paragraph includes at the end a set of cross-references to other sections in the same report.)

"C.6.2 The potential and sequencing of mitigation strategies to reduce GHG emissions will vary depending on a city's land use, spatial form, development level, and state of urbanization (high confidence). Strategies for established cities to achieve large GHG emissions savings include efficiently improving, repurposing or retrofitting the building stock, targeted infilling, and supporting non-motorized (e.g., walking, bicycling) and public transport. Rapidly growing cities can avoid future emissions by co-locating jobs and housing to achieve compact urban form, and by leapfrogging or transitioning to low-emissions technologies. New and emerging cities will have significant infrastructure development needs to achieve high quality of life, which can be met through energy efficient infrastructures and services, and people-centered urban design (high confidence). For cities, three broad mitigation strategies have been found to be effective when implemented concurrently: (i) reducing or changing energy and material use towards more sustainable production and consumption; (ii) electrification in combination with switching to low-emission energy sources; and (iii) enhancing carbon uptake and storage in the urban environment, for example, through bio-based building materials, permeable surfaces, green roofs, trees, green spaces, rivers, ponds and lakes (very high confidence) ("these examples are considered to be a subset of nature-based solutions or ecosystem-based approaches). {5.3, Fig. 5.7, Table SM5.2, 8.2, 8.4, 8.6, Figs. 8.21, 9.4, 9.6, 10.2}";

"C.6.3 The implementation of packages of multiple city-scale mitigation strategies can have cascading effects across sectors and reduce GHG emissions both within and outside a city's administrative boundaries. The capacity of cities to develop and implement mitigation strategies varies with the broader regulatory and institutional settings, as well as enabling conditions, including access to financial and technological resources, local governance capacity, engagement of civil society, and municipal budgetary powers (very high confidence). {Fig. 5.7, Table SM5.2, 8.4, 8.5, 8.6, 13.2, 13.3, 13.5, 13.7, Box 9}";

"C.6.4 A growing number of cities are setting climate targets, including net-zero GHG targets. Given the regional and global reach of urban consumption patterns and supply chains, the full potential for reducing consumption-based urban emissions to net-zero GHG can be met only when emissions beyond cities' administrative boundaries are also addressed. The effectiveness of these strategies depends on cooperation and coordination with national and sub-national governments, industry, and civil society, and whether cities have adequate capacity to plan and implement mitigation strategies. Cities can play a positive role in reducing emissions across supply chains that extend beyond cities' administrative boundaries, for example, through building codes and the choice of construction materials. (very high confidence) {8.4, Box 8.4, 8.5, 9.6, 9.9, 13.5, 13.9}"

All in all, the ECRM/SCRM is a conceptual core model of the assemble of all processes that take place in a city, i.e., a representation of how a city system works in physical terms. As such it provides a starting point for functional representations of a city-as-a-whole, or, to use the words from the literature on smart cities, a functional "digital twin", i.e., one that is indeed holistic and serves on the long term—not merely a descriptive 3D avatar, not just from one or two angles of view (e.g., the buildings and the land uses), and/or not just for a short-term interest related to a specific project or interest. For more about digital twins, see (e.g., [111, 256, 296, 297]).

Some of the features making up the HPECs have already been discussed in the book—because of their high-ranking in eco-city-related preoccupations, and as reflected in the literature. For example, resilience has been widely recognized to be crucial for urban sustainability. Similarly, the urban shape, urban footprint, and urban livability (at least as a general idea) are either directly discussed or simply presumed/implied in virtually most or all the literature about urban sustainability.

Perhaps unsurprisingly, all these popular topics are located at the beginning and the HPEC series. The middle of it, represented by connectedness, evolution, and potential has a more abstract character, but is equally important. Connectedness is indeed alluded in the discussion about the global network of cities and the double local-and-global character, but, aside from the idea of the "global influence" of a city (discussed in relation to the popular concept of "world cities" which refers mostly to the world's largest and influential cities [50, 51, 599, 685, 686]) the discussion remains a rather narrowly technical one (and usually separated from the big picture of what is going on inside a city itself). The same goes with the evolution and the potential of a city: intuitively, we can see what they refer too, but in-depth discussions can be very (even exclusively) technical and the connections with cities as whole systems tend to remain unaddressed in the usual type of study and publications.

Simple Mathematical Formalism for the Holistic Parameters in ECRM/SCRM

The holistic urban system descriptors in Fig. 3.14 and Table 3.3 are regarded as operational parameters because we need to be able to make quantitative comparisons (1) between the attributes of a city: between "before" and "after" an action or event; and/or (2) between cities in the process of working with each other and learning from each other. Obviously, this also enables analytically useful comparisons between cities-as-systems with different histories and identities. These same attributes are holistic because they refer strictly to the urban system-as-a-whole, not to other technical parameters that refer to system sub-components (those are the same as in the usual scientific and technical standards and require no treatment in this book).

Further, the relation between the holistic parameters is one of nested inclusion, i.e., each next one brings something new "on top of" the previous one, starting with *Urban Shape* and ending with the all-encompassing (sustainable) *Urban Livability*. Mathematically this can be summarized as follows (knowing that the symbol $\supset$ means "included in"/"it is the start-basis for", and $\subset$ means "includes"/"starts from"):

$$\boldsymbol{Shape} \supset \boldsymbol{Metabolism} \supset \boldsymbol{Footprint} \supset \boldsymbol{Connectedness} \supset \boldsymbol{Evolution} \supset \boldsymbol{Potential} \supset \boldsymbol{Resilience} \supset \boldsymbol{Livability}$$

or

$$\boldsymbol{Livability} \subset \boldsymbol{Resilience} \subset \boldsymbol{Potential} \subset \boldsymbol{Evolution} \subset \boldsymbol{Connectedness} \subset \boldsymbol{Footprint} \subset \boldsymbol{Metabolism} \subset \boldsymbol{Shape}$$

or

$$\boldsymbol{S_u} \supset \boldsymbol{M_u} \supset \boldsymbol{F_u} \supset \boldsymbol{C_u} \supset \boldsymbol{E_u} \supset \boldsymbol{P_u} \supset \boldsymbol{R_u} \supset \boldsymbol{L_u}$$

or

$$\boldsymbol{L}_u \subset \boldsymbol{R}_u \subset \boldsymbol{P}_u \subset \boldsymbol{E}_u \subset \boldsymbol{C}_u \subset \boldsymbol{F}_u \subset \boldsymbol{M}_u \subset \boldsymbol{S}_u$$

Thus, each Holistic Parameter of Eco-Cities (HPEC) can also be defined additively: the previous HPEC plus a set of new features, starting from $\boldsymbol{S}_0$ as the pre-shape of the city, and each next HPEC as the previous HPEC plus a characteristic set of new attributes (see Table 3.3 again).

This further means that the overall sustainable Urban Livability (i.e., the ultimate HPEC) can be defined as the final outcome of additions of new attributes down to the pre-settlement situation (pre-shape):

$$\boldsymbol{L}_u = \boldsymbol{h} + (\boldsymbol{g} + (\boldsymbol{f} + (\boldsymbol{e} + (\boldsymbol{d} + (\boldsymbol{c} + (\boldsymbol{b} + (\boldsymbol{a} + \boldsymbol{S}_0)))))))$$

Conversely, one can represent the evolution of an eco-city towards a given moment (as per Fig. 1.9), and so could potentially anticipate the series of steps towards a future desirable situation, say a golden age of an eco-city, by a series of progresses of the form:

$$(((((((\boldsymbol{S}_0 + \boldsymbol{a}) + \boldsymbol{b}) + \boldsymbol{c}) + \boldsymbol{d}) + \boldsymbol{e}) + \boldsymbol{f}) + \boldsymbol{g}) + \boldsymbol{h} = \boldsymbol{L}_u$$

This simple model has two principal uses. First, it allows us to represent, in an orderly fashion all actions towards an eco-city, from the most basic to the end (complete) set of systemic attributes. This allows us to locate each and every intervention made via an eco-city project (or/and wider program or/and policy) in this series of HPECs, which yields what I would call here (for lack of a better terminology and as I need a name for it) an "*Abstract Systemic Eco-City Operations*" (*ASECO*) model. Through this analytic structure, inherent systemic complexity is observed and the requirement of coherence at the scale of the city as-a-whole is served. In practice, we can thus talk about a holistic knowledge-action model that meets the need for a Big Picture perspective and is based on knowledge (instead of having the analyst or policy-maker, or/and a project team in general, being swamped in some "bazar" of disparate actions that would lead nowhere and would pull the projects down).

Second, this model complements the fundamental knowledge-action model for generating solutions for (here urban) sustainability (as introduced in Fig. 1.3, Sect. 1. 2), by providing a coherent series of work targets in the form of the HPECs series. In other words, while the model in Fig. 1.3 gives the "how" of principle for closing the unsustainability gap (a conceptual "sowing device" so to speak), the HPECs provide what mathematicians and engineers would call an "optimal substructure" of the gap (or problem) that needs to be addressed, in the form a logical series along which the "sowing device" can "zip" and address those different aspects of urban sustainability that require dedicated attention.

From an operational perspective (knowing that the field of Operations Research, while multi-disciplinary in its application, is a field of applied mathematics), this kind of abstract big picture is important to have because it distills a "mechanical" conceptual hard core that is independent from project contingencies and therefore can serve as a methodological default for eco-city projects from all contexts: as a start common denominator and reference for work within and between projects. Then, the details filling the ECRM model can vary as needed and can be adapted, e.g., project teams can (negotiate and agree to) establish their own list of definition of HPECs. What matters here is not so much how different those context-dependent definitions will be in each project, but that those definitions are agreed upon by all the participants in that particular project.

To this aim, the ECRM can also simply serve as a platform for kickstarting in-project conversations.

And, the same logic applies when the goal is just to learn from other projects (and other cities) through between project conversations.

The Connection Between HPECs/ECRM and the Delphi-Based Scenarios

The priority issues identified through the Delphi method basically take an empirical look as the status of an urban system and point to the top pressures. The HPEC series (and the simple mathematical model of it) inside the ERCM model provides the target against which those identified sore points (pressures/priorities) can be understood from a systemic perspective and mapped operationally (starting even from the diagnosis of the urban system), so as to know where the situation goes and to know what type of systemic outcome is desired for improvements.

In it, the series {a, b, c, d, e, f, g, h} represent the different functional "layers" (or "onion sheaths" if you will) of the status of a city at a given time. For example, if the priority issue of a city is identified to be related to its Footprint, then the model above tells us that (1) we look at layer "c" in the series, which also automatically tells us (2) where that issue is located with respect to the other main attributes of the city, and how they link (ergo, where and how to act in order to obtain the desired improvement instead of causing more harm than good). Concretely, in this example, the said series tells us that the root causes of that problem need to be sought in $(((\boldsymbol{S}_0 + \boldsymbol{a}) + \boldsymbol{b}) + \boldsymbol{c})$ and that whatever change we induce in layer "c" needs to be made sure it has a favorable (or at least neutral) impacts also in the layers {d, e, f, g, h} instead of creating other problems there. Importantly, once the detailed minimal lists and definitions are agreed upon in each layer by the project team (i.e., when there is a clear big picture), then the location of an issue identified through the Delphi method automatically shows what this means in the big picture.

At the same, time, one can use this model to determine what can be said about a given city when all we have is (as it is very often the case) a random and potentially incomplete set of disparate pieces of information about that city: by identifying where each bit of insight belongs, one automatically maps all those "puzzle pieces" along the ASECO model, and thus can see (1) how can they serve in advancing a city as an eco-city, and (2) what gaps remain to be addressed, and thus we gain visibility about the situation and we know better where to prioritize our efforts. And, this is of course a direct aid in the logic of prioritization used in the first parts of the Delphi-based procedure for scenario-generation as described in Module 2.

Put differently, the ASECO model is the result of a progressive addition of insights about urban feature from the moment of any initial settlement, when there is no clear urban form yet but there is a default shape (if only consisting in the physical features of the settlement) until the final end outcome which is the livability of that human settlement. Each step can be a bottleneck that can influence the trajectory of a city in the space of the possible (as per Fig. 1.9) and even push it over the border, which can result in the cessation or at least major dislocation of that human settlement (resulting in a sudden change of identity—so that one could talk about being another city altogether).

Then this model can provide a basis for the development of a set of functional scenarios that can help a community understand what is happening with the city, e.g., how different situations occurring or envisaged through foresight will impact the urban system of interest. For example, how something that influences resilience is influenced by the preceding HPECs, and how they influence the subsequent HPEC, which is Livability itself. In addition, this model allows explicit focus on the differences between different HPECs, i.e., what it takes to pass from one level to the next, so it helps feed relevant and coherent content into the scenario-generation process described in the previous Module.

Then, the ECRM/SCRM model as a whole (or via some of its components) can be used as a backdrop for all conversations during or outside the scope of scenario-generation. The USN sub-model can also be used for general or/and quick assessments of different situations.

Figure 3.15 illustrates how the USN model can help with a clear visualization and understanding the systemic challenges that can be caused by events that may appear unrelated directly with other topics but in fact are, e.g., a knot of several major urgencies instead of only one, some being already here, and some that can be easily be regarded as realistic scenarios.

Namely, one can see a security-energy-climate conundrum faced by cities (as part of their countries) but (1) many might not immediately see (and would discover empirically later—the hard way) that topics such as energy are actually part of a hard 4-nexus that automatically brings along a wider array of challenges, and (2) if this central core of topic is involved, then all the others topics will emerge as urgent, next in line being those from the 8-cloud of topics (notably, we already see different other issues related to food/hunger, health, population/refugees and so on, and the problems coming with them), and (3) a good deal of other analytical insights and that everyone can draw by simply analyzing it.

In short, the utility of USN and the related models is in encapsulating existing knowledge and thus showing a manager where to look first (and what to avoid omitting) to insure project work efficiency under constraints of time and other resources.

As an illustration of the importance of this matter, let me also point out that the new strategic concept of the *North Atlantic Treaty Organization* (NATO) adopted in June 2022, a document with 49 paragraphs spread over 8 pages of text [505], mentions the words "climate" or "climate change" 11 times, as follows: in paragraph 5 (at section Purpose and Principles); 11, 19 (at Strategic Environment); 36 (at Crisis Prevention and Management); and 43, 46 (at Cooperative Security). The Preface of the document also gives the general context of the matter: "The Strategic Concept emphasizes that ensuring our national and collective resilience is critical to all our core tasks and underpins our efforts to safeguard our nations, societies, and shared

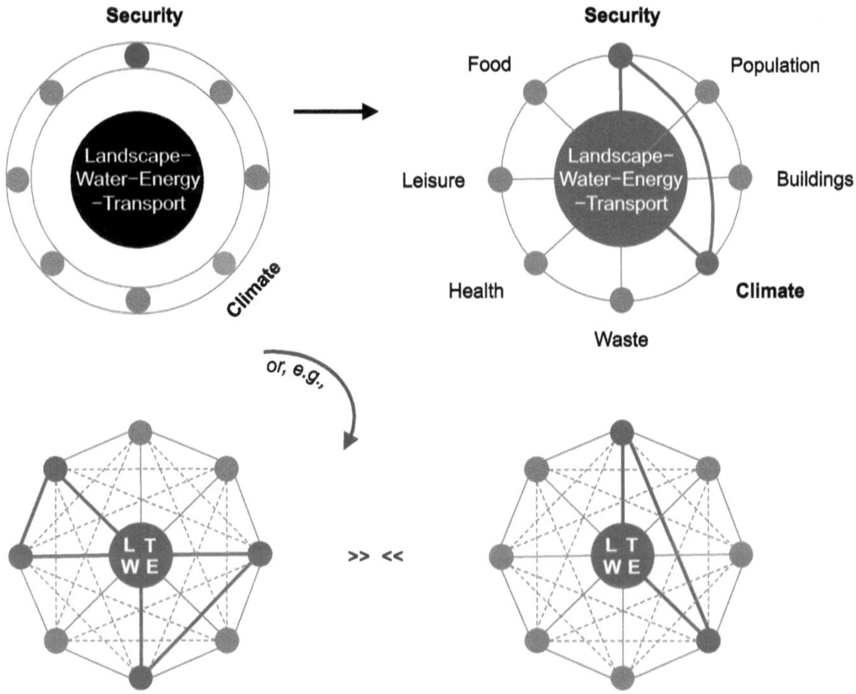

Fig. 3.15 Example of using the USN to represent a concrete type of challenge in the urban transition to sustainability: central interconnected topics and the knot (shown in blue) of present urgencies in European Cities in the context of Russia-Ukraine war (and not only), or some other combination of drivers that may be brought to front by events, and which a city must address based on a coherent and efficient understanding of the big picture. *Source* original

values. It also emphasizes the cross-cutting importance of investing in technological innovation and integrating climate change, human security, and the Women, Peace, and Security agenda across all our core tasks". To help translate this into practice, NATO has also established an Innovation Fund of over 1 billion Euros [506] for supporting early stage startups, thus recognizing the importance of developing new solutions through innovation.

3.5 Tapping on Global Pools of Experts: Brain Networking for Eco-city Projects

As already shown in Fig. 1.8 (Module 1, Sect. 1.3), every city can be functionally represented as groups of people interacting with each other in networks, while every city is also a node in regional and global networks of cities. This has direct consequences on the issue of human resources, which translates in whether or not a city is capable to attract and involve the best experts in the world, with direct consequences for the Urban Connectedness discussed in the previous section (Table 3.3).

An issue that is typically discussed in country contexts but is equally important for cities is that of the so-called "brain drain". Usually, this topic is discussed in the context of lesser rich countries (losing their best and brightest as they emigrate to richer or otherwise more attractive countries), but this phenomenon in fact affects cities and regions inside rich and poor countries alike, as more influential cities tend to attract human resources from anywhere in the world: "talent goes to global cities" [724].

However, this is only the beginning of the discussion, because brain drain is usually regarded as a disadvantage while in fact the mobility of the capable people is a phenomenon which (1) is both natural and legitimate, and (2) most relevantly for our discussion, not only it can be turned into an advantage but it can help cities (as well as universities and other institutions; and countries overall) to leap forwards through projects and programs that employ a strategy of Brain Networking between people that may live and work in different countries around the world but share a common culture and sentiment of a connection with the city (or country) in which they were born or where they studied or with which they have some other emotional ties [143, 147, 148, 154, 157, 158]. For further aspects related to the circulation of talent and the issue of intellectual resources for the development of different countries around the world, see also other examples (e.g., [85, 106, 126, 213, 412, 427, 499, 527, 581]).

Figure 3.16 shows a basic operational model modified for application in eco-city projects and programs, while the following paragraphs add a brief discussion about this concept and strategy.

For the immediate practice, the main idea is that, because all the flows of data and information (and the mobilization of knowledge across local, regional, and global networks of experts and actors) can be represented as interactions across networks of networks (as per Fig. 2.2 in Module 2, Sect. 2.2, and the discussions and references therein), each and every eco-city project is in fact a network of individual project participants that have complementary expertise and are united in that project by a set of rules. Now, the explicit rules are the official rules of the project itself, while the implicit rules are represented by the existing or to-be-achieved agreements on values, knowledge, goals, methods, etc., as described at the start of this Module, at Figs. 3.1 and 3.2. Therefore, because Internet and modern means of communication in general (remember the Internet-of-People/IoP), essentially anyone with relevant expertise can participate in a project in any city around the world if there is some connection between that person with that city; and, the emotional connections with a city can be the decisive factor that will make at least some of the people born in that city to seek and find how to make herself/himself useful in such projects—provided that they know about those projects in the first place, and that the person has at least a minimal level of trust in that project and in the project organizers. And, very important, having these people involved in an eco-city project/program, a city benefits not only from the specific expertise of its Diaspora and/or friends but also from the fact that these people have their own professional networks and thus can

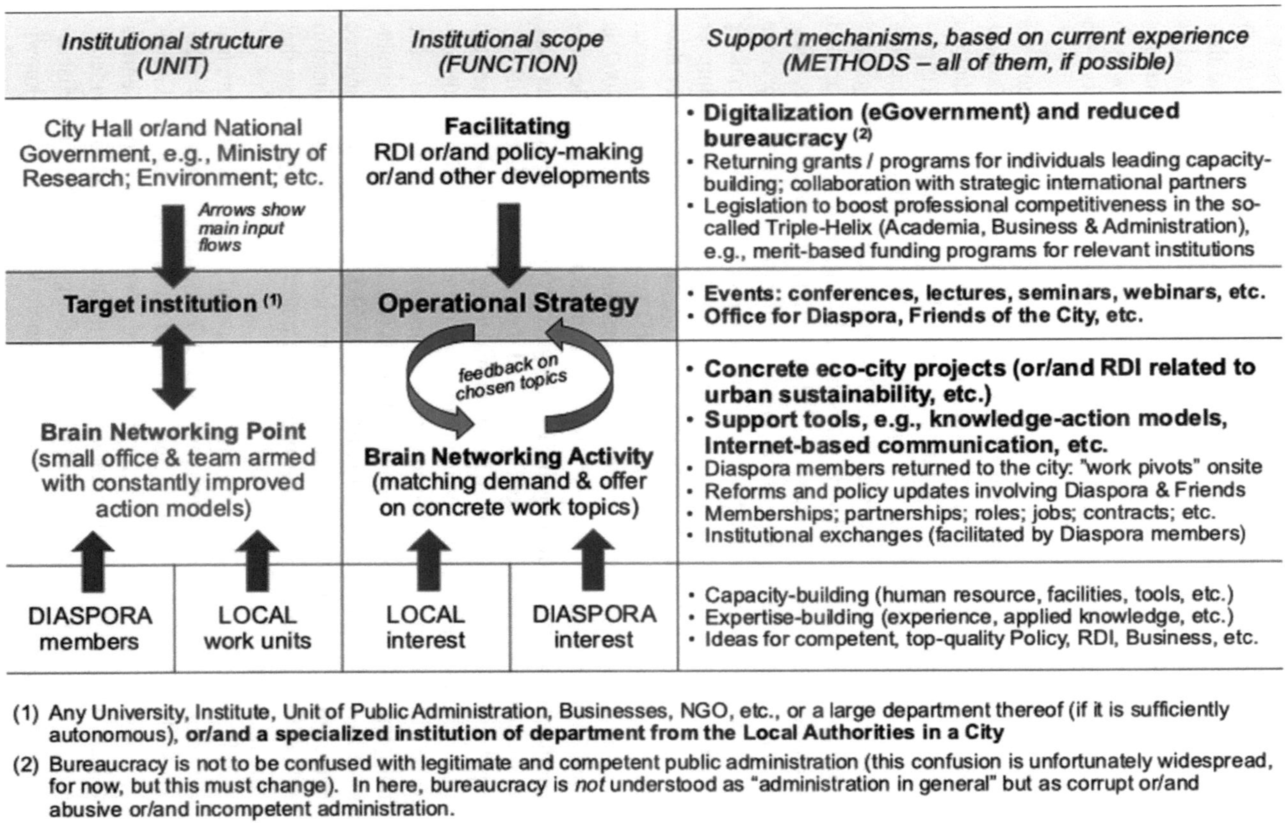

Fig. 3.16 Brain networking: basic operational model. *Source* modified based on [148, 154, 157]

serve as an interface for reaching out to other competent professionals (and various business actors and potential investors). One must also not forget that those members of a city's Diaspora have reached high levels of competence precisely because they went to other countries and institutions, i.e., to places where they were able to grow and fulfill their potential, and therefore it is a good thing that they are where they are and must therefore not be regarded as a loss for that city but as a gain and as a pragmatic connection of a city with world-level centers of expertise that may not easily exist in the 'home' city. In fact, no city can have top-level expert centers in all domains, and therefore, the existence of a professional Diaspora is—for any city—an immense asset and gate to the world.

In short, a first and foremost way for a city to have an intelligent approach (one oriented towards high-value generation for itself) is by working with its Diaspora and Friends, i.e., by applying Brain Networking. It goes without saying that, in this approach, universities can be of great help—not only by participating in triple-helix (university-government-business) projects but also by catalyzing the interaction with experts around the world. In general, universities use a brain networking logic naturally, because Science is global and collaborative by nature, so scientists around the world collaborate all the time and across geographies. Moreover, the decades-old phenomenon of electronic publication (through which print science reports are also made available in an electronic format) facilitates scientific exchanges, making communication more efficient and deeper, knowing that, since 2007 the number of publications that had already become accessible on databases was higher than the number of publications produced in the prior 100 years [33, 227]. To a keen observer, it might be more than an interesting coincidence the fact that 2007 is also, as mentioned in Module 1, Sect. 1.2, the year when Sustainability Studies, the transdisciplinary field *par excellence* gained a strong recognition as a field of its own, and begun to look more mature (as per Fig. 1.5 and the discussion related to it).

In the process, the electronic environment provided by the Internet and related technologies will also help address various other problems that require good solutions, notably (in the relation between scientists themselves, starting with the issue of terminological disconnects between different fields (e.g., [453]) and then the relations between scientists and media/journalists (e.g., [559]), and the issue of science literacy in the era of the Internet when so much information is available but is hard to understand because both the sheer quantity and fragmentation (e.g., [321]), so that "just google it" is not enough (e.g., [290]).

Together, Internet and Brain Networking throw a useful light on the fundamental relation between science and society in general ([559, 746], see also Figs. 3.24, 3.25, Table 3.7, and the related discussions in Sect. 3.6), and the broader matter of knowledge integration (or lack thereof) which has become overwhelming for scientists themselves and which calls for real synthesis, both at the level of available evidence and in terms of cross-domain (transdisciplinary) knowledge [33]. This same discussion also brings forward the notion of "knowledge about knowledge" (also known as "meta-knowledge") in science in general and in urban sustainability in particular [228, 611] and highlights the importance of university-business-government collaborations in the present knowledge-based economy [413], and the role of universities as "co-creators of sustainability" within this triple-helix and across the society (see, e.g., [697], which also includes lists of examples of partnerships on different continents). More specifically, Brain Networking-based approaches can help in what has been described as the challenges of "situating knowledge on an urban planet", in the context in which a top priority for Humanity as a whole is represented by the fast-growing cities in the parts of the world known as the Global South [547].

Obviously, in this very logic, businesses and other organizations can also facilitate Brain Networking, and we would not have enough space here to mention all of those (but see Fig. 3.21, Sect. 3.7), and the related discussions on public–private partnerships (PPP) for eco-city, knowing that, by their very nature, universities can be hosts and promoters of PPPs and eco-city projects in general). Hopefully, this short introduction here is just enough to kickstart relevant discussions about using Brain Networking in your projects no matter where you are in the world. If your city or town or village happens to be too small to have a significant Diaspora around the world, you can always make a local alliance of villages or towns that is big enough to generate useful projects.

Now about the background aspects related to Diaspora as a human resource. First of all, going to study and/or work in other countries is not a bad thing but constitutes the personal development of those people (one of the possible and legitimate ways).

De jure (and morally), it is everyone's right (to do what they think is appropriate with their own life, in order to develop and affirm their potential as human being). And, that doesn't mean that somebody "abandons" a place in favor of another. That place remains as part of the history and identity of a person, which means that the connection remains and is alive. When somebody is emotionally attached to a city, even if a person's life drives her or him somewhere else in the world, that person never really leaves that city emotionally. And this is part of what a city is and therefore part of what a city can be in the future. In his 1987 speech in Berlin (mentioned in Module 2, Sect. 2.6), US President Ronald Reagan highlighted this idea when talking about the spirit that ties the people of a city, citing the words of a 1960 song by German singer Lily-Marlen Dietrich: "*Ich hab' noch einen Koffer in Berlin*" (in English, "I still have a suitcase in Berlin"), meaning "I never really left that city".

De facto, and especially now in the era of global communication technologies, the international development of individuals is a positive phenomenon for every city, as it translates in a concrete development of the human capital as a central resource for a city that understands to look for its own future. Thus, the practical and logical concern should not be to "stop" this so-called "brain drain" phenomenon as if it were a loss (it is definitely not a loss and cannot be stopped anyways). Any approach that would regard this circulation of people as a loss would be short-sighted and false: those people who went in pursuit of their individual development would not have reached their potential if they didn't go to some specific places in the world where they could grow professionally according to their specific aptitudes and aspirations. Therefore, the human capital represented by this people (1) did not exist yet when those people relocated (what existed was only an unfulfilled potential, and this is not the fault of anyone), but (2) it exists right now and can be mobilized for concrete projects (instead of complaining about a presumed loss of things a city never had).

In a nutshell, lingering in false (pseudo-philosophical) debates about "brain drain" (instead of developing the capacity to apply Brain Networking) would be not only a very regrettable misunderstanding and immoral from a human rights perspective; it would be a form of ignoring a real resource while complaining about an imagined one, or as one popular saying goes: sitting on a treasure and begging for survival.

Even the term "brain drain" is no longer very appropriate in the twenty-first century. In the age of the Internet, "networking" is a proof of intelligence and "drain" is only a proof of impotence or disinterest. Furthermore, through its Diaspora, a city as a local community (or a nation) has a direct avenue towards a truly global horizon, strength, and relevance. As already discussed (starting in Module 1, Sect. 1.3) a city has a double local-and-global character, and becoming a true eco-city involves assuming this double character, and developing a voice in the world. Brain Networking is an obvious way to do that, and a logical strategy for eco-city projects.

3.6 Developing Shared Understanding and Choices: The DIKAR_process Cycles

A General Framework Model for Coordination Within and Between Projects

Once we agreed that eco-city projects must be based on evidence, we also have to acknowledge and manage all the divisions that exist between the different fields of expertise that are relevant for the project (or program or policy), lest that project finds itself in a situation of "managerial cognitive dissonance" [66] along underlying "collective cognitive dissonance" and a long list of closely related challenges [19, 314, 638]. To use a keyword from the precursor of cybernetics Stefan Odobleja, collectively knowing what we are doing is about building that "consonance" [249, 517–519, 711].

In this section, I present a model born in the practice of concrete pioneering projects for eco-innovation for/and eco-cities. The next section will add a set of complementary observations and visual summaries related to operational knowledge integration in projects dedicated to foresight, research, development, and innovation.

Since progress only happens when enough people are on the same page, and given all the issues in complex projects addressing complex systems, the central challenge is how to effectively insure a basis of common denominators between project participants (knowing that usually this basis is assumed to be sufficient but is not, and it is only a matter of time until misunderstandings and conflicts break any complex project apart). Beside general issues like cognitive and behavioral biases (e.g., overplanning or/and ideologies, i.e., sets of dogmas; or emotional or cultural tendencies that influence our cognition, objectives, preferences, and choices), competent action require preparation, starting with a basic understanding of what a project is and what it can do.

In the field of Organization Studies, projects are seen as "temporary organizations" [38, 433, 670, 677] emerging from the interaction of individual and group agents within wider social systems. As "permanent" institutions are typically (by their nature) too rigid, projects are being generated to put ourselves in the situation of doing things that current organizations cannot do (or not well enough). In other words, "temporary forms of organizing are an appropriate solution for dealing with organizational complexity" [623]. This reminds us the importance of systemic interdependencies (the nexuses in Sect. 3.4), and the fact that projects (and especially complex ones) might come as a surprise to (or simply as "unwanted" by) those who are used to work in a "silo-thinking" environment (like the "traditional institutions").

The crux of the matter is that our current institutions are borne in the strict division of labor that was the norm (along with the systematic separation between disciplines and fields) of the late nineteenth and the early twentieth centuries, i.e., during the industrialization era that lead to our current way of life, when most of today's technical universities and modern institutions in the most industrialized countries were founded. Thus, eco-city projects are not just "problem-solving machines" but also concrete opportunities for opening professional discussions about how institutions function, how they can contribute, and how they could evolve to accommodate the requirement of integrative thinking. The goal is to find the right balance between functional specialism and organizational versatility. This takes a lot of work and applied conversation, which, if too weak or inadequate, will stifle progress before it starts.

Figure 3.17 shows an iterative framework model called DIKAR_*process* which incapsulates how to proceed coherently, to insure low-risk and highly coordinated understanding and action in projects of virtually any degree of complexity and difficulty. Table 3.5 shows a basic user template of the model: an organizer structure to be used as such or as a guide for finding where project resources are needed to insure smooth work and coordination of efforts.

Of course, in discussions about Vertex 4/4′, the emphasis may fall—depending on the moment of the conversation—on the start knowledge (referred to as Vertex 4) or on the updated knowledge (referred to as Vertex 4′) but they obviously represent two instances of the same content (or "two sides of the same coin" if you prefer), and invoking one implicitly refers to the other too (Tables 3.4 and 3.5).

In terms of graph analysis of the model, an adjacency matrix can be used to visualize and store the relative importance (centrality) of each step, in terms of the total degrees TDC_{ij}, the in-degrees IDC_{ij} (reception), and the out-degrees ODC_{ij} (sending), where i and j denote the respective sets of categories in the basic coordination axes. In those, one observes the influence (weight) of each step among the other steps, and how "in" and "out" degrees add up:

$$\mathrm{TDC}_{ij} = \begin{pmatrix} 0\,2\,2\,0 \\ 0\,3\,5\,0 \\ 2\,8\,3\,2 \\ 4\,5\,0\,0 \\ 3\,3\,0\,0 \end{pmatrix};\ \mathrm{IDC}_{ij} = \begin{pmatrix} 0\,1\,1\,0 \\ 0\,1\,3\,0 \\ 1\,5\,1\,1 \\ 1\,2\,0\,0 \\ 1\,1\,0\,0 \end{pmatrix};\ \mathrm{ODG}_{ij} = \begin{pmatrix} 0\,1\,1\,0 \\ 0\,2\,2\,0 \\ 1\,3\,2\,1 \\ 3\,3\,0\,0 \\ 2\,2\,0\,0 \end{pmatrix}$$

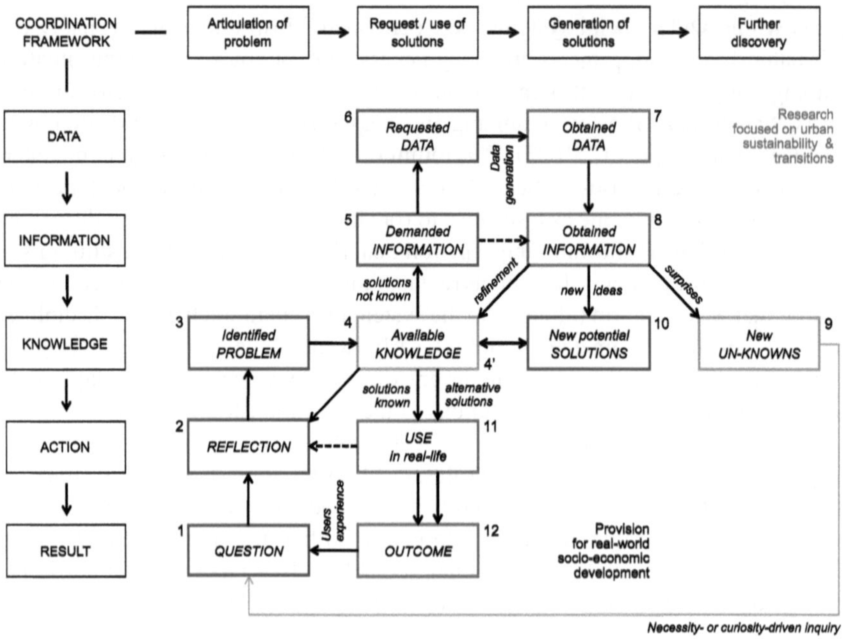

Fig. 3.17 The *DIKAR_process* model for structured co-work in a project or/and between projects or work packages. Case numbers denote steps operationalized as Negotiation Rooms (NRs; one can also understand them as "coordination rooms"), i.e., moments in the work process where project participants sit and check whether they use the same definitions of important project aspects, starting with shared values and knowledge, and continuing with knowledge, goals, methods, and anything needed (i.e., achieving agreements as per Figs. 3.1 and 3.2 in Sect. 3.1). Arrows denote the transformations that take place between steps in the full cycle. The full cycle starts with a Question (Step 1). The full cycle and the sub-cycles are iterative as needed: the optimal number of iterations is to be found empirically. *Source* modified after [153, 155]

Key Technical Observations and Discussions About the DIKAR_process Model

At this point, if you think you already understood the idea of this coordination model and prefer to go directly to the next Section (so, the next model), you can do so instead of continuing with the detailed discussions hereof. Or you can skip only this more technical sub-section, and go directly to the next sub-section—the one dedicated to the managerial benefits of this model. Later on, whenever you will have some questions and/or want to reflect a bit deeper about the model (for example, on how to apply-adapt it in a project of your own), you will probably want to also check this part.

In the basic terms of Graph Theory and related methods, the DIKAR_process model is basically a strongly connected direct graph (e.g., [537]): a set of vertices (nodes; representing categories) are connected by directed edges (morphisms/transformations/arrows; representing relations between categories) so that every vertex (category) can be reached from every other one. This type of representation

Table 3.4 Basic definitions of DIKAR categories (data, information, knowledge, action, result), as used in the *DIKAR_process* model in Fig. 3.17

Categories	Definitions	Example (from the topic of urban river water pollution)	Agent
Data	A measured or estimated value[a]	A measured or estimated concentration of, say, Atrazine, 2.5 μg per liter of water or sediment. (This is a herbicide that is also toxic to humans; notably, it is a hormonal disruptor)	Scientists; supervised lab team
Information	A result of scientific/ professional interpretation of data, in the sense of "the difference it makes"[b]	Telling whether the concentration of Atrazine is above or below a relevant threshold, e.g., respectively, 0.1, 2.0, or 3.0 by the EU (legal), the World Health Organization (recommended), and the USA (legal). The above concentration of 2.5 would give different interpretations, respectively: "polluted" ("not ok"), "alarming", and "not polluted" ("ok")	Scientists; other professionals
Knowledge	Ability to "connect the dots" (combine information items) in "the big picture" of a topic of interest, as described by theory and models	I.e., (or/and), – Identification of existing action options, – Improvement of existing theories/ models, – Update(s) of problem (co-) definition, – Request for new information items, – Generation of new solutions If none of the above, new unknowns emerge automatically, leading to new questions	Experts able to build and operate with theories and models

(continued)

Table 3.4 (continued)

Categories	Definitions	Example (from the topic of urban river water pollution)	Agent
Action	Choices and pursuit of chosen paths in the space of the possible (scenarios and procedures or adventures—when the space of the possible is still uncharted at action time t_n)	As per the example above, different actions will be taken: in the EU, proceeding along the protocol in place for the detection of pollution with that substance in the studied type of sample; in the US, nothing will be done until the next sampling and measurement In addition, the fact that the target problem (that pollution) is monitored and knowledge is updated also constitutes (a type of) action	All entities undertaking action[c]
Result	Any situation that is a basis for new experiences, questions, and evaluations	Legal pollution thresholds tend to follow to-date toxicological thresholds for humans and ecosystems; ergo, in short, the result will be a number of intoxications and disturbances	All entities experiencing a situation; stakeholders[d]

Source redrawn after [153]

[a] This refers to scientific data. But, depending on needs it can be adapted to mean/include other types, e.g., in information science and technology, it usually means all types of record. Relatedly, if the notion of "data type" (as well as "data model"/"data structure", "data management", and other terms from information and computer sciences) is (are) applied, then scientific data can be regarded as a particular type of data. Depending on the purpose, any other type of data can be used, but then the definitions must be detailed enough to avoid confusions and to facilitate functional mapping between scientific data (as grounds for evidence-based action) and other data types

[b] This roughly corresponds to Shannon information in science. However, depending on the chosen purpose, this definition can also (be adapted to) cover other meanings, e.g., algorithmic information and computational complexity. For supplemental discussions, see also [315–317]

[c] Individual and groups of citizens; any organization that is pursuing their interests and obligations. Agents have various degrees of freedom and can act cooperatively or not (or combine opposite choices in a series of situations). Thus, actions can be convergent or divergent, coordinated or not, coherent or not, amiable or hostile

[d] Stakeholder entities may have participated (or not) in the actions leading to a given result

allows the identification of steps, paths (sequences of steps), and loops, the totality of which (i.e., all grouping and path possibilities within the model) collectively describe all possible interactions that can take place in a group of agents (individual experts, teams, institutions, etc.).

It is also important to mention that this model is itself the result from the synthetic formalization of the workflows between (and feedbacks from) the ensemble of partners of a triple-helix (university-industry-government) cluster focused urban sustainability and generating eco-innovation therefore, led and hosted between 2009 and 2013 by the University of Versailles (UVSQ), when this institution was part of *PRES UniverSud* consortium of Universities in the Paris Region (in French, a "Pole de

Table 3.5 *DIKAR_process* user template for joint projects bridging fields; to apply and extend as a spreadsheet, text or/and diagrams to fit the chosen use purpose. Work steps are illustrated with two work packages. Each full cycle starts with a question

Negotiation Rooms		***Example Goal 1:*** *Building alternative paths towards affordable clean water in a Target City, given a climate change policy*	***Example Goal 1:*** *Developing a platform for real-time dialogue between local citizens and a network of actors and experts*
1	Question	Q1.1. How to improve running water safety, cost, and accessibility?	Q1.2. How to improve operational dialogue between relevant parties?
2	Reflection	In the real-world cities, water management is divided into separate sectors: supply; waste water; sanitation water, and stormwater [35, 514, 541]. To build and maintain valid options for sustainability, we need operational coordination	Project co-involvement of different types of stakeholders (as in the triple-helix of science-business-government) can elicit a useful convergence of perspectives [171, 715]. In crises, resilience involves effective co-work between sectors
3	Problem	Water governance failure, linked to a lack of tools, e.g., sensor network architectures for unified water quality monitoring and control	Science-society co-work failure, in a great measure due to a lack of tools for effective dialogues between experts and between them and citizens
4	Knowledge (if the real case is describable as "solutions known-and-available", jump to step 11. Else, continue to step 5)	Expert project team (focused on science), covering: ecology, analytical chemistry, toxicology and eco-toxicology, water management, engineering, etc., with a combination of specialized and holistic models	Expert project team (focused on society and business): social studies, communication, business management, informatics, law, arts, etc., with a combination of specialized and holistic/interfacing tools

(continued)

Table 3.5 (continued)

Negotiation Rooms		*Example Goal 1: Building alternative paths towards affordable clean water in a Target City, given a climate change policy*	*Example Goal 1: Developing a platform for real-time dialogue between local citizens and a network of actors and experts*
5	New information (demanded)	… (technical lists), e.g., – chemical, eco-toxicological tests; – loads of analytes (i.e., the studied substance) and related toxic effects – options for sampling and analysis – technical support options, e.g., based on automatic samplers and, relatedly, machine learning (ML) and artificial intelligence (AI)	… (technical lists), e.g., – infographics on risks and options; – visual demonstrations opportunities (places and access details) about scientific and engineering methods – relevant cultural references (local, national, global, plus comparisons) – issues (open discussions) about realities, perceptions, and policy
6	New data (demanded)	… (technical lists, as per project)	… (technical lists, as per project)
7	New data (obtained)	… (technical lists, as per project)	… (technical lists, as per project)
8	New information (obtained)	… (technical lists, as per project, i.e., as defined by the template user)	… (technical lists, as per project, i.e., as defined by the template user)
9	Potential solutions	… (details as per project)	… (details as per project)
10	New unknowns	… (details as per project)	… (details as per project)
4′	Knowledge (updated)	… (details as per project)	… (details as per project)
11	Use in real life	… (details as per project)	… (details as per project)
12	Outcome	… (details as per project)	… (details as per project)
NQ (new question/new iteration)		… (details as per project)	… (details as per project)

Source modified after [153]

Recherche et d'Enseignement Superieur"; now UVSQ is part of the larger University of Paris-Saclay (UPS) consortium). However the first formulation of this model in the present form (shown here) was made during a subsequent project called ACE-ICSEN, a transdisciplinary university cluster within the same UVSQ and UPS. The User Template shown in Table 3.5 was made as a response to an exploratory request by the participants in this project, notably in the work package dedicated to transdisciplinary (and "transversal") integration of three different thematic Modules, respectively focused on cities, climate changes and health (the relation between human health and environmental degradation).

Thought of as a universal map of activities and logical generator of work threads, the DIKAR_process model basically explicates the general path Data-Information-Knowledge-Action-Result (the DIKAR framework or concept from information sciences) [109, 723] along a work process carried out by project participants (under some form of managerial coordination) here understood as explicit agents (see, e.g., [76, 195, 474], for further discussions, see [153, 155]).

Essentially, this framework model is born from the Cartesian Product between two ordered sets of categories, the result of which is a bi-dimensional matrix within which one can identify a set of non-empty sets that can be connected to represent streams of work steps—the abstract flow shown in the model is constituted by the totality of those non-empty steps. Each of these non-empty categories then is regarded as necessary points of convergence for relevant project participants, i.e., they serve as conceptual spaces for meeting (i.e., "coordination rooms") for developing and managing a shared set of common denominators. In each of those, a set of dedicated definitions (especially, and starting from, the definitions concerning specific values and knowledge) are co-identified and agreed by expert interaction and expert collective decision. This can also happen in combination of experts and citizens, or of any/all types of participants that are required in a given project.

This coordination and decision is to be done through a set of (initially default, but then deliberately proposed and) commonly agreed qualitative and quantitative methods, such as expert (or/and citizen-expert) workshops and debate sessions, voting (and ranking) tables for quantifying weights of different parameters, as one can easily find in the management literature (e.g., [150, 541, 715]). The single most important goal is that those definitions and details are being agreed-on, and that the "what exactly is agreed-on" is defined inside the categories identified as non-empty cells of the project flow model above.

All items that have been agreed upon are then recorded as such, used in the target (and eventually, subsequent) projects. Their purpose is to insure semantic and operational alignment and coordination between project work packages (and any other work units). They are meant to be regularly updated (when needed, as needed), but the serious challenge here is that they first need to be produced.

In this general model, every work round begins with a chosen topical question and then continues along any of the steps, paths, and cycles that is logically possible within the model. Each such step, path, and cycle can be tested and then iterated for 2-n times until an optimal routine is found (i.e., when new iterations do not bring anything new) for the intended use (and there is no prescription about the number of iterations). Thus, the model allows efficient orientation and coordination towards the production of a set of co-action protocols, i.e., sequences and cycles of action types that can be pursued simultaneously, comparatively, and iteratively within any (large and) heterogeneous constituency of actors that are—at any one time—involved in building, say, resilience (or any other holistic/systemic parameter) in a given city.

Therefore, what matters is to find the proper routine, and then use that in the daily work instead of the DIKAR_process; then the DIKAR_process model can be used again for periodic (/as needed) recalibration of the project co-work routine. This is a key observation, as the general DIKAR_process model must not be confused with the routine of the project; instead, it is a tool for (1) finding a project-specific routine that is coherent with the requirement embodied by the DIKAR_process framework model and (2) for updating that project-specific routine based on new information and changes in a way that remains consistent with the underlying fundamentals of coherent (optimally-structured) work. In fact, once a project developed its routine and goes into updates, the project management team (and the teams) will have already become familiar with the model and re-calibrations will go fast. As my background is both in natural and social sciences, I think it might be useful to signal here that the notion of calibration and recalibration is common currency in natural sciences; to give one simplest example with which I am very familiar, in an analytical chemistry lab, the scale (employed for weighing the substances which are used in the preparation of reagents) is calibrated every morning before any weighing (/measurement) takes place. If this is not done, basically you can expect that all your subsequent work sits on grounds that are unverified (and therefore assumed false) and is invalid by default. This is why, you never forget to do that calibration first off. If you do, you worked a full day for nothing and you learned a lesson the hard way (and assuming you still have other reagents/resources left to do that work, this time correctly). In RDI and eco-city project management, given the complexity and the stakes of these projects, plus all the cognitive and behavioral biases that already well-documented, and knowing how easily the sum of institutional and managerial pulls and tensions can kill such projects, having and using a method for proper calibration and recalibration is a necessity.

As a colleague of mine used to remind us colloquially, there are thousands of ways in which a project can go wrong and (often only) one in which it can go right. Keeping a project on track is the hardest task of all, and eliminating unnecessary mistakes is a basic requirement. This is what calibration and recalibration is all about.

Obviously, a method such as this requires a high-enough level of reflexivity, but that should not be a problem for an eco-city project team that understands what is at stake and the nature of this kind of work. (Rather, the challenge here for a steering committee or initiator is to insure a high quality of the human resource—which goal can be aided via the Brain Networking discussed in the previous section).

From my experience, I can tell (a) that some people will quickly grasp the method (usually these are people who are capable of disciplined work and experimentation) and (b) that most will understand the logic of the model after due explanation. However, two basic conditions need to be met. Firstly, the project needs to include this model (or at least its basic logic) from the beginning, so that it is understood to be a part of the main work (not like a supplementary work load). Secondly, the method needs to be managed by a small team tasked with handling it and thus interfacing all project participants as necessary. At the limit this can be the project manager. This was originally my case in the Econoving cluster and master, and this is partly how I developed the method in the first place: by resolving mentally (at any moment when clarity was required before any daily managerial decision) the paired operational requirements shown in the coordination axes of Fig. 3.17; basically, using just a 2D mental model (for thinking "on the spot") that is structured as a 5 × 4 matrix, i.e.,

$$Spot\ Model = \{\mathcal{M}_{i,j}\}$$

in which $i = \{Data, Information, Knowledge, Action, Result\}$, and $j = \{Problem,\ Solution\ Demand/Use,\ Solution\ Generation,\ Further\ Discovery\}$

In other words, someone can mentally monitor where anything is in the cartesian grid composed by the two lists of requirements (applied as intuitive coordinates of "where you are" at any moment in the work of a given project). This sheer localization (when properly made) is highly informative about the kind of work that is being done and enables effective orientation along the natural paths to be followed in the work process. But this was the original case, and other projects can apply this model even better by anticipation and preparation. Normally, and especially in projects above a certain size (i.e., because of the sheer volume of all project-related information to be processed and the sheer work volume required to insure smooth interactions between project team members), the project manager will most likely need some dedicated support to properly use this method. In my case, I also had support logistic support from a managerial team, which freed me from a lot of daily administrative tasks and thus allowed me to focus on building a coherent big picture and dedicate my efforts on the related core actions.

As I noticed after the Econoving project closed and I moved on to other projects, there is another context detail that matters for eco-city projects more broadly: the basic logic behind the mental model and the more detailed DIKAR_*process* actually converges with the 3D Smart Grid Architecture Model (SGAM) co-developed by the European Committee for Standardization (CEN), the European Committee for Electrotechnical Standardization (CENELEC), and the European Telecommunication Standards Institute (ETSI) as a response to the standardization mandate M/490 by the European Commission [122, 544]. This makes that the DIKAR_*process* is complementary to SGAM and speaks about SGAM's "interoperability dimension" (which includes 5 layers covering different domains, from business to physical components). Conversely, the SGAM can be used in the structuring of (eco-city) project content details as operated by the DIKAR_process model, especially with respect to the goals related to generating eco-innovation for the energy transition, or/and to related to addressing energy-security-climate (and broader sustainability challenges) mentioned previously in this Module (Sect. 3.4, Fig. 3.15). Further, this can help in dealing with interactions between the different domains that are made explicit in the smart grid concept model of the National Institute of Standards and Technology

(NIST) of the USA [281]. In such instances, the taxonomy shown in Fig. 3.17 (and the extended taxonomy discussed in Sect. 3.9) can serve as a conceptual and operational connector, and as a way to maximize work efficiency and operational versatility.

Finally (for this sub-section), the reflexivity condition mentioned above and in the previous Module (Sect. 2.2; in the generation of solutions for problems, and transition to sustainability in general) is embodied in the iterative "reflexivity cycle" involving vertices {2, 3, 4} in its core version, but also vertices {11; 12, 1} in extended versions, and ultimately the entire {1, ..., 12} cycle. From an RDI and managerial perspective, this reflexivity and the model itself helps find the best balance between exploration and exploitation, i.e., insure managerial ambidexterity and deploy adequate weights of each side in given contexts and development stages (for example, in TRLs, exploration must be dominant in early stages but having in mind the end goal, while exploitation must be the dominant concern in the later stages but having in mind the location of the intended solution within the entire space of possibilities that has been (and still is) being researched). Without the explicit stages articulated in the DIKAR_process model, many aspects (from the importance of different work types and how they related with each other) would be hard to grasp.

In context, the reflexivity cycles (from minimal core to full cycle) give a concrete operational meaning to the expression "thought-through solution (/proposition)" in general: iterations of these (minimal, intermediate of full) reflexivity cycles can be executed until no new content is being added. In other words, when the conclusion simply repeats the same way, one can say that a given proposition has been thought through.

Exploring What is Possible: Work Scenarios Using the DIKAR_process Model

The utility of the model in the generation of scenarios resides (a) in its structuring of the process of interaction between experts (to facilitate the generation of scenarios based on the Delphi-based method described in Module 2), but also (b) in its iterations during eco-city projects aiming at advancing on the path traced by the scenario set generated with the Delphi-based method (i.e., without iterations, the model is just a sharp and potent tool but left mostly unused).

Concretely, through successive iterations, the bi-dimensional matrix described above naturally assimilates new inputs and thus adds variations in the space of options. Collectively, these variations give a depth dimension to each Negotiation Room and to the model as a whole, making it a de facto three-dimensional lattice that is charting and recording all identifiable options at each step and that is multiplying the number of paths inside the known space of options.

This means that every path taken inside the model itself is a de facto scenario, which further means two things. First, the value of a given scenario only exists because of the possibility to compare it with other scenarios. In other words, it is (in general) slightly misleading to talk about "scenarios"; it is much better (and it is technically correct) to talk about "scenario sets". Taken out of (or isolated from) its scenario-set context, a given scenario has no meaning and no practical use.

Second, inside the DIKAR_process model we talk about work scenarios as a knowledge/expertise integration basis for generating scenarios as path options for a city: in other words, the DIKAR_process model insures that the scenarios produced through the Delphi-based decision-trees described in Module 2 are indeed grounded in expertise and embody the idea of "to the best of our (collective) knowledge". Therefore, if the work is not done thoroughly in the DIKAR_process model to insure a good coordination and development of a shared and realistic understanding, then the value of the scenario sets generated through the decision tree, although following a logical method, remains based on a superficial (insufficiently studied, integrated) understanding of the target city (or of the target problem within a city).

Also, the fact of being a lattice model makes the DIKAR_process model highly compatible (and can easily be used in combination) with other models known from RDI and management. Essentially, lattice models are well-known in research and represent a usual "shared language" between problems and topics in both natural and social sciences, for the simple reason that they support realistic approximations of different processes and interactions between agents (or groups of agents), and between them and their environment.

Importantly, the DIKAR-process model makes clear the big picture so that one can start from its direct interest (often dictated by the professional specialty) and then see how this relates with different steps along the process.

Observations About the Benefit of Having Abstract Models More Broadly

While seeking to remain intuitively readable for the everyday work of anyone regardless of their professional background, the DIKAR_process model described above results from the concrete necessity (lived experience in such projects) to distinguish simultaneously between the categories in the DIKAR framework, i.e., {data; information; knowledge; action; result} and the principal stages along the process of problem resolution, so as to insure managerial coherence between work types (and value) in projects and situations that looked overwhelming. And, as the notion of agreement on terms (and as discussed in Sect. 3.1) stroke as a first priority for sound project work and management, the notion of "Negotiation Room" emerged naturally as the descriptor of what needs to happen in each stage identified in the workflow described by the DIKAR_process model.

Importantly, "negotiation" is here understood to comprise "all interactions leading to the establishment of common denominators". It is not about usual "bargaining" but about an effort made together to find the best and most substantial basis of common denominators, and thus it can involve anything that helps that purpose. The participants are meant to put on the table all the notions, definitions, and models that exist in different fields and disciplines and see whether they can work together, and when not, why and how to make them work together—which may take some adjustments or new developments in those models or/and the generation of new models/methods to bridge the divides between the work constraints of the partners. As a matter of fact, the DIKAR_process model itself is an example of bridging models: it resulted from an effort to cover unavoidable grand differences in a unified understanding of operations in a project, and it did so by mapping (the steps along) the work process as described above. In so doing, the model is also an example of Knowledge Map (as defined in Sect. 3.3). Thus, the notion of "Negotiation Room" (taken individually or collectively, i.e., the model as a whole) bears similarities with the notion of "Workspace" (and "Concepts" as a fundamental unit of thought) in analogy-based information processing [320, 480].

Within the frame of Negotiation Rooms (NRs), one includes also some conventions that the actors may agree to start using, and which can indeed involve some bargaining over specifics (although the usually sub-optimal results of bargaining, as evoked in Sect. 3.1, should be avoided), but—because the common purpose is to reach a shared understanding for operational purposes of common interest, and because the semantic range of the word "negotiation" includes but is not limited to "bargaining"—it cannot be reduced to it. For this reason, sometimes the term "coordination room" might be preferred by some, but even this is merely a highlight of a purpose and is not necessarily a better term overall; for example, one can easily imagine some forms of coordination that are not based on a common understanding—and which would not be a valid content for "coordination rooms". Alternatively, one might want to use a third term, e.g., "integration"; this can be adequate for some purposes, but it would obscure the fact that the most important aspect underlying everything is in fact the intense effort that is needed, i.e., that that clarifying interaction is a process not a given. The literature provides useful insights, e.g., when searched with keywords like Social Learning and Cultural Theory.

Also important to observe is that, because mathematically and operationally speaking, the DIKAR_process model is a directed graph, its components can be coded using the formalism from mathematics, logic, and computer science, e.g., Category Theory (so that every component of the model, and the model itself can be defined as a category), Set Theory (where the technical content of each NR can be defined as a set or set of sets) and Type Theory (where every component/term is being assigned a "type" which defines what it means and what operations may be performed on it). Further, the model can be applied through protocols and tools that use inter alia functional computation lambda calculus (i.e., bindings and substitutions of abstract terms derived from mathematical functions) and pi-calculus (i.e., process calculus: operating with entire process and sub-process paths understood as units), propositional calculus and functional programming.

This highlights the benefit of abstraction in such models. In dealing with the overwhelmingly complex projects addressing complex problems in complex contexts, abstraction of what is being done (the work itself) is necessary because abstraction allows efficient communication and understanding through existing methods, and facilitates connection with (and allowing support from) automation methods provided by information science and computer science. Since technology is part of our humanity and the modern way of life is embedded in science and reliant on a good and properly balanced humans-machines interaction, the future of our cities and society is in semi-automation (knowing full automation is rarely possible or/and desirable or/and useful in a complex system, because full automation creates rigidities, which in complex systems, inevitably yield sub-optimal or catastrophic results).

In this context, there is one key aspect of the DIKAR_process model (and benefit for project management): as per Type Theory, this model exists in a coordination framework which, as a Cartesian Product between paired sets of categories, is a typing environment in which every NR is an "assignment". Then, in lambda calculus, our model (a valid subset of relations between the non-empty pairs of the Cartesian product) is in fact an extension of what is known as "simply typed lambda calculus".

Obviously, from different disciplinary angles (within the spectrum of viewpoints from natural to social sciences) different authors may use different definitions or/and specifics for the word "negotiation". For example, the discussions about that can be different in management and in epistemology. But this model uses a most general conception that transcends domain boundaries and understands negotiation as any interaction leading to the development and agreement upon common denominators, the concrete purpose being to ensure that participants in an eco-city project "speak the same language", at least to a minimum level that allows useful co-work.

In that sense, considering a model of interaction as an iterative sequence of steps allows us to refer to the common properties of that model regardless of contextual particulars. And this is what this book is about, insuring (a) relevance for as many contexts as possible (virtually all; the goal of sustainability tells us that no context should be dismissed beforehand) and (b) an iterative character, so as to account for and assimilate change in every context; and (c) a common means for different cities and project teams to learn from each other.

In brief, the models themselves (not only their components) represent a basis of common denominators that can unite different project teams from anywhere in the world, which further means of course that these models can improve with time through application and interaction between those applying it, but this also means that a model that is being put on the table and proposed for this purpose must be striped of particular contextual details. This implies parsimony, which requires the people from those different teams to use abstraction as a natural way and platform of communication with other teams (see also the summary on the Scientific Method, provided in Sect. 3.7, and the applied knowledge-action model from Sect. 3.10). In other words, abstraction it is not only helpful for producing this book as a relevant tool for all teams—it is also helpful as means for eco-city projects to continue the dialogues with other eco-city projects and to cultivate the potential for attracting as many highly qualified experts as possible for serious, evidence-based collaboration.

Using an abstract model allows us to look at a procedure as an abstract algorithm and re-organize it (in general or as an adaptation to a given context) so as to improve its performance in terms of work efficiency (in a given time unit, e.g., how much has been done in a week or a month), work flexibility (which may allow for less efficiency on a project's short term, e.g., a week, in order to increase overall efficiency for medium and long term, e.g., a trimester, year or multi-annual), independently from particular implementations.

Thus, it is practical to abstract the idea of a knowledge integration procedure (used in the interaction between experts/project participants) as a model of data and information processing in a human social system. In effect, this can be understood as a form of "social computation" (including related terms like "human computation" and "crowd intelligence") (e.g., [130, 209, 357, 621]). In the context of cities, this is also directly related to the notion of "urban computing", i.e., "the process of acquisition, integration, and analysis of big and heterogeneous data generated by diverse sources in urban spaces, such as sensors, devices, vehicles, buildings, and humans, to tackle the major issues that cities face (e.g., air pollution, increased energy consumption, and traffic congestion)", and which "connects unobtrusive and ubiquitous sensing technologies, advanced data management and analytic models, and novel visualization methods to create win–win-win solutions that improve the urban environment, human life quality, and city operation systems" [765]. For further topical discussions and case studies, see, e.g., [69, 70, 141, 409].

Consequently, it is useful to regard every NR (an abstract space that can be implemented anywhere) as the central vertex (node) of a *Kahn Process Network (KPN)* (Fig. 3.18), which is a simple model of distributed computation in signal processing systems, an illustrative example of which is the notion of "social brain" as an emergent property of a network of actors that are relevant for and contribute to a topic [681], but also the notion of "decision tree model" (of computation in general) that is already used de facto in the Delphi-based method described in Module 2 (Sect. 2. 2) and (under the form of branching closed loops) in the DIKAR_process model described above.

The KPN model is also important because, given the huge amount of data that can be collected and managed with the help of computers for the benefit of the human actors involved in each NR, both humans and machines can be regarded as inputters to NRs and we need clear models for them taken both separately and together; different routines for combined man–machine interaction can be found in each project.

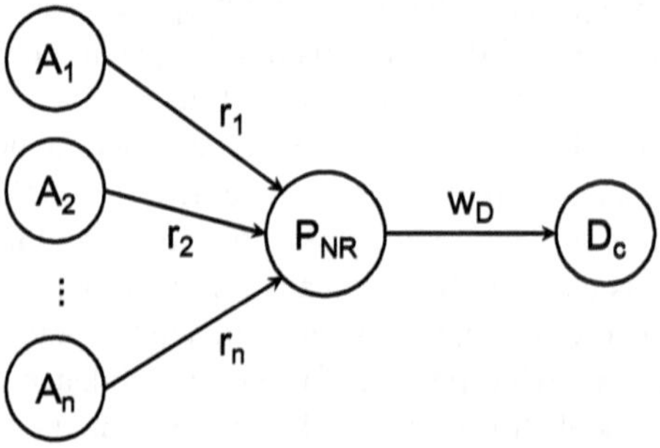

Fig. 3.18 An example of a *Kahn Process Network (KPN)* execution model of a Negotiation Room (NR). Work inside the NR is a process that reads from input processes (in our case, areas of expertise) and writes to an outcome vertex (in our case, the set of common denominators shared by the actors that participate in an eco-city project). In it, the process vertex P_{NR} reads from channels r1, r2 … rn coming from field processes A1, A2, …, An and writes to the channel W_D towards the outcome D_c. Observation 1: here, the word "network" refers to the graph shown in the model (not to the fact that source field areas can be represented themselves as part of a network, in the understanding of distributed knowledge; that is a different matter). Observation 2: this representation does not preclude the existence of intermediate levels of integration along the channels read by P_{NR}; it only indicates that integration must be sufficient for all actors to accept D_C as common denominators. *Source* original (based on current knowledge about KPN)

In a more elaborated description, the KPN model applies to NRs in two ways: (1) within-NR input processing, and (2) at the level of the whole model, between-NRs input processing because inputs into an NR from input NRs continues to flow in (in KPN, reading is blocking), which means that within-NR negotiation/integration does not ignore new inputs as they come (although some structuring delays can be agreed at the beginning, like integration is done once a month, or much more frequently, as needed, and incoming inputs are being collected constantly but their processing has its own time—unlike in machines systems where it can be done constantly as per established rules).

From the perspective of mathematics and operations research, the KPN model of input integration at the level of both NRs (as sub-units) and in the DIKAR_process model (as a whole) also relates naturally to the matter of concurrency, which is further related (at least in the main or start idea, and paying attention to avoid excessive interpretations of technical notions in different work contexts) to certain issues in innovation management and product development (CDI), such as (to use the consecrated terminology) the "concurrent design" versus "waterfall design" and their respective limits and complementarities (see Fig. 3.25 and the related discussions in the main text towards the end of this Section, a sub-section dedicated to Innovation).

At the abstract level in general (and in applications, notably in computer science and technology), beside equational reasoning and related approaches such as the lambda calculus and the pi-calculus mentioned above, this integration is also related to the key challenge of semantic coordination between definitions and models—a challenge that is known in computer sciences as "schema matching" or "ontology matching" or "ontology alignment" (for reviews, see, e.g., [248, 531, 536, 582]). Specifically, the DIKAR_process model in general (as a framework) and each NR in particular (as a focus point) provide context and locus for schema (and, its component ontology) matching. This model facilitates this work because it provides a common purpose: the need for a common definition as relevant for a very specific step in the work process to which every kind of work can be reduced.

A Wider Benefit—As Meta-Model for Transposing Other Useful Models from Their Scale of Origin to the Scale that Matters the Most: The Scale of the Project

One may observe that the general knowledge-action model introduced in Module 1 (Sect. 1.2) based on the systemic perspective on sustainability (and the related sustainability filters) represents a logical content for the Knowledge step (Vertex 4/4′) and related edges (links) in the DIKAR_process model. The same applies to all other models, methods, and insights discussed in this book.

In that sense, the DIKAR_process model represents also an operational model (or, perhaps a managerial "vehicle") for applying that basic model or principle for the concrete resolution of the unsustainability crisis in real projects. In other words, the DIKAR_process model is also a means for scaling/directing different models down (or up) to the scale of a concrete project: a meta-model of sorts.

This is important because, for a human community, there is no true resilience outside sustainability, and it is not acceptable to simply standby and witness how complexity overwhelms (or "paralyzes") society into "nothing significantly being done until crisis strikes". And even if some crisis will strike, it is unacceptable to waste it and fail to convert it into a "bouncing forward" moment. This means that good models need to be recognized wherever they are (in any domain and at any scale of study) and put to use in concrete projects.

And, this must happen in all contexts, so that projects learn from each other and cities learn from each other. As a holistic model, the DIKAR_process framework can coalesce and sustain meaningful and fruitful collaborations in projects carried out in multi-disciplinary clusters or networks (or both).

To take the example of water (strongly related to climate change, but not only to it), there is a systemic need for large programs that would combine (1) water research with science-society partnerships targeting "knowledge-to-action" [540] and (2) structural and non-structural measures towards effective reduction in flood risks, so as to achieve new multi-stakeholder governance arrangements, e.g., to coordinate flood-related measures with various aspects in land use, insurance, and urban revitalization, some inspiring examples coming, for example, from cities like Rotterdam (in the Rhine-Meuse-Scheldt delta) [194] and London (in the Thames estuary) [193].

A More Detailed Managerial Discussion on Uses and Benefits of the DIKAR_process Model

This section is dedicated to discussing this model in the context of current experience and literature, as a way to elicit further questions and ideas. So, this is where I delve a bit into some of the more frequent issues and important insights. If you just want to quickly pass to the next general model of this book Module, you can skip this sub-section for now and read it later when you need such details. My reasons to include this discussion as a sub-section here comes from experience. First, I know how easy is to someone pressed by time to make assumptions that may not hold at more detailed levels of analysis. Second, oftentimes the most useful insights for a problem we have in mind come from apparently small/supplementary observations. And third, such details tend to be the bread and butter of complex projects, and the reader might be able to corroborate my experience with hers/his to yield further key insights that can make a difference in practice. Ignoring that would be bad practice.

All in all, DIKAR_process is a synthetic model that is itself based on a structured state-of-the-art review, and provides the structure for future reviews of state of the art in the relevant fields of science, technology, and project management.

In the broad real-world situations describable in sustainability studies as Common Resource Pool (CRP) cases, e.g., the ecosystems on which cities and humanity as a whole rely, overexploitation leads to disequilibria, and ultimately disasters [149, 159, 166, 288]. This model however, can help partners (stakeholders) understand each other and move on in developing a common understanding and representation (mental model) of their shared resources, and of how it evolves and operates, so as to distill optimal rules and sustain reciprocity in a community of users [533]; see also [539].

In a specific project, like already discussed in the case of scenarios, this model is not a recipe but a tool for what has been called "complex project management as complex problem-solving" based on the understanding that knowledge management must inevitably be regarded as distributed across a plethora of expert networks [8] and/or for "creating resilient cities" [278] in spite of the many different purpose-determined ways that are possible to take when defining urban resilience [467] and knowing that as this (complex) issue is a component of the wider (complex issue of) urban sustainability (e.g., [11, 150, 210, 222, 466, 640]).

Based on this model, different project constituencies and expert may want to make in-depth trials, testing, adaptations, and related modeling. For the practical purpose of this book to make the logic of this model well understood, I would only mention that other relevant fields (types of expertise and corpuses of literature) are related to fundamental mathematics, notably Order Theory (e.g., [255, 484]), Category Theory (e.g., [561, 701]); and Graph Theory (e.g., [537, 668], see also Box 3.1, Sect. 3.2), Agent-Based Modeling and the related Multi-Agent Systems [76, 474]—all of which are part of the standard theoretical armamentarium in science, information science (as related to computerized modeling and automation) for crisis management, project management, and decision-aid methods.

In everyday practice, DIKAR_process is complementary to other models and approaches, e.g., problem structuring [159, 160, 567, 637], themselves related to notions and methods like "knowledge-structuring" (e.g., [392]) and "decision-structuring" (e.g., [308]), and various ways to structure data (e.g., [298, 699]), as well as comparisons of mental models (in order to check and further survey the shared understanding of a problem) [636]; topic/issue mapping [183]; (definitions and) measures of social learning by various actors [588, 715]; Delphi-based streamlining of priorities into scenarios [21, 150]; technological revolutions and new techno-economic paradigms [156, 556].

Importantly, the Negotiation Rooms (NRs; and the entire DIKAR_process model) can host a virtually unlimited range of views on a target work problem, and resolving disagreements in different contexts is a huge research theme per se and regards many and often intractable differences in terminology and methodologies. For this reason, one cannot realistically bridge semantic differences quickly. Instead, works in NRs need (1) to search for and select the best methods as suited for each concrete project, and (2) to aim at reaching not full theoretical clarifications and agreement on every detail, but hashing out a collective expert decision, e.g., as described with the Delphi method above (and the references already cited in here), so that different (even conflicting) views are collected and debated in rounds, until at least enough common understanding is reached to have an informed vote of the participating experts, and, ultimately, an aggregated expert decision. And, again, this decision must be accepted by all experts as a collective expert decision even if not everyone is "perfectly happy" with the details; in fact, this expert decision needs not (and should not) even be regarded as definitive (in fact, the exercise resulting in that collective expert decision can be re-iterated and the result of it refined and updated) but as a "to the best of our knowledge" decision in a given context. In other words, this kind of decision does not force a perfect or/and a common theory conciliating sectoral theories or/and preferences of different participants (individuals or institutions), but it helps with operations: to decide the best "what to do next" with regard to a target problem or issue.

Of course, this method can be used in complementarity with, or assimilate in its work, other approaches, e.g., measures of social learning whereby the world views of project participants (as per Cultural Theory, i.e., hierarchist, individualist, and egalitarian) converge or diverge during the course of a project/period of co-work and these changes can be measured through specific sociological methods [715].

A general and key observation is that this model (like other models representing processes in the real world) is based on analogy as a basis for understanding, because it comes down to models that allow us to predict, i.e., develop expectations (every concept being a "mental simulation" of reality). And, different models are further interconnected in higher order forms of understanding, from the simplest notion up to the Big Picture. The fundamental difficulty with such models is, therefore, a matter of representation, and most knowledge we have is in fact not in encyclopedias but in our minds and is based on our experiences in and about the world (books and research reports are just some passive records, beside the fact that they only represent a minority of all possible records).

Thus, the models of agreed knowledge are valuable for projects because they help generate a sufficient common perception of the target issue and its context. In so doing, but also because models like this one are representable with adjacency matrices (tables for storing, in coded forms, information about the relations between project-relevant categories), such models are compact-equivalent to the semantic networks that are being used in the current information science and technology, notably Internet, and specifically in relation to the Internet-of-People (IoP) and the Internet-of-Things (IoT). This similarity insures a broad basis for concrete operations in the world of today and of the future.

Then, depending also on how you decide to classify the depth variations at the level of its component vertices and edges, the 3D lattice can be generalized as multi-dimensional lattice ($nD, n \geq 4$), and thus linked with other mathematical models that you may want to use inside your project: a quick overview of reasons is provided Table 3.6, and a slightly more extended review is provided in the following paragraphs.

The main advantage (from which all the others derive) resides in the fact that it is a practice-distilled, operational (i.e., mathematical) formalization core of the interactions that take place in a project as regarded through the epistemic double-lenses of knowledge management and problem-solving process: it is as idealized as possible (i.e., stripped of the particulars of the original project and project management details) but at the same sufficiently sophisticated to capture all possible situations that may arise not only in the original (eco-city, as it happens) project but in virtually all possible projects.

From a similarly general perspective of physics and complexity sciences on project management, I would only mention that, once iterations have started, the space of options generated by this lattice can be understood as a form of implicate/explicate order—a concept with a wide range of applications, from quantum physics to living systems and social systems [82, 251, 543, 693], i.e., the total (virtual) space of options can unfold to display practical details as called upon by the specific set of priorities as per project details (also usually recorded and communicated as spreadsheets and

Table 3.6 Summary of advantages provided by the coordination lattice represented by the DIKAR_*process* model

Key reasons and benefits for eco-city project management	Suggested keywords further reviews of topical literatures
General/start design for project planning and managing work in complex projects	Disciplined experimentation; pioneering projects; multi-disciplinary and integrated projects
It structures work from initial questions to concrete solutions	RDI; project management; project work flows; value-added of research (and related activities); problem solving-oriented research
Strategic data acquisition	Big Data and data mining versus purpose-driven data acquisition; urban data, citizen science
Forcing solutions for social dilemmas	Creative problem-solving; breaking the vicious circle of non-collaborative behavior; social trust; game theory and (urban) sustainability
Versatility/suitability for any context; fighting attempts by hostile actors to gum up the project	Generalist models; collaboration platforms; ambidexterity (exploration vs. exploitation)
Avoiding semantic confusions provide common definitions of "slippery" terms such as data, information and knowledge	Cognitive dissonance (in an individual/groups of people); knowledge organization; knowledge systems; learning; empirical evidence
Helping us respect the necessary conditions for the transition to sustainability	Sustainability filters; reflexivity; finding relevant synergies and tradeoffs

other technical tables that can populate each NR as agreed-upon content). Obviously, the space of options is not fixed but remains dynamic as it evolves with every single variation in the criteria, definitions, and categories used inside each NR and along the paths connecting the NRs. For example, there are different possible definitions of data (e.g., kinds of scientific data, or/and as defined in information sciences) or/and various definitions and categories of information (e.g., Shannon information or/and algorithmic information or/and as labeled by classes of relevance and operation).

In daily works in complex projects with large and heterogeneous teams, the model gives the managers (and to all project participants) the benefit of a constant visibility on the whole project, namely, what matters, what is the status of different work packages and of the project-as-a-whole at any moment in time.

Moreover, the model provides an organized way to record all information about how the project (or a work package) arrived to be where it is (as it makes a logical or/and ontological description of the whole process and of its parts). In this sense, the model provides a type of functional benefit that is known from the blockchain methods (i.e., multiple records of the same agreement, automatically distributed to and owned by multiple agents: those who participated to the agreements in the NRs, provided, of course, that old records are not deliberately or inadvertently discarded).

And, the model is a logical definition of the space of options for the next paths that may be pursued in the project: the next coherent steps and trajectories that are possible between the chosen work question and the detailed requirements of the desired end-result.

A related and important observation is that the Questions (Vertex 1 in the model) are about experiences in real life, so they tend to emerge in formulations that are not dictated by some disciplinary habits: they are just "user questions" about currently employed goods, services, or situations in general. In other words, they are holistic by nature. It is later in the process (described by the full cycle of the model, and its iterations) that different perspectives (from across the triple-helix and across different disciplines or general fields of knowledge and expertise) contribute and are eventually melted together through integration in each of the 12 vertices. Questions emerged from the discovery of unknowns during research are also useful because they are the first step in the effort to expend the frontiers of knowledge.

This highlights the breadth of scope of this model, as it covers both research (exploration) in its upper half and business (exploitation) in its lower half, which further highlights the importance of Knowledge, as the NRs on "available knowledge" and "new potential solutions" (linked by the double arrow of "innovation") are sitting at the intersection of the two main domains (as does the entire row K from Knowledge in the DIKAR series), and NR4/4′ (Available Knowledge) is the main hinge.

This further means that the DIKAR_process model can be used recursively not only by iterations that look at the project-as-a-whole, but also as a work tool inside each NR: when people meet and ask "What do we have at this particular step, what are the differences between us, and how can we bridge them—are there available solutions for that, and if not what would we need to generate them?", this model is equally fit to help them. This functional self-similarity is a logical consequence of the fact that people work by systematically using their available knowledge to address any problem they face, anytime. As a fun fact (especially if your background covers mathematics and complexity science), this means that this model has an intrinsic fractal character. As a practical aspect, this also casts interesting lights on notions like finding the optimal substructure of a problem, modularity, and fractal manufacturing [153, 349, 555] that is important in engineering, industry, and business.

This brings us to the tension that exists in real-world projects between the often counter-intuitive character of science and the intuition-driven management in human organizations, which usually takes the form of *heuristics* and *biases*, known in the field of Managerial and Organizational Cognition (MOC) as "one of the most important psychological phenomena that influence the judgement and decision-making of actors at all levels of seniority, throughout the organizational hierarchy" [318]:

- *heuristics*: "rules of thumb that serve as potential aids to decision-making by focusing decision makers' attention on particular aspects of information";
- *biases*: "systematic deviations from rational judgement and thinking that can, and sometimes do, result from the adoption of heuristics".

In sum, current knowledge tells us that individual decisions tend to be based either on rough heuristics (that may or may not be based on experience) or on some combination of heuristics and causal understandings (which may lead to superior outcomes), but it also tells us that developing shared understandings and rules at organizational level is always challenging, and especially so in fresh situations.

This problem is known to be greatly amplified with project size and complexity, also known as "megaprojects" [241, 242]. In the specific case of "urban megaprojects", addressing such difficulties competently makes the difference between coordination and contradiction, and, given all variables involved, the difference between crisis and success; and can transform inherent ambiguity into concrete opportunity to make real progress in the transition to sustainability [101, 441–443].

Thus, looking at the inevitable (but also creative) tension between the (objective) scientific evidence and the (subjective) intuitive decision provides a rather straightforward insight: in a given complex project, the closer a shared understanding is to truth, the easier it is to apply that shared understanding in productive work, and, conversely, the more the individual (or small specialized group) understandings are shared via exchanges with the other actors in the project, the closer-to-truth that shared understanding will be (and the more refined and therefore valuable that understanding becomes). In short, the NRs and the DIKAR_process model as-a-whole are well positioned to address effectively the difficulties associated with cognitive and behavioral biases, and, this is crucially important in highly complex endeavors like eco-city projects. While heuristics may put speed above the quality of an analysis, they are prone to cognitive biases, and they only work (if at all) in an environment with very few unknowns—which, by definition is not the case of pioneering projects in which a city needs to explore and find some viable paths in a complex world and re-invent itself in the process.

Nevertheless, at the level of institutional participants in an eco-city project, for example, private companies involved in projects that take the form of public–private partnerships (PPPs; for details, see Fig. 3.23 and that entire sub-section dedicated to innovation), the relation between heuristics efficiency and quality level of analysis depends on the datums. For example, success of companies that cover the whole DIKAR_process model because they have both a business side and an RDI department doing research, e.g., Google, Amazon, and Tesla, has been related to the decision styles of their founders and CEOs, speed and innovativeness having a mediating role: reliance on heuristics favors speed over innovativeness while high decision standards favor innovativeness over speed [389]. The question for such a project actor (and for the project as-a-whole) then becomes: how to find the best balance in a given context for a given goal, and, subjacently, how to insure nimble change purposefully in context, i.e., when the situation demands it. Again, the DIKAR_process model (and the entire rational behind it) is a logical answer.

Furthermore, this model does not imply that high-quality levels of analysis are required all the time: it only requires the minimally high level that insures coherence so that the project does not fall apart before having a chance to make real progress in generating real solutions to real problems. In pioneering and complex projects, routines are not there and must be developed, the intermediate goal in the management of such projects being to arrive at the point when appropriate in-project routines begin to take shape—as a realistic basis for the development of solutions for the target problem(s). The DIKAR_process model helps project managers do exactly that: it can be iterated until a useful routine emerges. (I make some other, related observations about that in Sect. 3.7, dedicated to solution readiness levels.).

The idea is that, of course, when facing new and complex projects, higher quality decisions are much more adequate than heuristics, but once a new and useful routine has been developed, there is no reason to keep "re-inventing the wheel".

Basically, heuristics (like more detailed established action protocols) represent automated decision-making, which means that they are the product of a routine developed through practice, and therefore their utility is dependent on how close the context of application resembles (functionally) their original context of development. This is important because prioritizing heuristics basically means that a lot of information that sits in front of the CEO's eyes is simply ignored, and this is exactly the source of very consequential biases. This essentially means that the CEO (or any decision-maker) that applies heuristics uncritically doesn't know what they are doing and is deliberate about this ignorance. Operating (only) with clichés and other blunt tools equals not expertise but inadequacy and potentially even incompetence.

At this point, it is appropriate to be more specific about cognitive mismatches: because different types of actors and different individual professionals from different fields give different meanings to different notions, if the process of development of common denominators fails or is deliberately undermined (e.g., by actors pursuing some "divide and conquer" philosophy), then social dilemmas like the Prisoner's dilemma (with its different versions known from the literature on Game Theory in mathematics and its applications in social sciences) will delay or even prevent concrete resilience building in cities (and in society in general). Granted, a few common denominators emerge naturally by daily interaction between actors but this is obviously insufficient when we talk about systemic issues that require substantial agreement and effective coordination between many categories of people. In these cases, social dilemmas represent a situation of societal lock-in (a default vicious circle).

DIKAR_process helps us structure efforts towards (in projects aiming at successfully) solving such dilemma, and thus co-develop a common understanding as a necessary (but insufficient) condition for the development of useful scenarios. More often than typically recognized, project partners presume that they mean the same thing when they use this or that term (especially key notions, precisely because they think that, and because it is so basic, they surely mean the same thing). In reality, different professional fields assign different meanings to seemingly identical words or expressions, and those field-specific meanings may only overlap a bit or not at all. By failing to check the meanings of the terms, the project partners place themselves in a semantic trap and are up for unpleasant surprises: they will discover the mismatch later on, usually when it is already too late and the inevitable result is a significant loss of project resources, or assets they intended to protect, or all of that.

In short, the common (understanding of) denominators in terms of principles, concepts, methods, models, and tools must be verified to be sufficiently developed in advance of the projected main work, i.e., as part of the preparation of options for response (and before crises strike) or, in the case of an intended series of eco-city projects (i.e., a program), even as a fully fledged preparatory project in the first part of the program.

To this aim, and as an overall conclusion of this sub-section, the DIKAR_process model is general enough to be applicable in all contexts and types of projects; with any kind of question; it is adaptable to any kind of change in the objectives, structure and context of already-started projects; and can be used with virtually any software and hardware of choice. This generalism and versatility makes it useful also for the goal of scaling-up collaboration in and across organizations; and allows optimization of planning activities and/or work processes at any moment/in distinct phases, e.g., preparation, implementation, or/and evaluation.

Finally, a fundamental challenge in projects aiming at creating meaningful change is the fact that the necessity of change is proclaimed but not respected because the business-as-usual pressures upon the professionals are relentless, the end-result being that the very lack of change prevents the making of the change (in a vicious circle of inertia protecting itself), and that is how tensions keep accumulating as already discussed in Module 1 (in Tables 1.1 and 1.5, and the main text of Sects. 1.2 and 1.3). The operational conclusion one can draw from this reality is that, in order to break the vicious circle of inertia and create change dynamics, change must be learned by doing it not by over-planning it. To that aim, the DIKAR_process framework model provides a method for beginning that process with a robust procedure for dealing with knowns and unknowns, i.e., without having to fall to the natural temptation of wanting to "answer all questions from the beginning", and, instead, helping a team to rely on the trust in itself and its capacity to methodically process whatever input happens to emerge towards a good result in the desired direction.

3.7 Ground-truthing a Valid Space of Options: Data Versus Information Versus Knowledge

Terminological Traps and Tentative Clarifications

Data, Information, and Knowledge Are, of Course, not the Same Thing

An important issue in eco-city projects, like in all complex projects, is that there is a great deal of confusion in the literature about what is and constitutes Knowledge, Information, or Data. These three terms are often and wrongly used as if they were interchangeable, but, in the main model from Sect. 3.6, one sees that they are not. Not only is their respective nature different although (and it is no secret that) their meanings can also have field-specific and context-specific differences. This is why I find it important to make some supplementary clarifications about these terms and to discuss some other notions that are related or depend on these three, most notably Innovation and Science itself.

The kind of sloppy use of terminology mentioned above might be caused by daily pressures to stick to existing field-specific work protocols and/or appear to some of us to be just some harmless shortcuts that we should all be allowed to make because we are always in a hurry to meet some deadlines, and because it is (as this type of excuse goes) more important to be efficient *ceteris paribus*, i.e., all else being equal.

However, a usual problem in projects aiming at generating change is precisely the fact that all else is not equal, and the right distinctions must be made so as to stop perpetuating the confusion and start building healthier communication routines.

To break this vicious circle of inertial poor cognition and to infuse a productive learning dynamic, the DIKAR_process model introduced in the previous section is a good platform for operational clarifications. This starts with a first set of simple definitions of respectively Data, Information, and Knowledge (and of the other relevant notions) that make a clear distinction between these notions. Then the user of the model can further develop and adapt the definitions as needed in her/his specific context (with the condition, of course, that those new definitions are negotiated and agreed by all project participants).

Because a project team must start somewhere, the DIKAR_process model (with the definitions given here) can be used as a default/minimal platform for project design, management, and evaluation.

The Epistemic Background of Knowledge: A Summary

As per current philosophy, Knowledge is typically defined as "justified true belief" (JTB), to which certain caveats apply and are still under some debate. In common language, this means that "knowledge tracks the truth", which idea is said to ground an "epistemological practice of reason justification" [205]. Added to the JTB central idea is a "sensitivity condition" (or "Justification Condition") that is meant to cover some special situations that have been identified as being left uncovered by the JTB definition (or wrongly identified by it, like "lucky guesses"), thus resulting in a JTB+ [327].

In practice, knowledge is (1) basically the set of our best representations of reality, taking the form of a consolidated set of models (along with the reasons behind those models) and (2) it is the ground of prediction-based action (i.e., all the planning and the choices, e.g., for crises response and management) [153, 156]. Indeed, these models become better and more detailed over time, so much so that our modern life is tied to a "knowledge-based economy and society".

However, although prediction is central, it is not enough (i.e., it is a necessary but insufficient condition): knowledge must be operational. While normally not an issue at the small scales (as this usually implies only a limited expertise), it becomes a serious challenge in large systems and with multi- and trans-disciplinary issues like sustainability. In our efforts to address this challenge, different fields of study exist. Thus, the field of Knowledge Organization (KO) focuses on indexing, classifying, and representing all the knowledge that is captured in documents for the purpose of "information retrieval" (IR), electronic or physical browsing, and all the related tasks and processes executed by humans or computers. Obviously, IR is predominantly done by computers through different programs for "match techniques, popularity measures or personalization principles". However, the KO per se "attempts to reflect knowledge as reflected by science and contemporary scholarship" [317].

A related notion is that of Knowledge Systems (KS), which refers to the ways in which knowledge operates in our forms of social organization, starting with culture and institutions (for detailed discussions, see, e.g., [118, 119]). Given all that has been said in this book and in the references cited herein, the current challenges faced by our knowledge systems are formidable but surmountable. To give a general idea, here I would only mention how a very large team of authors that reviewed the issue in a recent paper [231] sums it up: "Formalised knowledge systems, including universities and research institutes, are important for contemporary societies. They are, however, also arguably failing humanity when their impact is measured against the level of progress being made in stimulating the societal changes needed to address challenges like climate change. (...) Future systems will need to be much more collaborative, open, diverse, egalitarian, and able to work with values and systemic issues. They will also need to go beyond producing knowledge about our world to generating wisdom about how to act within it. To get to envisioned systems we will need to rapidly scale methodological innovations, connect innovators, and creatively accelerate learning about working with intractable challenges. We will also need to create new funding schemes, a "global knowledge commons", and challenge deeply held assumptions."

In light of the definition of knowledge given above, both knowledge organization and knowledge systems seek to help the process of constitution and justification of knowledge. What matters in practice is the explanatory power of current models with respect to the available evidence, and the central approach to obtain it is the so-called "scientific method" or, more broadly, "scientific reasoning".

The Scientific Method: A Summary

These being said about knowledge itself, and given the existential stakes we have in the challenge of sustainability, the obvious question is: *How to maximize our understanding of the problem and of the real options in spite of all constraints?*

Gaining clear sight on problems and potential solutions is precisely where science comes to the fore as potential model of high-value social interaction. Concretely, science is characterized by what is called "the scientific method": a set of principles (necessary conditions) to be met in that search for truth, the main idea being that whatever is being proposed to be true must (a) be coherent with already-known facts and (b) pass the test of reality.

In the applied work of scientific modelling, this is often known under the name Maximum Entropy Principle (MEP), and thus means that a scientific model stay "as simple as possible but not simpler" ([124, 173, 341, 342, 648]; see also [226], and the knowledge-action model in Sect. 3.10). In other words, a scientific model must aim at finding the "sweet spot" between complexity and usefulness—an idea that is often referred to as the "Medawar Zone" of optimal benefit lying at an intermediate level of difficulty, or as an epistemic case of the "Goldilocks" principle of "not too cold and not too hot" as mentioned in Module 1, Sect. 1.3.

Usually the scientific method is applied in a sequence of actions, which in summary is: empirical observation, hypothesizing (i.e., building conceptual models), experimentation, analysis of results, and confirmation-or-rejection.

Without going into the debates on this topic (to start explorations on the topic, one could check [613]), I would only mention that the central requirement of the scientific method is a reliance on evidence through observation and experimentation: *a commitment to active and systematic sense-making.*

In the popular culture, this main idea has been formulated by Albert Einstein and often quoted in expert contexts (from management to psychology and in science in general) "It's not that I'm so smart, it's just that I stay with problems longer" (e.g., [182, 405, 663, 756, 761]) and which refers to the need of correctly understanding the process of problem-solving and to "failure as a catalyst for developing creative persistence" [658].

Now, lesser well-known is the sobering fact that this "persistence with the problem" (i.e., constant study and investigation) is highly demanding and unrewarding on the short term (unattractive for non-scientists, and hard for all scientists themselves: the more one persists the harder it gets, at least until a breakthrough). However, this is and is bound to remain the most efficient human activity on the long term because it allows us, humans, to actually know what we are doing and thus to achieve true civilizational progress, starting with the transition to sustainability. In that sense, the work of true scientists tends to be under-appreciated when it happens and, as a consequence, true scientists are often obliged by the social circumstances to either abandon science and start doing something else or to sacrifice themselves for the good of humanity by continuing to do science.

For our interest in eco-cities, the "active and systematic sense-making" means that the scientific method relies on the identification of patterns based on available empirical evidence (i.e., inductive generalization). Then, this objective conceptualization is followed by operationalization. Basically, it is a matter of discovering "what is" by imagining possibilities ("what could be"; i.e., hypotheses) and seeing "what works" (and what does not work) in a given set of conditions (i.e., limits).

The fact that actually doing this is much harder than how it sounds, is reflected by the existence of many and complex debates in the history of epistemology. Discussing those in great detail is beyond the scope of this book, but it is important to have at least a main idea about it. By their nature, as eco-city projects run regularly into sharp differences of perspectives between project participants (a typical example being the differences between natural and social scientists; another is the divide between different types of institutions). These hard conversations are necessary (as they are very informative about the kind of underlying divisions that need to be addressed in a project) but can also very easily slide into sterile ideological debates and traps that open inside the usual cultural mismatches, personal differences between worldviews and individual or group preferences that have nothing to do with objective professional operations. So, I encourage the interested reader to explore the literature on the matter but I would also urge the reader to pay attention in priority to those authors that are actually experienced scientists themselves. Based on experience in multi-disciplinary conversations, I regard real experience in science as a minimal condition of true competence in epistemology. Otherwise, it would be like, say, doing the philosophy of gardening without being a gardener (or, for that matter, like giving directives about innovation for lowering carbon dioxide emissions from current technologies without having ever practiced RDI; or offering lists of "should and must" about eco-cities without having any experience in real eco-city and eco-innovation projects). Of course, this brings us to another point: different types of experience are required, and a team must cover all of them as much as possible and then develop more of it by concrete practice in real-projects focused on hard, long-term objectives.

For the sake of knowing where we are in time, but also to highlight the deep, real-life origin of scientific inquiry, I would remark that the roots of the scientific method can be traced to Mesopotamia and Ancient Egypt, but the scientific inquiry per se started with the "natural philosophy" of the Ancient Greece, itself started with the pre-Socratic philosopher Thales of Miletus (640–546 BCE). Of all the personalities in the history of science, he was the first known to have postulated non-supernatural explanations for what we observe in the real world, i.e., without recourse to gods.

For the same orientational purpose, let us remind ourselves that mathematics is a companion of science and is employed in science but mathematics as such in not science. This is also trivial, and care must be taken in practice to avoid unnecessary confusions. Along with that, I would remind here that people in the Ancient Egypt were already using a decimal system. This system was introduced to Europe (and from there into modern science later on) only in the year 1202 CE, by Leonardo of Pisa (currently better known as Fibonacci) who noticed during a commercial trip to North Africa that people were using an arithmetic system which, due to the positional notation of those Hindu-Arabic numerals (which he called "*modus indorum*"), was much more efficient than the Roman numerals used in Europe at the time. Then, separately from the Ancient World, the Mayans in what is today the southern Mexico had developed their own system of mathematical notations, which also included an explicit symbol for zero.

Equally interesting, the roots of the scientific method itself are older than the Ancient Greece and trace back to the Ancient Egyptians. Beside the advanced mathematics that was used in the daily practice of people in their life environment, like land use and ownership in the flood plain of the Nile river, inquiry methods were also related to practical aspects, notably health, and disease. Most notably, we know from medical papyri from around 1600 BC (the Ebers papyrus and the Edwin Smith papyrus) that, in order to treat diseases (which they attributed to the body being invaded by "bad spirits"), the ancient Egyptians used a specific sequence of steps (*examination, diagnosis, treatment, and prognosis*) which anticipates the modern scientific method, and, as it has been argued (see; e.g., [425]; see also [619]), it contributed to its development because it was transmitted through learned individuals in the succession of civilizations—notably via the Ancient Greece, including the Hellenistic Egypt that has begun with Ptolemy I. But one can easily argue that this kind of continuity is also one of common sense: regardless of cultural peculiarities across space and time, any conscious attempt to address a problem must go through a logic series of basic steps (like the one mentioned above) that would be "rediscovered" any time it was needed.

Aside from the epistemic clarifications that matter in the practice of projects, there is a broader point of this short historical incursion: the scientific inquiry reflects human nature, and therefore geographical details only reflect some local circumstances at a given point in space and time.

For the purpose of this book, it is also practical to note that, technically, the review of primary research for the purpose of higher-order knowledge integration is as part of the research itself. Namely, the "review research" can be defined as "a class of research inquiries that employ scientific methods to analyze and synthesize prior research to develop new knowledge for academia, practice and policy-making" [393].

At this point, one remarks that the confusion mentioned at the beginning of this section between data, information, and knowledge is unfortunately still widespread, even among professionals; and this, per se, is one symptom of the current challenge represented by the combination between the fragmentation of science (lack of integration between fields) and the low awareness of the problem (particularly among specialists; by definition and usually also by professional necessity, they have a very limited understanding of the big picture). In a sense, the famous observation made by physicist Werner Heisenberg (1901–1976) that "*only a few know how much one must know to know how little one knows*" (which refers to the immense volume of knowledge that remains to be established by science; also known popularly as "the more you know, the more you realize how little you know") also applies to and anticipates the challenge of knowledge integration of today: the more you try to put things together (in one coherent picture for the practical purposes of problem-solving), the more you realize how much apart things are (and how much confusion this state of affairs is causing). So, discussions about the scientific inquiry are yet another type of instance in which knowledge integration comes to the fore as a key priority (and another relevant situation for the DIKAR_process model introduced in Sect. 3.6).

To cut a long story short, the scientific method tracks its origins in philosophical argumentation but relies on empirical observation thus achieving a great success in practice (because it works) in spite of what is known in philosophy as "the problem of induction"—which consists in the fact that a theory cannot be proven definitively right because, by definition, there is only a limited number of observation that we can make (and infer general rules from), and thus there is always the possibility that some future observation(s) will disprove the theory. In practice, this issue is resolved using the notion of probabilities: if a probability (/expectation) of something happening is deemed sufficiently low, it can be ignored (at least until new observations demands renewed attention).

Because of this apparent contradiction between induction in science versus induction in philosophy, induction has been cheerfully described in popular conversations as "the glory of science and the scandal of philosophy" [291]. But this kind of contradiction is actually productive from a scientific perspective. This is because, once an empirical pattern (a certain "what") has been identified, then the philosophical unease mentioned above becomes the motivation for further tests and clarification of the "why" (i.e., the current theory itself, i.e., the basic knowledge to date), and thus serves as a push for advancing the "how" (i.e., the current methods and technologies). Both the "why" and the "how" then provide feedbacks that further improve the "what" through a perpetual cycle of advancement in science and society.

Moreover, the problem of induction is important in the resolution of complex problems, and the solution to it is to aim for reasonable progress in the desired direction one step at a time. From a general perspective, this means that modern science as a disciplined inquiry is a necessary condition for staying on the path to success and maximizing avoidance of irremediable mistakes (and this is discussed in the next sub-section). From an applied project perspective, this also means that we need to continue to gather evidence as much as humanly possible and to base our decisions on current evidence, i.e., ultimately "to the best of our knowledge"—which further implies that we need to stay open to novelty and change because change is not only inevitable, it is a source of unavoidable dualities: risks and opportunities, problems and progresses.

The next two sub-sections address the matter of scientific inquiry from a social (and sociological) perspective before going into some technical details about Data and Innovation. Then, the next section is dedicated to discussing this approach from a methodological perspective, under the umbrella terms of disciplined experimentation and stepwise progress.

A Functional Social (and Sociological) Perspective on Scientific Inquiry

The previous sub-section briefly discussed the imperative of reliance on scientific evidence, terminology, and methodology, Science itself being a necessary condition for advancing on a real path to success and avoiding irremediable mistakes as much as possible.

In order to sum up (for all types of project contributors, not only scientists) how scientific evidence comes to see the practical light of the day (so to speak), this sub-section also discusses the social aspect of science and makes a resumé of how scientific reports (the primary research papers one can find in the scientific literature) arrive to be published through the normal filter of peer-review. These details only abbreviate the much more complex professional discussions and challenges, but references from the literature are used to suggest to the reader some entries into those wider conversations taking place out there. The overall goal here is to help all readers gain familiarity with the matter of scientific production and to do so in a way that highlights the deep relations between science and society.

The first point to make is that, once the core notion of scientific research is clear, the next step is to understand that different models of scientific inquiry co-exist and that finding the best one (or even at least a possible model) in a given social context depends on the field of inquiry and the context details.

Most importantly, the problem of induction mentioned in the previous sub-section makes that the way science can function in human systems has some intrinsic rigors which, can translate into (and often determine) certain methodological choices and operational constraints. Perhaps the best known of these is embodied by the critical rationalism of Karl Popper, according to which science necessarily relies on a procedure that consists in conjecturing hypotheses (based on available knowledge at a given moment in time), then deducing their potential consequences, and finally trying to empirically falsify those hypotheses. However, this is only the start of the discussion about real-world pursuit of scientific inquiry, and the practice of scientific research can be even "messier" from a philosophical perspective but operationally respectful of scientific rigor and useful for problem-solving.

Another important and well-known conceptualization comes from a sociological perspective on science: Thomas S. Kuhn famously studied the history of "scientific revolutions" in different fields and theorized that, as some abstract representations (general models) attain a wide acceptance in a group of scientists through psychosocial and social interactions, those dominant commonly-held views tend to become normalized: they emerge as dominant paradigms which unify entire generations of scientists (and societies), come to define different eras in the history of science (and in the development of the human society), and only change with them [391].

This perspective on "scientific revolutions" is not without critique, as, again, in reality things are even more complex than that. Interestingly however, if we look at this social theory of science from a current systemic perspective (which includes all natural and social sciences (for a discussion on 'systemic thinking', see, e.g., [6, 452]) and social network analysis) we realize that the social network dynamics (already known from the study of natural and human complex dynamic systems, from the industrial revolutions of the eighteenth century to present times) can be identified in the (social) history of science as well—for good or bad. Concretely, different issues that exist at present times at the intersection between science and some commercial interests or/and public policies can be objectively recognized as examples of network effects in social systems.

Specifically, we can see some undesirable effects in that the core of science (i.e., the advancement of knowledge through the pursuit of truth) is being trampled upon by short-term interests. As [205] summarizes it, "absence of the epistemic justification of reason defaults to the use of an unjustified form of reason that runs the play of an unrecognized and unchecked dialectic between epistemology and science". In this sense, different examples are being discussed throughout the book, where different pseudo-scientific notions are being used in public conversations and even in official policy-making, or are being given pseudo-scientific meanings—one example being even the term "resilience" (discussed in Module 1), two others examples being "circular economy" and "Industry. 4.0" (see also the last section of the book, notable around Fig. 3.40). Seemingly, this happens largely because of rhetorical reasons and other non-scientific agendas that ultimately result in some expressions becoming "buzzwords" [310] regardless of the evidence behind them. In other words, network dynamics in human systems in general and with regard to science-in-society itself is the same network mechanics known from Complexity Science (and studied in different areas of specialism, e.g., the field of statistical mechanics in physics) and can result in both desirable and undesirable effects as per how it affects us. From a managerial perspective, such harmful types of effects give us a reason to study thoroughly how different propositions that are made in the Agora would actually work in practice.

For further discussions on scientific inquiry and its challenges, you might want to check the further literature (e.g., [216, 512, 754]). For a dedicated recent general review and introduction, see, e.g., [644].

All in all, science is not a given "fallen out of the sky" for our benefit. The reliance of eco-city projects on real knowledge starts from the basic recognition of the natural limits of human cognition and is fundamentally based on the collective representations of local-and-global "communities that keep updating what they know". We do need to study and understand how scientific inquiry works in our human systems, and we need to help science help us. In the transition to sustainability and for the future of humanity in general, science is basically "the light" that makes it possible for us to see where we are and where we are heading: the process and the outcomes of our effort to obtain a practical representation of reality, and "our best set of reasoned predictive models grounded within communities of actors with complementary expertise" [153, 155, 156].

How Scientific Reports Get to Be Published and How Can This Be Used as a Model for the Wider Creation of Value in Society

Beside the fundamental Human Rights that constitute our core and guiding values, Modern Science provides the hard, objective start basis (a necessary but insufficient condition) for success. For starters, allow me to remind here that a necessary condition is something without which an intended outcome cannot be obtained (whatever you do, you need to have that onboard/solved), and attaining success may require (and usually it does) more than one necessary condition (so, in the end, all necessary conditions must be met, or there is no success possible). Only in relatively simple situations (i.e., very rarely) can we identify a "necessary and sufficient" condition (i.e., meeting that one requirement is enough). In the real world, most (and virtually all) of the times when we discern a necessary condition, it is a "necessary but insufficient condition". Moreover, it is very difficult to gain sufficient clarity about the entire set of necessary conditions that are required for attaining an objective. Usually, given all constraints, the best we can hope is to gain some partial visibility about some of them, so that (when the inevitable deadline comes), we can decide the best way forward based on that incomplete understanding ("partial visibility") we managed to have until then. This is important because the more complex a problem/situation is and the more incomplete is our understanding remains, the more likely is that our decision will result in failure.

Of course, through science, scientific advancements percolate in society as a whole, usually with a delay—which, depending on the topic and the technical details, can go from immediate impact (like it was, to give an easily-recognizable example, with some research on SARS-CoV-2 and COVID-19, say the discovery of a beneficial effect of a certain substance or clinical action in the treatment of the disease) to decades or even centuries.

Some discoveries and insights are so new and so fundamental that it takes scientists and society a very long times to digest and figure out how to use it (and usually this is not just like that but takes a huge amount of further work). For example, the basics of quantum physics have been established at the beginning of the twentieth century but applications like quantum computers are, as per 2023, still in their very early stages of development.

To give an idea, it takes on average about 10–20 years from the moment researchers make a clear-cut scientific discovery until the moment that discovery begins to have a significant impact in society. But the important thing is that it does. We are now living in a knowledge-based society and economy which is the result of the Scientific Revolution that has begun in the middle of the seventeenth century and generated the industrialization cycles of the eighteenth, nineteenth, and twentieth centuries.

Of course, like every other domain of human activity, science also has some major challenges that it needs to overcome, some of which have already been mentioned in this book module, like the problem of fragmentation between research fields. Some other challenges faced by science do not require dedicated discussion in this book but are worth mentioning, like the high difficulty of carrying out effective peer-review in inter- and transdisciplinary research, and the problematic business models of scientific publications which put scientists in rather difficult situations, along with the on-going new developments that may open some new ways.

These being said: scientific evidence established through the scientific method is generated through research projects and the concrete results of those projects are to be found in the professional literature produced with the scientific method, i.e., papers in peer-reviewed journals and books. The production of each scientific report depends on a trio of necessary conditions:

1. it starts from a critique of (and relies on) the knowledge until them; which further means that the authors of a scientific report must justify each statement they make in the content with scientific references, i.e., precisely logical technical citations of other previously-available scientific productions, according to standardized methods,
2. each of it brings a very new advancement obtained by research, i.e., application of the scientific method,
3. and each of it undergoes a process of peer-review during which other scientists read the manuscript and make a professionally objective (i.e., justified with evidence) critique of its content, a process that happens before being accepted (or rejected) for publication through the decision of an editor which is her/himself always an experienced scientist and bases his/her decision on the overall review reports of the reviewers. If a manuscript is rejected, its author(s) will go back to more research so as to renew its efforts (and the cycle resumes).

In concrete operations, what distinguishes scientific reports from non-scientific works is an iterative cycle that includes research, publication of research results, percolation of the new scientific contribution in the rest of science, and the provision of new resources for further research (Fig. 3.19). But, there is yet another important aspect that is inherent to science and which may translate in wider societal benefits. Namely, science is <u>not only</u> a source of knowledge <u>for</u> society: in spite of its internal challenges, the *modus operandi* used in science can provide a template for a *modus operandi* <u>in</u> science-society projects and in society in general. Put differently, because the challenge represented by the transition to a sustainable way of development is so formidable and our best asset at hand is science, *mutatis mutandis*, the best (and probably only) way through which eco-city and related societal projects can succeed

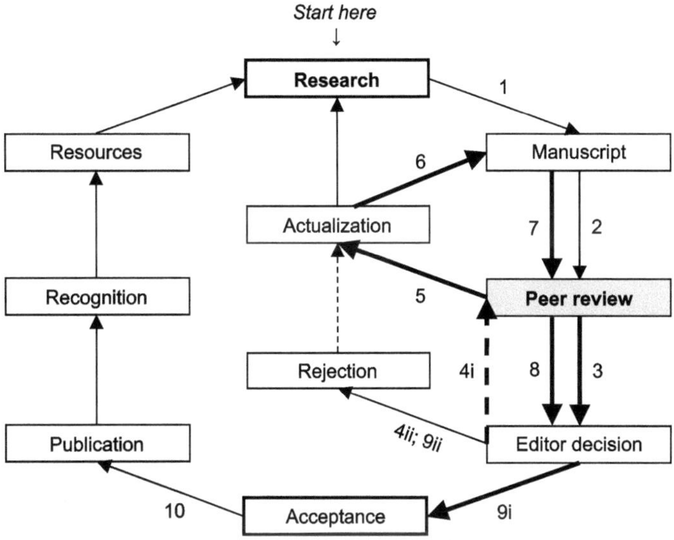

Fig. 3.19 A science production cycle (SPC) model, i.e., how scientific papers are being produced as a result of science's pursuit of truth. This visual is simplified at maximum. The terms in the box labels are further explained in Table 3.7. *Source* redrawn and modified after [274]

is by insuring that these very projects are grounded in truth. In other words, the requirement for us all, not just for scientists, is to make sure we know what we are doing even if we cannot know the future.

Concretely, Fig. 3.20 shows a generalization of the scientific knowledge-generation cycle procedure employed in science in the form of a societal value-generation cycle grounded in collaborative truth-pursuing work. Besides being inspired by how science itself works, this general model is also grounded in and informed by current practices in other science-society interactions, notably RDI, entrepreneurship and innovation, knowledge-based business, and evidence-based public policy. In this sense, this general model also (1) connects back with the DIKAR_process model as a model of interaction between science and society as intermediated by knowledge (as discussed previously in this Module), and (2) anticipates some of the discussions made in the next sub-sections and rest of the book.

A more detailed, project management-oriented model that incorporates the logic of this basic cycle of value-creation at the interface between science and society will be discussed in this Module of the book, Sect. 3.4.

Finally, to illustrate the amplitude and depth of the relation between science and society, I would cite the issue of misinformation as a case in point, and I cite here a relevant fragment from a recent paper [746]: "*Public engagement and understanding of science should be a priority for all scientists. This is not a matter of just teaching more astronomy or biology. Rather, it involves nurturing innate curiosity and teaching people to understand how science works, how to consider evidence when making conclusions, and how popular media distorts these conclusions.*" For further study,

Table 3.7 Components of the VGS model, as based on peer-review and user-review

	Science production cycle steps	General value production cycle, e.g., for an eco-city
1	Research based on the scientific method results in a research report, i.e., a "manuscript" (these days, it is usually an electronic document). The manuscript is submitted (sent) by the author to an academic journal. If there are more co-authors, one of them (usually the lead or/and coordinating author) must act as the corresponding author—handling all communications with the journal	Objective study and integration of what is known so as to gain clarity on the real space of options (what can/cannot happen) with respect to a question or a known problem
2	The journal evaluates the manuscript in two steps. First an editor checks that the minimal requirements are being respected (notably, the manuscript must be structured as a scientific report, and must fit the journal's topical focus (profile). If this is "ok", then the editor invites 2–3 peer-reviewers (which are other scientists who are known as experts in the topic because they have published in the past some scientific reports that are relevant for the topic). Normally peer-review is a voluntary, collegial work i.e., not paid; and, as a rule, they must remain unknown to the manuscript authors; one of the reasons why the manuscript handling is done by an editor is precisely because authors and reviewers must not have to communicate directly. The policy of the journal can be either "single-blind" review (when the reviewers can see the name of the authors but not vice-versa) or "double-blind" review (when the peer-reviewers do not see the name of the authors either—or any information that might "betray" their names). Those invitee who accept to conduct the peer-review read the manuscript and send to the editor a detailed evaluation report about the manuscript's merits/lacks thereof (with feedback for the manuscript authors and for the editor, and, based on those details, an overall recommendation: "accept as is", "conditional publication", i.e., accept for publication if some aspects are being further addressed (answers, errors, further work, etc.), or "reject"	Professional organizations (i.e., professional peers) and users in general (knowing that experts are also users—expert users) look at and evaluate the quality of an input (data, information, and knowledge) that is being communicated as potentially valuable with respect to question or problem of interest. Other aspects must be observed, notably the avoidance of any conflict of interest (and other moral aspects), and in general, the insurance of a good practice, so that the effort to resolve an issue does not create other issues of equal or higher gravity or that are otherwise unacceptable

(continued)

Table 3.7 (continued)

	Science production cycle steps	General value production cycle, e.g., for an eco-city
3	Based on the reports and recommendations provided by the peer-reviewers (as a rule, an editor uses at least two reviewers; sometimes three or more, especially in complex/multi-disciplinary papers), the editor makes the decision ("accept as is", which is very rare and is better considered as "never happens") or "conditional acceptance" (i.e., provided that each of the list of commends and demands for clarification or improvements provided by the peer-reviewers is being addressed satisfactorily by the manuscript authors) or "reject". As a rule, the editor respects the balance of recommendations provided by the reviewers. For example, if two out of three peer-reviewers recommend "reject", the editor will reject the paper; if two out of three peer-reviewers recommend "conditional acceptance", the editor will decide "conditional acceptance". If an equal number of reviewers make contradictory recommendations (e.g., one of them recommends reject and another "conditional acceptance"), the editor will seek the opinion of a third reviewer (or more) to see which way the balance of argumentation tips	Professional organizations make their own collective opinion about a new input proposed by some of the members/participants in that organization. Indirectly (through the intermediation made by the professional organizations) or/and directly, all users (citizens included) can and would make their opinion about propositions made in the society
4	The editor communicates her/his decision to the corresponding author of the manuscript (conditional acceptance, shown in Fig. 3.19 as case 4i, or rejection, shown as 4ii), and, usually, also to the reviewers themselves. Typically, all this information is handled in an electronic format and is available to the authors and the reviewers as per their role-determined type of access, i.e., anonymized, e.g., "comments by Reviewer 1", etc. If the editorial decision of the journal is "rejection", the interaction stops there (and the author will use the feedback obtained from this journal to try and improve the manuscript, normally by continued research before submitting it to another journal; this improvement is not just minimal but quite substantial—if a manuscript was rejected after peer-review, this means that some very serious issues need to be addressed before the authors try again at another journal with real chances of being accepted. The discussion here can be very complex for professionals, but, in principle, if those shortcomings for which the paper was rejected cannot be addressed successfully, then that manuscript will probably be rejected again and again	Professionals and users in general communicate their opinions about propositions made in the professional and general agoras. Collective opinions of experts and general users can be to accept a proposition right away (this is unlikely but possible). But, usually, the decision will be either to reject (at least for the time being) or to indicate that the position might be acceptable if a number of other aspects are being addressed so as to improve the first (tentative; intermediate; prototype) version(s) of a proposition
5	Normally, the decision is either rejection or conditional acceptance. The latter means that the Editor asks the authors to further improve the manuscript	One way or another, users ask solution proponents to integrate their feedbacks

(continued)

Table 3.7 (continued)

	Science production cycle steps	General value production cycle, e.g., for an eco-city
6	In this phase the manuscript goes through a round of improvements by the authors based on the comments of the peer-reviewers (plus those of the editor who can have her/his editorial comments on top of it)	Solution proponents (and generators) go through one or more rounds of improvements as needed
7	The improved manuscript is sent to the Editor (which sends it to the original peer-reviewers again), along with a separate document which contains a detailed response to each and every aspect raised by the peer-reviewers (point-by-point, not only what the authors want to respond to), and indicating how the original manuscript has been modified to respond to those peer-review and editorial issues. If the authors think that some comments or requirements made by the peer-reviewers are wrong or otherwise unjustified, they must state so and explain (justify) why they think so. But very often, the disputes are more a matter of clarifications (and what the peer-reviewers ask for more clarifications). So, at the essence, this is a form of debate (propositions made by authors inside a manuscript, followed by critique and response, also called "rebuttal") moderated by the Editor	The improved solution (based on the initial version integrating the user feedbacks) is being submitted again to the attention of the users, with modifications and explanations as needed (e.g., what is possible, what not, why, what is the actual space of options, and how can this be used concretely—what needs to be done according to the current analysis "to the best of our knowledge")
8	The Editor evaluates this technical response (in deed all technical, detailed responses contained in the overall response), and—as a rule—also sends this new input (both documents) to the original peer-reviews to see whether they estimate the authors responded satisfactorily in terms of modifications or/and justifications	The users look again at the updated solution and see whether at least the needed /expected updates have been made to a level that is satisfactory
9	If all is ok, the Editor will give the final approval for publication (case 9i in Fig. 3.19) and the manuscript goes into the final "production" phase (as described above) by the journal. If the updated manuscript does not succeed to meet the overall requirements, the manuscript is rejected (case 9ii). And, in practice, the authors are being thanked for their manuscript proposition, but, with regrets that this particular manuscript had to be rejected, and wishes for success with future works and publication in other journals). In principle, the authors can appeal this decision if they think some serious misunderstanding happens, but otherwise, the Editorial decision is final and the same manuscript will not be accepted for another round of peer-review (if the authors submit it regardless, it will be "desk-rejected" by the Editor, i.e., without being sent to peer-reviewers	The relevant institutions and persons managing the process of interaction between proposers and users follow a previously agreed procedure (recognized by everyone) to formally approve the use of a solution (after that, what remains to be done are just further measures that are required in the given social system, e.g., related to the legal, procedural, budgetary and other aspects)

(continued)

Table 3.7 (continued)

	Science production cycle steps	General value production cycle, e.g., for an eco-city
10	Once the manuscript is (definitively) accepted for publication (after the case 9i in Fig. 3.19), the manuscript will remain as it is at the moment of the decision in content but will still undergo normal updates in terms of grammar and visual outlook optimizations before publication	Once a solution is agreed by the users, its important specifications are "frozen" and further updates only relate to customary forms
…	Logically, the publication of a research paper brings to the authors recognition (new or enhanced) as an expert in the topic, and thus they help those authors secure more funding for more research (usually through competitions for funding, where they submit a research project proposal and the submitted propositions are being evaluated and ranked by a professional Jury of experts that are not in a situation of conflict of interests (e.g., an expert that participates in the competition as an author of a project proposition cannot be in the Jury too)	Normally, the successful generation of solutions brings to the authors of it a certain recognition, and this will help them attract further resources to generate other solutions in the future (in their domain of recognized expertise)

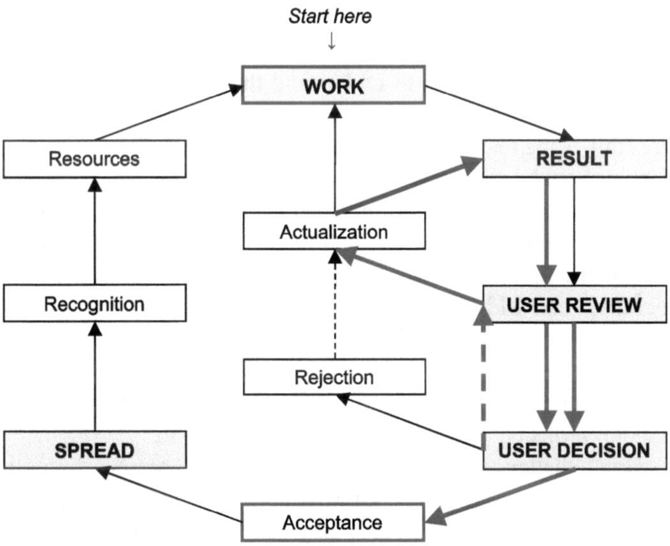

Fig. 3.20 A generalized truth-based value generation cycle (VGC) model using Fig. 3.19 as a template for understanding how real-life solutions can be produced based on truth/evidence. To facilitate comparison, changes are shaded and written with bolded capital letters. For further text explanation of the box labels, see Table 3.7. *Source* original

one might want to start with: [559] on the "gap between science and media" and the related role of "scientists as public communicators"; or [321] on the relation between science literacy, misinformation and Internet as an open source of information and communication; or [532] on "scientific mobilization of keystone actors" (e.g., for biosphere stewardship).

Of course, the role of scientists is made harder by the sheer explosion of research production that needs to be integrated into problem-solving-oriented understanding and knowledge-action models. In a recent paper in a professional journal dedicated to operations research, [33] sums it up as follows: "*Academic research has changed in recent years. It has entered the age of abundant scholarly information. New scientometric data shows impressive increases in both the quantity and quality of information researchers produce. Since 2007 about the same number of publications have become accessible on databases as more than the hundred years prior. At the same time, evidence synthesis has become key in making this wealth of information understandable and useful. Researchers need to be increasingly proficient in identifying relevant information—to be able to build on an increasingly comprehensive research base and to adhere to rising standards in evidence synthesis. Both these requirements make a 'true partnership between librarians and researchers' in demand like never before.*" To this I would only add that the 'true partnership' needs to go much further than strictly the interaction between librarians and researchers.

Then, the fact that we started to understand that "'just google it' is no longer enough" [290] is only the beginning, and this reality has been with us since decades; as it happens, this is the reason why, 10 years ago (right after the Econoving project mentioned in this book), I commenced a new work thread that was (and remains) dedicated to "interactive knowledge maps" and knowledge-action models (for supporting eco-innovation and eco-cities)—one which included among other things the development of an Internet platform with the motto "*Going beyond search engines*", and a graduate course (and wider research-innovation-education program) called *Transition to Sustainability: Human & Technological Systems* based on an annual series of lessons I taught between 2015 and 2020 at the University of Versailles (UVSQ)/ Paris-Saclay University (UPS) and the insights from a longer string of RDI and entrepreneurship projects.

But the overall challenge is even greater than what the present discussions about high-volumes and high-diversity of knowledge might appear at first look: most experts deal with this challenge by restricting their focus to narrower and narrower scopes of expertise (basically, this is a natural self-defense mechanism that helps people avoid being completely overwhelmed). Such adjustments of the individual work scopes are understandable but they further exacerbate the challenge of knowledge integration, and this is detectable in the bibliometric analyses: for example, terminology becomes more and more field-specialized, which (1) reduces the overall number of citations of individual scientific papers [453], and (2) undermines the professional participation of different scientists around the world through decreased visibility (or increasing "citation inequality") [509].

Ultimately, this self-maintained dynamic (a "vicious circle", or rather a "downward spiral" of) knowledge fragmentation means that current methods (current ICT-based methods included) are insufficient: instead, new, smarter, and more powerful qualitative approaches to using current knowledge for solving stringent problems (like urban and societal sustainability) remain to be developed. In a word, we need to become even smarter, and having computers pervading our lives is not enough to earnestly call ourselves, our projects, our cities, and our society "smart". Instead, we can use Internet and ICT to enhance interaction between experts and stakeholders across geographies, e.g., via Brain Networking (discussed in Sect. 3.5) and related as applied to concrete projects involving all relevant actors.

Those new and more powerful methods for managing processes, knowledge, and expertise will have to include individual and collective self-reflexivity of science and all other sources of knowledge and human capacity. Logically, this brings into the discussion the notion of "metaknowledge" (or knowledge about knowledge) in general and as applied specifically to cities— just to mention another notion which is decades-old but has not reached its potential in practice yet (see, e.g., [228, 611]).

Observations About Data Quality

The key question to be asked about the data being used in the process of deliberation and decision during the generation of scenarios is: *How good is your data?*

As visible from the DIKAR_process (Fig. 3.17) discussed in the previous section, the quality of the solutions being developed and proposed depends on the quality of the knowledge behind it, which depends on the quality of the information feeding into it, which is dependent on the quality of the data being interpreted (which further depends on what is asked, which further depends on what is known and not known).

Here, Figs. 3.21, 3.22, and Table 3.8 show a list of main properties that serve in checking that data employed in any project is good enough. In your projects, you might want to use it as a start list (and might add some other properties, and re-organize the whole set as needed). In brief, we must start by looking at how relevant, accurate, and timely the data is, with the observation that accuracy is related to precision. In fact, precision serves to enhance the overall quality but, technically speaking, precision is not equal to accuracy; and accuracy is in fact more important than precision, i.e., it is obviously better to be "about right" than "precisely wrong".

This latter observation is perhaps the most important for the project manager as s/he needs to make sure that project contributors avoid "popular confusions", e.g., that "science is about precise measurement".

This particular confusion is actually informative, because it seems to be related to situations when people try to refer intuitively to the idea of closeness to truth, without being familiar with the correct definition of terms and simply use the common words that come to mind. The point is that all team members need to master the basic technical vocabulary, and the project management must include adequate verifications and prompt clarifications.

At this point of the discussion, some other brief but important observations can be made. Because it is more useful to be "about right" than "precisely wrong", the order of analytical priority is this: what matters first and foremost is trueness, and

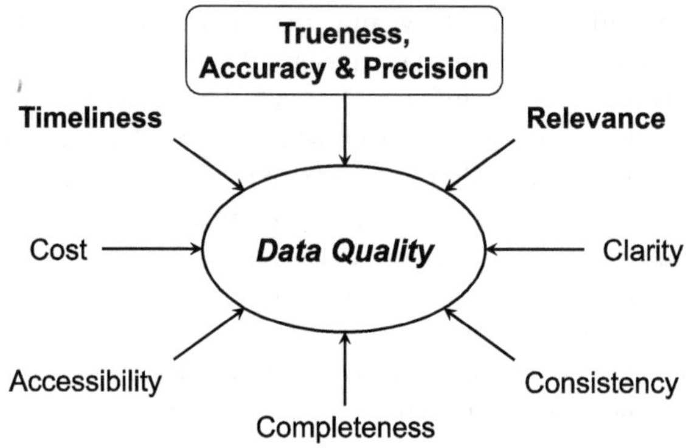

Fig. 3.21 Important properties of data contributing to Data Quality. Each of these properties must pass a minimal threshold (as defined by the user), but some of them (shown bold, on top) have an overwhelming influence: Relevance, Accuracy (and the technically related property of Precision), and Timeliness. *Source* modified after West and Fowler [747]

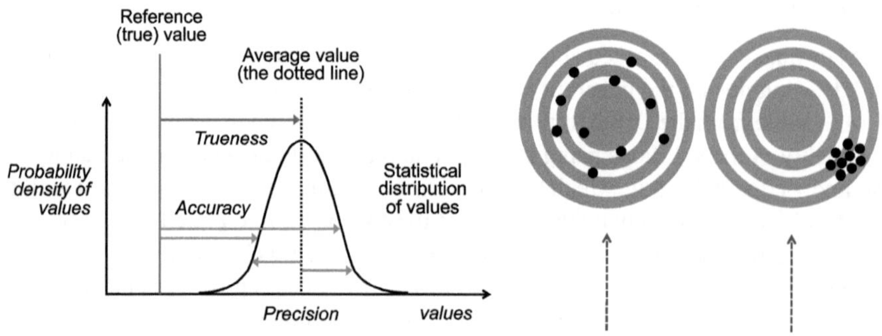

Trueness	The closeness of the mean of a set of measurements to the actual (true) value	Good	Poor
Accuracy	The closeness of a measurement to the true value	Low	Low
Precision	The closeness of agreement between the measurements (in a set of measurements)	Poor	Good

Fig. 3.22 Visual summary of the difference between data trueness, accuracy, and precision. The two targets illustrate the contrast between *About Right* and *Precisely Wrong*. *Source* original, based on ISO 5725 standard definitions and usual visuals in the public domain

this must not be confused with accuracy and precision, which are analytical details. This has certain managerial consequences: high trueness insures that you do have some evidence on which to base further decisions; low trueness does not insure that (exactly how low a trueness is "not good enough" represents a threshold which is usually either already-known from practice in that particular topic or it needs to be established).

Table 3.8 Important properties of data, as related to their definition or/and content

Properties contributing to data quality		As related to data's	
		Definition	Content
Relevance	Usefulness of the data in the target work context (project, business, policy, etc.)	x	
Clarity	Existence of a clear-and-shared definition for the data	x	
Consistency	Compatibility of data types from different sources of data	x	
Completeness	How much of looked-for data is available as per relevant parameters	x	x
Accessibility	Where, how, and to whom is the data available (or not)	x	x
Cost	Resources consumed in obtaining the looked-for data, and/or gaining access	x	x
Timeliness	Availability of as-actual-as-possible data at a target time (when needed)		x
Trueness	How close to truth (reality) is the mean of a set of data points (measurements)		x
Accuracy	How close to truth (reality) are individual data points (measurements)		x
Precision	How much do the data points (measurements) vary around their mean value		x

Then comes the matter of how intense (and therefore how costly) is to obtain that evidence. To insure a good trueness, one needs a minimal number of data points: "at least two" might not be enough to get an average that has a good trueness; looking at the target on the left, one understands why: if precision is low, then the trueness depends on which two data points you hit, so you need that minimal number of data points (i.e., corresponding to statistically representative sample when compared to the overall statistical distribution of the phenomenon under discussion); if precision is high, then even the lowest sample (two measurements) can be enough, and after optimization, and depending on the nature of the studied phenomenon (for example, in highly controlled and predictable settings), even one single measurement can suffice (in which case accuracy overlaps with trueness). However, obtaining a high precision might be very (or even too) costly; in that case, the managerially correct approach is to simply deploy a cheaper method to obtain sufficient data points.

To summarize it in everyday language, asking "How good is your data?" and then making the correct analytical and managerial decisions reflects whether or not a project team knows what they are doing (instead of applying a method blindly just "because others did it too"). Blind action will inevitably drive a project into failure, incur big losses on everyone involved, and in general will create more harm than good. This is to state the obvious, but evidence-based work often requires that for the sake of clarity, because, in complex projects involving all sorts of professional habits, all kinds of misunderstandings can easily happen.

Equally important for projects, the idea that "science is about precision" or about "precise measurement" is rather popular among non-scientists, presumably because a combination of lack of understanding of how science works and because of some rhetorical attributes of words like "precise" (and the related "exact"); we all know how often, in daily conversations, people say "exactly!" or "precisely!" to express beliefs about trueness of a statement.

Another popular (and equally mistaken) idea exists in the field of management, namely, that "what gets measured gets managed". The mistake consists in a hasty, naive reduction of science (and management) to some element in a scientist's (and a manager's) toolbox—measurement—that may be easier to grasp intuitively but which is only one component, it must be not confused with other aspects, and cannot possibly "speak" for the whole story (let alone the virtually unlimited possibilities of simplification and idealizations about unfamiliar domains).

Such undue extrapolations are in fact part of the long list of cognitive (and related behavioral) biases and heuristics that have already been evoked in this book and constitute a full array of research topics in and by themselves. Of course, one can study those biases and discuss in great technical details why they do happen, but the bottom line for project management is that we need clarity.

Similar discussions can take place with regard to the other properties listed in this table as contributing to answering the question of data quality, the basic conclusion is essentially the same: in a project team, we must insure that we know what we are doing, i.e., each of us must resist the temptation to believe that "if I know something, the others also know it". They might; but you have to check. Otherwise, you (we) effectively walk on untested grounds and misunderstanding will surely happen; some of those will be minor and some will be deadly for the project.

Contextualization: Some General Observations About Data Generation and Uses

The discussions about Data, Information, and Knowledge bring about other issues. For example, because of its integratory character, the DIKAR_process model is well-suited for "strategic data acquisition" for urban applications, i.e., it helps the operator have a well-defined and motivated purpose for data generation (e.g., [153–155, 294, 385]; see also [53, 502]), or, put differently, "strategically acquiring relevant data sets" [634], or "targeted data acquisition" [398], or implicitly-so inside a well-defined strategy for specific uses, e.g., strategic energy management or automation and "smart city" (briefly defined in Module 1, Sect. 1.3, Table 1.3), wireless communications, and various "data-driven business analytics" (e.g., [131, 339, 501]).

Now, general terms like "strategic" and "data acquisition" can be applied in many contexts and may be understood in many ways. But, a general goal in all projects is to avoid spending resources on acquiring just about any data, because this normally simply means low-quality and low-utility data; details provided below), the result being that valuable resources are being spent on the generation of large accumulations of useless-and-unused data that are known colloquially as "data graveyards" and "data swamps". These cannot represent a valid basis for eco-city scenarios and eco-city projects but unfortunately represents a typical problem. As [385] summed it up, we need "targeted and strategic data acquisition, provision and management (necessary information at the right time)", in order to "avoid the creation of 'data graveyards' and the often-mentioned 'DRIP phenomenon' (data rich, information poor)."

It is in this context that another popular term ("meme") that circulates in the Agora must be understood: "Big Data". To cut a long introduction short, the term is derived from the benefit of ICT, starting with substantial economic advantages that also scales up with the size: "big cities, big data, big economy" (see, e.g., [276]). This great potential often leads to attempts to generate and use "data lakes", i.e., large volumes of heterogeneous data being collected on a systematic basis and awaiting to be exploited (treated and interpreted) for the generation of useful insights [132, 470]. Its commercially-touted importance in society can be summarized by referring to public policy (local, national, or higher; for a more detailed discussion on policy see Sect. 3.9), and a recent paper sums it up, "Does big data serve policy? Not without context." (see, e.g., [283]). More often than not, data lakes turn out to be data swamps.

This clarification is important, because sources of (often innocent but even more often deliberate) confusion abound. For example, there are commercially catchy but scientifically (and socially) misleading metaphors like "Data is the new oil" [688] that purport the fundamentally-mistaken idea that Data (actually they usually talk about digital records, not necessarily scientific data) is a commodity: in reality, useful data is (and can only be) a limited product with a limited utility within the bounds within which it was generated. As one can see from Fig. 3.17, Data is generated for specific purposes (and it therefore requires a complex process and significant efforts to generate it) and its relevance is limited to that purpose, which gives its functional relation with Information and Knowledge.

Relatedly, the literature also includes conceptual propositions like "knowledge lakes as contextualized data lakes" [132]. Again, in certain contexts, this might be useful but must be used with care (such an expression might or might not be logical in the given context, so everything must be technically verified).

Observations About Advanced Automation Processes (A.I. and Data Science)

In the discussion about the DIKAR_process model's negotiation rooms (NRs) where different views of the world are being compared and conciliated, the entire model can be understood as a basic representation of (and start reference for) comparing and coordinating concepts (with their definitions, old and updated). Then, based on this, Data (i.e., Vertices 6 and 7 in the model) can be also understood as the "eyes and ears" of a project—a metaphor which reminds of an interesting concept from the literature: "analogy-making as perception" [480]. Seen through these lenses, data represents the shared perception that a project team has about the world and a target problem, simply because that data is generated as a result of a shared understanding and shared demanded information leading to that shared demanded (and then obtained) data.

To sum it up, in the end (as per the DIKAR_process model), knowledge indeed determines what information is demanded and further which data needs to be generated. In this sense, the DIKAR_process model of operation with Data, Information and Knowledge represents an instance of combining bottom-up and top-down approach (as you first use evidence and inductive reasoning to discern a model and then you impose your operational model onto the available evidence, while, and this is a crucial detail, you stay open and update the model as new evidence becomes available—or, to use an earlier employed expression in this book, you keep updating what you know).

In that sense, analogy is the key mental operation because you interpret data by comparing it with the model, which is a form of analogy, and which in the end means that the most important aspect for a good understanding of reality—and of the potential impacts of the actions of an agent in reality (i.e., for knowledge-action models) as per Fig. 1.6 (in Module 1) and the related discussion on complex dynamic systems—is the conceptual clarification of the employed notions per se and the clarifications about how notions relate to each other in a hierarchy of categories that describe reality aptly enough to make our problems tractable.

For the initial goal of the DIKAR_process model itself, analogy matters because it enables a common perception of the world through an idealized model that captures the essence of it. This is exactly what that a basis of common denominators is all about: the confrontation of different view angles naturally results in hashing out the essence of a problem and using it as a common reference for further study and action.

Another interesting and directly related topic comes from the engineering field of statistics-based automation, which is better known outside science and engineering under "scene name" (to use a popular urban vocabulary): "*Artificial Intelligence*" or AI. Because of certain, occasionally spectacular applications of it, the field of AI has recently attracted a lot of attention and is a familiar topic in technology forecasts published by both private and public sectors. Some commentators even venture to call this a "revolution" and compare this with the advent and spread of use of electricity about a century ago, or with the steam engines three centuries ago, which for our interest here tends to make suggestions like "our cities will look completely differently" and so on. However, a sober look at the evidence shows that the grounds for overexcitement are not there: for the time being, the so-called AI only involves a very small number of types of automation that are essentially reducible to simple tasks (and types of data), combined with speculations and assumptions. In short, there are too many 'if's, and the current situation, objectively speaking, does not rise to the level implied by the imagination-flaring name used in public agoras. The fact that most people were not (and still aren't) aware of the underlying mechanics and technicalities of AI and are therefore easily impressed by different applications (especially when advertised to that effect by commercially interested parties) does not make it in any way a revolution. At most, it is just another hype, and hypes per se constitute a research object and topic in innovation studies.

Automation is what technology does in general, since the Paleolithic (i.e., the stone age) to the agricultural instruments we made in the Neolithic and modern era machines. Of course, this is not to diminish the merits of the works being done in ICT and related fields, or to deny the importance of different applications for our daily life. It is only about making the right distinctions based on current knowledge and placing novelties in their context.

Given the reality that we live in complex dynamic systems (as discussed since the beginning of this book, starting with Fig. 1.6 and the entire Sect. 1.2), the potential of those statistics-based automations to substantially assist in the generation of solutions for eco-cities and sustainability is real. However, far from being "revolutionary", it is some functional optimization of those processes that involve a constant high throughput of data of the same kind, and the spectacular side of it comes from a quantitative aspect (the sheer speed given by computing power), not from a truly qualitative leap.

Indeed, other benefits of computing and automation (that have now become familiar) are much more profound for us. Specifically, the impacts of Internet are still unfolding and are truly shaping present and future cities and the human society as a whole. Even the fact that the Internet is already being taken for granted is a direct proof of its transformative character for the human society (i.e., a leap to a next level), and the science and engineering behind the set of automations that are part of how Internet works is much more important. In this context, virtually all those automations that are claimed to be part of the A.I. movement are in fact identifiable as a type of content of what is known as the Internet-of-Things (IoT), while the central aspect about Internet is the Internet-of-People (IoP; and Brain Networking as a key resource for cities and projects as already discussed in Sect. 3.5). More interesting from the CDS perspective are other phenomena, an example of which is the notion of "Social Brain" as a generator of innovation during major socio-economic transformations like with the so-called "industrial revolutions" in the past [681] and potentially with the transition to sustainability that is now beginning; and related themes like smart cities and societies [155, 156]; see also [9, 190]. But again, we should avoid rhetorical traps: change is more related to evolution, and what are being called "revolutions" are in fact either periods of fast evolution or sudden systemic readjustments that occur when tensions have been accumulating for long periods inside human systems.

Based on current evidence, one would argue that humanity will come anywhere close to developing veritable AI (that is safe and worthy of its name) when entire human communities (i.e., cities, towns, villages, projects, etc.) will have begun to take extensive advantage of the fundamental and truly immense potential of Brain Networking and IoP combined, of course, with support methods from computer-driven automation and IoT, and inside the broader idea of "network thinking" behind our understanding of complex systems from biomolecular interactions to cities and the Internet [479].

Relatedly, I would also highlight that the notion of "smart city" only refers to (and reflects) certain ICT-based types of methods that might (or might not) be integrated into a real eco-city. For further discussions on smart cities and potential relations with AI (and broader "data-intensive" approaches) and urban sustainability, see also (e.g., [15, 208, 595, 653, 762]).

Another worthwhile highlight is that, in more engineering and commercial-driven applications (which is the usual case), AI typically refers to Data as digital records (not necessarily scientific data; and those digital records are used for a small set of automations). But we must not forget that you can only automate repetitive tasks within very narrow bounds, and, by definition, one cannot automate real-world complex dynamic systems. Therefore, what may appear to most people as a "very complex technology" is better described as a string (of strings) of automations, and the complexity here is not what most people would expect: it comes not from the number of components linked together but by the inherent dynamics of a system that is composed of a network of subsystems and components. For such a "marvel of technology" to work, real complexity (which comprises non-linear dynamics and unpredictable outcomes) must be reduced to zero because technology is about predictability, and the harder to produce (and the more expensive) a piece of technology is, the more critical it becomes to eliminate unpredictability sources and possibilities. In this case too (in line with the discussions already made in Module 1), the fact that "complex" sounds nicer or "cooler" than "complicated" does not imply that the two words are essentially synonyms (they are not); relatedly, the fact that in daily life we use these words more freely does not translate into interchangeability in professional contexts. Then, it follows that the word "intelligence" in the name of the field "Artificial Intelligence" is a misnomer, because it tends to create more confusion than clarity and reflects more a distant aspiration than a reality. There is no generally accepted definition of what constitutes intelligence, and, moreover, intelligence does not "automatically" imply sentience or consciousness.

In fact, more than its achievements in the optimization of goods and services the field of AI (not to forget the blunders too, and also not to dismiss them too hastily, because failures are in fact part of true learning (see, e.g., [658]), has actually served to ask very important questions within the interdisciplinary "hexagon" program of Cognitive Sciences, composed of at least six main fields: Philosophy, Psychology, Computer Science (not just AI), Neurobiology (which implies Biology/Life Sciences in general), Anthropology and Linguistics [176, 473, 515].

In this context, one observes that if urban systems are being discussed in computer science (as already mentioned in this book) in terms of "smart cities", then the corresponding term in the overall cognitive sciences is "cognitive cities" [238, 273, 436, 493]. Consequently, all knowledge-action models discussed in this book actually fall within the scope of the notion of "cognitive eco-city", which further (a) includes the notions of "cognitive city" which includes the notion of "smart city"; and (b) is included in the notion of "eco-city". In basic logic formalism, this can be noted as follows:

$$Smart\ City \subset Cognitive\ City \subset Cognitive\ Ecocity \subset Ecocity$$

All in all, "AI is harder than we think", for numerous of reasons [481], but the works in the field of AI have the merit of highlighting the importance of data in science and society, notably the fact that solutions need to be based on interdisciplinary knowledge and knowledge is based on hard evidence (i.e., data and information derived from it) and long-term and difficult efforts of integration across disciplines.

Also, such discussions highlight the importance of making the right distinctions between issues and goals in different fields, because science, business, and policy have their own logics that must remain autonomous and this requirement itself generates conflicts of priorities and semantic mismatches that must be observed and resolved so that the end-result of the interaction between science, business and government is dominated by synergies rather than chasms, and by the creative tensions between complementary perspectives rather than the burden of unresolved systemic tensions.

These being said, I would reiterate an earlier summary of it [153]: "by anchoring itself in all (of) science, AI could widen its relevance and deepen its potential beyond this initial hypes and distrust, and achieve a more balanced discussion between qualitative and quantitative aspects. This can provide useful avenues for discussions about the connections between smart cities, urban metabolism models and sustainable cities/eco-cities, and how AI can play a catalytic role in managing risks, crises and transitions in human communities. (...). AI-aided holistic models can enable solutions to problems that have been impossible or excessively hard to address". A first example of such problems is the type of social dilemmas described in the introduction of this book (Sect. 1.2), notably those over common-pool resources (CPR; or "the tragedy of the commons"). Thus, the works around AI can assist us in preventing the long-term collective destructions of our shared resources, both the natural resources (e.g., air, water, ecosystems) and the products of civilization (e.g., common urban spaces and infrastructure).

Another "popular" but problematic term (at least when used uncritically) is the term "Data Science". This seems to be another fancier name for statistics (perhaps combined with engineering aspects about data generation in computer science), motivated by the commercial interest of some companies to avoid using the term "statistics" which is considered to be "unpalatable" (or even repellent) for the customers and the wider public in general. However, because data is part of what Science itself is about, it only makes sense to talk about "data science" if the immediate concern is meta-science (the "science of science" or SciSci), i.e., when science itself and how it works is the object of study (in line with the discussion about science from the previous sub-sections of this Section): "with a deeper understanding of the factors that drive successful science, we can more effectively address environmental, societal, and technological problems" [244].

It would make perhaps sense to discuss about "Data Science" in the context of epistemology. But, usually, the term "data science" is used outside of those concerns and more in relation to commercial interest of inducing to potential customers the idea that the meme "big data" (presumably something that certain big companies have a lot and can present as an asset and measure of competence) is grounded in science. In reality, because the scientific method is all about evidence (and data is at the heart of evidence), all science fields work with data (otherwise they wouldn't be science fields).

Unfortunately, this is not the only problem of this type, and care must be taken to avoid confusing commercial claims with science. The uncontested necessity and benefits of interaction between science and business means not that things can be mixed up. And, the same applies to the relation between science and public administration and the relations between public administration and business (to complete the triple-helix).

A related and perhaps deeper problem with the term "data science" is that it implies that "everything starts with data", which brings us back to the idea of purposeless collection of data with the vague ideal of generating a big resource ("data lakes") while such pools of data can only be a resource for the purpose for which that data was generated. So, to use the hastily-concocted expression of *The Economist*, many things can look like "oil" (or any other commodity) and not be.

Of course, one can occasionally find some data that can be reused for other purposes than the initial one; but such situations exist inside a lucky convergence of so many factors that it can only be very rare, i.e., clearly not a commodity. This risk exists concretely for eco-cities, because some discussions in the literature are about "data-driven smart eco-cities" (see, e.g., [68–70] and the references therein). Such concepts are dangerously close to the claims (in fact speculations and undue extrapolations) and mistakes discussed earlier about the "data-driven" type of solutions (vs. those based on real knowledge, and knowing what we are doing).

The background lesson here is similar to the earlier discussion (in Module 1) about the naivety and dangers of looking for "silver bullet" solutions to complex problems in general: in eco-city projects, searching for "data bonanzas" is a non-starter. Some reliable sources of data may exist, but, arriving in a position to have good data requires generating those data (as described in relation to the DIKAR_process model), and this takes knowledge and hard work, and not just the work of some people but the co-work by the project teams as teams. This is what ground-truthing of projects is essentially all about (although, in different fields, it may translate in different practical aspects).

Then, at the levels of a city, country, and humanity, systemic ground-truthing of option-developments for present and future times require reliance on science and scientists through what (already a quarter of a century ago) has been called "a new social contract for science and scientists" [27, 269, 430, 431] (see also [532]). This is a profound long-term necessity and it is materialized, I contend, through the cutting-edge works of eco-city projects.

Automated Scenario-Generation Using the DIKAR_process Model, in Addition to the Delphi Method

The fundamental need of strategic data acquisition in RDI projects and "data-driven business analytics" is coded in the DIKAR_process model by the second half (shown in green and pink) of the circuit (vertices 7–12) but should be understood and managed in the integrated (and integrative) context of the full model.

In terms of the space of options for eco-city projects, this further means that, once the broad scenarios have been produced and consolidated with the Delphi-based scenarios described in Module 2, data-based scenarios for in-depth technical explorations and optimizations can be generated directly with the DIKAR_process model in the form of different solution options being produced through model iterations that use different (purpose-driven) data sets, based on different data sources or/and types of interpretation of data or/and different configurations of the resulting insights into novel solutions or/and knowledge updates. This is an important aspect during the preparatory phases of crisis management (i.e., for the development of contingency plans and emergency protocols).

Further, one notices that DIKAR_process model supports process automation in both machines, in teams of human, and (probably most importantly) as an interface between human and machine agents. If combined with AI methods and tools, the DIKAR_process can assist even more in facilitating certain types of activities and tasks.

In the context of the high throughput of data, the main advantage of this general model is that it allows efficient identification/formulation and therefore automation of routine searches (of solutions) and activities. This can bring down coordination costs and thus free resources for qualitative interaction.

Observations About Workflow Automation and Potential Utility in Projects

Another observation about data, information, and knowledge concerns the tension between the complex nature of projects and the deterministic nature of the automation protocols provided by computers when automation is possible, i.e., workflow automation and related families of methods like the Robotic Process Automation (RPA). The latter topic is popular in engineering and somewhat popular in relation to digitalization and smart cities [3, 319, 437, 586, 650, 714, 737]. But I propose an exercise: try to find in the literature some studies that do both RPA and Eco-innovation. Just for the sake of it (and to check), I already made this exercise and it would seem that there is essentially nothing at the intersection between the two. Given the current state of the art, this is not surprising (the two notions are at the two opposite ends of a gradient between predictability: eco-innovation is about pushing the frontier of the unknown through unexpected and ground-breaking developments, while RPA is about finding situations with full predictability. However, this does not mean that the two cannot be complementary: they are and the DIKAR_process model is compatible with both and can make this bridge.

For clarity, we first need to make the proper distinctions: to be useful, computational models must be deterministic. However, this is not a problem for our purpose here, because we are talking about models that are used for structuring a target problem when that problem is the (in-project) process of interaction itself, i.e., not the object of that project (sustainability and cities). So, using this model simply provides a template for helping a project team advance in their work in a structured way, i.e., to define and map options and therefore pathways (scenarios of action and progress in the project as a project) without actually contesting the complex nature of a project or of the human system in which the project is embedded. Put simply, these models are just tools, i.e., embodiments of automation (predicted outcomes) within given limits within complex dynamic systems. To use an analogy, we use predictive (deterministic) models to talk about a well-equipped house in which people need to carry on with their activity successfully—not about what people choose to do in that house and with that equipment.

A Brief Discussion About Innovation in the DIKAR_process Model (Vertex 10 and Its Links with Vertices 8, 4/4′, 11 and 12)

This second part of this Section includes a few quick points that I estimated to be useful for starting the conversation about innovation in eco-city projects. To that aim, the main text is kept at a minimum and the visual summaries are largely left to speak for themselves.

As per definition, eco-innovation entails everything that innovation (in general) does, plus the need of compatibility with sustainable development and the transition to it. For the purpose of this book, two general (start) requirements can be identified right away: (1) diminishing the impact upon the environment, and (2) helping cities perform as eco-cities. The first one relates directly to the full set of sustainability filters discussed in this book (since Fig. 1.3 in Sect. 1.2), i.e., not only the "Ecologic Sustainability Filter"; as already explained, we talk about systemic innovation, not just some optimization by maximizing or minimizing one parameter or another. The second one highlights the fact that an eco-city itself stands as a system and therefore as city-scale example of eco-innovation.

Another point is that smaller scale (eco)-innovations may not all strike like a truly systemic innovation, but appearances rarely reflect systemic realities; notably, small-scale innovation can become or/and help generate other small-scale or larger scale (eco-)innovations which ultimately can transform a system—the relations between different innovations at different scales and of different sizes (depending on how they are measured) is basically just as complex as any complex dynamic system is. Notably, some innovations that do not strike at a first look as significant (or even as an eco-innovation at all) may actually be key for the generation of a larger eco-innovation, and vice-versa. In fact, innovations that may appear to an observer to be an "obvious" eco-innovation actually require confirmation (that this is the case of an eco-innovation) by in-depth systemic analysis: it might very well turn out that, when all its life-cycle is considered, an innovation generates a higher (not lower) burdens upon the environment (and so on). Put differently, an innovation can easily appear "eco-" within a narrow set of considerations but actually be harmful to the environment when the entire chain of production, use, and disposal is considered. Or/and, an innovation might have a low environmental impact on the short term but create serious socio-economic problems which later on will result in an even higher ecological footprint ("with a vengeance"). This is why (as a reminder) those three sustainability filters must be taken together.

The question then becomes: *How to generate large, system-scale (i.e., true and truly transformative) eco-innovations?*

The next (and last) three Sections of this Module contain some concrete models for a general approach in eco-cities as complex dynamic systems, starting from the notion of "disciplined experimentation". Meanwhile, this sub-section continues the discussion about the in-project deliberative process in the DIKAR_process model with focus on some key aspects.

In that model, particularly important is the negotiation room NR number 10, i.e., the "Innovation" vertex: a moment in the entire process when dots are connected in new ways and some new potential solutions are being generated. Thus, an important operational objective in the management of a project is to understand when would an innovation from Vertex 10 have good chances to flow successfully along the pathway described by vertices V4′-V11-V12.

Of course, current knowledge about innovation in general provides us already with a good start platform for eco-innovation. What distinguishes the latter from other innovations is its comparative multi-dimensionality and systemic transformational character, which makes that, on a basic gradient from the simplest to the most complex innovations, eco-innovations tend to be at the highest complexity end.

Table 3.9 illustrates the best kind of concise representation ("Big Picture" synthesis) I found in the literature: a compact synthesis (and abstract decision tree) of logical narrative options. It is a modified and structurally updated version of an older model (by [438]) for critiquing any innovation of interest (or a novelty in general) by filtering it through those instances that are known to be causes for failure (i.e., situations to be avoided on the path for success). For discussion and comparisons, see also [178].

In the abstract terms of Category Theory and the earlier discussions made in this book about what can or cannot happen in the real world, this model charts a space of main limitations (what can usually go wrong) and thus implicitly charts an abstract space of options, and represents it as a (another) 2D matrix: a Cartesian product of *Features* (of a given innovation; rows) and *Application Cases* (columns). In this form (essentially a text-based decision tree), (a) this table can serve as a minimal default for general orientation (to see and anticipate what can block the wider adoption and success of a new solution), and (b) can also be updated and extended by an active user for her or his specific purposes and additional insights.

(Technical note: this above is of course a separate 2D matrix and another cartesian product, and must not be confused with those shown at the DIKAR_process model; these are different models. Now, 2D matrices and cartesian products are very often used in practice, although they may not always be called explicitly what they are mathematically. They are so often used because this is a most efficient way for us, humans, to represent reality: by drawing categories on a surface and checking how they correspond with each other. In professional contexts, the practice of using the established mathematical vocabulary and formalisms simply insures that this entire process of abstract representation is clear and correct).

Further, Fig. 3.23 provides a concise "Big Picture" of the financial and institutional context that characterizes the development of a solution from science (i.e., laboratory research) to business (i.e., commercial success validating the fact that users/buyers appreciate the end-result).

To give a quick example, in a study that compared "smart city" developments in the cities of Amsterdam (in the Netherlands), Hamburg (in Germany), and Lingbo (in China), beside showing tables that illustrate perfectly of the kind of work outcomes which (I would observe) would result from negotiation rooms (NRs) of the DIKAR_process model, [585] noted—based on their own experience and analysis—that private–public partnerships (PPPs) represent a natural solution at the very heart of urban experimentation. For example, Amsterdam adopted a PPP approach simply because private funding was the only solution that was available for insuring the

Table 3.9 An integrative model of basic factors limiting the adoption of innovation (all novelty)

<table>
<tr><th colspan="2" rowspan="2">A new solution (a technology or good or service or policy) fails to replace older ones when</th><th colspan="3">In the case of</th></tr>
<tr><th>The individual user</th><th>A community of users</th><th>A given market/ industry</th></tr>
<tr><td rowspan="3">Its (general features)</td><td>Utility</td><td colspan="2">… is perceived to be less/lower than that of the older ones [191, 486, 763]</td><td>… fails to exceed the old ones' measurable specifications [618]</td></tr>
<tr><td>Complexity</td><td colspan="2">… focuses attention on (obliges the user to prioritize on) the overall effectiveness, not on the newest features per se [440, 488]</td><td>… renders really new innovations less frequent overall [662]</td></tr>
<tr><td>Complementarity</td><td colspan="2">… of older technology with the already existing solutions results in higher total utility of old versus new solutions [649]</td><td>… does not lead to a (new) dominant design [1]</td></tr>
<tr><td rowspan="3">Its (social situations/ dynamics in which it is embedded)</td><td>Context</td><td>… creates material/ operational limits to access [368, 386, 641]</td><td>… supports social divisions to access [125]</td><td>… restricts access, on behalf of proprietors or of the state [295, 684]</td></tr>
<tr><td>Orientations</td><td>… towards its use are negative [98, 363, 490]</td><td>… are towards the older technology [93, 605, 744]</td><td>[a]</td></tr>
<tr><td>Contagion</td><td colspan="2">… is not strong enough to displace (incumbent-favoring) existing community norms [98, 104, 597]</td><td>… is low due to weak marketing and/or operations [211, 485, 672]</td></tr>
<tr><td rowspan="3">Its users' learning</td><td>Capacity</td><td>… or cognitive ability limits learning [174]</td><td>… to access relevant education is limited [475]</td><td>… of resources or guidance is inadequate [299]</td></tr>
<tr><td>Capability</td><td>… generated by the old solutions helps not in using the new solutions [174, 215]</td><td>… of users has not created a (relevant) community of expertise [17, 89]</td><td>… to experience the solution is diminished [10]</td></tr>
<tr><td>Costs</td><td colspan="2">… related to switching are (too) high [488, 646]</td><td>… as required by the new solution are prohibitive [243]</td></tr>
</table>

[a] Means that a competitive market or industry is already oriented towards innovation diffusion by definition

Reading examples: (1) "In the case of the individual user, a new solution fails to replace older ones when its utility is perceived to be lower than that of the older ones"; (2) "In the case of a given market, {it fails when} its users' learning costs as required by the new solution are prohibitive". *Source* modified after [438]

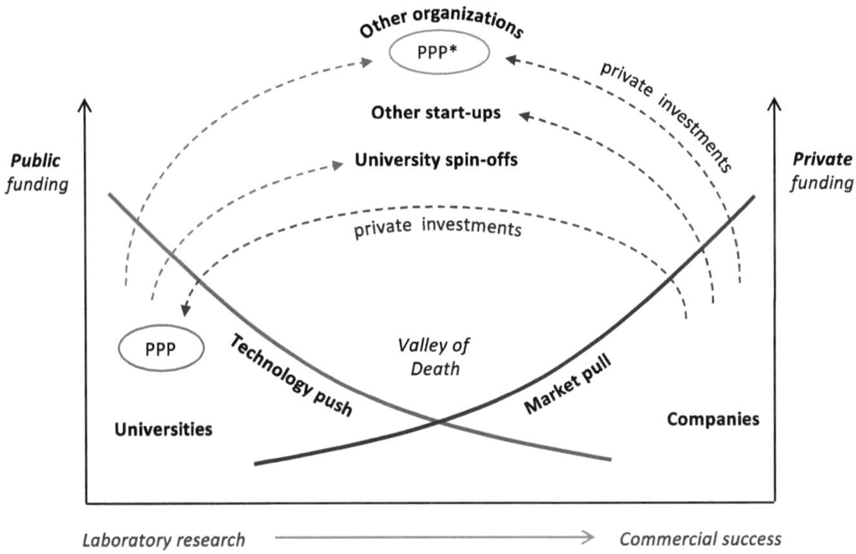

Fig. 3.23 Funding of public and private projects. *Source* original, based on general literature on innovation and entrepreneurship as covered by the metaphor "Valley of Death", e.g., [264, 369, 450], and the author's personal professional experience in such projects

continuation of their urban projects during the difficult context of the 2008–2010 financial crisis.

The next logical question is *How can eco-innovation survive the Valley of Death?* To help find an answer in concrete projects, Fig. 3.24 provides a visual summary of the basic types of actions that can be taken (or/and policies that can be put in place) for facilitating the funding of new solutions from lab to market.

Then, a related question is *How to generate large, system-scale eco-innovation?* As this requires a more complex answer, a dedicated discussion is provided in the next, dedicated book section, but the main idea is "disciplined experimentation", i.e., an evolution from a start emphasis on exploration (of potential solutions) to an end emphasis on exploitation (of developed and available solutions), while institutions try and cultivate the capacity to maintain a long-term dynamic equilibrium between the exploration and exploitation, i.e., "ambidexterity".

Yet another key question is *How to organize the work of solution development?* Essentially the answer involves the predominance of one of the two basic models: "waterfall" and "concurrent" (Fig. 3.25).

In startups and pioneering projects, the work tends to collapse in the form of a small (central) team that does everything that needs to be done "at the same time" but following an internal logic that insures coherence. Essentially this means a concurrent model where the work cycle is iterated very quickly and frequently, as needed. This can also be done inside large companies that want to escape the traditional "waterfall" routine and promote innovation through intrapreneurship, e.g., in dedicated special

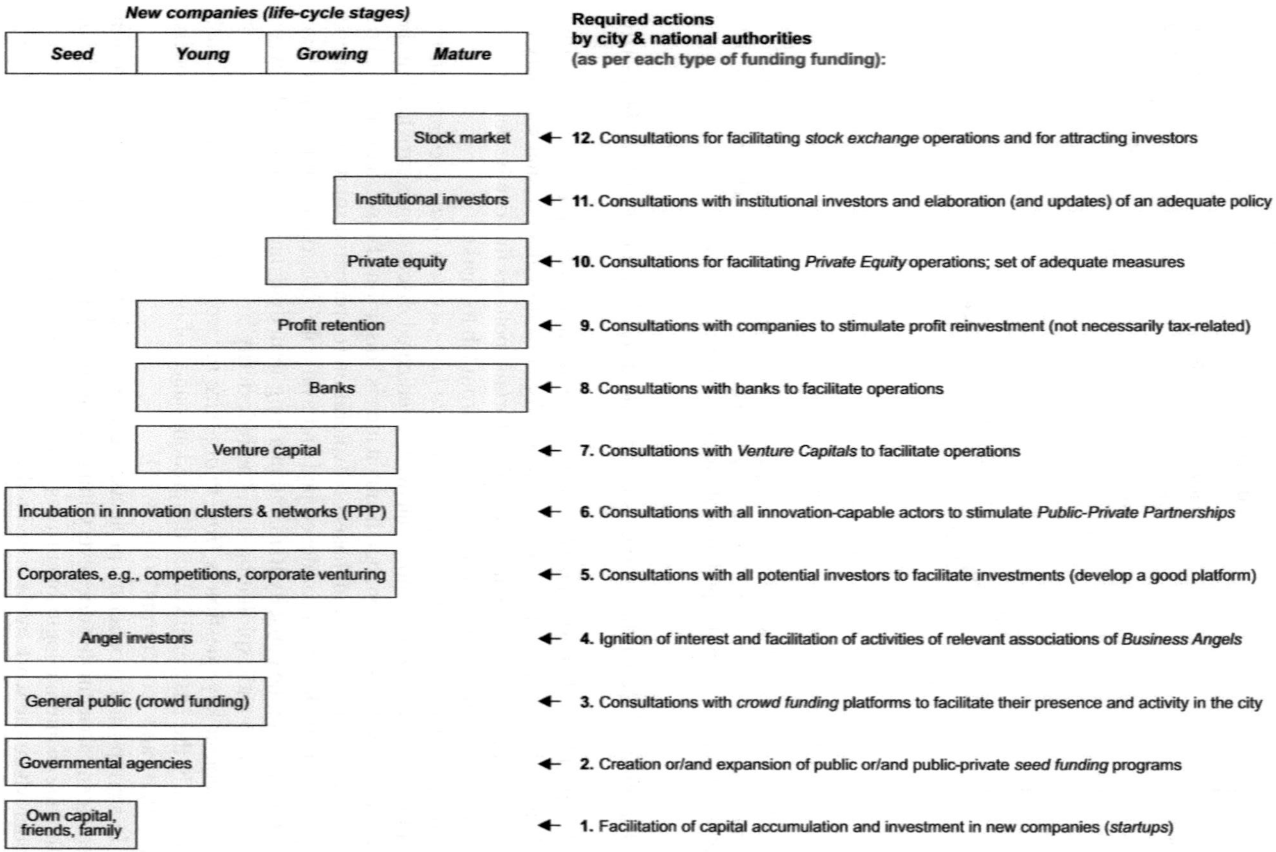

Fig. 3.24 Basic actions for facilitating the financing of new solutions from lab to market. *Source* modified and expanded after [79]

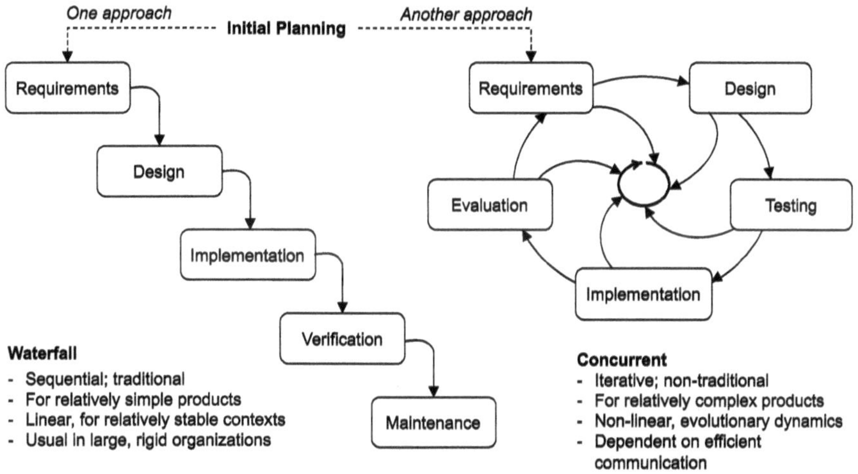

Fig. 3.25 Comparison of typical Waterfall versus Concurrent models of solution development. *Source* original, based on general literature on the matter

units that apply the concurrent approach in some specific internal initiatives (projects) for specific purposes, or/and through participation of some of their units (usually the RDI department, typically in collaboration with the commercial department) in PPP projects.

One observes of course that both kinds of organization exist in a city, but we know that small-and-medium enterprises (SMEs) are more innovative and have a particularly important role in generating the innovativeness that is required for a city to escape the business and usual and re-invent itself as an eco-city. For further discussions on the role of SMEs in "urban sustainability interventions", see, e.g., [748], see also [102, 103, 135, 735]. For another way of looking at it, see also the (equally old) kind of discussion that conceptualizes different approaches to solution development like "rational plan", "communication web", and "disciplined problem-solving", while at the same time making the distinction between product effectiveness and process performance [95, 129, 676] and "dynamic capabilities" of a firm (or, for that matter, any organization that is developing solutions), i.e., "a set of specific and identifiable processes such as product development, strategic decision-making, and alliancing" [212], see also [201, 219, 417, 729].

It any case, for our concern here, it is important to keep in minds that, in most projects, the term "eco-innovation" is used to refer to some specific technologies, products or/and services, i.e., small-scale solutions that are being proposed within larger systems (city districts or hubs of activity, whole cities, manufacturing, and industrial plants, parts of or entire supply chains, etc.). In such cases, and given the relatively fast pace of changes on the market, the concurrent approach is the dominant (and usually inevitable) approach applied in the most innovative organizations, e.g., startups and pioneering complex projects; and this may create tensions with those project contributors that fall in the category "large organizations" like governmental institutions or/and large companies. While such tensions may appear as (and indeed are) serious obstacles, in light of the systemic perspective on sustainability discussed earlier in the book, they are also concrete tests of viability and therefore should be regarded as part of the systemic problems to be solved during the systemic transformations that are implied in the notion of eco-innovation as "innovation for sustainable development".

This idea is also a segway to another important observation: when we talk about the complex products envisaged by PPPs or other initiatives covering several domains of expertise (i.e., not most of the goods and services on the market but the systemic transformations that involve multiple stages of development along the path between scientific research/exploration to ready-to-use mature solutions, each stage requiring a different complex discussion), a third approach becomes necessary: one that combines the two approaches mentioned above in the form of a sequence of stages from idea to end-result, each stage including specialized expert methods and general managerial strategies that exhaust a detailed list of technical requirements under the classic project constraints of "performance, schedule and budget". This stepwise approach is best known through current readiness levels models, the most prominent of which is the Technology Readiness Levels (TRL) scale from RDI [446]. In the next sections I develop on this managerial approach in relation to eco-city and related policy, programs, and projects. Before that, however, the last sub-section of this section addresses the matter of scale from small to large and across systems.

One may observe a connection with the Delphi method described in the previous Module. Namely, in any project that uses this method to generate a map of potential futures, (1) the spread of the method along a series of stages is an example of a waterfall approach, while (2) the sessions of direct interactions between experts are instances of concurrent work. This means that the overall method is itself an example of the combination of the two approaches in a structured-and-agile strategy.

Supplemental Visual Summaries About Knowledge Integration for Generating and Maintaining a Valid Space of Options

As we move on, it also becomes obvious that dealing with a broad (and multidimensional) range of insights from science in general (on the one hand) and from a specific city/local community (on the other hand), means that the number of variables that must be factored in by project teams and project managers is immense: in the absence of highly effective tools, this massive challenge will dwarf and overwhelm any expert or project. (Not to mention other issues that compound the overall challenge of a true project, like false expertise and hidden agendas).

In an attempt to provide additional "big picture" support for project management, Figs. 3.26, 3.27, 3.28 and 3.29 are meant to serve as visual resumés about the notion of Space of Options.

Figure 3.26 shows how a concrete and dynamic space of options (namely, the adaptive capacity of a city) exist between opposing but reciprocally-influencing forces and what this gives as an end-result—in this case, the lower/higher vulnerability of a city to adverse events and phenomena. Figure 3.27 generalizes this idea to a notion

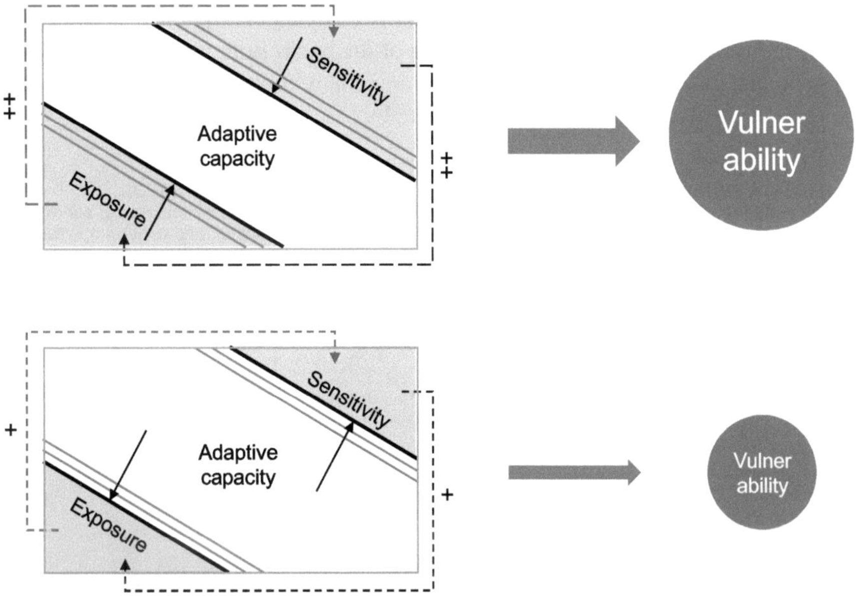

Fig. 3.26 Vulnerability of a community to a certain phenomenon is a function of exposure, sensitivity and adaptive capacity. *Source* modified after [220, 689]

of abstract space of options that may exist (or not), and needs to be cultivated and defended as vital for the generation of solutions.

Figure 3.28 sketches the basic features of the hard process of generation of a space of options though knowledge integration. Relatedly, Fig. 3.29 sketches the evolution of situations and roles in a project, from the abstract perspective of agent-based

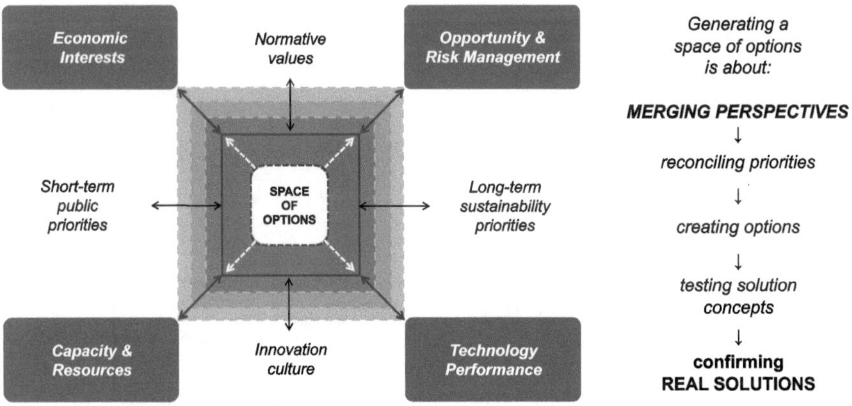

Fig. 3.27 Visual summary of the reasons why holistic models are needed to generate and expand a Space of Options between conflicting factors. *Source* original

modeling. In this sense, this holarchy model is complementary to the agent-based model provided in Fig. 1.6 at the beginning of the book. Both models help making the point that, in complex dynamic systems, any policy co-evolves with its target and the roles of involved agents change accordingly.

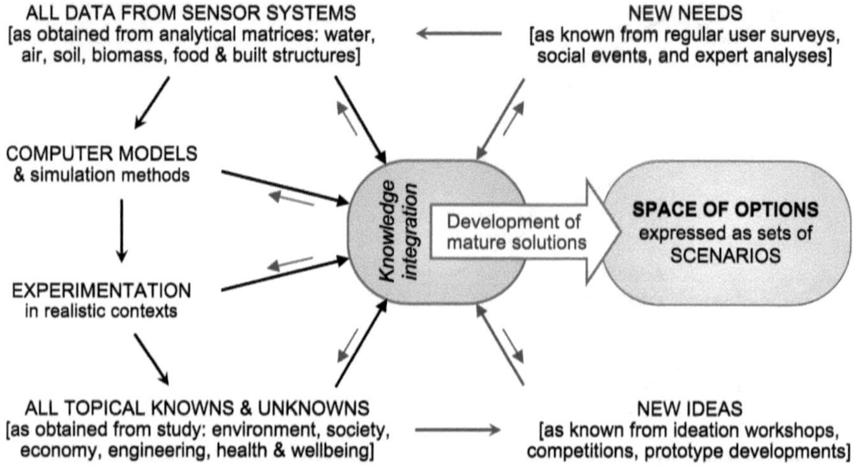

Fig. 3.28 General input–output model of problem-driven knowledge integration. This is also an example of knowledge map, one depicting the "big picture" of the process of development of solutions from initial idea to end-result. *Source* modified based on [153]

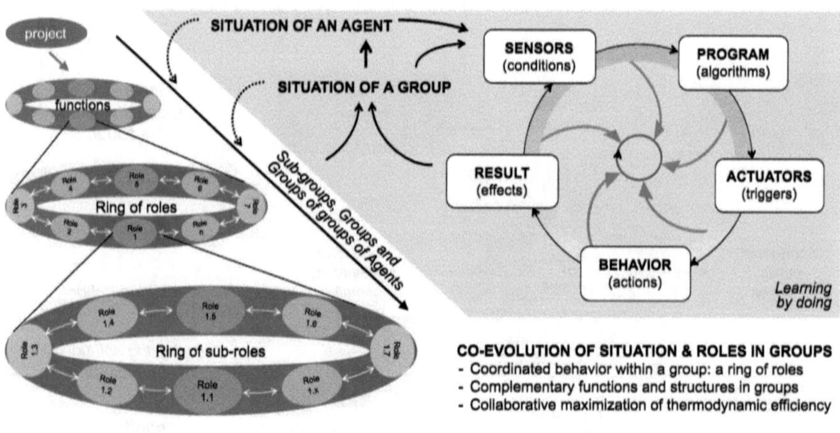

Fig. 3.29 Abstract summary of the evolution of roles (division of labor) in-project teams or/and communities of actors-stakeholders understood as groups of agents engaged in a project and learning their way by doing (interacting between themselves and with the problem). Operationally, an agent can be a person, a machine, or a group thereof. *Source* original

3.8 Stairway to Heaven by Disciplined Experimentation: Solution Readiness Levels

A General Model for Structuring Eco-city Pathways

An Introductory Discussion: the Concrete Challenges of Public Sustainability-Oriented Policies/Programs

Having a good policy-making process (as a context and basis for a good program) for the transition to sustainability is a true endeavor which faces some fundamental challenges. Firstly, as discussed so far in the book, knowledge integration across scientific and other field boundaries is both essential and hard; and it involves different levels of research (from pushing the boundaries of knowledge to coordination between domains), and this is a never-ending work by definition. The reality is that the importance of knowledge integration is being systematically underestimated, and the reason for that happening is primarily the fact that most specialists (be it in scientific research or in other professions) are usually (and, again, by definition) focused on in-depth details about their/chosen specialty topic and methods but are (with some exceptions) unfamiliar with other fields. While this situation is unlikely to change (and does not need to change completely, because we do need specialist depth of focus), a bigger but fortunately solvable problem is that the overwhelming majority of specialists are also not familiar with transdisciplinary approaches. To be fair, this might be the case because, among other causal contributory factors, it is not enough to have a general grasp of what transdisciplinary work is (i.e., holistic problem-solving-driven): also required are enough serious efforts in the direction of developing-by-doing methods and models to make *Transdisciplinarity* operational. The observation above is of course just a reminder (so, not a new discovery) that the operational level is another key requirement (and limitation). In the literature on sustainability transitions this is well-recognized since a long time: even the older theorizations on the multi-level "transition management" movement from over two decades ago were already acknowledging the "operational level" (e.g., [360], Figure 1 therein).

For further practical aims, the periodic reports of the IPCC (probably the best known and arguably one of the most advanced practical efforts towards sustainability) synthesize primary scientific research and provide some executive summaries. However, even those reports are only the beginning of the discussion, and there is a long way from those reports (which document the problem and explore some general potential paths forward) to truly systemic eco-innovation as concrete advancement (i.e., changing how we generate the goods and services which we consume, so as to lessen the overall ecological footprint, and knowing of course that this is not only about the environment, but also about the society and the economy).

In the big picture, there are many science-policy interfaces that need to be recognized and dealt with, as well as numerous "effectiveness factors and impacts on policy-making" [731]. At present, the situation can be summarized like this: we already did a relatively good job in documenting the problem but the work on solutions is still at beginnings—in spite of occasional hypes and commercial or political short-term interests that claim the contrary. In fact, many of us have already signaled in the past in different projects and programs or other conversations that such irrational hypes and "silver bullet" promises have become a most serious risk in our transition to sustainability, because they channel public attention towards unachievable goals and divert resources away from the problems that need resolution and from the real works that are required to address them. From a "policy coherence" perspective, this situation has been described as policy being "lost in action" [128].

All in all, policy-making (and programs for implementing policy) requires special attention. Even though, in principle, a policy is a means to promote solutions, not a solution per se, projects aiming at developing a city-scale policy (or, for that matter a national or other scale) require a functional awareness of (a) existing knowledge gaps and (b) potential synergies and/or tradeoffs with other topics (as illustrated in Fig. 1.10 from Module 1 in the case of Climate Change and Sustainable Development Goals). Effective policies, programs, and projects also require intelligent use of existing infrastructures (especially Internet) for communication across relevant networks of experts, actors, and stakeholders [507].

However, the context of a policy-making exercise has an impact on what is being done (or not): traditions and overall capacity of (urban or national or other) public administration can effectively help or can actually hamper the process of policy-making [72], and the details of such positive or/and negative effects have something to do with the concrete peculiarities that are embedded in the place-specific functional division of labor between elected agents (i.e., politicians), and non-elected agents, (i.e., professional administrators, sometimes called "the bureaucracy" [463]).

Insuring Citizen Participation: Democracy is a "sine Qua non" Condition

Now, what may appear to be just a collateral semantic detail is in fact reflective of an issue which is key for the whole discussion about premises for policies towards systemic solutions: the "bureaucracy".

I will start with the immediate practical aspect (and then I will continue with the reason behind it): in this book and in general, I recommend the use of the term "public administration" instead of "bureaucracy". The fact that many people from all walks of life use the two terms as interchangeable denotes a certain confusion that lingers in public and professional conversations (not completely unlike the "popular" confusion between Data, Information, and Knowledge signaled previously in this book).

As I understand it based on current knowledge, "bureaucracy" is not synonymic to "public administration" but, instead, represents the situations where/when public administration is corrupt or/and grossly incompetent. Essentially, democracy means that (or describes the social reality in which) the power (or "cratos") belongs to the people (or "demos"), while bureaucracy means that (or describes the social reality in which) the power belongs to (those holding) administrative "bureaus"/offices (an English word coming from the French word "bureaux"). In fact, the term "bureaucracy" refers to the context from before the French Revolution (of 1789–1799), when the power indeed belonged to the offices bestowed by the king (who used to say "The

State is me", in French, *L'état c'est moi*) along with numerous privileges that were at a point abolished during the French Revolution and through the subsequent social transformations. Except that, the term bureaucracy simply continued to be used in modern times in France, and from there it passed to other countries and languages, although the absolute monarchy had been toppled a long time before. Of course, one may wonder whether bureaucracy too has been toppled (or not), but that would be another discussion, for another kind of book.

What matters here is that using the term "bureaucracy" without understanding its meaning simply denotes a lack of reflexivity about word semantics. In the daily life of a person this might not be important, but, in the context of policy-making, and supposing that we want a responsible relation between people and the public administration, this kind of terminological misuse should trigger an alarm bell and operational clarifications. At best, the term "bureaucracy" could be used properly as an allegation (or accusation) against some specific deed or sub-department of public administration that requires urgent reform (or other due action).

These being said, the fact that eco-city projects require higher levels of competence also means that eco-city projects can provide some concrete benefits in this direction: they can help detect shortcomings in public administration and ring those alarm bells (to which, needless to say, competent public administrators will not fail to listen). Obviously, this general observation is also true for all the organizations that are involved in the triple-helix of eco-city projects; in the end such projects will not only generate solutions for some problems "out there" but will also help organizations shape up and raise to the challenge. If the participating organizations succeed, the whole project is in a good position to succeed. Of course, our collective experience already shows that the opposite outcome can also emerge: entrenched interests and organizations that prefer the status quo (and business-as-usual) have a natural tendency to re-enforce each other's actions and stifle any project that may challenge their preferences.

On the one hand, we have to overcome inertia and insure that the public administration does not turn into a bureaucracy—and, when it does, we need to act (within the professional logic of projects) to keep it on track and fit for the mission of serving the public interest of today (not for reproducing habits from past centuries). But on the other hand, bureaucracy is not at all the only danger for a project, program, or policy pursuing the goal of urban/national/global transition to sustainability. Eager and over-heated activism (i.e., hypes that ignore evidence and seek to impose a set of beliefs just because those beliefs are enthusiastically shared in some closed circles) is at least as toxic and unprofessional (if not plainly subversive) in the process of generating real solutions.

Indeed the discussion about bureaucracy brings us to the fundamentals: democracy is one of the the necessary (but insufficient) conditions for sustainability, a part of the social sustainability filter introduced in Fig. 1.3 in Module 1. In other words, any solution concept that is being proposed as compatible with sustainable development must, by definition, be coherent with the democratic standards of the society from the beginning—or is no valid solution at all.

In democratic countries, public administration exists as a consequence of a basic principle called "the consent of the governed", which emerged (together with the related idea of representative government) in the European cultural space during a long historical process and gained global recognition of as a morally and socially justified and therefore sustainable dynamic equilibrium in terms of balance of power in the human society (because, I should add, the roots for human dignity and democracy exist in all human communities and therefore all cultures around the world, and human emancipation in every set of historical circumstances is just a matter of time) (for a review of historical details, see [666]) and is contained in documents defining our modern civilization, like the US' Declaration of Independence (in 1776; in the second paragraph "… Governments are instituted among Men, deriving their just powers from the consent of the governed"), and UN's Universal Declaration of Human Rights from 1948 (in Article 21: "The will of the people shall be the basis of the authority of government").

To echo and complement in this context the earlier book sections discussing the issue of urban (and societal) resilience, let us observe that, democratic resilience that characterizes consolidated democracies is the capacity to withstand systemic shocks by absorbing challenges and thus avoid the erosion of democracy [81, 432]. This characteristic is supported by the predominance of emancipative (vs. authoritarian) values in the cultural background of a people [742].

In this sense, it is essential to also note that, although global trends appear to indicate that democracy is losing ground (a phenomenon known as "deconsolidation thesis" [270, 407, 742, 743]), the big picture is a little different. First, when compared with long-term historical data, the situation of democracy around the world seems to be at or near an all-times high [696]. Second, a deeper and comprehensive regard at the issue shows that democracy backslides into authoritarianism do not happen everywhere but predominantly in societies in which emancipative values are still under-developed, and that "the prevalent generational profile in people's moral orientations exhibits an almost ubiquitous ascension of emancipative values that will lend more, not less, legitimacy to democracy in the future" [742].

The overall stable (or robust) status of democracy is, in principle, a good premise for eco-city projects, but we should not forget that the majority of people around the world still live in predominantly non-democratic societies, and this is a very serious concern and drawback for the global transition to sustainability.

In other words, the fundamental aspiration of inter-generational equity embodied in the very definition of sustainable development given by the United Nations (as discussed at the beginning of this book) remains as valid as ever and even more so. And, given the high stakes (represented by "our common future"), any form of democratic backsliding (wherever it happens) remains bad news that should soberly inform our efforts to succeed in the transition to sustainability. By definition, the loss of democracy as a universal aspiration of humanity (hypothetically speaking, in a thought experiment) would be fatal to the transition to sustainable development.

This is why, to make a cross-reference to the earlier discussions on the Eco-City/Sustainable Community Reference Model (Sect. 3.4), (1) citizen involvement at all practical levels is a necessary condition for sustainable human systems, from cities to countries, to continents, and to the global community of nations, and (2) Democracy must be actively cultivated and defended every day and with all occasions (instead of being taken for granted). If you allow a small digression, I would mention that I have lived the first 14 years of my life in a dictatorship: until the fall of the communist regime in Romania (and the rest of east-Central Europe), in late 1989. And, I lived and witnessed the long and profound pains of the societal recovery—this is why I am not taking democracy, nor anything else, for granted. I know progress is hard-won and easily lost, and I am always comparing alternatives to have a balanced perspective: freedom and tyranny, richness and poverty, wisdom and naivety, good work and a waste of time (and other resources), responsibility and complacency. Equally important to say is that, because so many people around share this view and underlying set of core human aspirations, "our common future" has a chance.

Given the key role that cities have in our lives in practical terms, local communities are the quintessential loci of democratic life and collective pursuit of evidence-based policy, and eco-city projects have a critical role to play in this respect too.

In this sense, it is important for eco-city projects and related works to highlight, to use the words of [253], "the democratic relevance of cities" and the ideals of the democratic city such as "the city as a school of democracy, the urban cosmopolis, the city as a commons and the sustainable city", as well as related concepts like "democratic action" [16, 253, 664] and "deliberative democracy" [36, 206, 221, 292, 293, 403, 563] in the form of experimental problem-solving within the city, and a way to achieve a "spatialized understanding of democracy" at city scales and towards the development of new and further "pathways for democratizing urban space". For further discussions, one can also check John Dewey and his intellectual legacy on democratic action, see (e.g., [606]); especially for the importance of deliberative democracy in education, community building in general, see also [16, 651, 663].

All these notions relate easily to the 17 UN's Sustainable Development Goals, starting with SDG16 (Peace, justice, and strong institutions) and SDG17 (Partnership for all SDGs) (also shown in Fig. 1.10), and with the whole discussion that is made in this book Module plus the available literature out there.

These being said, it is no surprise that cities and local governments are often more advanced than national/state in certain topics, including climate institutions because they are closer to the citizen.

Last but not least, the state of democracy around the world is directly connected to global affairs, starting with security (of persons, of energy, of food, of water, and so on) through the "geopolitical rivalry between free and repressive world order visions" [707]. From a strictly operational perspective, this means that reaching global sustainability (a) is probably not going to be done through general collaboration between countries, even if only because many countries fail the social sustainability requirement (because they are not democratic, and the many consequences of this situation, starting with non-cooperative behavior by definition), but (b) instead, and realistically speaking, progress towards sustainability will be achieved like with all (or nearly all) other instances of progress in the past: superior solutions will be adopted by different countries at their own pace (not simultaneously). In the last sub-section of the last section of this Module, I discuss the mechanics of this transformation as anticipable based on current knowledge about how change happens, namely, in the form of "disruptive innovation".

For now, I only observe that the imperative of democracy as a necessary condition for sustainability further means in practice that (1) our major objective as Humanity is that sustainability-compatible solutions and systems emerge as superior and then as winner, and (2) because cities have a key role as engines of progress, the protection of democratic life and democratic participation of citizens in the making of their city is of the utmost importance for (and must be pursued by) all eco-city policies, programs and projects. In other words, this is very much about a healthy democracy in all its forms, not only correct and lively elections but also constant consultations and involvement of citizens in the making of the systemic solutions for their own cities through projects that not only bring together experts from different disciplines but which bring together experts and citizens. "Concocting" "solutions" in "expert circles" that ignore social and societal realities (thus being no expert at all but pseudo-expert) and then trying to deceive or coerce the public into adopting those "expert" "solutions" is a non-starter and a sure path to failure. Autonomous work of true experts is necessary (like with all professions) but this work must involve the rest of society on honest bases from the beginning, through sound dialogue and appropriate methods and tools that respect our common fundamental values.

Insuring Competent Government: An Illustrative Discussion Using the Topic of Climate-Related Policy

A competent (local or national) government is more than the sum of good individual public administrators: it rests on coherent action as consistent with the reality of complex dynamic systems and with the systemic perspective on sustainability as discussed in this book. Competent government requires a "whole-of-government" (WOG) approach, i.e., a functional coordination between public administration departments [140]. To give an example, WOG was adopted by the US administration to "assess and address" the vulnerabilities (for the national economy) that relate to the dependence (in the sector of renewable energy sources) on critical minerals such as rare earth elements (as current technology requires those in manufacturing permanent magnets, which are used in electric vehicles and wind turbines). In that respect, an effective policy includes (a) growing supply volumes and (b) the so-called de-risking supply chains; these goals must be pursued through separate (dedicated) policies because they have distinct underlying dynamics (in other words, each is a topic in its own right), but the two policies are still sufficiently interconnected to require functional coordination (for details on critical minerals, see, e.g., [272]).

Another challenge to keep in mind is the fact that unresolved differences between different governance levels (e.g., EU, national, and regional/local) naturally result in insufficient "vertical" coordination between, e.g., between country and city levels, but also horizontal coherence, e.g., between cities or/and between organizations.

Because of these reasons (and many others which we cannot possibly discuss in detail here) policy-making tends to be less like a streamlined flow process from knowledge to legislation, and more like a disorderly chain of reactions to different pulls and tensions popping up in a local or national community as the most pressing (the "biggest burning fires" for the public administrators to extinguish). Equally important to acknowledge is that, so far, the policy-making process has been studied mostly retrospectively (the focus was "what has been done") although a more explicitly prospective understanding ("how policy might successfully reach its objectives") would be more helpful in practice [237].

In the case of the status of climate policy (and keeping in minds that climate change is perhaps the most widely recognized topic related to the sustainability), it is important to understand that, although there is a great richness of climate studies documenting the problem, there is overwhelming paucity of policy studies. Overall, at this point, the situation is characterized by a lack of realism, big "implementation challenges", and a "weak delivery". One (and perhaps the main) problem is that many national climate policy (as well as strategies and plans) objectives tend to be limited to "importing" global agendas without due adjustment to national, regional, and local contexts and vulnerabilities. Unsurprisingly, public participation and support and the policy-evaluation mechanisms are still very limited. However, these issues are common around the world (or at least at regional levels) and could therefore occasion communication and learning between experts and constituencies in different countries and localities (for more detailed discussions, see, e.g., [18, 44]).

Yet another serious challenge is with the requirement of appropriate mobilization of funds (public and private), as "it is crucial to understand which policies are effective" (for more detailed discussions, see, e.g., [570]). A central problem is that most investors rely on current climate models without giving enough feedback to help develop those models, the result being that climate models are weaker (in terms of that "to the best of our knowledge" supporting technical model assumptions). In the end, financial risks are higher than they could be.

One such risks and challenge is an overall "disorderly transition" (as it has been called by the Network for Greening the Financial System (NGFS) (for an in-depth discussion, see, e.g., [48]). Another (but related) risk and challenge is the generation of inherent systemic conflicts: for instance, the potential "carbon lock-in" created by short-term quantitative easing, i.e., large asset purchases by central banks (to sustain liquidity on the markets), which contradict and undermine efforts by the same institutions to encourage financial markets to "better account for climate-related risks" (see, e.g., [113]). Such contradictions suggest insufficient scientific study of the systems of interest, and insufficient coordination between different policy-packages, an additional major effect being a higher risk for countries and businesses to end up with large stranded assets (for a discussion about stranded assets, see, e.g., [112] and the technical summary in the next sub-section).

These complex challenges make that, the "raison d'être" (i.e., 'reason to exist') of a policy-making process is (in a nutshell) to make the right distinctions and "find the appropriate short-term action options that are compatible with long-term goals and keep future options open" ([604], see also [87, 706]).

When it comes to identifying the general trends in a society, this can be done through (more or less usual) consultations. However, this is only the beginning. A sustained engagement of a community on a path of systemic transformation towards sustainability requires a community of actors and stakeholders to keep asking questions that go down to operational details. In other words, thorough policy-making requires sufficient analytic granularity, clarity of understanding and realism, starting with the acknowledgement and correct understanding of the reality of complex dynamic systems (CDS; as per Fig. 1.6, in Module 1), and the fact that we are not (and have never been, and cannot be) separate from nature but, our own existence is occurring within those "coupled human and natural systems" (CHANS) that I briefly introduced in the beginning of this book [235, 307, 308, 395, 421, 422].

This also means that policy-makers do not have the freedom to dismiss scientific knowledge, and must constantly strive to avoid any of the situations (e.g., carelessness, imprudence, ideology/priors, or spiraling rhetoric) that would make them ignore evidence. Failure to meet this basic requirement will inevitably have negative consequences throughout the social systems (as detailed in the sub-section on risks, in Module 1, Sect. 1.3): aggravation of current problems, cascading disasters, and a (logical) collapse of the social system.

From the field of "organization studies", we already know that the kind of failure I mentioned above actually happens much more often than we realize and more often than we might want to admit: there is an already well-known, long list of cognitive biases and (consequently) an equally important list of behavioral biases. These are routinely distorting our individual decisions (even when we consciously try to avoid them, because most of them are deeply engrained in how we are wired as humans).

Group decisions are prone to similar biases but there is a key difference: the inner diversity of perspectives that occurs in (sufficiently large) groups provides the grounds and balance for obtaining high-quality decisions.

Such biases decisions distort resource allocations: to the disadvantage of the truly-best projects and favoring project proposals that "look best on paper"; and from relevant and necessary long-term investments to predominantly short-term investments. In other words, we often ignore key information items even when that information is right in front of our eyes ([242, 348, 638], see also [318]).

From this understanding, one can derive two key facts about the nature of good policy-making:

1. in practice, any planning aspiration (legitimate or not) is limited by misalignments between chosen goals and between actions, and by complex (intended or unintended) effects of different actions, by contexts, and by just the intrinsic properties of all CDS.
2. consequently, any policy that is strong enough to have an impact will (logically, have to) co-evolve with its moving target (because, if there was an impact of that policy, the situations changed and the change in CDS is always complex, so the goals of the policy in place must adjust to this new situation).

This means that priorities, goals, tasks, and methods need to be updated all the time, based on the new, incoming, and updated Questions, Knowledge, Information, and Data items—as per project work cycles described by the DIKAR_process model in the previous book section.

For current policy-making, this also means that the essentially political negotiations in the COP meetings of the UN's Framework Convention on Climate Change (UNFCCC) represent only a necessary but insufficient condition. Cities and city networks have an increasing role in the ensemble of these negotiations, but these global negotiations remain, at least in legal terms, a matter between national governments. However, it is already widely acknowledged that any solution must coming out from any international negotiation must be accepted by those living with the most direct (and impactful) climate changes, i.e., the local communities. This further means that the real content of the COP meetings must include relevant operational details which cities (and local communities in general) have agreed or can agree on, i.e., we talk about policies that must be coherent across scales.

Equally important to acknowledge is that the sets of expertise and skills (and therefore the profiles of people/team members) that are required in the description of the problem versus the generation of solutions are different. Projects in the respective endeavors rarely involve the same individuals and/or institutions. There is some overlap of interest but not much more, and the relation between the two kinds of work is one of necessary complementarity: these kinds of people have to work with each other but they need to organize their work in their own respective ways.

Unfortunately, here too reigns a lot of confusion, in the sense that, for example, there is sometimes a strange expectation from public (and private) administrators that "climate experts" who know the problem also have the solution to the problem. They do not, and that is normal; they may participate in some of the projects aiming at generating solutions, but whether or not they can and want to participate in this or that solution-generation project depends on many factors.

Then, there is yet another important issue in policy: the chronic (and oftentimes acute) policy accumulation because the rhythm in which legislation is generated exceeds the rhythm of (older) policy termination). One easily understands that this particular problem signals that there is a lack of systematic evaluation of the complex causal contributions (and therefore of the conditional effects) that this or that policy or policy component may have [4].

The underlying question, therefore is: *Can we find a general (and generalist) way to make logic updates of existing policy and legislation instead of just adding new policy "patches" over old ones?*

Fulfilling the necessity of having a smart and coherent (instead of heavy and chaotic) policy and legislation is important because this can "make or break" public support for existing or/and would-be policies regarding climate change and the transition to sustainability in general [204, 449]. This issue has already been documented by policy studies: around the world, numerous departments and units of public administration have devised various sectoral policies and "took action" within their purview (i.e., assuming that coordination between sectors is unnecessary, and that different measures would just "add-up" and contribute to a general progress forward), only to realize—after those policies started to have effects—that many of those sectoral ("in my department of responsibility") measures were contradictory in practice, have unexpected effects, notably reciprocally canceling each other across scales or even collectively generating in overall negative effects [403]. In simple words, "doing something, anything" does not represent progress: in fact, this tends to worsen the problems.

This issue is not marginal but central in climate policy, and climate policy anticipates many basic issues of sustainability policy in general (even though it is only a part of it). As already shown in this book in Fig. 1.10 (Module 1), IPCC documents this issue in its most recent (the sixth) assessment report.

The bottom line is that we need a policy-making process that makes sure that the decision-makers "think ten-n times before cutting once" (be it with a new policy or with a policy update), starting of course with the national level (because nations are the central actors in the international negotiations over climate) but in coordination with cities and all the actors that are locally relevant.

This nation-city duality is very important: cities cannot act outside the authority of national governments, but, at the same time, national governments cannot ignore what happens in the daily context of citizens, i.e., in cities and towns.

Concretely, engaging in superficial short-term "solutions" (that may help extinguish some immediate political "fires" without addressing the root problem and thus setting other fires in other places) inevitably exhausts public resources and support, wastes precious time and undermines trust in public institutions, and hinders cooperation between people and institutions.

In short, in the whole policy-making process (that is concerned with either "hot" topics climate changes or the all-encompassing challenge of urban and national development that is compatible with sustainability goals), the idea is that we need to learn from past experiences but also to start with an emphasis on more prospective work, i.e., to try and answer this question: "*How to generate a good sustainability-oriented policy (instead of just critiquing what was already done)?*".

Recently, I had the opportunity to do just that, as a participant in a national policy-making exercise that was focused on climate change (mitigation of and adaptation to), in the context of national authorities coordinating with the rest of the European Union. Even though that exercise did not explicitly deal with cities (because it was meant for the use of a national government), the issue of local context and sustainability seen from the level of a city came regularly in the conversations, and several of us stressed with multiple occasions that we need to talk to cities (if you want to see this country-level report, check [573], in Romanian, with a summary in English]).

Relatedly, I should observe that cities have also featured highly in earlier works in which I participated, for example, right from the beginnings (in 2010) of *Climate KIC* (*Climate Knowledge and Innovation Community*), one of the several topic-driven network projects (bringing together universities and companies, start-ups included, from several European countries, notably France, Germany, The Netherlands, UK and Switzerland) of the *European Institute of Innovation and Technology (EIT)* funded by the European Commission. Similarly, cities represented and represent a frequent direct or indirect topic in other networks in which I participated, e.g., in conversations that took place decades ago about the security and sustainability of energy in Europe (see, e.g., [639]).

Especially the insights from that recent national exercise have determined me to start develop a general model for a step-wise solution development that can be applied to any scale: that of a project, of a city, of a country, or of a continent, with the goal to learn from past and current insights to develop a mature policy (with inputs specific to this goal) or a more classical item as in RDI, e.g., an applied technology, or a public policy, or a commercial good/service for specialized or wider use. This model is presented and discussed in the remainder of this section.

Making Complex Problems Operationally Tractable: A Brief Technical Summary

In complex dynamic systems, a space of options for addressing a complex problem comprises one or more of (mostly yet-unknown) possibilities, and the longer you look in the future, the lesser visibility you have. When the long-term goal is to take the shortest overall path, the logical approach is to seek the best short-term action that keeps future options open.

Concretely, and as illustrated in Fig. 3.30, the prospective work (and the overall managerial challenge) of finding the best path of a city towards sustainability is an instance of what is known in mathematical optimization as "dynamic programming", with the additional observation that the qualifier "best" in the expression "best path" refers to the shortest amongst all potentially viable paths (and knowing that longer paths have a lesser viability and that they are more exposed to unexpected effects and disruptions).

One should also not forget that the word "shortest" in "shortest path" means a path that is both viable and as short as possible (but not shorter than that, as this would become "non-viable").

The situation described in Fig. 3.30 is a typical problem in the fields of Operations Research and the broader Mathematical and Computational Optimization. The way

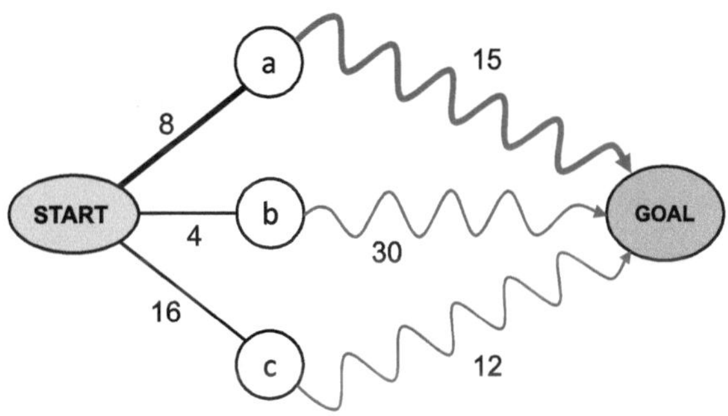

Fig. 3.30 A basic abstract model of the challenge of finding the shortest path from start to goal in a complex dynamic system, i.e., identifying the optimal substructure of a path-finding problem (as known from dynamic programming in mathematical optimization). "Start" refers to the given situation of any city. "Goal" refers to sustainability and in practice means full compatibility of a target city with sustainable development. Straight lines denote known (clear) sub-paths between known "stepping stones" on short term. Wavy lines denote yet-unclear (but real) sub-paths. Thick lines indicate the overall shortest path. Numbers illustrate comparative efforts that may be required by alternative paths. The values displayed in here are chosen to show that what can appear to be the shortest path on the short term can still turn out to be the longest overall (if still viable at all)

to deal with it in "dynamic programming" is as follows: a multi-period planning problem is broken into simpler steps at (different points in time), i.e., to obtain a stepwise gain of visibility onto what is best to do next. This approach also insures a gain in operational tractability of the target problem (like minimizing travel distance, minimizing cost/effort, maximizing utility, and so on).

In the next paragraphs, I will summarize the technical aspects of this approach. So, if you are already familiar with it, or if you are otherwise not directly concerned with the method details, you can skip it (I only include this technical part because I think it is important to have it at hand if needed).

At the beginning, the simple fact that future steps are unknown can be formalized as an infinite-horizon decision problem of the form

$$V(x_0) = \max_{\{a_t\}_{t=0}^{\infty}} \sum_{t=0}^{\infty} \beta^t F(x_t, a_t),$$

where $V(x_0)$ is the optimal value obtainable by maximizing the objective starting from an initial state x_0 (and continuing through the state x_t at time t), assuming that the state x changes to a new state $T(x, a)$ when action a is taken, under the constraints $a_t \in \Gamma(x_t)$ and $x_{t+1} = T(x_t, a_t)$; $F(x_t, a_t)$ denotes the payoff of taking action a, and $0 < \beta < 1$ is a discount factor, i.e., the impact of other factors, notably unexpected/undesirable effects and imperfect execution (e.g., impatience).

Under the Bellman's Principle of Optimality ("an optimal policy has the property that whatever the initial state and initial decision are, the remaining decisions must constitute an optimal policy with regard to the state resulting from the first decision" [55, 56, 203, 275]), this problem can be broken into smaller sub-problems. Namely, by separating the current decision from future ones and re-writing the problem as a recursive definition of the goal value function (which also implies that each decision in the series is made by acknowledging explicitly that all remaining-future decisions will be made optimally), one obtains the Bellman equation:

$$V(x_0) = \max_{a_0}\{F(x_0, a_0) + \beta V(x_1)\}$$

or, as simplified by referring directly to the value of the next state:

$$V(x) = \max_{a \in \Gamma(x)} \{F(x, a) + \beta V(T(x, a))\}$$

By solving this "functional equation" (i.e., finding the unknown best-possible value V of the objective as a function of the state x), one also finds the "policy function" $a(x)$ describing the optimal action given the state x. As an observation, this is a mathematical term in which the word "policy" can be used in a mathematical sense (i.e., independently from managerial concerns related to a "policy" in general), but it does correspond to the same core idea.

Put differently, in finding the shortest path (as per Fig. 3.30) the aim is to identify the lowest number of sub-paths, i.e., sub-problems into which a path (problem) can be broken to accomplish the shortest path from Start to Goal.

In practice, this says what we intuitively know if we have enough experience: basically, the purpose of a sustainability-oriented policy or long-term program is to make the right distinctions and identify those short-term action options that are (1) appropriate with respect to current knowledge, (b) compatible with the long-term goals, and (c) keeping future options open ([604], see also [87, 706]).

With a view to the exploration of the literature, I would also remind here that the Bellman principle and equation find applications in virtually all fields, both in the form described here (for discrete-time problems) and as partial differential equation (for continuous-time problems; also known as the Hamilton–Jacobi–Bellman equation). Therefore, discussions on the concrete issues that need to be addressed in the context of cities will inevitably include some of these applications, both in natural-sciences-and-engineering and in social sciences, especially in economics. But one can also find studies that discuss the Bellman equation in explicitly urban settings, e.g., the management of urban traffic lights (e.g., [40]), the management of urban drainage systems (e.g., [516]), *et cetera.*

At this point, the reader might have already noticed that Bellman's Optimality Principle is implicit in the work process discretization made by the DIKAR_process model discussed in the previous two sections, but also in the notion of sustainability filters discussed in Module 1, and in the "maturity models" known from RDI, starting with but not limited to Technology Readiness Levels (TRLs).

This optimality-driven stepwise approach is the essence of the generalized model described in the next sub-section and the models discussed in the last two sections of this Module and book and applying in practice the idea of acting "to the best of our knowledge" inside and upon complex dynamic systems.

A General Stepwise Approach to Complexity: Solution Readiness Levels (SRL)

Figure 3.30 and the related discussion in the previous sub-section reminds of the intuitive managerial stepwise approach described in the Delphi-based method for generating scenario sets as described in Module 2. To this conceptual convergence adds the idea of a city's trajectory as described in Fig. 1.9 (Module 1) and the idea of "sequential decision making" as a logical approach to decision-making (itself derived from the same body of knowledge provided by fundamental physics and science in general (see again [529]), and the numerous insights coming from eco-city projects, RDI and policy-making as discussed so far.

The result of this entire synthesis is summarized in the model that is presented and briefly discussed in the remainder of this Section.

For eco-city and related eco-innovation projects, the notion of disciplined experimentation plays a central role [155]. Concretely, not only is this important in technology developments and at the front-end of innovation when "product parameters are still flexible" [78], it also matters in the general development of systemic solutions, when new paths (and different scenarios) are still "alternative solutions" that can be modified through interaction between participating experts and stakeholders, i.e., before becoming the "known solutions" mentioned in the DIKAR_process model (in Fig. 3.17, Sect. 3.6).

In the current political and economic context, the idea above also resonates with older signals about the imperative of building versatility into our techno-socio-economic systems, starting with energy and urban systems (e.g., [150, 151, 639]). The main point to highlight here is that the current state of human global affairs (geopolitics) shows again that "indulging in long-term "blindness" for the sake of short-term ease logically leads to a rude awakening" [155].

In projects, instead of generating uncontrolled and resource-consuming chains of insufficiently thought-through propositions towards solving complex problems, collaborative work can flow coherently from basic research to sophisticated modeling, tests, and refinement, and on to validation by use in relevant real-world contexts.

For this, Fig. 3.31 shows a synthetic framework model for applying the disciplined experimentation and the related stepwise approach discussed above: a "parent scale" called Solution Readiness Levels (SRL) and two particular applications to special kinds of concern: the well-known TRLs ([445, 446, 521], see also [155, 156]; further discussed in this context in the following paragraphs) which are here understood as the Technology-focused SRLs, and the PRLs (Policy-focused SRLs) (an on-going work and publication related to PRLs is currently under peer-review and expected to be published later on).

This model relies on the notion of Dynamic Programming described above and the rich body of experience and insights provided by the empirically-developed TRL scale since almost 50 years to operationalize the core idea that the general development of solutions (or any kind: a technology, or a policy or anything else) happens through advancement along a gradient from exploration to exploitation, in a dynamic equilibrium (at every step) between relevance and efficiency and acknowledging the reality (illustrated by the TRLs) that the concrete work conversation at the different levels along that gradient go very differently and therefore require a dedicated "room" in the form of a distinct "level".

In management, this good equilibrium between exploration and exploitation (or "ambidexterity") refers to the challenge of maintaining a good equilibrium between the efforts directed at finding new ways to fulfill a target demand (i.e., RDI) and the ways to obtain commercial success (i.e., ROI—return on investment). By the SRL approach described here, an eco-city project positions itself to maximize project success rate and minimize opportunity (and transaction) costs.

The relations between these three scales are being discussed in the following paragraphs. Similar relations can be identified or explored with the other maturity concepts and models from the different fields; for example, the notion of Minimum Viable Product (MVP) or Minimum Viable Service (MVS) that is well-known from entrepreneurship may correspond to SRL 6 or 7, depending on the context.

Development and Benefits of the SRL Model: A Summarizing Discussion

In practice, the idea of maturity scale first emerged empirically and consolidated through practice in the form of Technology Readiness Levels (TRL) model: a project

Solution Readiness Levels (SRL)

9	Solution set consolidated through successful operation
8	Mature solution set assimilated refined feed-backs from real world
7	Basic solution set demonstrated utility in relevant real-world localities or sectors
6	Solution options demonstrated in relevant context (pilot use in limited settings)
5	Details (solution parts) tested in realistic complex models (including simulations)
4	Details (solution parts) verified in simple logic models
3	Real-life (analytical) data appear to be supporting the formulated possibility
2	Solution possibility (concept) or/and direction formulated
1	Basic principles observed & reported

(6-9)*: solution deployment institutions

(1-6)**: solution generation institutions

(1-3-4)***: convened groups of experts

Technology Readiness Levels (TRL)

9	Actual system "flight proven" through successful mission operations
8	Actual system completed and "use qualified" through test and demonstration
7	System prototype demonstration in space environment
6	Sub-/system model or prototype demonstration in a relevant use environment
5	Component and/or breadboard validation in relevant environment
4	Component and/or breadboard validation in laboratory environment
3	Analytical & experimental critical function and/or characteristic proof of concept
2	Technology concept or/and application formulated
1	Basic principles observed & reported

(7-9): final operational validation

(1-6): demonstration of potential value

Policy Readiness Levels (PRL)

9	Policy packet consolidated through successful operation
8	Mature policy packet assimilated refined feed-backs from real world
7	Basic policy packet demonstrated utility in relevant real-world localities or sectors
6	Policy options demonstrated in relevant context (pilot uses in limited settings)
5	Details (policy components) tested in realistic complex models (including simulations)
4	Details (policy components) verified in simple logic models
3	Real-life (analytical) data appear to be supporting the formulated possibility
2	Policy possibility (concept) or/and direction formulated
1	Basic principles observed & reported

(6-9)*

(1-6)**

(1-3-4)***

Fig. 3.31 Solution Readiness Levels (SRL) framework model with its versions for the development of Technology and Policy, respectively, as applicable for eco-city and eco-innovation projects. Shaded levels show the middle zone where push (topical experts) and pull (need/market) factors meet. TRL levels are the "classic" ones already known. The definitions of the SRL and PRL levels are original and have been derived analytically from the TRL as explained in the main text management tool used in RDI to develop the potential encapsulated in some initial science-and-technology ideas and to achieve practical value in the form of mature products, e.g., the "ready to fly" equipment by NASA, where TRLs originally emerged as a method, but also in RDI projects funded by the European Commission and other institutions, and in both public and private sectors [445, 446, 521].

As inspired by TRL, other solution maturity models have been developed and are being applied with very different purposes, thus confirming the underlying general logic of the original TRLs [100, 727, 751], the key goal of connecting all these applications being the achievement of maturity of a desired solution concept.

Thus, it is no surprise that different maturity scales (similar to the TRLs) exist. For example, the U.S. Accountability Office uses a set of readiness level scales as complementary to TRLs, as a practical aid in mapping key "decision points" along an internal process of public funds allocation—notably a Manufacturing Readiness Levels (MRL) model, an Integration Readiness Levels (IRL) model, and a Systems Readiness Levels model (SRL; not to be confused with the "solution readiness levels" discussed in here) [709].

Another good example is provided by the last IPCC-WGIII report [332], from which I cite in Box 3.4 a short summary in full for the readers to be able to see the details and draw their own insights.

Box 3.4 Text snippets from the IPCC-WGIII report (2022, Sect. 16.2.1.4).

As an observation, be informed that this is from the pre-final report, and the final edits of that document might introduce some modifications like punctuation or word choices; so all eventual text errors are mine and I invite the reader to check the definitive text version of that IPCC-WGIII report once available.

"*Technology Readiness Levels (TRLs) are a categorization that enables consistent, uniform discussions of technical maturity across different types of technology. They were developed by NASA in the 1970s* [445, 446] *and originally used to describe the readiness of components forming part of a technological system. Over time, more classifications of TRLs have been introduced, notably the one used by the EU. Most recently, the IEA extended previous classifications to include the later stages of the innovation process* [328] *and applied it to compare the market readiness of clean energy technologies and their components* [328, 520]. *TRLs are currently widely used by engineers, business people, research funders and investors, often to assess the readiness of whole technologies rather than single components. To determine a TRL for a given technology, a Technology Readiness Assessment (TRA) is carried out to examine program concepts, technology requirements, and demonstrated technology capabilities. In the most recent version of the IEA* [328], *TRLs range from 1 to 11, with 11 indicating the most mature (see Table 16.2).*

The purpose of TRLs is to support decision making. They are applied to avoid the premature application of technologies, which would lead to increased costs and project schedule extension [708]. *They are used for risk management, and can also be used to make decisions regarding technology funding and to support the management of the R&D (Research and Development) process within a given organization or country* [198].

In practice, the usefulness of TRLs is limited by several factors. These include limited applicability in complex technologies or systems, the fact that they do not define obsolescence, nor account for manufacturability, commercialization or the readiness of organizations to implement innovations [224] *and do not consider any type of technology-system mismatch or the relevance of the products' operation environment to the system under consideration* [446]. *Many of these limitations can be eased by using TRLs in combination with other indicators such as System Readiness Levels and other economic indicators on, for example, investments and returns* [328]."

For the reasons discussed above, and to formalize (for the purpose of this book) the general the need for a structured allocation of resources in the complex projects and in the overall process of generating eco-innovation, I applied the classic principles of scientific modeling to articulate the Solution Readiness Levels (SRL; this abbreviation is not to be confused with the "systems readiness levels" mentioned above) for broader use in eco-city projects.

Essentially, the SRL model can be understood as "a succession of core requirements to be met from ideas to end-result" but—and this is an important distinction—without excluding or ignoring the feedback loops that exist between levels in the form of updates when new items of data, information, and knowledge become available or, for that matter, when the required features of the aimed-at solution have been updated. Indeed, such an update of the target automatically triggers a full or partial re-evaluation of the location of the solution under development on the SRL scale and, consequently, of the practical conclusions regarding what remains to be done to achieve full maturity of that solution (Figs. 3.32 and 3.33).

This model basically details the blue arrow ("development of mature solutions") in Fig. 3.28 (on knowledge integration, in Sect. 3.7) and the pink arrows (development of alternative solutions) in Fig. 3.17 (on DIKAR_process, Sect. 3.6). This scale applies to any solution as defined according to a target problem. Thus, once the transition of a city to sustainability has been defined as a problem to be solved, and this problem has been studied enough to be re-defined as having an optimal substructure, the generalized scale can be applied iteratively to each and every problem and subproblem, from the development of a certain technology or service or infrastructure (which in the end is also a service for citizens) to the city itself. In this sense, this model provides a universal method for organizing projects pursuing eco-city/eco-innovation goals.

Concretely, each stage is described by a list of expert-determined, purpose-driven questions (of the format: "*Has requirement or standard X been met already by the solution draft D?*") with a binary "yes or no" answer. When (almost) all questions

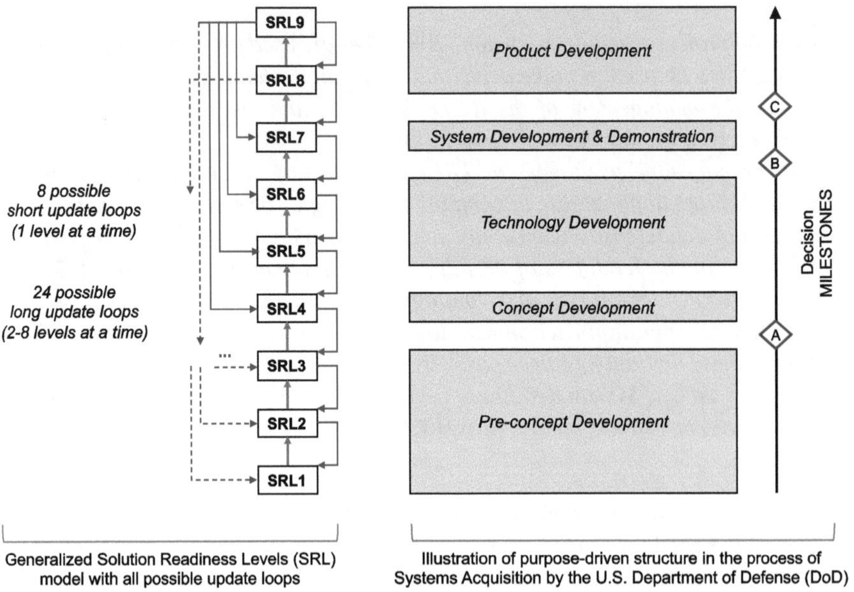

Fig. 3.32 General Solution Readiness Levels (SRL) model for any eco-innovative product/service, business, or policy, illustrated (from an operational regard, and bearing in mind the very origin of scenario analysis in the military domain) with a de facto use case of defence equipment. *Source* original; illustration based on [521, 709]

of a stage have been answered "yes", development can pass to (or rather "grow" to include) the next stage with its own, different requirements. The entire process can be monitored through a table (e.g., in MS Excel, as it usually happens with TRLs) that instantaneously displays the status (state-of-the-art technical requirements, what has been achieved, what is left—and why). Thus, any team/constituency knows the situation anytime and can interact with anyone to the point.

Intuitively, a mature solution (e.g., a technology or policy) can be imagined as a sphere-shaped fruit with a central heart (SRL1; represented in all cases by Science or Knowledge in general) wrapped up in 8 successive layers (SRL2-9). Then, qualifying a would-be solution from the core upwards is equivalent to adding one or more new layers (each layer being a list of technical, expert-determined requirements). Reassessment of a solution is therefore bound to be similarly orderly: one knows precisely where on the scale is required for the solution developer/updater to look again and how the required update progresses to recover solution maturity.

From a procedural perspective, as the promotion from one step to the next is based on an expert yes/no decision, the pathway from start to goal can be easily formalized by regarding it through the lens of computational Complexity Theory, i.e., the SRL set of scenarios (mapping that path to an expert-defined aim) can be recasted as an decision problem, so that the procedure is an algorithm and the expert-determined content of each step is the input. This inputted content can be any data, information, and knowledge proposed/agreed upon by a project team.

General phases			*STEPS in orderly succession*
Test, launch & operations of solution set	END	9	Solution set consolidated through successful operation
	Solution set or subset development	8	Mature solution set assimilated refined feed-backs from real world
Demonstration of potential solution		7	Basic solution set demonstrated utility in relevant real-world localities or sectors
		6	Solution options demonstrated in relevant context; pilot programs in limited settings
	Potential solution development	5	Details (solution parts) tested in realistic complex models (including simulations)
Research to prove feasibility		4	Details (solution parts) verified in simple logic models
		3	Real-life (analytical) data appear to be supporting the formulated possibility
	Path-finding research	2	Solution possibility (concept) or/and direction formulated
START		1	Basic principles observed & reported

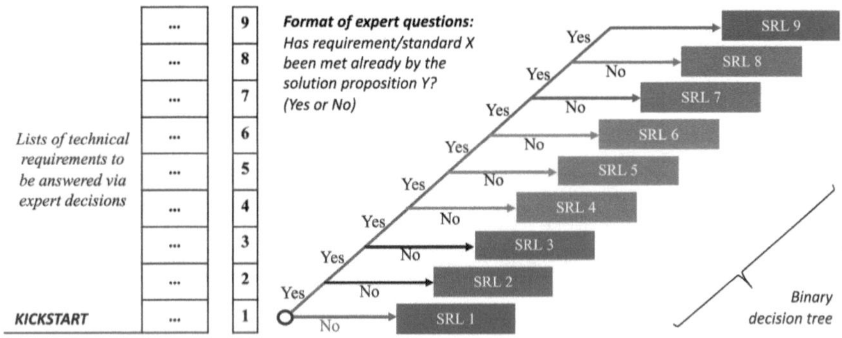

Fig. 3.33 Solution Readiness Levels (SRL) as a generalized scenario set for stepwise, lowest-risk advancement in the generation of mature solutions to the target problem. It is a general scale for structuring the always complex and usually costly process of solution development. "..." denote level-specific lists of expert requirements. *Source* original

In practice, this means that the content of each step (e.g., the list of requirements and other technical specifications) functions as an expert-determined filter and that the achievement of each of those requirements (i.e., allowing a "yes" answer) can itself represent the outcome of a string of steps.

Thus, a city's viable path to sustainability is formalizable as an *Eco-city solutions Readiness Levels (XRL)* framework model (Fig. 3.34), and the fulfillment of each of the many requirements that are listed inside an XRL step can be regarded as an end-result of successfully running the path from an eco-innovation idea to a mature eco-innovation (i.e., one that helps a city advance on its path to sustainability)—which itself, can be formalized procedurally as an *Eco-innovation Readiness Levels (ERL)* scale (Fig. 3.35).

Logically, this means that the outputs of ERL serve as inputs for XRL, so that the advancement of an eco-city towards sustainability can be imagined intuitively as a river with nine sectors in succession from spring to mouth, each sector receiving the combined "waters" from the earlier river sector and a set of smaller rivers represented

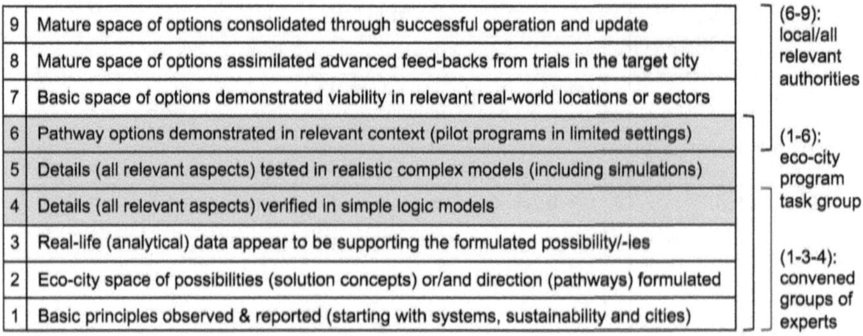

Fig. 3.34 Eco-city solutions Readiness Levels (EcsRL or XRL) framework model for structuring an eco-city program; and a start basis for coordinating (and integrating results from) eco-innovation projects. Shaded levels show the middle zone where push (topical experts) and pull (need/market) factors meet

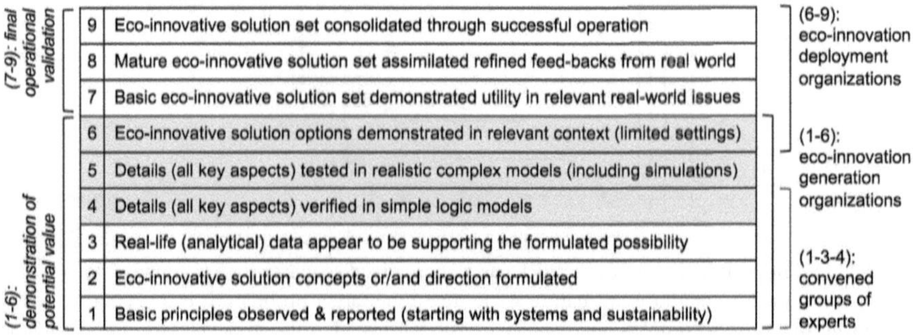

Fig. 3.35 Eco-innovative solution Readiness Levels (EisRL), or Eco-innovation Readiness Levels (ERL) for short: a model for structuring the generation of eco-innovation for eco-city projects. Shaded levels show the middle zone where push (experts) and pull (need/market) factors meet

by mature-enough eco-innovations flowing into a given segment of the river (those eco-innovations being distinct by scope and features but sharing the fact of being generated through the ERL). Further, the XRL and the ERL models share the fact that they represent the most efficient respective pathway that is possible from their respective "start" to their respective "goal".

As an observation, this relation between eco-cities and eco-innovation(s) does not mean that a city is an eco-city only if it uses mature eco-innovations (quite on the contrary, the fact that a city is an eco-city can trigger eco-innovation generation, starting from ERL1, and the fact that there is an eco-innovation process feeding the eco-city is part of the proof that a city is engaged on a path to sustainability and thus being demonstrably an eco-city. In addition, many or most eco-innovations are being developed in the context of an eco-city from the beginning, as the city is their motivator and test bed). It only means that the advancement from one XRL level to the next requires a sufficient number of mature-enough eco-innovations (say, from ERL7 or at least ERL6, depending on how the minimum viable product (MVP) or service (MVS) is defined in each concrete program).

Further, this complex interaction (but still logical and straightforward at a close look) between eco-city and eco-innovation highlights the importance of cities as loci of solution generation for sustainability (as discussed already in the first part of the book), and the benefit of the general SRL scale (taken together with its daughter scales) for "knowing what we are doing" and disciplined experimentation: it serves for comparing solution concepts and options between different projects and constituencies. Or, to use a previous expression from [629], talking about cities as 'living labs' and related concepts, it serves for "shaping real-world laboratories by comparing them" (see also [115]).

All in all, the scales and their feedbacks and inter-scale interaction pathways give an operational mapping of the Space of Options, and thus provide concrete details that matter in the generation and the implementation of the general directions and scenarios discussed in Module 2 of the book. Obviously, these scales serve for the generation and implementation of detailed "solution development scenarios" at the scale of concrete projects. In other words, these "technical scenarios" (if you will) serve for answering the "why" and "how" types of questions about the grand policy scenarios that give the general trajectory of a city as an eco-city.

Supplementary Discussions

In the rest of this section, I use these three models from an RDI and project management perspective to outline some of the issues sitting at the crossroads between natural and social sciences and experiences.

Observations About Policy-Making and PRLs

The stepwise logic of the Policy Readiness Levels is a way to bring, fit and adjust together many concepts that are known to have been proposed towards overcoming gridlocks (from geopolitical to national to local and back) in climate changes and the other issues related to the goal of sustainable development, namely, to address the global "tragedy of the commons" and other social dilemmas, e.g., polycentric and multi-scale developments [14, 284–285, 361, 534, 535, 706, 726]. The PRL scale model can do that because it recognizes and maps the continuity between bottom-up and top-down dynamics.

Good scientific and science-society conversations are not just "good to have" but are in fact necessary especially in situations of acute (and hardened) economic and political competition [118], which is what we have right now in many regions of the world: a complex climate-energy-security nexus. In that sense, the PRL scale model is a knowledge organization template to use in concrete projects and can serve as a methodological common denominator for engaging semantic ambiguities through coherent science-policy interaction [732]; and as a framework for policy-relevant communication between countries, between cities, and between a country and its cities, i.e., to operationalize what [387] called "the concept of shared climate policy assumptions" and the developments based on them. For RDI in general, the

PRL model serves to channel the very many and very different types of specialized experience and models towards the end goal (PRL9), by integrating all those in an evidence-based stepwise manner.

In short, because it combines conceptual simplicity and operational versatility, the PRL model (like the TRLs and the entire SRL family described above) addresses the need of a project to meet the classic requirements of "performance, schedule, and budget" [446]. As a side remark, I hope this will also help address some outstanding climate-related policy issues at a global scale (as these have a strong influence on what happens in cities). To give one example, at the time of finalizing this book manuscript, there is still a lingering absence of an operational framework for the Green Climate Fund (GCF) of the United Nations Framework Convention on Climate Change (UNFCCC)—a problem which combines with the frequently insufficient administrative capacity, transparency and accountability at national and local levels to delay the development of solutions [525].

Another problem that can be targeted with the PRL model and method is the still-poor understanding (and recognition) of the potential value of RDI for different emerging technologies which may be highly beneficial (and, given the current state of affairs, might become badly needed very soon) but which right now also have some very high (even critical) opportunity costs [300] or/and information-processing costs (see also [226, 529]). With respect to such difficult issues, PRL helps us pursue non-naive climate policy developments.

In the case of renewable-source energy, a non-naive approach begins with acknowledging that energy systems are complex dynamic systems (CDS) as well and cultivating (and integrating) a set of required attributes of these systems: resilience, reliability, security, economic investability, affordability, versatility (also flexibility), and sustainability. Arent et al. [25] sums it up as follows: "*The intelligent design of these increasingly complex, multidimensional energy systems is a significant challenge. Innovations required include approaches to formulate and optimize the complex, dynamic, multiscale interactions among energy sources, electricity generation and distribution, energy services, energy-intense processes and products, and markets*". So, the PRL model (and the SRLs in general) is a framework that can help policy-makers (and solution developers) in managing the exponential growth of uncertainties in complex systems and thus fight the "negative learning" i.e., "the phenomenon of new information leading to increasing or sustained divergence of current belief from the true outcome" [41, 526].

Relatedly, this also means the "traditional" approach to science-policy interactions consisting in a small group of political representatives making top-level decisions on climate and sustainability only on the base of "advice" from a few experts that they happen to be acquainted with (i.e., not following a due process of development but just obtaining some/any opinion) is not realistic and very dangerous (and, I should say, given the high societal stakes, irresponsible). Another problem beside naive policy-making is "token" (i.e., pretend) actions. Tokenism is the situation when consultations are purely formal: ideas are gathered from citizens or experts just to make them believe they have a voice (and to be able for the policy-makers to pretend that they have consulted the experts and the people), but there is no impact of those ideas on the process of policy-making (for some discussions about tokenism in sustainability and/or city context and the origin of the term, see, e.g., [28, 87, 234, 416, 504, 654, 655]). In a particularly bad approach, tokenistic consultations are being made in the sense that some top-level political levels want to remove for themselves the risk of being accused that they didn't consult with the society and the experts, and they try "a bit" to use some of the feedbacks obtained from consultations and hope that this would be enough for the policy to work (presumably, they do this because they are not prepared for this kind of challenge, do not understand the system they are dealing with, and/or they do not know or do not want to do more than they usually do).

But naive or/and tokenistic "policy-making" approaches to the complex and grave problems of today can only have catastrophic consequences, and we must avoid it. In a professionally serious and non-tokenistic process of solution development based on evidence, the expert advice (as located-identified in Module 1, Sect. 1.2, Fig. 1. 1. top loft corner) is only the beginning and it actually involves a large network of contributors (basically, as many as needed to cover all required types of competence and work capacity) that interact via real projects and use truly integrative methods and tools, i.e., they are not just giving some personal advice to a "policy-maker", regardless of how experienced or "prestigious" the advice giver it (as good practice tells us, in the concrete work of real solution development, prestige, social position, and other personal descriptors are irrelevant).

In the global context, it is worth observing that the progress coming out of the Conference of Parties (COP) meetings (of the UNFCCC) is extremely slow (and tends to appear "much too little, much too late") although, of course, there are important systemic insights to be obtained (e.g., see [632]). But such meetings, while necessary, remain ineffective precisely because they are not accompanied sufficiently by a solid-enough bottom-up evidence-based process of policy/solution development. Of course, the latter is not an easy process but it is the due process. The PRL scale (and the SRL model more broadly) is a way to actually do it.

As summed up a few years ago by Dave Snowden (cited by [87]), "*having a good sense of direction and an understanding of what is possible, and which keeps options open, is more important than creating some messianic vision of the future; the latter whatever the intention, just gives a cult-like excuse for not dealing with present reality.*"

Observations About the General Scale Represented by the SRLs, via Contrasting TRLs and PRLs

A first typical issue in pioneering projects is that the idea of comparing policy readiness with technology readiness could be perceived by some, as "uncomfortably technological", because "people aren't machines". When voiced (this is the fortunate situation; many times, such concerns remained "hidden"), this concern indicates some fundamentally legitimate questions about basic human rights in the modern society in which technology has such a pregnant role. However, this one issue is easy to address: the plain answer to this concern is that none of the two models (TRL and PRL) is a "machine-making tool". Both scales are in fact operational manifestations of a strategy to manage great volumes of data, information, and knowledge (typically coming from many disciplines). These two scales are only two forms of discerning "what must come before what", by means of a series of constraints that reflect universal patterns in pioneering projects, i.e., endeavors to develop new and advanced products (be it the product of an RDI program or the product of a policy-making effort). In this light, the difference between TRLs and PRLs is given by the chosen project goals: a mature policy or a mature technology.

Because they share a common reality (the general dynamics of requirements succeeding themselves in all complex projects), the TRLs and the PRLs map to each other (the levels, one to one), but each of the two scales has its own lists of specific parameters and questions to be addressed (in each level) when their turn comes in the course of a project. Even inside the TRL scale, which has a multi-decade history behind it (for reviews, see., e.g., [304, 521]), the content specifics of each maturity level vary greatly between topics and are determined by each project. Policy, of course, comes with its own details.

Thus, because PRLs and TRLs share a fundamental logic of "end-product development" (which can be expressed, in generalized form, as SRLs), the long history and proof of the utility of TRLs means that the overall structure and resemblance between these maturity scales (and with the other ones known in the literature, as mentioned previously) is an advantage. Of course, this advantage should not be overestimated either. No model whatsoever is to be used uncritically.

Relatedly, one notices that notions like "policy maturity" and related concepts of maturity may be used in different contexts, sometimes consistently and sometimes not (instead, it may simply be evoked to suggest whether a policy exists at all with respect to a given topic, i.e., not necessarily discussing how well-developed that policy is). Needless to say, the PRL and all SRLs discussed here are meant to serve as a coherent method.

However, the idea of "maturity" of policy or other desired outcome are becoming more and more important in the generation of solutions towards sustainable development. To illustrate this point, Box 3.5 contains some examples coming from the last (2022) IPCC-WGIII report [332].

Box 3.5 Text snippets from the latest IPCC-WGIII report [2022], showing (as underlined) several mentions of the expression "policy maturity".

As an observation, be informed that these paragraphs are from the pre-final report, and the final edits might introduce some modifications like punctuation or word choices; so all eventual text errors are mine and I invite the reader to check the definitive text version of that IPCC-WGIII report once available. The section numbers and display items mention in the text, refer to parts of the cited IPCC report (I kept them in order to respect that complex nature of that report, and to thus avoid inducing any form of over-simplification of the messages in the cited report). The highlight of the words "policy maturity" and other constructions including "maturity" belongs to me.

Extract from page TS-82 (of the full report):

"The level of policy maturity and experience varies widely across the mitigation options (high confidence). Energy efficiency is a well-established policy field with decades of experience from voluntary and negotiated agreements, regulations, energy auditing and demand side-management (DSM) programs. In contrast, materials demand management and efficiency are not well understood and addressed from a policy perspective. Barriers to recycling that policy could address are often specific to the different material loops (e.g., copper contamination for steel and lack of technologies or poor economics for plastics) or waste management systems. For electrification and fuel switching the focus has so far been mainly on innovation and developing technical supply-side solutions rather than creating market demand. {11.5.2, 11.6}"

Extract from page TS-94 (CDR refers to Carbon Dioxide Removal):

"CDR is a key element in scenarios that likely limit warming to 2°C or 1.5°C by 2100 (high confidence). Implementation strategies need to reflect that CDR methods differ in terms of removal process, timescale of carbon storage, technological maturity, mitigation potential, cost, co-benefits, adverse side-effects, and governance requirements. (Box TS.10)"

Extract from page SPM-47 (section C.11.1 on CDR; DACCS refers to Direct Air Carbon Dioxide Capture and Storage):

"CDR refers to anthropogenic activities that remove CO_2 from the atmosphere and store it durably in geological, terrestrial, or ocean reservoirs, or in products. CDR methods vary in terms of their maturity, removal process, timescale of carbon storage, storage medium, mitigation potential, cost, co-benefits, impacts and risks, and governance requirements (high confidence). Specifically, maturity ranges from lower maturity (e.g., ocean alkalinization) to higher maturity (e.g., reforestation); removal and storage potential ranges from lower potential (<1 Gt CO_2 yr^{-1}, e.g., blue carbon management) to higher potential (>3 Gt CO_2 yr^{-1}, e.g., agroforestry); costs range from lower cost (e.g., 45–100 USD/tCO_2 for soil carbon sequestration) to higher cost (e.g., 100–300 USD/tCO_2 for DACCS) (medium confidence). Estimated storage timescales vary from decades to centuries for methods that store carbon in vegetation and through soil carbon management, to ten thousand years or more for methods that store carbon in geological formations (high confidence). The processes by which CO_2 is removed from the atmosphere are categorized as biological, geochemical or chemical. Afforestation, reforestation, improved forest management, agroforestry and soil carbon sequestration are currently the only widely practiced CDR methods (high confidence). {7.4, 7.6, 12.3, Table 12.6, Table TS.7, Cross-Chapter Box 8 in Chapter 12, WG I 5.6}"

With respect to the lead topic of climate changes, one also observes that, to this date, the concrete transition to an economy (and techno-economic paradigms) based on renewable energy sources is still incipient. This is because (not all but) most technologies that could be employed for mitigating climate changes by reducing the emissions of carbon dioxide (and other emissions greenhouse gases) are still in the early TRL stages and maturity also concerns the internal transformation/transition processes of producers (the companies, and their technology/product/service development) (see, e.g., [394]) and the supply chains (see, e.g., [179]), and therefore policy planning horizons depend on the current and future learning curves of those technologies that may become mature on medium- and long-term (e.g., [607] for CO_2 utilization, [225, 545]).

Relatedly, one can also think in terms of "industrial maturity" to refer to refer to technological maturity (the TRLs) combined with the additional requirement of applicability in terms of "economy of scale", "industrial scale" or "industrial levels" [225], which is important for all cities, and especially when one of the priority issues of a city is deindustrialization (like with the so-called "rust belt" in the USA; but also in Eastern Europe after 1989, and in other places around the world) and the potential re-industrialization based on technologies that are compatible with sustainable development.

And, of course, analyses in terms of maturity levels of potential solutions can be conducted to articulate policies and programs addressing many other environmental (and related social and economic) issues, from complex diffuse pollution and solid waste in urban and densely-populated land areas to ecological issues and indicators across the Earth's oceans (e.g., [159, 160, 166, 189, 598, 719]), which, let us not forget, also constitute a major source of inputs for urban systems and must be factored in.

Another important aspect to highlight is that it is always practical to make explicit the feedback loops (i.e., to describe in what consists the interactons) between the distinct levels of a readiness scale (as shown in Fig. 3.32 for the parent SRL scale, and as replicable for all daughter scales) and between different scales as needed—first because we are of course talking about complex dynamic systems, and second because there are some significant, purpose-driven technical differences between the scales which further command correct understanding of each scale and maximum clarity over how different scales and models employed by a team project "talk to each other", especially when they are employed by people from different fields which may assign different practical meanings and may draw different conclusions.

Thus, as per Fig. 3.32, depending on how frequent (and in which way) updates of a scale are required, the content of the mathematical connections between readiness levels can be very different (also quantitatively, besides the obvious qualitative specifics). Namely, the dynamics of updates required in an eco-city policy (which is very much dependent, for example, on the 4-year work cycle in updating of local and national public policy) is rather different than the more technical updates in an eco-innovation policy (which needs to give at least equal priority to the science production cycles (which tends to be about 7 years in terms of major breakthroughs but is unpredictable in terms of novelty updates because hundreds of research/academic work

9	Institutional reforms consolidated through successful operation
8	Mature set of institutional reforms assimilated refined feed-backs from real world
7	Basic reform packet demonstrated utility in relevant real-world localities or sectors
6	Institutional reform options demonstrated in relevant context (pilot use in limited settings)
5	Details (organizational reform) tested in realistic complex models (including simulations)
4	Details (organizational reform components) verified in simple logic models
3	Real-life (analytical) data appear to be supporting the formulated possibility
2	Institutional reform possibility (concept) or/and direction formulated
1	Basic principles observed & reported (starting in the field of organization studies)

(6-9): Institutional reformation authorities

(1-6): Institutional reform task group

(1-3-4): convened groups of experts

Fig. 3.36 Institutional Readiness Levels (IRL) framework model generated based on the SRL scale (as parent scale model), as inspired by the idea of institutional readiness [713], for the benefit of generating institutional support for eco-city and eco-innovation projects. As with the other scales above (Figs. 3.31, 3.34 and 3.35), shaded levels show the middle zone where push (experts) and pull (need/market) factors meet

results are published every day, this kind of uncertainty being what research means by definition: you do not know in advance what you fill find) and the business cycles in the private sector (which are mostly trimestral and annual). And, the dynamics in hard-technology developments can also be very different from the more transdisciplinary goal of eco-innovation, although the two types of work share many general features.

Then, there is the issue of the context, as the progress along the scale of readiness of a solution (and then the dynamics of its adoption and deployment) also depends on something called "institutional readiness" which requires its own discussion (see, e.g., [713]). Here I only limit myself to observe that this can also be derived from the SRL scale, by defining Institutional Readiness as yet another type of Solution Readiness, as shown in Fig. 3.36 as a matter of principle. As per current knowledge in organization studies and related fields, this Institution Readiness Levels (IRL) model applies to the question of how to better and/or transform our institutions on the path to sustainability (as part of the wider societal transformation that we need to achieve). In terms of definitions, such a scale replaces the word "solution" in the SRL scale with the word root "institution". In terms of content, the IRLs refer to all relevant aspects that need to be addressed (as determined in the lists of requirements that each project/program team can build as needed), for example, as related to institutional reform or institution building (in the public or the private sector). This is another concrete exercise that you and all the readers of this book can do, when relevant.

Epistemic Observations About the SRL Approach

The SRL scale is a model of knowledge organization and knowledge management, one which follows from using scientific modeling to meet the need of any manager of a complex project to prioritize objectives (and related tasks) from the most consequential ones downwards (which usually means from the most general to the most specific). In practice this consists in raising and addressing the question "what comes first" (or "what comes before what") in a systematic (not only occasional) fashion. This insures that knowledge-to-action theoretical mappings and concrete work flows respect the Maximum Entropy Principle (MEP) in scientific modeling (see the details from in Sect. 3.7).

The stepwise approach that is built in the SRL scale addresses coherently the issue of the limited relevance of planning in complex systems described in Fig. 1. 6 (Module 1) specifically through that "*finding the appropriate short-term action options that are compatible with long-term goals and keep future options open*" [604] (as mentioned above in the technical summary around Fig. 3.30), i.e., finding the way through "disciplined experimentation" and allocation of resources to logical priorities in any given set of circumstances, all under the basic constraint of efficiency as derived from thermodynamics and the related MEP ("maximal non-committal with regard to missing information") in scientific modeling ([155, 529]; see also the applied model in Sect. 3.10).

This knowledge-action ordering by prioritization and parsimony is essential for meeting the practical imperative of transdisciplinary problem-solving [367], because of the very nature of the problem, as described in Module 1 Sect. 1.3: complex dynamic systems are challenging because tension tends to build up undetected or only showing subtle warning signals that may pass as unimportant or not urgent) before a/the "burst", i.e., a crisis that catches a community of people without a mature solution to the issue that just broke havoc. Although a system has some resilience that can buffer small shocks so that the entire human system continues to function seamlessly, there comes a moment when delaying due action (for the protection of the system against the important stresses that undermine its capacities) is toxic and does not represent an option anymore. In such situations, famously encapsulated in the expression "post-normal science", "facts are uncertain, values in dispute, stakes high and decisions urgent" [258, 259], and what needs to exist at hand is a set of knowledge-action models that are sufficiently versatile-and-potent in the "situation as is". Because in a situation of crisis decisions are taken under time and resource constraints, decision cannot wait for more scientific research and there is little time to think about: one must use whatever competence exists and what evidence has been gathered until that moment.

The SRL family of models anticipates such situations and, at the same time highlights the importance (1) of disciplined experimentation while there is time and (2) of the readiness to take decisions at any moment in time based on whatever previous preparation has been done until that moment and any knowledge that is available at the time of decision.

In an exercise of epistemic reflexivity, this also reminds of what has been called the "eco-cognitive model" of abduction, which articulates the direct experience as a source of functional cognition: a decision-maker has no choice but to look around to get an intuitive grasp of what happens, and this has a value that transcends (but complements) explicitly articulated knowledge: it enhances the understanding of the scene where things happen, and confers a minimal (some would say "surviving") basis for at least well-oriented decision even when the available explicit knowledge is not formally defined (or not yet) as "inference to the best explanation" [439]. On the temporal gradient from the immediate to the long term, this becomes part of the human resource underlying eco-city foresight and project management in the form of "experience as horizon" [562], and therefore "disciplined experimentation" can be understood as a strategy for broadening the cognitive and operational horizon. For eco-city projects, this has the double benefit of simultaneously enhancing, metaphorically speaking, the two "electrodes" between which good projects happen and good solutions are being generated: concrete experience and explicit knowledge. Lastly but first as importance, this experience-knowledge interaction is a way, when deliberately used, to increase the effectiveness of our efforts towards proper "problem structuring in transitions" as discussed in this book.

As supplementary resources for problem structuring in sustainability transitions, and related notions (e.g., "knowledge structuring", "structuring decisions under deep uncertainty"), I suggest to check these references too: [118, 160, 311, 392, 633, 637] (see also [58, 119, 183, 309, 313, 476, 567]). Complementarily, on the matter of knowledge-action models which pervades this book, I can suggest the following references (obviously a very limited, but open list; and also concerning the matter of knowledge representation in networks and the related issues applied to cities as discussed in Module 2, Sect. 2.2, Fig. 2.2): [232, 497, 498]. For related explorations of the "epistemology of the urban", one might want to start with some topic-dedicated papers (e.g., [91, 177]).

Observations About the Connections of the SRL Approach with Different Disciplines

The basic mathematical formalism of the SRL model of structuring collaborative multi- and transdisciplinary (i.e., multi-angled and problem-driven) projects allows smooth connection with virtually all other mathematical models across relevant domains and topics that can contribute inputs to eco-city projects. This makes this approach not only fully compatible with the systemic perspective in sustainability studies (e.g., [160, 271, 282, 739]) (which should not be a surprise, as this entire work has started from that systemic perspective), but it makes it conceptually coherent and operationally tractable. This later aspect is also natural because all these models are thought from fundamentals up, namely, by integrating—as already discussed earlier in this book—the notion of entropy as a basis for realistic management (see also [529]) and the sheer reality of feedbacks in the "coupled human and natural systems" (CHANS) as interrelated complex dynamic system (see also, e.g., [124, 337, 459, 616]). Notably, the model links naturally with other entropy models in the literature concerned with the historical patterns and dynamics of innovation and related socio-economic transformations during the industrial revolutions in the past 250 years and potentially towards sustainable cities and societies in the next decades and centuries (e.g., [240, 554, 681]).

Further, the SRL model is coherent with mathematical graph analyses of trends in sustainability studies per se, some remarkable topics being (inter alia) sustainable manufacturing, sustainable supply chains, and renewable energy sources [156, 471]; and with methods addressing the issue of uncertainty, e.g., min/max multi-criteria decision analysis (e.g., [229]); participatory policy-making for generating efficiency/

robustness tradeoff scenarios (e.g., [150, 759]). Above all, entropy-minded models are important in the generation of solutions in general, and in policy-making in particular, because the "chaotic" policy-making is currently more the rule than the exception and it tends to generate "complex socio-political constructs" and "complex human systems and socio-political constructs perpetuate vulnerability" [358].

From the perspective of innovation management (specifically with regard to the desiderate of organizational ambidexterity, i.e., relevant capacity of an organization to do exploration AND exploitation), this process insures an ordered flow of data and information from source(s) to sink(s), i.e., from exploration to exploitation. In this matter, and as illustrated in the previous book sections, the discussions can be very complex, but a first requirement is to avoid silo-thinking and a practical way to do so is to explore all options, and eventually take more seriously (if this didn't happen yet) some of those "Cinderella" approaches, e.g., Open Innovation. Specifically, being based on Intellectual Property (IP) sharing, this latter approach is not focused only on technological innovation but seeks broader value-creation through interactions among all actors that are relevant in a given market or industry [134]. This approach has a great (and much under-appreciated) potential to contribute to eco-innovation for cities and human society in general.

Finally, a common feature in all work aiming at developing readiness is the need for structured models of interaction in the creation of viable and reusable solutions (i.e., "stable" at least for a useful period of time). One would expect this need to be deeper in the case of human systems (when compared to technological systems), the lack of which models being a major causal contribution to higher (and potentially overwhelming) disorder in projects and in the transition to sustainability in general.

Potential Epistemic/methodological Questions (and Anticipatory Answers) About the SRL Family of Models

Some might fear that the use of a structured scale like SRLs (or/and any of its daughter models) carries a non-insignificant risk of premature disqualification of new propositions as "immature". However, this concern is also easy to address: the logic of a readiness scale is not a priori disqualification of work inputs, but sound and timely development of a solution for the chosen goal. The practice of insuring high-quality standards is necessary in the context of constant pulls and tensions that characterize the development of new solutions. At the same time, the application of judicious evaluations does not mean that a readiness scale needs to be some sort of blunt instrument—quite on the contrary: solution readiness scales are not meant to discourage that effervescence of ideas and propositions that is also necessary for the generation of new and better solutions; they are meant to make sure that good ideas come along with arguments, so that their merits are being tested and do not remain just good ideas that never reach utility in the real world. Using a cultural analogy, I could say that SRLs are a matter of building a stairway to heaven precisely because we already know that "the road to hell is paved with good intentions". The difficulties one encounters upwards only serve discerning value from non-value, or, as we like to say in Latin: *per aspera, ad astra* (in English: through hardships, to the stars).

The set of scenarios that are derivable from SRL (with all of their knowledge-based, expert-determined requirements) provide exactly that: the evolutionary steps which good ideas can climb to success, i.e., utility in real life.

In the same vein, it follows logically that readiness scales must be protected not to be (in any way) corrupted to serve as a method for suppressing new ideas which incumbents may not like; in fact, as an expert tool, they are by definition hard to corrupt—as long as society knows it needs competent experts to do their work. The SRL scale is a prioritization tool for insuring focus, coordination, and overall value-creation in complex projects. Attempts to use it as a pretext for restricting research and exploration or to limit the reach of different actors or sectors to a given program would have short legs. By contrast, the intrinsic logic of the SRLs gives the managerial flexibility that is required in complex projects to optimize work and find the best balance between emerging needs and emerging ideas.

In the broad picture of eco-cities and eco-innovation, SRLs are in fact an argument for stimulating new ideas and an effervescent creativity that expands the volume of possibilities and maintains a healthy ambidexterity between the exploration and the exploitation of options. In that sense, SRLs and the related scenarios are needed to address the widening gap between what we now know and what we now do.

In fact, this methodological structural quest for equilibrium is what characterizes the history of the best-known SRL model: TRL. The current form of the TRLs dates from 1995 [445] but was historically preceded by other constructs that explored the idea of stepwise prioritization inside NASA: the "technology readiness review" (in 1969) and the general "readiness levels" (in 1989). These were favored also by the "triple-helix" way of thinking: a stepwise structured process was regarded not only useful for work streamlining inside existing programs and projects, but also as a means to encourage cooperation between NASA itself and partners from research institutions, industry, and government. Then, between the general idea from half a century ago and the TRLs scale as-we-know-it since (as it happens) a quarter of a century ago, there were some schematic, earlier TRL set and versions, which were meant "to allow more effective assessment of, and communication regarding the maturity of new technologies" [304, 446, 624].

I made this brief evocation of the history of the TRLs because the whole world of RDI project experience of TRL use provides an important practical lesson for eco-city developments, and for the path to sustainability ahead of us: this historical evolution of TRL was also driven by practice, namely, the general and operational imperatives of projects that are big, complex, costly, and have a great societal impact. While, at the time when TRLs were born, the challenge undertaken at NASA (and other institutions) was winning a technological race (which was part of the Cold War at the time), the question we must answer nowadays is multi-dimensional: How to gain a systemic capacity to transition to sustainable development without losing important achievements of the human civilization so far.

Another important issue (albeit partly touched upon in the previous observations) is the number of readiness levels: Why exactly nine levels and why the same number in all these readiness scales?

First of all, the original number of levels of the TRL model (the oldest and most tested SRL scale) was initially 4 (a general list of phases), then 6, and then the scale gradually grew until it stabilized at 9 divisions—simply under the pressure exerted by practice. Arguably the most important factor is the necessity to make sure that allocating money (resource in general) follows some verification that progress (1) happened, (2) in the right direction. Or, in other words, the "permanent question" of a project manager is: *When is it best to actually stop and decide what next?* This was important for TRLs because, in RDI, the cost of the next level tends to be at least one order of magnitude higher, ergo the double verification mentioned above is vital for a good anticipation of needs that will be faced by each specific task, project and program (for some more discussions, see [304, 446]).

This raises another important issue that needs to be well understood in such projects. Once the need to use a stepwise process has been recognized, the establishment (through practice) of the number of levels is just an operational detail that is optimized based on lessons from use. This in turn is a matter of abstract modeling, and the literature—along with practice—provides some important lessons for anyone interested in the competent use of abstract models for project management—no matter how simple those models appear to be (and that is because the simplicity of a good model is not there by chance: it is found by hard work and their value multiplies when the user understands at least the basic reality behind that model).

Thus, it is relevant to know and remember that the stable TRL scale published in 1995 (with the number and basic definition of levels) was derived abductively (i.e., by abstract "sense-making") from practical data and requirements, and then it was consolidated (validated empirically) in time. That is, each TRL level exists because it was proven to be needed in practice. For the related SRL family of models, this has two implications:

1. the number and definition of the readiness levels in a given model does not need to be definitive: concrete project management is going to tell us whether and what modifications will be required.
2. the nine-level template of SRL is just a practice-justified basis to begin with as the logic behind the SRLs (starting with the TRL model) encapsulates the core features of any complex process of developing a complex product in complex social systems functioning in complex contexts.

These constitute the two facets of a realistic premise for the SRL family of models, as consistent with the requirement of justification "to the best of our knowledge": using all available data, information, and knowledge at hand (and the knowledge management skills developed though practice) to find the best course of action in eco-city and related projects.

This is implicitly recognized by the technology readiness assessment guide provided by the US-GAO, which uses those "sister scales" mentioned above as a tentatively-holistic way of regarding the readiness of target features (technology, manufacturing, or systems) from separate managerial angles. For example, the latest US-GAO report notes about the Manufacturing Readiness Levels scale: "Although the MRLs are numbered, the numbers themselves are unimportant. The numbers represent a non-linear ordinal scale that identifies what maturity should be as a function of where a program is in the acquisition life-cycle." [709, p. 122].

This further highlights why it is important to respect the natural evolution of stages (as defined by priorities) in the process of making and deploying an eco-city policy: (1) a mature (i.e., viable) policy cannot be obtained otherwise (the alternative would be a series of unstudied, inevitably discretionary initiatives that are doomed to fail), and (2) the panoply of effects of a bad policy towards sustainability (inevitably affecting all socio-economic sectors) would be catastrophically wide.

Last but obviously not least, some others might want to understand whether juridical and financial considerations would only be solicited in the last PRL stages. The answer is "no". In general, aspects related to laws and the funding should inform all solution development stages and, because this would contribute to detecting unfeasible avenues (or new and interesting openings) as soon as possible. The fact that policies are made definitive through legal and budgetary decisions does not mean that legal and budgetary factors are reserved to some final fine-tuning. Nor does it mean, in the broader context of eco-city (or national) policy-making for sustainable solutions, that institutional legacy and inertia are inevitable. Public organizations in charge with law, finances—as well as those related to national security, democracy, free market and other key area will have the final word before the process of a policy-making is (provisionally) ended, and their input can also involve important requirements, like with any other relevant domain. And this is not (and must not be seen as) an impediment on a healthy overall process of good policy-making—quite on the contrary: experts from all domains interact in order to create a set of options for the elected (i.e., democratically accountable) actors to choose from.

Even more importantly, all relevant inputs (from any domain) must be made available as soon as possible in the process of generating solutions for eco-city advancement. The fundamental benefit of a step-wise model comes from their integration power: many specialized inputs are very narrow and may indeed concern only one or several steps along the SRLs, and most individual expert contributors may not be (given in their job) a good position to "see" the entire path through which their (or others') specific inputs trickles down towards implementation.

Moreover, not all solution sub-/components can possibly advance equally fast along the overall work process. But, a spelled-out standard core process model articulating the path from knowledge to action, as the SRL family of models does, provides that necessary common reference and scale which, when properly used, reduces substantially the possibilities of misjudgment and miscalibrations.

In a tentative nutshell, taking the fundamental perspective of physics (and related mathematics), the Delphi-based scenarios and the SRL family of models supporting (and feed into) them can be abstracted as flows of Shannon information ("surprisal", i.e., novelties) from the initial and collateral sources (represented by scientific discoveries and practical insights) through a set of managerial-operational filters (i.e., constraints reflecting distinct types of requirements) that facilitates conceptual and operational integration along a list of priorities identified in practice. This overall flow also translates in the utilitarian terms of information sciences as an increase in algorithmic complexity, and further translates in the terms of economics, business and project management as an ordered approach to feasibility and cost.

In a Nutshell Illustration of the Core Idea of SRL with the Case Study that Motivated Its Generation

All the above can be better understood if placed in the operational context of the whole process of policy-making, the knowledge/technology push provided by science and technology and the demand pull exerted by different sectors of activity and society in general, without forgetting the existing knowledge-action gaps that need to be managed.

In this sense, the SRL scale and family of models introduced above reflects a push–pull (demand-offer) type of interaction between problems (the demand for solutions) and problem-solving capacity based on knowledge and expertise. And, in the context of current policy-making practice and research, this corresponds to the relation of complementary between bottom-up and top-down approaches that predominate respectively in public and private domains (to simplify, for the immediate analytic purpose of this sub-section, the complex discussion started at the beginning of this Section).

Figure 3.37 illustrates this point and shows the operational mechanics connecting top-down and bottom-up forces in the process of development of a concrete evidence-based climate policy at a national level (of an EU member state: Romania). The model itself emerged from the experience in this policy-making exercise to which I have participated during 2022 and 2023 and is informed by the literature on the development of policies for sustainability and related aspects like sustainable supply chains, re-industrialization, knowledge management, innovation management, and entrepreneurship (see also: [155, 156, 304, 399, 727, 738]; and the other references used throughout this book).

This national climate policy exercise mentioned above reflected exactly this push–pull dynamic: once the Head of State (in this case the elected President) explicitly formulated a country's priority need to develop a comprehensive and integrated set of policy options (which in Fig. 3.37 corresponds concretely to the content of PRLs 7–9), those experts which were contacted and responded positively have been convened and became a source of inputs for PRLs 1–4.

In this particular case, the two forces have been brought closer together by a small team of organizers missioned by the President of Romania, and the works have been conducted in a way that was in line with what is needed in the general triple-helix model of interaction mentioned already in this book (and covering academic, private and public institutions).

Importantly, one of the start requirements (or even the first one) for this exercise and the participating experts was that, while adequately meeting the needs for an effective climate policy, any individual policy proposition and the entire policy packet must account fully for the specific national context, i.e., there was no room for "copy-paste" measures imported into national legislation uncritically. This is a key detail, because this is, unfortunately, exactly what happens in many locations around the world [18] (but see also, e.g., [358, 449]).

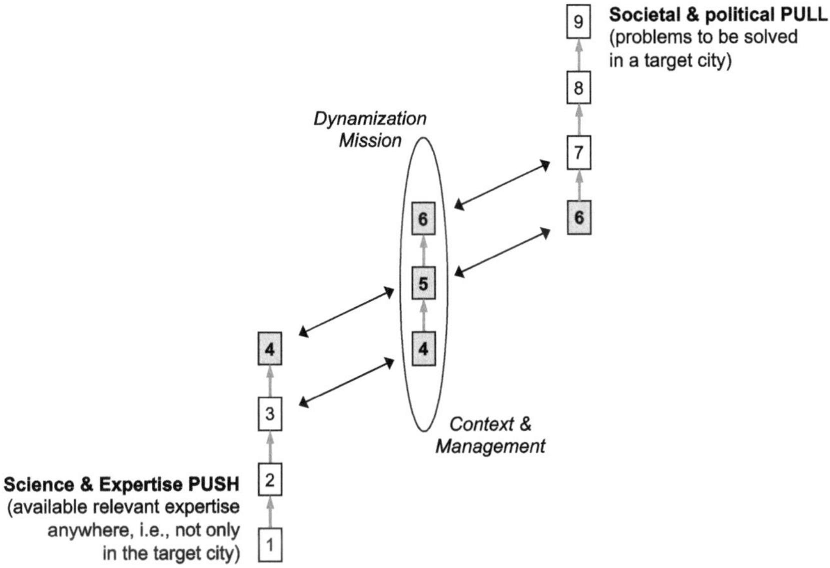

Fig. 3.37 Push–pull dynamics along the SRL scale, as applied as PRL or XRL: offer of, and demand for, knowledge and expertise. This also stands as a general model for structuring the process of policy-making based on current knowledge and practice, and as informed by Category Theory (and related areas) from fundamental mathematics. In the whole process of solution development, SRL5 is a point of inflection and therefore an intermediate target to reach during the early phases of policy (or more solution) development

This specific requirement imposed a standard for "no nonsense", in-depth discussions, and has placed the emphasis on realism. In this context, the need for public–private partnerships (PPPs) at the implementation level emerged naturally, in full resonance with current knowledge and experiences from pioneering and pragmatic projects, and in view of further action by the Government, notably via an Inter-Ministerial Committee on Climate Change (established by mid-2022 and headed by the Prime Minister).

The packet of policy propositions (main options/directions) that came out from the work of this expert group (and then consolidated by further work after a minimum period of one month in which the initial report was made public and subjected to a wider public consultation) reached a level corresponding to PRL4 (or SRL4 in the parent scale). Then, the small team missioned by the President took it from there and conducted all conversations that were needed to advance and develop the set of measures to PRL5 (simulation models; the halfway mile stone), then PRL6 (the positive confirmation of an advanced set of realistic policy options). This was the point when the Government was able to take (and indeed took) over, the packet of policy propositions undergoing now the legal and administrative shaping and development that can result in a mature national policy (PRL9) vis-a-vis climate change, which also includes due coordination within the European Union.

At this point, one notices that the PRL model can serve as a reference and tool all along the process of policy development, for cross-communication and integration all relevant levels, as discussed in the previous sub-section that compared SRLs, TRLs, and PRLs. As cities are one of the key scales for generating solutions, and as climate is only one of the many sustainability sub-topics, a next step in this national study case (and in any other case) is to use the PRL model and SRL family of models for organizing projects at city levels and for coherent coordination between relevant actors, across scales and topics.

Supplementary Mathematical Formalism and Background for Nested Models

SRLs and their dynamics can be expressed formally as abstract sets of entities and relations using the basic logic and mathematics Category Theory, as follows.

First off, the SRL model represents a nested set of sets (i.e., a hierarchy)

$$SRL_1 \subset SRL_2 \subset \ldots \subset SRL_9 \Leftrightarrow SRL_9 \supset SRL_8 \supset \ldots \supset SRL_1$$

where each level is a set containing two distinct objects, namely, a set of sets corresponding to all subjacent SRLs and a set of new level-specific requirements (i.e., a list of attributes that emerge from a concrete, applied project are uniquely new to that SRL level). In this hierarchy, SRL_9 is the original parent and SRL_1 is a self-contained set. Obviously, one can repeat this formalism by replacing SRL with any daughter concepts mentioned in this section, e.g., TRL or PRL.

Further, because PRLs and TRLs are epistemic instances of SRLs, we can clarify the formal logical relations between the three notions and the empirical evidence available so far. Accordingly, one recognizes TRLs as the first instance of SRL that emerged from the necessities of experience. This means that the SRL model M_{SRL} can be recognized as being derived from the TRL model M_{TRL} through conditional generalization, the qualifier "conditional" meaning that the generalization rests on the assumption that the total recognizable volume of past experience (i.e., recorded evidence) E grounding the widely-consecrated TRL scale is sufficient to be representative for the general (i.e., <u>core</u>) dynamics taking place in solution development projects (not just in technology-development projects), i.e.,

$$E \vdash M_{TRL}, \vdash M_{SRL}$$

Further, recall that the need to have a readiness scale specifically for the "policy" type of solution development can be derived from SRL by deduction but it was also inspired, like with the other types of maturity scales mentioned previously in this Section (notably by the US Accountability Office, the wider literature and other applied works like the IPCC report and RDI in general), directly by the experience and the rationale behind the TRL model, and this is an instance of abductive reasoning. That said, any direct one-to-one comparison between the scales in the SRL family (and, for that matter, between the levels of the same scale) is a form of "reasoning by inference from one particular to another particular", i.e., analogy.

Therefore, for the sake of formal clarity, the relation between TRLs and PRLs (as SRL daughter scales of the same order) is also denotable by a formal union between two cognitive models M_{PRL} and M_{TRL} which (1) are independently consistent with classic scientific modeling principles (see especially Sect. 3.7), and (2) are simultaneously consistent with each other and empirical evidence, i.e.,

$$M_{PRL} \cup M_{TRL} \models Observations$$

We can also define M_{SRL} as a cognitive model denoting a virtual union between these two models above and the ensemble M_{rRL} of all the other empirically justifiable readiness models that are reciprocally coherent, i.e.,

$$M_{SRL} \equiv (M_{PRL} \cup M_{TRL}) \cup M_{rRL} \models Observations$$

For further modeling purposes, one observes that M_{rRL} can also be expressed as a sum of all existing readiness (maturity) models M_{aRL} supported by empirical evidence minus M_{PRL} and M_{TRL}:

$$M_{rRL} \equiv \sum_{m=1}^{n} M_{aRL} \setminus (M_{TRL} \cup M_{PRL}) \models Observations$$

Further, we can employ basic Category Theory to define maturity models (and the levels inside a maturity model) as mathematical structures, as follows.

The three models discussed above can also be represented as related mathematical objects, i.e.,

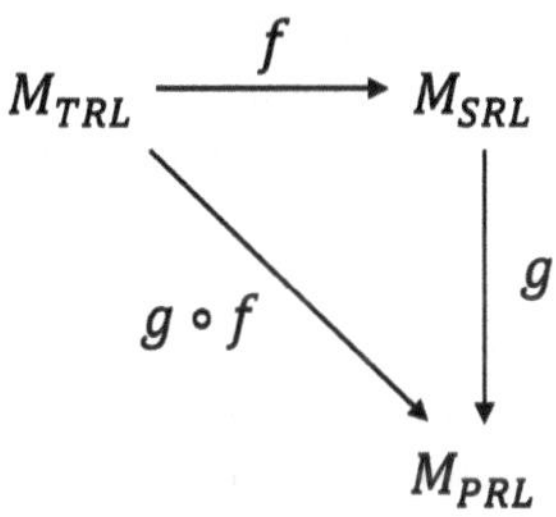

where morphisms f and g denote generalization and deduction, respectively.

In perspective, this brief formalization can help for further research and clarifications in projects, including the development of new models as needed. In this sense, it is important that M_{PRL} can be envisaged by abduction from M_{TRL}, as this is a form of "hypothesis generation via discovery of similarities" [523] (this reference also provides an in-depth discussion about abduction in science). One further observes that the model generation described above in relation to empirical observations also illustrates abduction as "inference to the best explanation", and that making sense of empirical and lived experience (which may also be called "studies of human practices") is essential in the entire process of sense-making, modeling and solution generation. This highlights abduction as a realistic approach for "finding grounds, means or anticipations for transformative practices" [538]. This can take the usual formal aspects known from science but can also be simply implicit in transdisciplinary narratives in real-world communities faced with concrete risks and challenges (see, e.g., [717, 718]).

Finally, within any readiness scale model, each and every Readiness Level RL_k can be represented as a mathematical object, and thus each promotion (or demotion) from one level to an adjacent one becomes a between-level morphism. The overall result is that each flow of technical information between the different levels of the same scale (as shown in Fig. 3.37) becomes a composition of morphisms $h_{n,n+1}$, knowing that, in the case of these 9-level models, n takes values from 1 to 8:

$$RL_9 = h_{8,9} \circ \left(h_{7,8} \circ \left(h_{6,7} \circ \left(h_{5,6} \circ \left(h_{4,5} \circ \left(h_{3,4} \circ \left(h_{2,3} \circ h_{1,2}\right)\right)\right)\right)\right)\right)$$

3.9 Allocation of Resources by Activity Categories: The SRL*DIKAR_p Taxonomy

In this short technical section, I combine two models discussed in the previous three sections, namely, the SRL and the DIKAR_process (or DIKAR_p for short), to develop a general taxonomy of activities in eco-city (and virtually all types of) projects regardless of specifics. Of course, differences also matter. For instance, it matters how different team sizes fit for different purposes, e.g., "large teams develop and small teams disrupt science and technology" [760]. Another example comes from the topic known as "urban living labs" (ULLs): they are very important and can provide useful insights about "participation, collaboration, experimentation, learning, and governing in cities" but it is hard to find uniform definitions and they can be very different in terms different study "dimensions", e.g., "geographical embeddedness, experimentation and learning, participation and user involvement, leadership and ownership, and evaluation and refinement" [728].

Overall, the underlying issues related to project management represents a convergence point and a common platform for finding ways to succeed (rather than fail). In the following sub-section, I summarize this work-organization method and discuss a few issues to help visualize the interactions and draw additional insights.

Mapping Specific Issues and Experiences with RDI in Eco-city Projects

The DIKAR_p model can be used, *inter alia*, to effectively code the steps and components of different models (virtually any of them), by using its vertices (boxes) or/and edges (arrows) as words of a core general language for describing anything that is part of a work process-oriented models, based on the formalisms and tools provided by fundamental mathematics, starting with Graph Theory and Category Theory (the basic terms of the latter have been introduced in Sect. 3.4, Box 3.2).

Figure 3.38 illustrates this idea by coding the Value Generation Cycle (VGC) from the discussion on ground-truthing a space of options (Sect. 3.7, Fig. 3.25 and Table 3.7) with the step labels from DIKAR_process (Sect. 3.6, Fig. 3.17): each and every step (category) in the VGC model can be mapped onto (i.e., can be expressed as) a composition of steps (categories) in the basic DIKAR_process model.

In this case, one observes that the reciprocal is also true: the steps (categories) in the DIKAR_process model steps can also be coded (mapped onto) the VCC model. Whether the reciprocal is true (or practical) with other models must be checked on a case-by-case basis (I expect that it depends on the level of generality of those other models and on the level of granularity used therein). When it is, the direction of the useful mappings will depend on the context, objective, and the questions of interest.

In practice, the model is meant to help clarify the scene whenever a question like "*What are we actually doing in this project?*" emerges. A project manager or a small team of 1-n people can code all main work activities identifiable in a project and then use the model to check for coherence; or to take an activity of interest and see where it fits in the template and derive the consequences of that categorization.

One also observes that the taxonomy above helps further orientation and definition of "who does what and why", and agreements in the Negotiation Rooms (NRs) of the work coordination process.

For example, the NR "Available Knowledge" (Vertex 4) involves establishing what is being known by different participants by organizing inputs along the SRL scale—including all relevant topics, e.g., including technology (TRLs), policy (PRLs), etc., so as to not only understand the different definitions from different fields/ perspectives, but also to structure those usefully along an idea-to-result streamline.

General Remarks

Technically speaking, each user of this method can exercise it by coding her/his own experiences. The details in Table 3.10 are just a very brief illustration with my own experience and details in the Econoving cluster, meant to give a start idea. But the minute details are to be defined for each project of application, by each team or team member undertaking this coding.

Those of you who are familiar with or/and willing to learn about Graph Analysis and Category Theory will see additional uses of this work-coding method, e.g., by simply comparing path lengths and centrality values of different types of activity, one can identify gaps and then prioritize project tasks and allocate resources accordingly. Further, this coding template can be used as a general framework for linking with and combining other methods, and as a common coder of actions and courses of action, thus gaining much-needed versatility in complex projects that otherwise risk sliding into blockage or inefficient (sub-optimal) muddling-through.

At a first look, doing this kind of coding can appear as an unwanted "supplementary work". And this might be in fact true for projects that are not complex enough to overwhelm intuition-based management, although we should also not forget that, because we are anchored in our own background specialties, complexity is easy to underestimate and our real competence is easy to overestimate. All in all, however, eco-city projects worthy of that name cannot by-definition be "simple-enough". The

Cartesian Product			SRLs								
			Basic science (principles)	Solution concept or direction formulated	Analytical / functional proof of the possibility	Solution parts verified in simple logic models	Solution parts verified in realistic complex models	Solution options demo in a relevant context	Basic solution set demo in target context	Solution set assimilated refined real-world feedback	Solution set consolidated by successful operation
			1	2	3	4	5	6	7	8	9
DIKAR_process steps	Question	1	A11	A21	A31	A41	A51, B51	A61, B61	A71, B71	B81	B91
	Reflection	2	A12	A22	A32	A42	A52, B52	A62, B62	A72, B72	B82	B92
	Identified problem	3	A13	A23	A33	A43	A53, B53	A63, B63	A73, B73	B83	B93
	Available knowledge	4	A14	A24	A34	A44	A54, B54	A64, B64	A74, B74	B84	B94
	Demanded information	5	A15	A25	A35	A45	A55, B55	A65, B65	A75, B75	B85	B95
	Requested data	6	A16	A26	A36	A46	A56, B55	A66, B66	A76, B76	B86	B96
	Obtained data	7	A17	A27	A37	A47	A57, B57	A67, B67	A77, B77	B87	B97
	Obtained information	8	A18	A28	A38	A48	A58, B58	A68, B68	A78, B78	B88	B98
	New unknowns	9	A19	A29	A39	A49	A59, B59	A69, B69	A79, B79	B89	B99
	New potential solutions	10	A110	A210	A310	A410	A510, B510	A610, B610	A710, B710	B810	B910
	New/updated knowledge	4'	A14'	A24'	A34'	A44'	A54', B54'	A64', B64'	A74', B74'	B84'	B94'
	Use in real life	11	A111	A211	A311	A411	A511, B511	A611, B611	A711, B711	B811	B911
	Outcome (of each action)	12	A112	A212	A312	A412	A512, B512	A612, B612	A712, B712	B812	B912

Fig. 3.38 A managerial taxonomy of activity types in eco-city projects, generated by pairing the Solution Readiness Levels (SRLs) and the DIKAR_*process* steps. A-codes and B-codes denote Academia-led and Business-led activities, respectively (details in Table 3.10). Some observations made in the text are marked in the figure as well: the grey shade indicates the two sets combined in the Cartesian Product; the yellow shade indicates activity types that belong to the critical and challenging developments between SRL5 to SRL7; the thick blue border is used to locate the initial preparatory phases of the project; and the thick orange border between "Requested Data" and "Obtained Data" separates the two halves (respectively "ascendant" and "descendent" arrows) in the DIKAR_process model in Fig. 3.17 (Sect. 3.6). *Source* modified after [155]

Table 3.10 Types of activities in a given (target) project (in occurrence, the Econoving cluster from UVSQ, as described previously in the book, notably in Module 2), as a general illustration of content of the A-codes and B-codes from the matrix of correspondences (in Fig. 3.38 above) between the steps of the DIKAR_process knowledge-action model and the SRL model of stepwise solution development from initial idea to "ready-to-use" maturity

Main actions in a medium size train station: Versailles Chantiers (which is a district of the city of Versailles, in the Paris region of France)	***A. Academia-led activities, along several main themes***:
	• **Project writing**: convening the initial set of participants (institutions and people) in an eco-innovation cluster (PPP), composed of 6 universities and 6 private companies • **Further development of human capital and networking**: an international graduate (Master) program fully dedicated to eco-innovation—a world's first as a central point of concrete collaboration in the cluster; PhD students and postdocs • **Active participation in international professional networks**, e.g., the European Institute of Technology's Climate KIC funded by the European Commission; and Science|Business, a network of universities and businesses (e.g., [151, 639]) • **Research, Development and Innovation (RDI)**: diagnosis of the energy performance of the target (Versailles Chantiers) train station; generation, use of and refinement of market-relevant solutions (new algorithms) based on the NIALM technology [336, 508]; energy modeling of the building(s) [695] and exploring complex engineered systems, notably smart grids [7, 398, 397] and round-the-clock patterns of use of electrical vehicle (EV) batteries plugged into the electric grid [312]; • **Development of scenarios** for the concrete re-development of the Versailles Chantiers train station and of its neighborhood (Versailles Chantiers city district) [150, 151]
	B. Business-led activities:
	• City-district (and wider) scale developments: public–private complex projects of refurbishing the Versailles Chantiers train station and the urban regeneration around it, resulting in a larger, greener (parks included), and multi-functional station, city neighborhood, and a better city-as-a-whole within the larger conurbation of the Paris Region (Ile-de-France); details available online by [142]

method described above emerged precisely from experience at the bleeding edge, as a necessity in the daily sense-making effort and result-oriented management of projects. While the use of different methods needs to be decided in each project and context, the common point to recognize by all of us is that competent action requires preparation—and this is what this method is about. It can be used to check whether used assumptions about common understanding (like "we already know that" or "everybody knows that") hold, and whether different actions are compatible in the big picture of the work process from initial question to solution generation and validation. Briefly put, coding of project activities can also be done as a minimal-version exercise by an individual manager (or small managerial team) as a way to check coherence and to spot aspects that required increased attention or recalibration. In a broader perspective on contexts, let us keep in mind that some (or many) of the joint crises of today have grown from certain past collective complacencies.

Therefore, and given the overwhelming volumes of scientific and general work outputs and the high level of heterogeneity of technological state-of-art (which is not bad, just very hard to deal with in complex projects in which multi-dimensionality increases exponentially), the challenge we are facing with eco-city projects is not primarily related to specialized expertise (the competition for the highly skilled is a different matter), but consists in the integration of available knowledge in novel solutions, under the constraint of compatibility with sustainable development goals (resource- and time-efficiency included).

Lastly, precisely because it is grounded in fundamental mathematics, this coding method can help with further potential objectives, like those known in engineering and management as "workflow automation" and "robotic process automation" (RPA), both in terms of software and users (and including human–machine interfacing).

Supplementary Discussion (with Some Illustrative Details from Past Projects)

Again, depending on your immediate purposes, you may wish to go and read this details-focused section or skip it and read it at a later time when you want to delve deeper and find some fuel for your own works. In either case, know that I have kept the list of details mentioned here to a minimum that I estimated necessary for eliciting some initial reflections that can be then directed by the reader to the context and the goals of specific projects—basically, just some food for thought.

In principle, many aspects and details can inform the reflection of a project team about the model presented in this section, both from the projects leading to it and the insights provided by the extant literature.

For example, one aspect worth highlighting as potential beneficiary concerns the phenomenon of social learning that happens one way or another inside a project (but which needs to be serving the goals of the project). In fact, some complex projects like innovation clusters may include formal education programs or/and other forms of non-formal training (internships, thematic workshops or/and courses, indoors and outdoors, and so forth) (e.g., [20, 23, 162, 163, 169, 180, 181, 223, 279, 345, 353, 447, 496, 530, 622, 657, 661, 683, 690, 767]). Thus, social learning is a natural resource in the development of the human resource (and can be further discussed, in given contexts, in terms of human capital).

In the Econoving cluster mentioned in this book, this was indeed the case, as many of the teachers involved in the formal courses of the cluster's master program were also experts in its industrial program. Moreover, all our main partner institutions and several other organizational collaborators from the private sector had teaching contributions (at least one morning or afternoon, delivered either by single experts or by teams of several experts; and all have been instructed to have an interactive style and to reserve time for answering questions, and—as a Director of that master program—I personally checked and made sure things happened as needed; and, while this form of collaboration with partners was not without difficulty, ultimately went excellently). And, as already mentioned in the book (in Module 2, Sect. 2.2, Figs. 2.4–2.6 and Table 2.5) but worth reminding in this discussion here, the students themselves also participated (by program design) directly in the works of this eco-city-oriented eco-innovation cluster. In addition, most students had internships at some of these partners and collaborators, and the works also involved exchanges of experience and structured conversations (exchange workshops focused on students also involving the teachers) between our program and a master program dedicated to architects (devoted to sustainable construction and eco-districts; original title in French "*Construction durable et éco-quartier*"), a joint program of our university and the National School of Architecture of Versailles (in French, *Ecole Nationale d'Architecture de Versailles*, or *ENSAV*).

The details about this particular inter-program exchange are relevant for the above method because it illustrates well some of the many benefits of exchanges between project teams (and between expert constituencies in general). This relation between an eco-innovation program (essentially focused on RDI and management, with a systemic and transdisciplinary/holistic perspective) and a program specialized for architects (sharing the general urban sustainability aim) was obviously motivated by a shared topic and helped me develop a visibility upon of the target issue from different practical view angles (especially epistemic, institutional, technologic and managerial). As I had to insure coherence of the entire work process, I engaged in all types of work organization that were required, and this helped me further develop and refine the Big Picture understanding of the interactions between the relevant human and institutional agents. For example, I co-moderated the live interactions between the two groups of students (a type of work that bears a lot of similarity with the moderation of the Delphi-based method and the moderation of interactions in the Integration Seminar described in Module, 2 Sect. 2.2); and I participated as a jury member and provided constructive feedback in the intermediate and final presentations (and work evaluations) of the architecture students. In all instances, it was for me required to insure work coherence according to the logic behind the SRL model and the logic behind the DIKAR_p model at the same time. By the way, in return, I also gave feedback to help the students and the other involved experts step out of the specific boundaries of their field, as part of the general process in which we all learn how to operate effectively across domains (as we are required to do in any sustainability- and complex systems-related work).

I am sure many of the readers of this book may recall some experiences that they had as students, experts or/and teachers and which are similar to those I mentioned above. The point is that this kind of work is more important that it may appear at a superficial regard, because the fact of internalizing different angles upon a problem helps us (and it helped me) see and operate with the different perspectives between expert fields with respect to cities (as already mentioned in Module 2, that Table 1.2 with lists of disciplinary perspectives). The challenge therefore is to develop enough work discipline for maintaining the operational capacity to convert a great diversity of work experiences into a high level of project management efficiency (as a way to maximize added value and to keep a hard, complex project on track). The model and template described in this section is meant to help with addressing that precise challenge.

It might also be worth noting that the overall experience of all participants in the original Econoving project (a pioneering project at its time and still remaining one of the very few examples of holistic projects and, at the same time, a unique learning program with many insights for current and future eco-city and/or eco-innovation projects) has touched on many of the issues that are already known in the literature about the essentially "unavoidable" topics already discussed in this book such as the Cultural Theory (and the related social learning (e.g., [715])) and Project Management (and the related issues with cognitive and behavioral biases (e.g., [242])). In this sense, it is equally worth mentioning that many practical lessons "had to" be learned by doing (or, as many would say, "the hard way"). The overall importance of "working together to learn how to work together" was summarized at an end-of-project meeting (in 2013) by one of the main institutional partners and main beneficiary in the Econoving cluster (citing from memory): "it is only now, after 3 years, that we really learned how to work with each other". In the literature on inter-/transdisciplinary sustainability studies, this has been adequately called "learning to collaborate while collaborating" [250], and the relevant institutions are being called to facilitate this process, e.g., as summed up by [114]: "research institutions devoted to sustainability should focus more on creating the conditions for experimenting with multiple kinds of knowledge and ways of knowing to foster sustainability-oriented learning".

Yet another aspect worth highlighting in this supplementary discussion concerned with the SRL*DIKAR_p method links back to the more technical discussion from Sects. 3.6 and 3.7 in this Module, namely, that there is a fundamental need for "strategic data acquisition" (notably mentioned in Table 3.6 and detailed in the subsection dedicated to in-context data generation and uses from Sect. 3.6, and the references therein). This fundamental need exists both for RDI in general and for the so-called "data-driven business analytics" [131, 339, 501]. This latter field is coded by the second half of the template in Fig. 3.38 (below the thick orange line, i.e., vertices 7–12) and it is important to highlight that it needs to be managed interactively, i.e., in the context of the full DIKAR_p model (and of the derived SRL*DIKAR_p model template). In a practical sense, this model informs the so-called "data-driven approaches" (DDA for short) that they can add value through complementarity with targeted data acquisition.

Concretely related to DDA, in the Railway Station Program discussed in Module 2 (Sect. 2.2, around Figs. 2.4–2.6 and Table 2.5) and briefly evoked above in this section, the DDA-related approach called "knowledge discovery" (which infers basic patterns through statistical treatments of large amounts of data) could not be applied in the context of target train station because there was not a great deal of data available, for historical and other reasons that are usual in real-world projects. In that case of ours, even some of the most basic information items were hard to find (because of lost or otherwise unavailable original plans of the site and building, a number of later and often undocumented modifications of the building, plus legal issues about information sharing and aspects about technological decisions and sheer cost of generating new data in the given context). For the project, this meant that some of the initial objectives had to be abandoned and new objective established and pursued. For project planning in general (activities A51 to B74 in Fig. 3.38), this means that one must anticipate that there will be some "impossible situations" and therefore planning (a) should allow as much flexibility as possible (i.e., only what really needs to be pre-decided from the beginning must be pre-decided) and (b) make sure resources are reserved to sustain a lot of necessary future modifications of the project. In our concrete case, the foresight exercise carried out in the cluster had to include a redoubling of efforts so as to compensate this loss in the space of options with the creation of novel perspectives and approaches (based on every bit of information that could be gathered), and we managed to provide SNCF (the French Railways, a public company) with knowledge-based options for their (and our collective) objective of urban regeneration. The latter was achieved through another program that followed and was implemented successfully, resulting in an enlarged and upgraded train station, a new city district re-development around the train station, and the re-development of different functions at the level of the city-as-a-whole, and within the Greater Paris Region, especially in terms of land uses, transportation, energy, green areas, culture, tourism, housing and business.

Further, the SRL*DIKAR_p model addresses an important aspect of work in real-world projects: depending in which stage of development is a solution that is desired to a concrete problem, the conversation can be very different. Concretely, scientific discussions at lower-end SRLs are very different from the business discussions at the higher-end SRLs. Moreover, each maturity level of the SRL scale (and its daughter scales like the TRL model) requires its own, distinct type of work and conversation (otherwise, there would be no distinct level and no scale at all, as already discussed in Sect. 3.8). This explicit multi-layer model, and the rationale and the empirical support behind it shows clearly how easy it is for experts from different institutions and with different operational concerns to talk past each other. And, the overlapping code labels corresponding to SRL5-7 shows where the critical dialogues between academic and business agents happen. Thus, correct identification of work activity types in the SRL*DIKAR_p taxonomy is essential for avoiding grossly sub-optimal use of resource, project time, and the time of experts which also have competing demands from different projects and may be tempted (or obliged) to cancel their participation in those projects that they see as less than a good use of their time (knowing that the more complex a project is the higher this risk is, and that eco-city projects are inevitably complex, as the more they go to the root of problems the more complex they have to be).

To give some more elements about institutional learning process in the Econoving cluster and Railway Station program focused on Versailles Chantiers as pilot station and test bed for collaborative work, I can point out the differences between the research pursuit by different PhD students and post-doctoral fellows. For the PhD students whose research topic dealt with fundamental aspects of smart microgrids for the potential integration of renewable energy sources in and around a train station (Versailles Chantiers or others), the greatest interest and support came from academic research laboratories. For those whose research focus was on optimizing round-the-clock uses of electric vehicles (and their batteries) in relation to diurnal variations in electricity demand (and prices), the greatest support came from private companies. The same pattern was replicated in the case of postdocs, respectively those focused on understanding the intrinsic complex dynamics of the embedding of a train station in its urban environment versus those concerned with modeling current energy fluxes in the train station. Interestingly and tellingly, the main beneficiary (SNCF) was interested in all of those, because it had an interest in a combination of short- and long-term perspectives.

In this context, it would seem fit to add that, by its nature, the research conducted by PhD students is bound to be located more at the bottom of the TRL (and SRL) scales, because they need to obtain and publish scientific results, while the RDI conducted by private companies (including by experts having doctorates already) is bound to aim at the top end of the TRL scale. Therefore, the profile of an eco-city-related project team also reflects the composition of the project participants and their interest. Now, that doesn't mean that the work of PhDs and postdocs cannot go very high on the TRL/SRL scale—it simply depends on the objectives, contexts, and organizational details of each project and whether or not the project is the first in a multi-project work thread or can build on progresses made in previous projects—which again, highlights the importance of long-term thinking and programs. For instance, my own doctoral research (in another country and about a decade before Econoving) involved the development of an immunosensor prototype for on-site semi-automated screening of environmental (water and soil) pollutants, and covered all levels from TRL 1 to TRL7, and a main factor in this was a fruitful collaboration with people from different domains (biology, chemistry, engineering, business), different sectors (universities and research institutions and private companies, plus appropriate support from the government), and gathering experience from different locations in Germany (Munich, Mainz, Leipzig, Augsburg; and with contributed experience and materials from the USA), which in the end made it possible to overcome the usual and inevitable obstacles in such projects [144, 145, 167, 168, 252, 371, 372, 379–384].

The broader point is that institutions are called to facilitate experimentation, as this is how we can learn our way forward in the development of solutions to complex problems, and the involvement of different categories of young and more experienced experts is an integral part of the whole process. When institutions do that job of effective facilitation, real progress is possible.

Finally, the more an institutional (and personal) agent trusts that there is a long-term perspective for the project, the more that agent will be inclined to invest in work types that bring a higher but later value (ergo it will feel comfortable with work activities which, no matter where they are in DIKAR_p stage, are in the lower SRLs (closer to scientific research or even typical science itself). And, the more that agent loses trust in the project (e.g., if the project seems to not be advancing or if it believes it didn't benefit a fair share from the project advances) the more it will be inclined to discount long-term value and to insist on short-term benefits.

3.10 Cost-efficient Eco-city Pathways: An Entropy-Wise Knowledge-Action Matrix

A Synthetic Model for Project Efficiency and Verification of Proposed Avenues

As our general aim is to distill those aspects and dynamics of the relation between knowledge and action that can help us stay on track in the transition to sustainability, this section operationalizes the systemic perspective on sustainability (described in Sect. 1.2, Figs. 1.1–1.3 showing the sustainability filters and the general model for solving the apparent conflict of priorities between human needs and sustainability requirements) by combining it with the generalized model of Solution Readiness Levels (SRL) from Sect. 3.8 under the constraint of thermodynamic efficiency from Sect. 1.3 (related to "the space of the possible").

First, let us recall that the systemic perspective on sustainability basically means (1) that a social system can exist without having an economic system inside itself but that an economic system cannot exist independently from (i.e., outside of) a social system in which it is embedded, and (2) that a natural (ecologic/biophysical) system can exist without humans inside itself but we (Humanity as a whole) cannot exist independently from (i.e., outside of) the natural support system in which we are embedded (as biological beings, we are part of nature and the Earth' Biosphere in which we emerged as a species, not separate from it).

Then, let us also recall that, in operational terms (and as per Fig. 1.3 in Module 1), this nested inclusion relationship means that the resolution of the above-mentioned conflict of priorities in each and every relevant context must follow (as a necessary but insufficient condition) a certain sequence of steps along two phases: first the problem-definition phase that begins with immediate human needs (i.e., in order from economic to social to environmental issues), and then the phase of articulation of potential solutions, which begins with the fundamental sustainability requirements (i.e., in order from environmental to social to economic aspects).

This sequence follows logically from the fact that all three systemic sustainability filters (i.e., lists of indicators corresponding to the three main types of systems under discussion in the systemic perspective) must be consistently respected, and knowing that the economic aspect is the natural entry to the entire cycle because this is what motivates (or not) the adoption of any potential solution in the first place and then insures (or not) its validity on longer term.

That said, it is also important to keep in mind that this particular sequence is not about "putting the immediate economics above sustainability". Why not—because, in this sequence, the proposed solutions are not validated if they do not follow the order of development required by our systemic understanding of sustainability. So, this sequence is the best (and arguably the only) way of operating the "problem definition" and the "problem resolution" as distinct phases with their own internal logic, and has the immediate benefit of escaping the conflict of priorities that appears when this distinction is not being made.

This set of requirements is synthesized and resolved in the model from Fig. 3.39: a knowledge-action matrix which maps the series of 3 + 3 steps (of the full cycle of the basic resolution model from Module 1, Fig. 1.3), against the SRL steps "from idea to end-result" (as per Sect. 3.8 in this Module, namely, Fig. 3.33) and then adds the requirement of thermodynamic efficiency of scientific models (as per the well-known Maximum Entropy Principle) according to which any statistical or conceptual model agrees with the existing data but remains as structureless as possible (i.e., a "maximally non-committal with regard to missing information" [173, 341, 342, 648]).

The result of this synthesis of requirements is a state-space model for assessing the advancement of eco-innovative solutions (i.e., the set of all possible states of the development of solutions) where columns {a, b, c} represent the problem sub-space and columns {d, e, f} represent the solution sub-space.

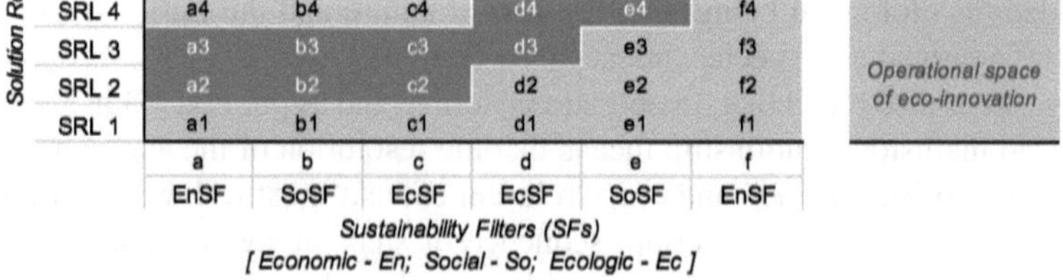

SRL 9	Solution set consolidated through successful operation
SRL 8	Mature solution set assimilated refined feed-backs from real world
SRL 7	Basic solution set demonstrated utility in relevant real-world localities or sectors
SRL 6	Solution options demonstrated in relevant context (pilot use in limited settings)
SRL 5	Details (solution parts) tested in realitic complex models (including simmulations)
SRL 4	Details (solution parts) verified in simple logic models
SRL 3	Real-life (analytical) data appear to be supporting the formulated possibility
SRL 2	Solution possibility or/and direction formulated
SRL 1	Basic principles observed and reported

Fig. 3.39 Knowledge-action matrix for generating eco-innovation for urban systems and wider human systems. *Source* modified after [156]

In it, the most efficient path from an initial idea (case *a1*) to a systemic solution that works in a test environment that is relevant for sustainable development (cases *f6; f7*) corresponds mathematically to the Manhattan distance

$$d_{ij} = \left|x_i - x_j\right| + \left|y_i - y_j\right|$$

where (x_i, y_i) and $\left(x_j, y_j\right)$ are the lattice graph coordinates of the vertices i (= a1) and j (= *f7*) of the state space [156].

Added to this is the constraint (prior insight from the field of innovation management and from the initial basic model in Fig. 1.3) that "problem definition" should not extend on the SRL scale higher than the 4th level of specifications (i.e., the SRL4), because we need to avoid premature specialization to first-considered variables (in other words, we must not make decisions about which potential solution avenue to follow before we have completed the justifications of all details in the phase of problem definition; if we do, we inevitably exclude some parts of the space of options for potential solutions before having a valid justification for doing so—or, to use a colloquial expression, we risk throwing away the bath water with the baby in it). In this state sub-space, SRL2-3 can be regarded as an optimal compromise between the need to develop an in-depth understanding of the target problem and the need to keep future options open (to avoid that premature specialization). Beyond this point (i.e., starting with the first step of solution development), progress on the SRL scale happens step-by-step: as per the requirement of thermodynamic efficiency quantified by the Manhattan distance, moving from left to right affords more freedom to advance upwards, i.e., directly towards the end-result target in the upper-right corner of the matrix, knowing that the last SRL steps consist in operational refinement in practice (optimization for smooth application), so the core solution is basically already defined at SRLs 6 and 7.

In the model from Fig. 3.39, the dark-green states represent the most efficient states and together constitute a thermodynamic efficiency front and thus the shortest path (in terms of effort) from idea to end-result. Then, and equally important, the states shown in light-green above the best-efficiency front/path designate situations of transition to (undesirable) high opportunity costs; these are not the most efficient but are still quite acceptable because (1) they are not a very large diversion from the most thermodynamically efficient path and (2) they represent situations that are potentially useful from an epistemic perspective because they may provide supplementary information and practical questions that can compensate for the moderate loss of efficiency. In fact, the more complex a project is, the more stringent is to stay close to the efficiency front but equally important is to have such "scouting" above the efficiency front because this can help with the correct assessment of what is efficient and what is not, and thus can ultimately increase the overall efficiency of a project or program. Put differently, efficiency goals are well served by sacrificing enough efficiency to ensure that visibility without which the overall project efficiency could be lost.

The entire model can also be formulated as an adjacency matrix

$$A_{ij} = \begin{pmatrix} 0\,0\,0 & 0\,0\,1 \\ 0\,0\,0 & 0\,0\,1 \\ 0\,0\,0 & 0\,0\,1 \\ 0\,0\,0 & 0\,1\,2 \\ 0\,0\,0 & 1\,2\,2 \\ 1\,1\,1 & 2\,2\,1 \\ 2\,2\,2 & 2\,1\,1 \\ 2\,2\,2 & 1\,1\,1 \\ 1\,1\,1 & 1\,1\,1 \end{pmatrix}$$

where columns i designate SRL1-9 and rows j designate the 6-series (3 + 3) of successive steps along the entire cycle and process of problem definition and resolution across sustainability filters. Such matrices can be used to store symbols designating practical values; in our case, "0" denotes "not ok", "1" denotes "OK", "2" denotes "best".

In other words, the most efficient path from idea to a validated solution goes along the cases marked with 2 (and shown above in dark green) but valid paths can take any trajectory inside the sub-space marked with "1" or "2" (both light and dark green).

In this form, the model is compact-equivalent to the semantic networks used in semantic technologies and knowledge graphs (e.g., [133, 137, 153, 377, 444, 574, 765, 766]), and allows operational coherence with a very broad range of other models that a project team may want to use, through the usual formalisms provided by fundamental and specialized mathematics (i.e., Set Theory, Order Theory, Graph Theory, Category Theory, vector spaces and so forth).

In addition, experts from applied fields may have noticed some direct connections with other usual models (as intermediated by these abstract formalisms). For example, this entire model is equivalent to a reverse Kalman filter known from statistics, engineering, and RDI more broadly [32, 553, 569], in the sense that you already know the desirable (target) states and you use the available information to find which states in the state-space are efficient.

Then, it is also worth mentioning another broad advantage: the model is reciprocally compatible with other explicitly entropy-minded approaches to and methods for solution development, including *inter alia* the requirement of strategic (targeted) data acquisition discussed in the previous sections of this Module. This allows a straightforward mapping of (and identification of correspondences between) concepts from across different fields and specialized bodies of literature as needed.

For example, this analytical system can help us see how far on the SRL scale can a concept or an entire eco-city project be efficient (all things considered), i.e., it helps us visualize and understand when the opportunity costs will begin to outweigh the benefits. In that sense, one can check some examples in Table 3.11. Obviously, one can do other similar tables using other concepts of interest, and with different details (for some start discussions, see the examples in Box 3.6).

In this sense, it is less important what location is being estimated for a concept by a project team (or individual expert) doing this exercise. After all, different (and teams)

Table 3.11 Location of a selection of concepts of interest on the analytic backdrop provided by the SRLs–SFs model in Fig. 3.39 and the related adjacency matrix (semantic network)

Concept	Range	Home
Industrial revolutions (18–20 centuries)	a5–a9	a9
Industry 4.0 (as such; 2011 to present)	a1–a7	a7
Climate changes	c5–c9	c9
Urban resilience	f1–f6	f6
Energy transitions Sustainability transitions Eco-innovation (as defined in this book)	f1–f7	f7
Eco-cities (as defined in this book)	f1–f8	f8
Industry 5.0 (as currently declared) Society 5.0 (as currently envisioned) Sustainable development Complex dynamic systems Coupled human and natural systems Technobiosphere (and the CHANS) Anthropocene (as proposed in the literature) Knowledge-action models (in general)	f1–f9	f9

"Range" defines the state sub-space of all possible locations leading to home, and "Home" denotes a stable target location of the basic/core meaning of a concept. *Source* modified after [156]

people can compare and see how their assessments differ and why; and they can eventually discuss and clarify the differences (i.e., reach a common understanding, with all due justification based on available data, information, and knowledge), the obvious benefit being a better understanding of the concepts thus-tested (compared to what we can hope for when such concepts are employed in silos or/and outside a whole-picture operational framework. Then, even more important than the review of concepts per se is whether or not this kind of analytical comparison is being done at all: because such exercises help a project constituency test, shake untested presumptions, and ask useful questions about the real space of options that may or may not be available. And, this overall mapping of concepts (ideas, theories, models, etc.) will help a project team with the articulation of a logical basis of common denominators for coherent analysis and action later on in the project.

Box 3.6 Brief illustrations of start discussions about a few concepts "dropped" in the knowledge-action matrix from Fig. 3.39 as listed in Table 3.11.

One example is the concept of eco-cities. As any reader of this book would have already understood by now, the notion of eco-city, it by definition spanning the entire path from idea to sustainability-compatible solutions. So, this is the easiest example possible. The only reason I assigned it in Home f8 instead of f9 is because f9 represents the end of the road and the notion of the eco-city refers to the road itself; f9 is simply urban sustainability. However, this can still be debated, since even sustainable cities in the future will not be a definitive and immutable situation but the continuation of a dynamic equilibrium and evolution, so the notion of 'sustainable city' is in fact an eco-city sub-category (special case). Anyways, as we are not there yet, I prefer to keep the term eco-city reserved for all situations except that special case.

A related example mentioned in the table is Urban Resilience (one of the holistic parameters discussed in the eco-city reference model in Sect. 3.4 *and a popular topic already introduced in Sect.* 1.3*, ranking just under Urban Livability and above Urban Potential). The reason I assigned its Home to case f6 is because f6 is the upper part of the maximum efficiency front and corresponds in the SRL scale to pilot demonstration in limited setting, i.e., the minimum core of a demonstrable solution, and thus indicate that resilience is the hard rock on which livability must sit. However, this is also debatable, as some may bring good arguments in favor of placing the Home of Urban Resilience in case f7. I suspect it depends on the context and different teams might reach different decisions.*

Another example of a popular topic is Industry 4.0, a broad, "umbrella" policy concept which refers to "the interoperability of all functions/services that are based on the new cyber-physical systems, including those in open control loops", and which "necessarily goes through semantic technologies" ([347]; *see also* [156])*—a concept that I have recently reviewed and studied precisely for the purpose of epistemic and operational clarification (as it is often treated as a promising avenue for urban sustainability).*

In this case, the problem is, unfortunately, that popular speculations about the benefits of Industry 4.0 methods continue unabated by various (and too many) specialists who seem to know enough in their specific field to believe they are right about a multi-domain issue (like sustainability; it seems that everybody knows what is sustainability nowadays, but only a few understand what it means in concreteness) but not enough to have a minimal understanding of the issue of sustainability as-a-whole. This seems to be the phenomenon behind this trend in which many ideas are launched and actively promoted although they are by definition invalid. (Of course, everybody could gain clarity on the matter if they would review the extant literature, but many do not do that). Concretely, the concept of Industry 4.0 had many lavish promises (like how fast carbon dioxide emissions would be reduced, how much resource-efficiency would be gained, and how many social problems would be solved) that were in fact only speculations about potential benefits of efficient engineered systems upon social and ecological aspects. Those environmental and social promises never materialized and, in light of the current knowledge in sustainability studies, cannot possibly materialize. Instead, and more pragmatically, any close observer of this trend can notice that many concepts and techniques used in or in relation to Industry 4.0 can be integrated separately in other concepts and approaches to generating sustainability-compatible solutions, e.g., multi-modal transportation, smart grids (as discussed in this book already), fractal manufacturing and functional modularity [156].

Putting the concept of Industry 4.0 in the broad picture of the knowledge-action matrix in Fig. 3.39 (and thus in the general context of sustainability studies) shows that (and how) we can recuperate many of Industry 4.0 constitutive elements in truly future-oriented eco-innovative solutions for smart and sustainable cities and societies.

A good place to start this re-analysis of Industry 4.0 is the framework of complexity sciences and 'complex dynamic systems' as described in this book (since Chap. 1). This already reveals some useful pathways. For example, those 'fractal manufacturing systems' "allow the introduction of resilience requirements (capacity to react to changes in a turbulent environment) and to reduce the complexity of its structure, operation and management", precisely through modularity [555], *see also* [350] *and in relation to analyzable network properties (and other aspects of complex systems) that are studied in engineering through cluster-detection methods like the Louvain algorithm* [77, 268, 448] *or/and (hyper)graphs-based methods, or/and usual approaches like ABM and MAS.*

For the time being, the kind of misunderstandings mentioned above in the case of the Industry 4.0 seem to be more the rule than the exception. For instance, there are numerous engineering-only discussions about "greening" or "greenifier" ideas that have no connection with the complex reality of sustainability and eco-innovation and are just unstudied (and undue) extrapolations based on some newly discovered or hypothesized method for gaining efficiency in some technological system or subsystem. Of course, many of these ideas could be properly studied (and often updated or/and adequately developed ideas can emerge from those initial intellectual shots). Nonetheless, a condition for such useful repurposing of individual and specialized efforts is that such works happen in truly multi-disciplinary projects using adequate, experience-based approaches, strategies and methods. And, because this kind of work is usually not done (or is not enough to count) for reasons related to business-as-usual, most of the resources that are now or will be in the future nominally allocated to "sustainability" projects will be in fact simply wasted from the outset. This logically means that we are blindly incurring huge opportunity costs, and we are of course talking about limited resources and a transition to sustainability that is in reality much taller an obstacle than it is usually thought of (or admitted), and there is no guarantee whatsoever that we will actually be successful (knowing that failure will inevitably lead to—in lack of a better word—a nasty race to the bottom).

It is interesting to observe that the notion of "modularity" discussed in Box 3.6 is also directly relevant for the model from Fig. 3.39, and it is partly connected (via category definition and problem structuring) with the topics discussed in the four previous book sections, i.e., including the DIKAR_process model and the SRLs, and is a basis for addressing other "open research issues" related to interoperability within complex dynamic systems [429].

In this sense, the model in Fig. 3.39 allows a mathematical and computational regard on RDI management in the context of eco-city projects in a way that I have summarized somewhere else [156] and can restate here for brevity reasons, namely, "by representing the process of generating eco-innovation as an optimization problem that can be broken up in sub-problems (e.g., the lists of yes/no questions constituting the SFs and the TRLs) that can be decided by expert decisions, i.e., by defining each project as a problem with optimal substructure and then using the Bellman equation and related techniques to facilitate understanding of, and resource allocation in, complex projects". In this managerial sense, the thermodynamic efficiency front represents the minimal cost function that is linked to the maximum utility in the end-result.

Equally important to observe is that the knowledge-action matrix described in this final section also closes the circle with the first model presented in this Module, the CIARPO/systems table (Sect. 3.2), and provides a similar benefit: while that other model served for operational orientation in a project's preparatory phases, this model here serves for operational orientation towards the desired end-result. In addition, the common goal of these two models (also shared by other technical models discussed in the middle parts of this Module) is to avoid those many instances and situations in which project contributors with deep (but typically narrowly specialized) expertise can lose sight of the big picture; and, thus they can help to avoid that a project slide into a situation where the true contribution made by those specialist actors lessens and, consequently, that the total added value of the project diminishes.

The same requirement of intellectual clear sight and managerial adequacy applies across the board and implies constant verification and minimal justification of the knowledge we may think we have.

Based on the knowledge-action matrix in Fig. 3.39, Table 3.11 and the related discussions, different concepts can be tested for compatibility with sustainable development. Figure 3.40 shows such an analytic list of concepts. In it, one observes that, out of all concepts considered, only eco-city and eco-innovation are (fully and therefore effectively) compatible with sustainable development. However, and in the same vein with the discussion made at Table 3.11, even more important than the assessment of individual concepts is that every concept must be clearly defined and confronted with the concrete operational necessities required in the transition to sustainable development of humanity (and then the preservation of it).

To illustrate the kind of discussions behind the notions listed in Fig. 3.40, I will use an example already used in Table 3.11 and Box 3.6. Thus, Fig. 3.40 not only shows (perhaps in an even more straightforward way) why the concept of Industry 4.0 (per se) is outside possible paths towards sustainable development (it fails the environmental and social sustainability filters; and, although it possibly retains some hope of passing the social sustainability filter in the future, no such hope exists for the environmental filter) while many of its components can in fact independently help

GENERAL CONCEPTS	Compatibility of principle with systemic Sustainability Filters EcSF (Econ.)	SoSF (Social)	EnSF (Environ.)	Overall compatibility with sustainable development
Paleolithic / stone age	yes/no	yes/no	yes/no	yes/no
Neolithic revolution (12-8th centuries BCE)	yes/no	yes/no	yes/no	yes/no
Industrial revolutions (18-20th centuries)	yes	yes/no	no	no
Neoclassical (synthesis in) economics	yes	yes/no	no	no
Ecological Economics (accounting for nature)	yes	yes/no	yes	yes/no
Economic "degrowth"	no	no	yes/no	no
Green Growth (as per the Club of Rome)	yes	yes/no	yes	yes/no
Progress of Economy (GDP per capita)	yes	yes/no	yes/no	yes/no
Progress of Science (frontiers of knowledge)	yes/no	yes/no	yes/no	yes/no
Progress of Education (human development)	yes/no	yes/no	yes/no	yes/no
Progress of Social Life and Institutions	yes/no	yes	yes/no	yes/no
Progress of Environmental Protection	yes/no	yes/no	yes	yes/no
Reduction & elimination of toxic pollutants	yes/no	yes/no	yes	yes/no
Renewable energy sources	yes/no	yes/no	yes/no	yes/no
Non-renewable energy sources	yes/no	yes/no	yes/no	yes/no
Circular economy (and "circular cities")	yes/no	yes/no	yes/no	yes/no
Smart society (and smart cities)	yes/no	yes/no	yes/no	yes/no
Technobiosphere / CHANS	yes/no	yes/no	yes/no	yes/no
Blue and green infrastructure	yes/no	yes/no	yes	yes/no
Eco-districts	yes/no	yes/no	yes	yes/no
Urban agriculture (hydroponics included)	yes/no	yes/no	yes/no	yes/no
Urban sprawl or densification	yes/no	yes/no	yes/no	yes/no
Energie transition (in German, "Energiewende")	yes/no	yes/no	yes/no	yes/no
National autarchy ("self-sufficiency"/isolationism)	no	no	no	no
Mitigation of Climate Changes	yes/no	yes/no	yes	yes/no
Adaptation to Climate Changes	yes/no	yes	yes/no	yes/no
Industry 4.0 (as such; 2011 to present times)	yes	yes/no	no	no
Industry 4.0 (components of it, e.g., smart grid)	yes	yes/no	yes/no	yes/no
Industry 5.0 (as declared in tentative papers)	yes	yes	yes/no	yes/no
Society 5.0 (as currently envisioned)	yes	yes	yes/no	yes/no
New materials, bionics & biomimetics	yes/no	yes/no	yes/no	yes/no
Recycling & additive manufacturing (3D-printing)	yes/no	yes/no	yes/no	yes/no
Other new technologies (e.g., miniaturization)	yes/no	yes/no	yes/no	yes/no
Smart sensors & actuators; Internet of Things (IoT)	yes/no	yes/no	yes/no	yes/no
Internet of People (IoP); Brain Networking	yes/no	yes/no	yes/no	yes/no
Advanced Systems Engineering (ASE)	yes/no	yes/no	yes/no	yes/no
Machine Learning & Artificial Intelligence	yes/no	yes/no	yes/no	yes/no
Embodied Intelligence; Servitization & Digitalization	yes/no	yes/no	yes/no	yes/no
Open innovation (sharing Intellectual Property)	yes/no	yes/no	yes/no	yes/no
Eco-cities (as defined in this book)	yes	yes	yes	yes
Eco-innovation (as defined here)	yes	yes	yes	yes

Fig. 3.40 Concise epistemic status of different concepts (a non-exhaustive list) based on the systemic perspective on sustainability and the operational compatibility with it (using a set of necessary but not sufficient conditions) as encapsulated by the notion of Sustainability Filters (defined at the beginning of the book and used in the knowledge-action matrix in Fig. 3.39) and the logic outcome to be expected. Basic rule (as per Fig. 1.3): all SFs must be passed successfully. If at least one filter is failed (i.e., "no"), the overall result is "no". The "yes/no" means it can be "yes" or "no", depending on considered details and real-life context. *Source* modified and extended after [156]

in the making of those solutions towards sustainability. This figure also shows, in a concise way, how the notion of Industry 4.0 stands in comparison to other notions that can be found in sustainability-related works and professional literature.

A second example is the notion of "circular economy". The discussion about this notion boils down to one issue: this idea originates with environmental engineering and refers to reuse of available resources as a way to avoid further impact on the environment, but its potential in practice has objective limitations, and therefore can only be a contributor to sustainable development, not "the solution".

As summarized by the last IPCC-WGIII report) ([331], Module 5, sub-section dedicated to circular economy, and the references therein), "in line with the principles of SDG12 (responsible consumption and production), the essence of building CE (i.e., circular economy) is to retain as much value as possible from products and components when they reach the end of their useful life in a given application [411, 415, 418, 665]". But, very importantly, "there are three key concerns relating to the effectiveness of the CE concept. First, many proposals on the CE insufficiently reflect on thermodynamic constraints that limit the potential of recycling from both mass conservation and material quality perspectives or ignore the considerable amount of energy needed to reuse materials [184]. Second, demand for materials and resources will likely outpace efficiency gains in supply chains, becoming a key driver of GHG emissions and other environmental problems, rendering the CE alone an insufficient strategy to reduce emissions [60]. (…) Third, cost-effectiveness underlying CE activities may concurrently also increase energy intensity and reduce labour intensity, causing systematically undesirable effects. (…) Moreover, the prevalent CE practices such as reuse, refurbishment, and recycling can differ substantially between developed and developing economies, leading to highly context-specific mitigation potentials and policy approaches [461]. (…) Besides, CE concepts that extend the life-time of products and increase the fraction of recycling are useful but are both thermodynamically limited and will remain relatively small in scale as long as demand of primary materials continue to grow, and scale effects dominate. In spite of presenting a large body of literature on CE in general, only a small but growing body of literature exists on the net effects of its strategies from a quantitative perspective, with key knowledge gaps remaining on specific CE strategies."

Unfortunately, terms such as "circular cities" which (like circular economy) may have some valid initial scientific justification (and limitations) but in most public conversations are being used lightly, even randomly or "freely", i.e., without those scientific justifications. Usually this happens with people which have no competence relevant to the issue and obviously do not understand the issues and the professional vocabulary behind such idea (otherwise they wouldn't use the words the way they do), but clearly and explicitly have some political or commercial interest in promoting this or that term, often simply because it sounds good. Terms like "circular economy" and "circular cities" can only be used correctly in a technical context, with a lot of technical qualifiers (i.e., conditions of validity, consisting in explicit expression of the context and the limits within which some "circular economy" mechanism works). The problem is that, when it passes in non-scientific contexts, these qualifiers are quickly abandoned and what remains is just rhetoric (without science) and the generation of false promises. The important idea here is that recycling can have some benefits but is far from being more than a marginal solution.

In addition, as it is well-known (and already discussed in this book), cities cannot be sustainable per se because they are dependent on inputs (and all the interactions) from outside the limits of a city. Therefore, talking seriously about "circular cities" without those many qualifiers is completely impossible and dishonest and therefore professionally inacceptable (essentially, it is irresponsible). The best one can do is to use the term in a purely metaphorical sense and with very many technical qualifiers as mentioned above.

Obviously, eco-city projects need support from all sectors, political and commercially included. But due clarifications must be made in every circumstance we can. For example, in the case of "circular economy", the right way I see possible is as follows: one place to start (or help feeding) conversations about eco-cities is within the original limits of "circular economy" from engineering, which essentially means recycling ([635] and [332]: Chap. 5 in that reference, as mentioned above), but on the condition that those conversations and the related actions (projects, policies, etc.) do not lead to toxic extremes that feed political or commercial "fires" while damaging future options for eco-innovation and eco-cities [156]. On the one hand, undue generalizations must be avoided (because they tend to stimulate politically and/or commercially driven promises and exaggerations. On the other hand, real work is not just a matter of just "box-checking"; simply covering a set of known or/ and desirable requirements does not equal progress yet. Important to remember is the summary on 'circular economy' made the last IPCC-WGIII report, and to follow future reports on the matter in the scientific and RDI literature and by IPCC itself. I also discuss it in a Module of a separate book [156]. For our purposes here, it is enough to also mention this key point from IPCC about the practical utility of the current academic literature (which is broader than the RDI and scientific literature) on this "popular term" ([332]: Chap. 5 as mentioned above): "only 10% of those provide insights on how the CE can support mitigation, and most of them found only small potentials to reduce GHG emissions [116]. Recycling is the CE category most investigated, while reuse and reduce strategies have seen comparatively less attention [116]. However, mitigation potentials were also context- and material-specific".

Figure 3.39 also shows clearly that, in the complex dynamic systems (and coupled human and natural systems; CHANS) in which we live, "box-checking" individual classes of indicators (economic, social, or environmental) technically means (a) that sustainability can still be missed (the transition to sustainability "isn't over until it's over"), and (b) that sustainability per se represents a dynamic equilibrium that can be lost again if we fail to know what we are doing (with obviously devastating consequences: not only immense opportunity costs but also societal demotivation). Instead, if all these technical details about sustainability are integrated in a systemic perspective, projects can have the structural coherence that is required for success.

How is Urban and Global Sustainability Likely to Be Obtained? A Perspective

At this point, a final idea is important to summarize and highlight in this Module and book. Namely, there is a slim (virtually zero) chance that all cities around the world will realistically join in some sort of planetary collaborative concert, but that is ok, for the following reasons: firstly because it is extremely unlikely that progress will happen otherwise than it always happened in the past of humanity (and nothing in the past happened through such a grandiose event), and secondly because this is actually not needed (in fact, I would contend that something like that would be very worrying). In reality, progress happens naturally through the competition between different actors (people, institutions, cities, countries, etc.) that seek to do better than others. Therefore, the idea is that what we must seek is not some illusionary "total cooperation" (an idea that strikes more like some propaganda of totalitarian regimes, which, by definition cannot be sustainable) but a functional agreement on a minimal set of aspects that are shared by a majority of countries.

Fortunately, this reality is already recognized at the level of our current international relations by the existence of major international organizations like the United Nations and the World Trade Organization (and some other which are well-known and require no reminder), plus some future basic agreements as needed. But the development of real solutions happens through concrete projects in real communities. Some of those solutions are or will be scalable at higher, national and international levels via natural relations between people and between institutions.

All in all, what we can realistically expect is that good systemic eco-city solutions will emerge from the "coo-petition" between cities (i.e., cooperation on some issues and competition on other issues).

In this complex reality, those cities that appear to be the most advanced today (basically the incumbents in terms of capacity or current advancement) are not necessarily the ones that will emerge as the best eco-cities of the future. Instead, smaller and less influential local communities can disrupt the market of eco-city solutions and take the lead at least in some aspects. The best solutions and combinations of solutions will then spread from city to city, because every local community wants to improve itself and gain a higher profile in the world.

Various types of solutions or/and patterns of combinations of solutions will be better fit for different regions of the world. Some solutions will eventually become quasi-global (at least for a while, as few things last "forever") but others would be optimal for specific contexts. After all, nobody expects that an eco-city in sub-polar regions would look just like eco-cities in tropical regions although some similarities will exist, and nobody in their right mind expects to see local cultural preferences disappear (nor should they).

Therefore, the transition to urban and wider societal sustainability of humanity as embodied in the concept of eco-city consists in a complex process of transformation that identifies what can stay, what must change and what real options exist or can be explored.

To illustrate the dynamics of how change can be expected to happen in the real world, Fig. 3.41 explains the interaction between cities (as providers of Livability for their citizens) starting from the phenomenon known in the field of management as "disruptive innovation" (details in Box 3.7 after Fig. 3.41).

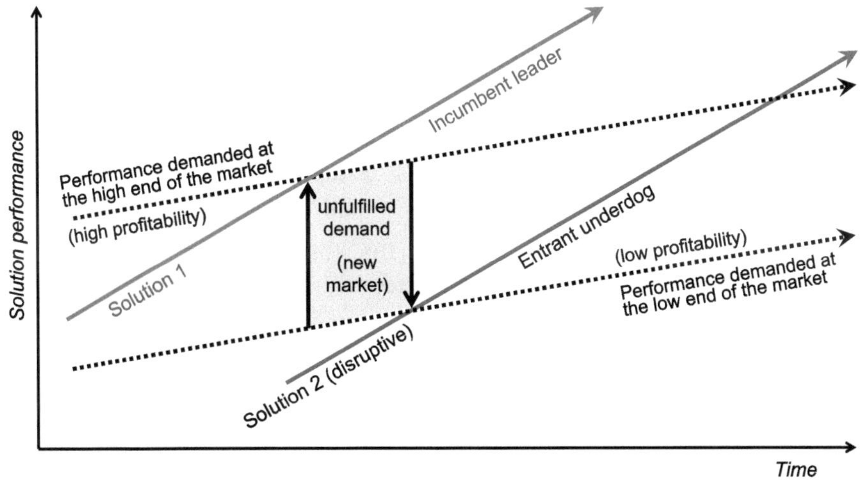

Fig. 3.41 A generalized model of disruptive eco-innovative/eco-city solutions for meeting demand unfulfilled by business-as-usual/outdated/otherwise-insufficient present providers. This shows how opportunity emerge for virtually any new project or/and any city around the world to disrupt the regional (and global) “market place” for ideas, people, and organizations competing to provide the best solutions for urban liveability. Depending on scale being used, an “Incumbent leader” or an “Entrant underdog” can be any individual or collective agent pursuing the development of sustainability-compatible solutions, from a small start-up team to a large industrial/organizational alliance and through to an entire city or alliances of cities. The low end of the market is defined by the minimal criteria insuring compatibility with sustainability. The high end of the market needs not to be defined, as it results naturally from aggregate regional and global demands (i.e., the highest level of solution quality for which there is enough demand/buyers to justify the investment in producing those solutions). *Source* original redrawing and re-conceptualization based on the classic notion of “disruptive innovation” by Christensen [138, 139]

In brief, this concept has an already long history and still represents a very active research topic. Essentially, it evolved “from a technology-change framework—essentially descriptive and relatively limited in scope—to a more broadly explanatory causal theory of innovation and competitive response” ([139], see also [138]). The main idea is that, precisely because newer (and smaller) organizations lack the great capacities of the dominant organizations, they know it is hard to compete with the incumbents. Instead, they have good reasons to invest their little resources in some new solution concepts which—at least at the beginning—tend to be less performant for the typical user but which may have other qualities that give them an advantage.

For example, these new solutions tend to be cheaper and can be fit for some non-mainstream use cases which are ignored (i.e., remain under-served) by incumbents because those lesser (or even “marginal”) use cases are regarded as being “too small to bother”. But once a newcomer provider has entered the market, i.e., once there is enough demand for the new concept to begin to be adopted, the entire set of demanders and providers begins to slightly adjust (and re-organize) and the market will start to look different. (Here, the word “market” denotes any interaction between any demand for solutions and any offer of solutions, i.e., including the public services provided by public authorities and other public organizations, not just the good and services that are sold and bought in commercial exchanges).

Box 3.7 A summary of the classic notion of "disruptive innovation" as relevant for eco-cities and eco-city projects aiming at generating change.

Originally observed in the private sector [138, 139], disruptive innovation is a set of complex dynamics where change results from the interaction of three main phenomena:

(1) in many industries (type of problem), the pace of (technological or general) progress outstrips the demand of customers for higher performance, the result being that incumbent organizations (providing a type of good or service) can "overserve" the market (by producing more advanced and sophisticated products than actually needed by most customers) and thus underserving the bottom part of the market, thus beginning to leave a gap at the bottom end of the market which represents an opportunity for new entrants;
(2) in any industry/sector of activity, a distinction can emerge between different technological and business model innovation, i.e., between (i) the usual "sustaining innovations" which simply keep improving products and services for the mainstream customers (because this enables the company to sell more products and services to their best customers at higher profitability), and (ii) the rarer "disruptive innovations" which are inferior to incumbent products/services in terms of usual performance parameters but offer a new mix of attributes that attracts fringe customer groups, for example, by being cheaper, more accessible or more convenient in other respects;
(3) established customers/users and business models tend to disincentivize those established firms to invest in novel solutions, which such investments can actually be attractive to the new market entrants who have no established customers and few other opportunities to manifest themselves, the result being that the well-established (incumbent) organizations tend to be unmotivated to develop disruptive innovations because those (at least initially) inferior products and services would be useless for their existing customers, and because those kind of innovations only promise some lower profit margins and smaller markets (i.e., "too small to bother").

All in all, "no innovation is inherently disruptive"; instead, organizations "make strategic choices to position an innovation in a disruptive way" [139], most often by targeting new potentials.

This further means that roles are gradually being updated and the demand for the new solution concept may increase (or "concepts", at plural, because there is no restriction and there can be many; the model in Fig. 3.41 is simplified for two actors only for clarity reasons). So, because there are many propositions out there, complex interactions exist between all actors, the overall result being that some of the newer solutions will become dominant and some of the older solutions will become obsolete or marginal.

Also important to stress is that a disruptive solution (in terms of relevance for eco-cities and urban sustainability in general) is not just any change or solution that promises to be the "next big thing", but a solution that holds and proves a potential to change the market-for-solutions itself. Put differently, some cities can change the world and will change the world—at least in some respects. The end-result is the global network of cities influencing each other, some being more influential than others. But each and every local community can have an influence on the world's network of cities (at least in some ways) while having access to and benefitting from the global network of cities.

In other words, every city in the world has a fair chance to lead the way, provided that it pursues eco-city ideals in earnest. Moreover, once a solution is well-proven, it gets adopted by many or even most (or all) cities. In this, cities that have the most experience as eco-cities will not only have the advantage of leading in some ways but will also be well-placed to understand and adopt the new solutions that come from somewhere else, and thus they will be able to secure for themselves a greater space of options and to enhance their capacities in the global network of cities. In this respect (and again, matching what we know from the practice and study of innovation and entrepreneurship), even the failures that a city inevitably will have in seriously undertaken endeavors are at least as valuable in this process of capacity-building because these failures come along with valuable experience and therefore represent a practical resource for development on the long term.

In a nutshell, (1) competition is necessary for solution generation (as part of that coo-petition mentioned above, and (2) in these complex dynamics, each and every city (and local community in general) around the planet has a chance to develop and maintain for itself a decent and meaningful place in the world.

Like in the discussion with large vs. small companies (and for that matter, with all types of organizations), larger cities might have a lesser interest in disrupting the business-as-usual (which they know well and in which they thrived). But even when they want to look seriously into the future (as many of them already do, in fact), they might have much less flexibility than small cities and therefore will find themselves at a comparative disadvantage (in spite of having more resources than smaller cities). The great advantage of smaller local communities is a combination of local uniqueness and superior agility in the pursuit of sustainable prosperity.

All these mean that, contrary to some appearances, it might be easier for small and medium size cities to engage in eco-city programs and be more successful or at least advance faster as eco-cities. Put differently, the play field is much more even than it looks like when looked at through the lenses of time. Interesting solutions will emerge from "somewhat expected" or "mostly unexpected" places. While the pre-Internet era was dominated by "world cities" like New York, Tokyo, and London and powerful brands like Paris, Rome, and Shanghai (more recently, also cities like San Francisco, Rio de Janeiro, Singapore, and Dubai, to cite only a few), the emerging new era of the mature Internet, dynamic RDI landscape and a deeply reconfigured (and reconfiguring) global supply chains breeds a great diversity of urban concepts and solutions.

References

1. Abernathy WJ, Utterback JM (1978) Patterns of industrial innovation. Technol Rev 80(7):40–47. https://teaching.up.edu/bus580/bps/Abernathy%20and%20Utterback,%201978.pdf
2. Aflaki A, Mirnezhad M, Ghaffarianhoseini A, Ghaffarianhoseini A, Omrany H, Wang ZH, Akbari H (2017) Urban heat island mitigation strategies: a state-of-the-art review on Kuala Lumpur, Singapore and Hong Kong. Cities 62:131–145. https://doi.org/10.1016/j.cities.2016.09.003
3. Afriliana N, Ramadhan A (2022) The trends and roles of robotic process automation technology in digital transformation: a literature review. J Syst Manage Sci 12(3):51–73. https://doi.org/10.33168/JSMS.2022.0303
4. Adam C, Steinebach Y, Knill C (2018) Neglected challenges to evidence-based policy-making: the problem of policy accumulation. Policy Sci 51:269–290. https://doi.org/10.1007/s11077-018-9318-4
5. Addai K, Serener B, Kirikkaleli D (2022) Empirical analysis of the relationship among urbanization, economic growth and ecological footprint: evidence from Eastern Europe. Environ Sci Pollut Res 29(19):27749–27760. https://doi.org/10.1007/s11356-021-17311-x
6. Agazzi E (2019) Systemic thinking: an introduction. In: Urbani Ulive L (ed) The systemic turn in human and natural sciences, pp iv–xvii. Springer, Cham. https://doi.org/10.1007/978-3-030-00725-6
7. Ahat M, Ben Amor S, Bui M (2013) Agent based modeling of ecodistricts with smart grid. In: Nguyen NT, Do TV, Thi HAL (eds) Advanced computational methods for knowledge engineering, pp 307–318. Springer, Heidelberg. https://doi.org/10.1007/978-3-319-00293-4_23
8. Ahern T, Leavy B, Byrne PJ (2014) Complex project management as complex problem solving: a distributed knowledge management perspective. Int J Project Manage 32(8):1371–1381. https://doi.org/10.1016/j.ijproman.2013.06.007
9. Al Sharif R, Pokharel S (2022) Smart city dimensions and associated risks: review of literature. Sustain Cities Soc 77, article 103542. https://doi.org/10.1016/j.scs.2021.103542
10. Alba JW, Hutchinson JW (1987) Dimensions of consumer expertise. J Consum Res 13(4):411–454. https://doi.org/10.1086/209080
11. Alberti M, Marzluff JM (2004) Ecological resilience in urban ecosystems: linking urban patterns to human and ecological functions. Urb Ecosyst 7(3):241–265. https://doi.org/10.1023/B:UECO.0000044038.90173.c6
12. Alberti M, Marzluff JM, Shulenberger E, Bradley G, Ryan C, Zumbrunnen C (2003) Integrating humans into ecology: opportunities and challenges for studying urban ecosystems. Bioscience 53(12):1169–1179. https://doi.org/10.1641/0006-3568(2003)053[1169:IHIEOA]2.0.CO;2
13. Alberti M, Wang T (2022) Detecting patterns of vertebrate biodiversity across the multidimensional urban landscape. Ecol Lett 25(4):1027–1045. https://doi.org/10.1111/ele.13969
14. Aldy JE, Kotchen MJ, Stavins RN, Stock JH (2021) Keep climate policy focused on the social cost of carbon. Science 373(6557):850–852. https://doi.org/10.1126/science.abi7813

15. Allam Z, Dhunny ZA (2019) On big data, artificial intelligence and smart cities. Cities 89:80–91. https://doi.org/10.1016/j.cities.2019.01.032
16. Allsup RE (2003) Mutual learning and democratic action in instrumental music education. J Res Music Educ 51(1):24–37. https://doi.org/10.2307/334
17. Aggarwal P, Cha T, Wilemon D (1998) Barriers to the adoption of really-new products and the role of surrogate buyers. J Consum Mark 15(4):358–371. https://doi.org/10.1108/07363769810226000
18. Alves F, Leal Filho W, Casaleiro P, Nagy GJ, Diaz H, Al-Amin AQ, de Andrade JBSO, Hurlbert M, Farooq H, Klavins M, Saroar M (2020) Climate change policies and agendas: Facing implementation challenges and guiding responses. Environ Sci Policy 104:190–198. https://doi.org/10.1016/j.envsci.2019.12.001
19. Ambos TC, Birkinshaw J (2010) How do ventures evolve? An inductive study of archetype changes in science-based ventures. Organ Sci 21:1125–1140. https://doi.org/10.1287/orsc.1090.0504
20. Amin MS, Permanasari A, Setiabudi A (2019) The pattern of environmental education practice at schools and its impact to the level of environmental literacy of school-age student. In: IOP conference series: earth and environmental science. IOP Publishing, vol 245, no 1, p 012029. https://doi.org/10.1088/1755-1315/245/1/012029
21. Anastasiadou K, Gavanas N, Pyrgidis C, Pitsiava-Latinopoulou M (2021) Identifying and prioritizing sustainable urban mobility barriers through a Modified Delphi-AHP approach. Sustainability 13:10386. https://doi.org/10.3390/su131810386
22. Andersson E (2021) Reconnecting cities to the biosphere: stewardship of green infrastructure and urban ecosystem services—where did it come from and what happened next? Ambio 50(9):1636–1638. (This article is from Ambio's 50th Anniversary Collection. Theme: Urbanization). https://doi.org/10.1007/s13280-021-01515-z
23. Ando T, Noda M (2017) Non-formal environmental education in Japan. Jpn J Environ Educ 26(4):39–44. https://doi.org/10.5647/jsoee.26.4_39
24. Antonova N, Abramova S, Polyakova V (2020) Cities are good for us: assessment of urban attractiveness by students. IOP Conf Ser Earth & Environ Sci 459(3), article 032090. https://doi.org/10.1088/1755-1315/459/3/032090
25. Arent DJ, Bragg-Sitton SM, Miller DC, Tarka TJ, Engel-Cox JA, Boardman RD, Balash PC, Ruth MF, Cox J, Garfield DJ (2021) Multi-input, multi-output hybrid energy systems. Joule 5(1):47–58. https://doi.org/10.1016/j.joule.2020.11.004
26. Arnfield AJ (2003) Two decades of urban climate research: a review of turbulence, exchanges of energy and water, and the urban heat island. Int J Climatol J Roy Meteorol Soc 23(1):1–26. https://doi.org/10.1002/joc.859
27. Arnott JC, Kirchhoff CJ, Meyer RM, Meadow AM, Bednarek AT (2020) Sponsoring actionable science: what public science funders can do to advance sustainability and the social contract for science. Curr Opin Environ Sustain 42:38–44. https://doi.org/10.1016/j.cosust.2020.01.006
28. Arnstein SR (1969) A ladder of citizen participation. J Am Inst Plann 35(4):216–224. https://doi.org/10.1080/01944366908977225
29. Asăndulesei A (2017) Inside a Cucuteni settlement: remote sensing techniques for documenting an unexplored Eneolithic site from northeastern Romania. Rem Sens 9(1), article 41. https://doi.org/10.3390/rs9010041
30. Asmussen CG, Nielsen B, Goerzen A, Tegtmeier S (2018) Global cities, ownership structures, and location choice: foreign subsidiaries as bridgeheads. Compet Rev 28(3):252–276. https://doi.org/10.1108/CR-09-2017-0063
31. Attwater R, Derry C (2017) Achieving resilience through water recycling in peri-urban agriculture. Water 9(3), article 223. https://doi.org/10.3390/w9030223
32. Auger F, Hilairet M, Guerrero JM, Monmasson E, Orlowska-Kowalska T, Katsura S (2013) Industrial applications of the Kalman filter: a review. IEEE Trans Indus Electron 60(12):5458–5471. https://doi.org/10.1109/TIE.2012.2236994

33. Avenali A, Daraio C, Di Leo S, Matteucci G, Nepomuceno T (2023) Systematic reviews as a metaknowledge tool: caveats and a review of available options. Int Trans Oper Res 30:2761–2806. https://doi.org/10.1111/itor.13309
34. Awoyemi AG, Ibánez-Álamo JD (2023) Status of urban ecology in Africa: a systematic review. Lands Urb Plann 233, article 104707. https://doi.org/10.1016/j.landurbplan.2023.104707
35. Bach PM, Rauch W, Mikkelsen PS, McCarthy DT, Deletic A (2014) A critical review of integrated urban water modelling—urban drainage and beyond. Environ Model Softw 54:88–107. https://doi.org/10.1016/j.envsoft.2013.12.018
36. Bächtiger A, Dryzek JS, Mansbridge JJ, Warren M (eds) The Oxford handbook of deliberative democracy. ISBN 978-0-19-180969-9. https://www.worldcat.org/oclc/1057358164
37. Bahers JB, Athanassiadis A, Perrotti D, Kampelmann S (2022) The place of space in urban metabolism research: Towards a spatial turn? A review and future agenda. Lands Urb Plann 221, article 104376
38. Bakker RM, DeFillippi RJ, Schwab A, Sydow J (2016) Temporary organizing: promises, processes, problems. Organ Stud 37(12):1703–1719. https://doi.org/10.1177/0170840616655982
39. Balany F, Ng AW, Muttil N, Muthukumaran S, Wong MS (2020) Green infrastructure as an urban heat island mitigation strategy—a review. Water 12(12), article 3577. https://doi.org/10.3390/w12123577
40. Baldi S, Michailidis I, Ntampasi V, Kosmatopoulos EB, Papamichail I, Papageorgiou M (2015) Simulation-based synthesis for approximately optimal urban traffic light management. In: 2015 American Control Conference (ACC). IEEE, pp 868–873. https://doi.org/10.1109/ACC.2015.7170843
41. Bamber JL, Oppenheimer M, Kopp RE, Aspinall WP, Cooke RM (2019) Ice sheet contributions to future sea-level rise from structured expert judgment. Proc Natl Acad Sci 116(23):11195–11200. https://doi.org/10.1073/pnas.1817205116
42. Banerjee S, Yik SK, Dzyuban Y, Crank PJ, Yi RPX, Chow WT (2022) Analysing impacts of urban morphological variables and density on outdoor microclimate for tropical cities: a review and a framework proposal for future research directions. Build Environ article 109646. https://doi.org/10.1016/j.buildenv.2022.109646
43. Barros Moreira de Carvalho G, Bueno da Silva L (2023) The microclimate implications of urban form applying computer simulation: systematic literature review. Environ Dev Sustain (online first, 11 August). https://doi.org/10.1007/s10668-023-03737-5
44. Baniya B, Giurco D, Kelly S (2021) Changing policy paradigms: how are the climate change mitigation-oriented policies evolving in Nepal and Bangladesh? Environ Sci Policy 124:423–432. https://doi.org/10.1016/j.envsci.2021.06.025
45. Barot S, Abbadie L, Auclerc A, Barthélémy C, Bérille E, Billet P, Clergeau P, Consales JN, Deschamp-Cottin M, David A, Devigne C (2019) Urban ecology, stakeholders and the future of ecology. Sci Total Environ 667:475–484. https://doi.org/10.1016/j.scitotenv.2019.02.410
46. Barrera PP, Carreón JR, de Boer HJ (2018) A multi-level framework for metabolism in urban energy systems from an ecological perspective. Resour Conserv Recycl 132:230–238. https://doi.org/10.1016/j.resconrec.2017.05.005
47. Bassett R, Young PJ, Blair GS, Samreen F, Simm W (2020) The megacity Lagos and three decades of urban heat island growth. J Appl Meteorol Climatol 59(12):2041–2055. https://doi.org/10.1175/JAMC-D-20-0059.1
48. Battiston S, Monasterolo I, Riahi K, van Ruijven BJ (2021) Accounting for finance is key for climate mitigation pathways. Science 372:918–920. https://doi.org/10.1126/science.abf3877
49. Bauwens T, Reike D, Calisto-Friant M (2023) Science for sale? Why academic marketization is a problem and what sustainability research can do about it. Environ Innov Soc Trans 48, article 100749. https://doi.org/10.1016/j.eist.2023.100749
50. Beaverstock JV, Smith RG, Taylor PJ (1999) A roster of world cities. Cities 16(6):445–458. https://doi.org/10.1016/S0264-2751(99)00042-6
51. Beaverstock JV, Smith RG, Taylor PJ (2000) World-city network: a new metageography? Ann Am Assoc Geogr 90(1):123–134. https://doi.org/10.1111/0004-5608.00188

52. Beck HE, Zimmermann NE, McVicar TR, Vergopolan N, Berg A, Wood EF (2018) Present and future Köppen-Geiger climate classification maps at 1-km resolution. Scientific Data 5(1):1–12. https://doi.org/10.1038/sdata.2018.214
53. Bega M, Sapel P, Ercan F, Schramm T, Spitz M, Kuhlenkötter B, Hopmann C (2023) Extension of value stream mapping 4.0 for comprehensive identification of data and information flows within the manufacturing domain. Prod Eng Res Dev. Published online: 7 July 2023. https://doi.org/10.1007/s11740-023-01207-5
54. Bell D, Jayne M (2009) Small cities? Towards a research agenda. Int J Urban Reg Res 33(3):683–699. https://doi.org/10.1111/j.1468-2427.2009.00886.x
55. Bellman R (1952) On the Theory of Dynamic Programming. Proc Natl Acad Sci 38(8):716–719. https://doi.org/10.1073/pnas.38.8.716
56. Bellman RE (1957) Dynamic programming and the numerical solution of variational problems. Oper Res 5(2):277–288. https://www.jstor.org/stable/167357
57. Beloin-Saint-Pierre D, Rugani B, Lasvaux S, Mailhac A, Popovici E, Sibiude G, Benetto E, Schiopu N (2017) A review of urban metabolism studies to identify key methodological choices for future harmonization and implementation. J Clean Prod 163:S223–S240. https://doi.org/10.1016/j.jclepro.2016.09.014
58. Belton V, Stewart T (2010) Problem structuring and multiple criteria decision. In: Ehrgott M, Figueira JR, Greco S (eds) Trends in multiple criteria decision analysis, pp 209–240. Springer+Business Media. https://doi.org/10.1007/978-1-4419-5904-1_8
59. Benatar SR, Gill S, Bakker I (2011) Global health and the global economic crisis. Am J Publ Health 101(4):646–653. https://ajph.aphapublications.org/https://doi.org/10.2105/AJPH.2009.188458
60. Bengtsson M, Alfredsson E, Cohen M, Lorek S, Schroeder P (2018) Transforming systems of consumption and production for achieving the sustainable development goals: Moving beyond efficiency. Sustain Sci 13:1533–1547. https://doi.org/10.1007/s11625-018-0582-1
61. Benito BSA (2023) A step-by-step experimental procedure for water quality assessment of Blue Lagoon: comparison to socio-demographic and economic profile for a teaching model. ASEAN J Sci Educ 2(2):107–120. https://ejournal.bumipublikasinusantara.id/index.php/ajsed/article/view/245
62. Benson LV, Pauketat TR, Cook ER (2009) Cahokia's boom and bust in the context of climate change. Am Antiq 74(3):467–483. https://doi.org/10.1017/S000273160004871X
63. Berger T, Grabert S, Kempa B (2016) Global and country-specific output growth uncertainty and macroeconomic performance. Oxford Bull Econ Statist 78(5):694–716. https://onlinelibrary.wiley.com/https://doi.org/10.1111/obes.12118
64. Beumer C (2017) Sustopia or cosmopolis? A critical reflection on the sustainable city. Sustainability 9(5), article 845. https://doi.org/10.3390/su9050845
65. Bhayunagiri IBP, Saifulloh M (2023) Urban footprint extraction derived from Worldview-2 satellite imagery by random forest and k-nearest neighbours algorithm. IOP Conf Ser Earth Environ Sci 1200(1), article 012043. https://doi.org/10.1088/1755-1315/1200/1/012043
66. Bhimani H, Mention AL, Salampasis D (2023) Disengagement in open innovation: a cognitive perspective. Br J Manag 34(1):241–258. https://doi.org/10.1111/1467-8551.12594
67. Bi C, Little JC (2022) Integrated assessment across building and urban scales: a review and proposal for a more holistic, multi-scale, system-of-systems approach. Sustain Cities Soc 82, article 103915. https://doi.org/10.1016/j.scs.2022.103915
68. Bibri SE (2019) On the sustainability of smart and smarter cities in the era of big data: an interdisciplinary and transdisciplinary literature review. J Big Data 6(1):1–64. https://doi.org/10.1186/s40537-019-0182-7
69. Bibri SE (2021) Data-driven smart eco-cities and sustainable integrated districts: a best-evidence synthesis approach to an extensive literature review. Eur J Fut Res 9(1):1–43. https://doi.org/10.1186/s40309-021-00181-4
70. Bibri SE (2022) Eco-districts and data-driven smart eco-cities: emerging approaches to strategic planning by design and spatial scaling and evaluation by technology. Land Use Policy 113:105830. https://doi.org/10.1016/j.landusepol.2021.105830

71. Bibri SE, Krogstie J (2020) Smart eco-city strategies and solutions for sustainability: the cases of Royal Seaport, Stockholm, and Western Harbor, Malmö, Sweden. Urb Sci 4(1):11. https://doi.org/10.3390/urbansci4010011
72. Biesbroek R, Peters BG, Tosun J (2018) Public bureaucracy and climate change adaptation. Rev Policy Res 35:776–791. https://doi.org/10.1111/ropr.12316
73. Binkley S (2003) The seers of Menlo Park: the discourse of heroic consumption in the 'Whole Earth Catalog'. J Consum Cult 3(3):283–313. https://doi.org/10.1177/14695405030033001
74. Blanke L, Walmsley A (2022) Resilient cities: renewal after disaster in three late antique towns of the East Mediterranean. In: Martinez Jiménez J, Soulsby SO (eds) Remembering and forgetting the Ancient city, pp 69–108. Oxbow Books, Oxford. Digital ISBN 978-1-78925-816-5. https://library.oapen.org/bitstream/handle/20.500.12657/54112/external_content.pdf?sequence=1#page=320
75. Bleischwitz R, Spataru C, VanDeveer SD, Obersteiner M, van der Voet E, Johnson C, Andrews-Speed P, Boersma T, Hoff H, Van Vuuren DP (2018) Resource nexus perspectives towards the United Nations sustainable development goals. Nat Sustain 1(12):737–743. https://doi.org/10.1038/s41893-018-0173-2
76. Bloembergen D, Tuyls K, Hennes D, Kaisers M (2015) Evolutionary dynamics of multi-agent learning: a survey. J Artif Intel Res 53:659–697. https://doi.org/10.1613/jair.4818
77. Blondel VD, Guillaume JL, Lambiotte R, Lefebvre E (2008) Fast unfolding of communities in large networks. J Statist Mechan 10:100–108. https://doi.org/10.1088/1742-5468/2008/10/P10008
78. Bocken NMP, Farracho M, Bosworth R, Kemp R (2014) The front-end of eco-innovation for eco-innovative small and medium sized companies. J Eng Tech Manage 31:43–57. https://doi.org/10.1016/j.jengtecman.2013.10.004
79. Bocken NMP (2015) Sustainable venture capital—catalyst for sustainable start-up success? J Clean Prod 108(part A):647–658. http://www.sciencedirect.com/science/article/pii/S0959652615006460
80. Boeing G (2019) Urban spatial order: Street network orientation, configuration, and entropy. Appl Netw Sci 4(1):1–19. https://doi.org/10.1007/s41109-019-0189-1
81. Boese VA, Edgell AB, Hellmeier S, Maerz SF, Lindberg SI (2021) How democracies prevail: democratic resilience as a two-stage process. Democratization 28(5):885–907. https://doi.org/10.1080/13510347.2021.1891413
82. Bohm D (1980) Wholeness and the implicate order. Routledge, London. ISBN 978-0415289795. https://www.routledge.com/Wholeness-and-the-Implicate-Order/Bohm/p/book/9780415289795
83. Boivin M, Tanguay GA (2019) Analysis of the determinants of urban tourism attractiveness: the case of Québec City and Bordeaux. J Destin Mark Manag 11:67–79. https://doi.org/10.1016/j.jdmm.2018.11.002
84. Bolund P, Hunhammar S (1999) Ecosystem services in urban areas. Ecol Econ 29:293–301. https://doi.org/10.1016/S0921-8009(99)00013-0
85. Bonilla K, Arrechea S, Velásquez Pérez LG (2022) Connecting scientists residing abroad: a review of converciencia as a practice to engage the Guatemalan scientific diaspora from 2005–2020. Front Res Metric Anal 7, article 37. https://doi.org/10.3389/frma.2022.898496
86. Boschken HL (2003) Global cities, systemic power, and upper-middle-class influence. Urb Affairs Rev 38(6):808–830. https://doi.org/10.1177/1078087403038006003
87. Boyes B (2020) Case studies in inappropriate responses in the face of complexity. Real KM, December 1st. https://realkm.com/2020/12/01/case-studies-in-inappropriate-responses-in-the-face-of-complexity
88. Bozzo R, Conca A, Marangon F (2014) Decision support system for city logistics: literature review, and guidelines for an ex-ante model. Transp Res Proc 3:518–527. https://doi.org/10.1016/j.trpro.2014.10.033
89. Brand MJ, Huizingh EKRE (2008) Into the drivers of innovation adoption: what is the impact of the current level of adoption? Eur J Innov Manag 11(1):5–24. https://doi.org/10.1108/14601060810845204

90. Brandt P, Ernst A, Gralla F, Luederitz C, Lang DJ, Newig J, Reinert F, Abson DJ, Von Wehrden H (2013) A review of transdisciplinary research in sustainability science. Ecol Econ 92:1–15. https://doi.org/10.1016/j.ecolecon.2013.04.008
91. Brenner N, Schmid C (2015) Towards a new epistemology of the urban? City 19(2–3):151–182. https://doi.org/10.1080/13604813.2015.1014712
92. Broto VC, Allen A, Rapoport E (2012) Interdisciplinary perspectives on urban metabolism. J Ind Ecol 16(6):851–861. https://doi.org/10.1111/j.1530-9290.2012.00556.x
93. Brown JS, Duguid P (1991) Organizational learning and communities-of-practice: toward a unified view of working, learning, and innovation. Organ Sci 2(1):40–57. https://doi.org/10.1287/orsc.2.1.40
94. Brown PG, Garver G (2009) Right relationship: building a whole earth economy. Berrett-Koehler Publishers. https://www.bkconnection.com/static/Right_Relationship_EXCERPT.pdf
95. Brown SL, Eisenhardt KM (1995) Product development: Past research, present findings, and future directions. Acad Manag Rev 20(2):343–378. https://doi.org/10.5465/amr.1995.9507312922
96. Browne D, O'Regan B, Moles R (2009) Assessment of total urban metabolism and metabolic inefficiency in an Irish city-region. Waste Manage 29(10):2765–2771. https://doi.org/10.1016/j.wasman.2009.05.008
97. Brozovsky J, Gaitani N, Gustavsen A (2021) A systematic review of urban climate research in cold and polar climate regions. Renew Sustain Ener Rev 138, article 110551. https://doi.org/10.1016/j.rser.2020.110551
98. Bruland K (1995) Patterns of resistance to new technologies in Scandinavia: an historical perspective. In: Bauer M (ed) Resistance to new technology. Cambridge University Press, Cambridge. ISBN 9780521599481. http://eprints.lse.ac.uk/id/eprint/39607
99. Bruneckienė J, Zykienė I, Stankevičius V (2016) Critical analysis of city attractiveness factors in Lithuania–Poland cross-border regions: the viewpoints of businessmen and youth. J Geogr Polit Soc 6(2):45–58. https://www.ceeol.com/search/article-detail?id=522484
100. Bruno I, Lobo G, Covino BV, Donarelli A, Marchetti V, Panni AS, Molinari F (2020) Technology readiness revisited: a proposal for extending the scope of impact assessment of European public services. In: Proceedings of the 13th International Conference on Theory and Practice of Electronic Governance, 1–3 April, Athens, Greece, pp 369–380. https://ec.europa.eu/isa2/sites/isa/files/technology_readiness_revisited_-_icegov2020.pdf
101. Bruns-Berentelg J, Noring L, Grydehøj A (2022) Developing urban growth and urban quality: entrepreneurial governance and urban redevelopment projects in Copenhagen and Hamburg. Urb Stud 59(1):161–177. https://doi.org/10.1177/0042098020951438
102. Burch S, Andrachuk M, Carey D, Frantzeskaki N, Schroeder H, Mischkowski N, Loorbach D (2016) Governing and accelerating transformative entrepreneurship: exploring the potential for small business innovation on urban sustainability transitions. Curr Opin Environ Sustain 22:26–32. https://doi.org/10.1016/j.cosust.2017.04.002
103. Burch S, DiBella J, Wiek A, Schaltegger S, Stubbs W, Farrelly M, Ness B, McCormick K (2022) Building urban resilience through sustainability-oriented small-and medium-sized enterprises. Urb Transform 4(1):1–7. https://doi.org/10.1186/s42854-022-00041-9
104. Burt RS (1987) Social contagion and innovation: Cohesion versus structural equivalence. Am J Sociol 92(6):1287–1335. https://doi.org/10.1086/228667
105. Butera FM, Palme M (2021) The city as a complex thermodynamic system. In: Palme M, Salvati A (eds) Urban microclimate modelling for comfort and energy studies. Springer, Cham, pp 13–22. https://doi.org/10.1007/978-3-030-65421-4_2
106. Butler DL, Del Castello B, Elliott S, Goldenkoff ER, Warner IA, Zimmermann AC (2022). Recognize and alleviate a resource management conundrum facing science diaspora networks. Front Res Metric Anal 7, article 898770. https://doi.org/10.3389/frma.2022.898770
107. Buzási A, Csizovszky A (2023) Urban sustainability and resilience: what the literature tells us about "lock-ins"? Ambio 52(3):616–630. https://doi.org/10.1007/s13280-022-01817-w

108. Byrne LB (2022) Ecology with cities. Urban Ecosyst 25(3):835–837. https://doi.org/10.1007/s11252-021-01185-5
109. Bytheway A (2014) Investing in information: the information management body of knowledge. Springer, Heidelberg. https://doi.org/10.1007/978-3-319-11909-0
110. Cai Y (2020) Renaissance of resilience: A buzzword or a new ideal? Manag Organ Rev 16(5):976–980. https://doi.org/10.1017/mor.2020.46
111. Caldarelli G, Arcaute E, Barthelemy M, Batty M, Gershenson C, Helbing D, Mancuso S, Moreno Y, Ramasco JJ, Rozenblat C, Sánchez A (2023) The role of complexity for digital twins of cities. Nat Comput Sci 3:374–381. https://doi.org/10.1038/s43588-023-00431-4
112. Caldecott B, Clark A, Koskelo K, Mulholland E, Hickey C (2021) Stranded assets: environmental drivers, societal challenges, and supervisory responses. Annu Rev Environ Resour 46(1):417–447. https://doi.org/10.1146/annurev-environ-012220-101430
113. Campiglio E, Dafermos Y, Monnin P, Ryan-Collins J, Schotten G, Tanaka M (2018) Climate change challenges for central banks and financial regulators. Nat Clim Chang 8:462–468. https://doi.org/10.1038/s41558-018-0175-0
114. Caniglia G, Luederitz C, von Wirth T, Fazey I, Martin-López B, Hondrila K, König A, von Wehrden H, Laubichler MD, Lang DJ (2021) A pluralistic and integrated approach to action-oriented knowledge for sustainability. Nat Sustain 4:93–100. https://doi.org/10.1038/s41893-020-00616-z
115. Caniglia G, Schäpke N, Lang DJ, Abson DJ, Luederitz C, Wiek A, Laubichler MD, Gralla F, von Wehrden H (2017) Experiments and evidence in sustainability science: a typology. J Clean Prod 169:39–47. https://doi.org/10.1016/j.jclepro.2017.05.164
116. Cantzler J, Creutzig F, Ayargarnchanakul E, Javaid A, Wong L, Haas W (2020). Saving resources and the climate? A systematic review of the circular economy and its mitigation potential. Environ Res Lett 15(12), article 123001. https://doi.org/10.1088/1748-9326/abbeb7
117. Carréon JR, Worrell E (2018) Urban energy systems within the transition to sustainable development. A research agenda for urban metabolism. Resour Conserv Recycl 132:258–266. https://doi.org/10.1016/j.resconrec.2017.08.004
118. Cash DW, Clark WC, Alcock F, Dickson NM, Eckley N, Guston DH, Jäger J, Mitchell RB (2003). Knowledge systems for sustainable development. Proc Natl Acad Sci 100(14):8086–8091. https://www.pnas.org/https://doi.org/10.1073/pnas.1231332100
119. Cash DW, Belloy PG (2020) Salience, credibility and legitimacy in a rapidly shifting world of knowledge and action. Sustainability 12(18):7376. https://doi.org/10.3390/su12187376
120. Castellani V (2013) The lost world of old Europe: the Danube Valley, 5000–2500 BC. https://doi.org/10.1080/10848770.2013.765105
121. Castle D, Culver K (2013) Getting to 'no': the method of contested exchange. Sci Publ Policy 40(1):34–42. https://doi.org/10.1093/scipol/scs118
122. CEN-CENELEC-ETSI (2012) Smart Grid Coordination Group. Smart grid reference architecture. https://ec.europa.eu/energy/sites/ener/files/documents/xpert_group1_reference_architecture.pdf
123. Chaisson EJ (2015) Energy flows in low-entropy complex systems. Entropy 17(12):8007–8018. https://doi.org/10.3390/e17127857
124. Chang YF (2020) Development of entropy change in philosophy of science. Philos Stud 10(9):517–524. https://doi.org/10.17265/2159-5313/2020.09.001
125. Chatman EA (1996) The impoverished life-world of outsiders. J Am Soc Inform Sci 47(3):193–206. https://doi.org/10.1002/(SICI)1097-4571(199603)47:3%3c193::AID-ASI3%3e3.0.CO;2-T
126. Chávez JEG (2016) Institutional reform through innovative diasporas in the knowledge economy. Reforma institucional a través de diásporas innovadoras en la economía del conocimiento (Doctoral dissertation). Universidad Complutense de Madrid. https://hdl.handle.net/20.500.14352/26744
127. Cheek W, Chmutina K (2022) Measuring resilience in the assumed city. Int J Disas Risk Sci 13(3):317–329. https://doi.org/10.1007/s13753-022-00410-9

128. Chen C, Loft L, Matzdorf B (2023) Lost in action: Climate friendly use of European peatlands needs coherence and incentive-based policies. Environ Sci Policy 145:104–115. https://doi.org/10.1016/j.envsci.2023.04.010
129. Chen L, Wang M, Cui L, Li S (2021) Experience base, strategy-by-doing and new product performance. Strateg Manag J 42(7):1379–1398. https://doi.org/10.1002/smj.3262
130. Chen Y, Ghosh A, Kearns M, Roughgarden T, Vaughan JW (2016) Mathematical foundations for social computing. Commun ACM 59(12):102–108. https://doi.org/10.1145/2960403
131. Chen YT, Sun EW, Lin YB (2020) Merging anomalous data usage in wireless mobile telecommunications: business analytics with a strategy-focused data-driven approach for sustainability. Eur J Oper Res 281:687–705. https://doi.org/10.1016/j.ejor.2019.02.046
132. Chen Z (2022) Observations and expectations on recent developments of data lakes. Procedia Comput Sci 214:405–411. https://doi.org/10.1016/j.procs.2022.11.192
133. Chen Z, Wan Y, Liu Y, Valera-Medina A (2024) A knowledge graph-supported information fusion approach for multi-faceted conceptual modelling. Inform Fus 101985. https://doi.org/10.1016/j.inffus.2023.101985
134. Chesbrough H, Lettl C, Ritter T (2018) Value creation and value capture in open innovation. J Prod Innov Manag 35(6):930–938. https://doi.org/10.1111/jpim.12471
135. Chien F, Ngo QT, Hsu CC, Chau KY, Iram R (2021) Assessing the mechanism of barriers towards green finance and public spending in small and medium enterprises from developed countries. Environ Sci Pollut Res 28(43):60495–60510. https://doi.org/10.1007/s11356-021-14907-1
136. Chirisa I, Bandauko E (2015) African cities and the water-food-climate-energy nexus: an agenda for sustainability and resilience at a local level. Urban Forum 26:391–404. https://doi.org/10.1007/s12132-015-9256-6
137. Cho S, May G, Kiritsis D (2019) A semantic-driven approach for industry 4.0. In: The 15th international conference on distributed computing in sensor systems (DCOSS). IEEE, pp 347–354. https://doi.org/10.1109/DCOSS.2019.00076
138. Christensen CM (1997; 2011) The innovator's dilemma. Harper's Business, New York. ISBN 978-14221-8998-6
139. Christensen CM, McDonald R, Altman EJ, Palmer JE (2018) Disruptive innovation: an intellectual history and directions for future research. J Manage Stud 55(7):1043–10781197. https://doi.org/10.1111/joms.12349
140. Christensen T, Lægreid P (2007) The whole-of-government approach to public sector reform. Public Adm Rev 67:1059–1066. https://doi.org/10.1111/j.1540-6210.2007.00797.x
141. Chuo HSE, Tan MK, Chong ACH, Chin RKY, Teo KTK (2017) Evolvable traffic signal control for intersection congestion alleviation with enhanced particle swarm optimisation. In: 2017 IEEE 2nd international conference on automatic control and intelligent systems (I2CACIS), pp 92–97. IEEE. https://doi.org/10.1109/I2CACIS.2017.8239039
142. City of Versailles (2022) Versailles Chantiers (in French). https://www.versailles.fr/actualites/toute-lactualite/detail-dune-actualite/actualites/versailles-chantiers-un-nouveau-quartier-en-2019/. Accessed on 31 July 2022
143. Ciumasu IM (2005) Environmental science and technology in Romania: the need and the possibilities of cooperation between the Romanian universities and the Romanian students and researchers at the European universities. Talk and book chapter. University in Society—UNISO 2005, 18–25 July. University of Caen Basse Normandie, Cherbourg. In: Zaharia SE, Ivan MS (eds) UNISO 2005. University and labour market. ISBN 973-596-305-1, pp 309–312. https://www.ad-astra.ro/library/papers/Ciumasu_UNISO_2005_Diaspora_Plan.pdf
144. Ciumasu IM (2006a) Development of a miniaturized immunochemical flow-injection system for on-site analysis of selected nitroaromatics and pesticides in water. Doctoral dissertation, Technische Universität München. https://mediatum.ub.tum.de/?id=603680
145. Ciumasu IM (2006b) Sustainable development needs biosensors. In: The 9th world congress on biosensors, 10–12 July, Toronto, Canada. https://www.elsevier.com/events/conferences/world-congress-on-biosensors/about/previous-events/biosensors-2006

146. Ciumasu IM (2007c) International circulation of talent—a solution for the development of Romania. Talk & book chapter. University in Society—UNISO 2007, 9–12 July. University of Versailles Saint-Quentin-en-Yvelines, Versailles, France. In: Zaharia SE, Ivan MS (eds) UNISO 2007. Lifelong Learning and Qualifications in Higher Education, pp 207–217. https://citeseerx.ist.psu.edu/document?repid=rep1&type=pdf&doi=2fd5e4f57e599ebf184acb04827a285eab325154
147. Ciumasu IM (2008) Brain networking—how to do it? Talk and book chapter. University in Society—UNISO 2008, 16–20 July. "Alexandru Ioan Cuza" University of Iasi, Romania. In: Zaharia SE, Ivan MS (eds) UNISO 2008. Competence-based higher education: challenges and solutions, pp 197–203
148. Ciumasu IM (2010) Turning brain drain into brain networking. Sci Publ Policy 37(2):135–146. https://doi.org/10.3152/030234210X489572
149. Ciumasu IM (2012c) Ecological disturbance theory for future ecocities: towards annular cities? Talk at EcoSummit 2012—Ecological sustainability. Restoring the planets ecosystem services, 30 September—05 October, Columbus-Ohio, USA. https://ecosummit2012.org/newsletter-eco-sustainability.html. Booklet: https://surumer.uni-hohenheim.de/fileadmin/einrichtungen/surumer/Booklet_EcoSummit_2012.pdf
150. Ciumasu IM (2013) Dynamic decision trees for building resilience into future ecocities. Technol Forecast Soc Chang 80(9):1804–1814. https://doi.org/10.1016/j.techfore.2012.12.010
151. Ciumasu IM (2013b) We need large-scale experimentation platforms for cross-sector innovation in Europe, Science|Business, 16 January 2013. https://sciencebusiness.net/news/76003/We-need-large-scale-experimentation-platforms-for-cross-sector-innovation-in-Europe
152. Ciumasu IM (2014) Challenges of systemic innovation: Cities as experimentation units? In: Deltas in Time of Climate Change II, Rotterdam, The Netherlands, 24–26 September, pp 278–279. http://www.climatedeltaconference2014.org/
153. Ciumasu IM (2018a) A coordination lattice model for building urban resilience. In: Proceedings of the 15th Information Systems for Crisis Response & Management (ISCRAM) conference, Rochester, NY, USA, 23–25 May, pp 419–427. http://idl.iscram.org/files/ioanmciumasu/2018/2119_IoanM.Ciumasu2018.pdf
154. Ciumasu IM (2018b) Networking brains: how Horizon Europe can foster excellence in central European science. Science|Business, 23 October 2018. https://sciencebusiness.net/framework-programmes/viewpoint/networking-brains-how-horizon-europe-can-foster-excellence-central
155. Ciumasu IM (2023a) Smart solutions for sustainability: RDI for urban and societal transitions requires cross-sectoral experimentation platforms. In: Da Silva Portela CF (ed) Sustainable, Innovative and Intelligent Societies and Cities. Springer, Cham. https://doi.org/10.1007/978-3-031-30514-6_10
156. Ciumasu IM (2023b) Mapping Industry 4.0 onto eco-city transitions: a knowledge-action matrix. In: Da Silva Portela CF (ed) Sustainable, Innovative and Intelligent Societies and Cities. Springer, Cham. https://doi.org/10.1007/978-3-031-30514-6_13
157. Ciumasu IM (2023c) Brain networking: model operațional de bază. In Smart Diaspora—diaspora în învățământ superior, știință, inovare și antreprenoriat. Diaspora și prietenii ei, 10–13 April, Timișoara, Romania
158. Ciumasu IM, Bos D, Bakker J (2007c) The effect of cattle grazing upon plant diversity in salt marshes: the importance of the sample size. In: The British Ecological Society's Annual Meeting 2007, University of Glasgow, 10–12 September, Glasgow, Scotland
159. Ciumasu IM, Buzdugan L, Stefan N, Culver K (2012a) Ecosystems: planning. In: Jorgensen SE (ed) Encyclopedia of environmental management, four volume set. CRC Press, pp 632–642. eISBN 9781351235860. https://www.routledgehandbooks.com/https://doi.org/10.1081/E-EEM-120047256
160. Ciumasu IM, Costica M, Costica N, Neamtu M, Dirtu AC, De Alencastro LP, Buzdugan L, Andriesa R, Iconomu L, Stratu A, Popovici OA, Secu CV, Paveliuc-Olariu C, Dunca S, Stefan M, Lupu A, Stingaciu-Basu A, Netedu A, Dimitriu RI, Gavrilovici O, Talmaciu M, Borza

M (2012) Complex risks from old urban waste landfills—a sustainability perspective from Iasi, Romania. J Hazard Toxic Radioact Waste 16(2):158–168. https://doi.org/10.1061/(ASCE)HZ.2153-5515.0000090
161. Ciumasu IM, Costica M, Secu CV, Gurjar BR, Ojha CSP (2013d) Adapting to climate change: technologies, perceptions, education, and perspectives. In Surampalli RY, Zhang T, Ojha CSP, Gurjar BR (eds) Greenhouse gas emissions and climate change, ASCE Press (American Society of Civil Engineers), pp 490–512. ISBN 978-0-7844-1271-8. https://doi.org/10.1061/9780784412718.ch. Also available at: https://www.researchgate.net/publication/292784855_Adapting_to_Climate_Change_Technologies_Perceptions_Education_and_Perspectives
162. Ciumasu IM, Costica N (2008) Environmental education: education for transition to sustainable development. Ann Sci | A.I. Cuza University of Iasi, Romania, vol. LIV (fasc. 1, section Plant Biology), pp 146–152. http://cercetare.bio.uaic.ro/publicatii/anale_vegetala/issue/2008F1/20-2008F1.pdf
163. Ciumasu IM, Costica N (2010) Impacts of air pollution on the ecosystem and human health. Health and environmental impacts. In: Gurjar BR, Molina LT, Ojha CSP (eds) Air pollution. Health and environmental impacts, pp 447–492. CRC Press (Taylor & Francis), Boca Raton. eBook ISBN 9780429075599. https://doi.org/10.1201/EBK1439809624
164. Ciumasu IM, Culver K (2011a) Eco-cities in a globalized future—from Constantinopolis to "Cosmopolis"? In: The 9th International Conference of the European Society of Ecological Economics (ESEE), 14–17 June, Istanbul (2011). https://www.academia.edu/26227428/Eco_cities_in_a_globalized_future_from_Constantinopolis_to_Cosmopolis
165. Ciumasu IM, Culver K (2011b) The city as disturbance of the local ecosystems. In: The 9th International Conference of the European Society of Ecological Economics (ESEE), 14–17 June, Istanbul (2011). https://ecolecon.eu/call-for-papers-for-the-esee-2011-special-issue-of-epg-deadline-november-14/
166. Ciumasu IM, Culver K, Costica M, Vanderlinden JP (2021) Ecosystems: planning and trade-offs. In Fath BD, Jorgensen SE (eds) Handbook of environmental science, 2nd edn, vol 2. Managing biological and ecological systems. CRC Press, Boca Raton, pp 105–124. https://www.taylorfrancis.com/chapters/edit/https://doi.org/10.1201/9780429346170-12/. https://doi.org/10.1201/9780429346170
167. Ciumasu IM, Krämer PM, Weber C-M, Kolb G, Tiemann D, Windisch S, Frese I, Kettrup AA (2004) Single-use immunosensor for environmental pollutants. Proof of principle for nitroaromatics and pesticides. In: Proceedings of the Eighth World Congress on Biosensors, Granada, Spain, May 24–26 (Abstract BS67). https://www.elsevier.com/events/conferences/world-congress-on-biosensors/about/previous-events/biosensors-2006
168. Ciumasu IM, Krämer PM, Weber CM, Kolb G, Tiemann D, Windisch S, Frese I, Kettrup AA (2005) A new, versatile field immunosensor for environmental pollutants: development and proof of principle with TNT, Diuron, and Atrazine. Biosens Bioelectron 21(2):354–364. https://doi.org/10.1016/j.bios.2004.10.027
169. Ciumasu IM, Lupu A, Costica N, Netedu A, Stratu A, Miftode V, Stefan N (2008c) Integrating social perception of nature with economic development expectations in a peri-urban area in Iasi, Romania. Talk and paper. In: ISEE 2008: applying ecological economics for social and environmental sustainability (10th Biennial Conference of the International Society for Ecological Economics), 7–11 August, Nairobi-UNEP
170. Clancy JS (2008) Urban ecological footprints in Africa. Afr J Ecol 46(4):463–470. https://doi.org/10.1111/j.1365-2028.2008.01041.x
171. Clark WC, van Kerkhoff L, Lebel L, Gallopin GC (2016) Crafting usable knowledge for sustainable development. Proc Natl Acad Sci 113(17):4570–4578. https://doi.org/10.1073/pnas.1601266113
172. Coelho A, Bairrada C, Simão L, Barbosa C (2022) The drivers of the city brand equity comparing citizens' and tourists' perceptions and its influence on the city attractiveness: the case of the city of Coimbra. Int J Hosp Tour Adm 23(2):242–268. https://doi.org/10.1080/15256480.2020.1746215

173. Cofré R, Herzog R, Corcoran D, Rosas FE (2019) A comparison of the maximum entropy principle across biological spatial scales. Entropy 21:1009. https://doi.org/10.3390/e21101009
174. Cohen W, Levinthal D (1990) Absorptive capacity: a new perspective on learning and innovation. Adm Sci Q 35(1):128–152. https://doi.org/10.2307/2393553
175. Collins JL, Wellman B (2010) Small town in the internet society: Chapleau is no longer an island. Am Behav Sci 53(9):1344–1366. https://doi.org/10.1177/0002764210361689
176. Contreras Kallens P, Dale R, Christiansen MH (2022) Quantifying interdisciplinarity in cognitive science and beyond. Top Cogn Sci 14(3):634–645. https://doi.org/10.1111/tops.12609
177. Cook M, Horne R, Potter S, Valdez AM (2019) Exploring the epistemicpolitics of urban niche experiments. In: Jensen Stissing J, Cashmore M, Späth P (eds) The politics of urban sustainability transitions: knowledge, power and governance. Routledge, Abingdon, Oxon, pp 125–147. https://files.core.ac.uk/pdf/86/161882888.pdf
178. Cooper RG (2019) The drivers of success in new-product development. Ind Mark Manage 76:36–47. https://doi.org/10.1016/j.indmarman.2018.07.005
179. Correia E, Carvalho H, Azevedo S, Govindan K (2017) Maturity models in supply chain sustainability: a systematic literature review. Sustainability 9(1), article 64. https://doi.org/10.3390/su9010064
180. Costica N, Ciumasu IM, Costica M (2006) Partnership for human resources development in Romania and the European Union: training the trainers in environmental education. In: Proceedings of the 2nd European Fair on Education for Sustainable Development "Promoting Education for Sustainable Development in Europe", September 13–15, Hamburg
181. Costica N, Stratu A, Costica M (2012) The key role of higher education institutions in implementing education for sustainable development (ESD). In: Conservation of plant diversity, pp 524–529. https://ibn.idsi.md/sites/default/files/imag_file/524-529.pdf
182. Cowen PJ (2010) Not fade away: the HPA axis and depression. Psychol Med 40(1):1–4. https://doi.org/10.1017/S0033291709005558
183. Cronin K, Midgley G, Jackson LS (2014) Issues mapping: a problem structuring method for addressing science and technology conflicts. Eur J Oper Res 233(1):145–158. https://doi.org/10.1016/j.ejor.2013.08.012
184. Cullen J (2017) Circular economy: theoretical benchmark or perpetual motion machine? J Indus Econ 21:483–486. https://doi.org/10.1111/jiec.12599
185. Dabaieh M, Iarkov I, Rodil K (2023) The 'Z-Free'home: a circular thinking and eco-cycle design practice. Energies 16(18), article 6536. https://doi.org/10.3390/en16186536
186. Dafermos Y, Nikolaidi M, Galanis G (2017) A stock-flow-fund ecological macroeconomic model. Ecol Econ 131:191–207. https://www.sciencedirect.com/science/article/pii/S0921800916301343
187. D'Amico G, Taddeo R, Shi L, Yigitcanlar T, Ioppolo G (2020) Ecological indicators of smart urban metabolism: A review of the literature on international standards. Ecol Ind 118:106808. https://doi.org/10.1016/j.ecolind.2020.106808
188. D'Amico G, Arbolino R, Shi L, Yigitcanlar T, Ioppolo G (2022) Digitalisation driven urban metabolism circularity: a review and analysis of circular city initiatives. Land Use Policy 112, article 105819. https://doi.org/10.1016/j.landusepol.2021.105819
189. Danovaro R, Fanelli E, Aguzzi J, Billett D, Carugati L, Corinaldesi C, Dell'Anno A, Gjerde K, Jamieson AJ, Kark S, McClain C (2020) Ecological variables for developing aglobal deep-ocean monitoring and conservation strategy. Nat Ecol Evolut 4(2):181–192. https://doi.org/10.1038/s41559-019-1091-z
190. Da Silva Portela CF (ed) (2023) Sustainable, innovative and intelligent societies and cities. Springer, Cham. https://link.springer.com/book/9783031305139
191. Davis FD (1989) Perceived usefulness, perceived ease of use, and user acceptance of information technology. MIS Quart 319–340. https://doi.org/10.2307/249008
192. Davies A, Manning S, Söderlund J (2018) When neighboring disciplines fail to learn from each other: the case of innovation and project management research. Res Policy 47:965–979. https://doi.org/10.1016/j.respol.2018.03.002

193. Dawson RJ, Ball T, Werritty J, Werritty A, Hall JW, Roche N (2011) Assessing the effectiveness of non-structural flood management measures in the Thames Estuary under conditions of socio-economic and environmental change. Glob Environ Chang 21(2):628–646. https://doi.org/10.1016/j.gloenvcha.2011.01.013
194. De Graaf R, Van der Brugge R (2010) Transforming water infrastructure by linking water management and urban renewal in Rotterdam. Technol Forecast Soc Chang 77(8):1282–1291. https://doi.org/10.1016/j.techfore.2010.03.011
195. De Haan FJ, Rotmans J (2018) A proposed theoretical framework for actors in transformative change. Technol Forecast Soc Chang. https://doi.org/10.1016/j.techfore.2017.12.017
196. De Noni I, Orsi L, Zanderighi L (2014) Attributes of Milan influencing city brand attractiveness. J Destin Mark Manag 3(4):218–226. https://doi.org/10.1016/j.jdmm.2014.06.001
197. de Oliveira JAP, Bellezoni RA, Shih WY, Bayulken B (2022) Innovations in urban green and blue infrastructure: tackling local and global challenges in cities. J Clean Prod 362, article 132355. https://doi.org/10.1016/j.jclepro.2022.132355
198. De Rose A, Buna M, Strazza C, Olivieri N, Stevens T, Peeters L, Tawil-Jamault D (2017) Technology readiness level: guidance principles for renewable energy technologies. Publ Off Eur Union Luxembourg. https://doi.org/10.2777/577767
199. Dearborn DC, Kark S (2010) Motivations for conserving urban biodiversity. Conserv Biol 24(2):432–440. https://doi.org/10.1111/j.1523-1739.2009.01328.x
200. Diachenko A, Menotti F (2017) Proto-cities or non-proto-cities? On the nature of Cucuteni-Trypillia mega-sites. J World Prehist 30:207–219. https://doi.org/10.1007/s10963-017-9105-8
201. Dias ÁL, Manuel EC, Dutschke G, Pereira R, Pereira L (2021) Economic crisis effects on SME dynamic capabilities. Int J Learn Chang 13(1):63–80. https://doi.org/10.1504/IJLC.2021.111662
202. Dijst M, Worrell E, Böcker L, Brunner P, Davoudi S, Geertman S, Harmsen R, Helbich M, Holtslag AA, Kwan MP, Lenz B (2018) Exploring urban metabolism—towards an interdisciplinary perspective. Resour Conserv Recycl 132:190–203. https://doi.org/10.1016/j.resconrec.2017.09.014
203. Dreyfus S (2002) Richard bellman on the birth of dynamic programming. Oper Res 50(1):48–51. https://doi.org/10.1287/opre.50.1.48.17791
204. Drews S, Van den Bergh JC (2016) What explains public support for climate policies? A review of empirical and experimental studies. Clim Policy 16:855–876. https://doi.org/10.1080/14693062.2015.1058240
205. Dreyer VM (2022) The epistemology and science of justified reason. Philosophia 50(2):503–532. https://doi.org/10.1007/s11406-021-00399-3
206. Dryzek JS, Bächtiger A, Chambers S, Cohen J, Druckman JN, Felicetti A, Fishkin JS, Farrell DM, Fung A, Gutmann A, Landemore H, Mansbridge J, Marien S, Neblo MA, Niemeyer S, Setälä M, Slothuus R, Suiter J, Thompson D, Warren ME (2019) The crisis of democracy and the science of deliberation. Science 363(6432):1144–1146. https://doi.org/10.1126/science.aaw2694
207. Duque JC, Lozano-Gracia N, Patino JE, Restrepo P (2022) Urban form and productivity: what shapes are Latin-American cities? Environ Plann B: Urb Anal City Sci 49(1):131–150. https://doi.org/10.1177/239980832199930
208. Duygan M, Fischer M, Pärli R, Ingold K (2022) Where do Smart Cities grow? The spatial and socio-economic configurations of smart city development. Sustain Cities Soc 77, article 103578. https://doi.org/10.1016/j.scs.2021.103578
209. Dworkin L, Kearns M (2015) From "in" to "over": behavioral experiments on whole-network computation. Proc AAAI Conf Hum Comput Crowdsourc 3:52–61. https://doi.org/10.1609/hcomp.v3i1.13223
210. Eakin H, Bojórquez-Tapia LA, Janssen MA, Georgescu M, Manuel-Navarrete D, Vivoni ER, Escalante AE, Baeza-Castro A, Mazari-Hiriart M, Lerner AM (2017) Urban resilience efforts must consider social and political forces. Proc Natl Acad Sci 114(2):186–189. https://doi.org/10.1073/pnas.162008111

211. Easingwood C, Koustelos A (2000) Marketing high technology: preparation, targeting, positioning, execution. Bus Horiz 43(3):27–27. https://doi.org/10.1016/S0007-6813(00)89198-3
212. Eisenhardt KM, Martin JA (2000) Dynamic capabilities: what are they? Strateg Manag J 21(10–11):1105–1121. https://doi.org/10.1002/1097-0266(200010/11)21:10/11%3c1105::AID-SMJ133%3e3.0.CO;2-E
213. Ekwo U (2011) Diaspora media and citizen engagement in the digital age: analyzing the impact of transnational journalism on democratization. In: Diaspora strategies: encouragement, evolution, and engagement. The Clinton Institute of American Studies, University College, Belfield, pp 9–11. https://www.mycmpi.org/wp-content/uploads/2012/07/Diaspora-Media-and-Citizen-Engagement-in-the-Digital-Age-Analyzing-the-Impact-of-Transnational-Journalism-on-Democratization.pdf
214. Eliade M (1957) Das Heilige und das profane. Rowohlts Deutsche Enzyklopädie Rowohlt Taschenbuchverlag, Hamburg
215. Ellen P, Bearden W, Subhash S (1991) Resistance to technological innovations: an examination of the role of self-efficacy and performance satisfaction. J Acad Mark Sci 19(4):297–308. https://doi.org/10.1007/BF02726504
216. Ellis G, Silk J (2014) Scientific method: Defend the integrity of physics. Nature 516(7531):321–323. https://doi.org/10.1038/516321a
217. Elmqvist T, Fragkias M, Goodness J, Güneralp B, Marcotullio PJ, McDonald RI, Parnell S, Schewenius M, Sendstad M, Seto KC, Wilkinson C (2013) Urbanization, biodiversity and ecosystem services: challenges and opportunities: a global assessment. Springer Nature, p 755. ISBN 9789400770874. https://doi.org/10.1007/978-94-007-7088-1
218. Elnabawi MH, Hamza N, Dudek S (2015) Numerical modelling evaluation for the microclimate of an outdoor urban form in Cairo, Egypt. HBRC J 11(2):246–251. https://doi.org/10.1016/j.hbrcj.2014.03.004
219. Ellström D, Holtström J, Berg E, Josefsson C (2021) Dynamic capabilities for digital transformation. J Strateg Manag 15(2):272–286. https://doi.org/10.1108/JSMA-04-2021-0089
220. Engle NL (2011) Adaptive capacity and its assessment. Glob Environ Chang 21(2):647–656. https://doi.org/10.1016/j.gloenvcha.2011.01.019
221. Ercan SA (2014) Deliberative democracy. In Phillips D (ed) Encyclopedia of educational theory and philosophy, pp 214–216. SAGE Publications, Thousand Oaks. https://doi.org/10.4135/9781483346229
222. Ernstson H, Van Der Leeuw SE, Redman CL, Meffert DJ, Davis G, Alfsen C, Elmqvist T (2010) Urban transitions: on urban resilience and human-dominated ecosystems. Ambio 39(8):531–545. https://doi.org/10.1007/s13280-010-0081-9
223. Eshach H (2007) Bridging in-school and out-of-school learning: formal, non-formal, and informal education. J Sci Educ Technol 16:171–190. https://doi.org/10.1007/s10956-006-9027-1
224. European Association of Research Technology Organisations (2014) The TRL scale as a research & innovation policy tool. http://www.earto.eu/fileadmin/content/03_Publications/The_TRL_Scale_as_a_R_I_Policy_Tool_-_EARTO_Recommendations_-_Final.pdf
225. Evangelopoulou S, De Vita A, Zazias G, Capros P (2019) Energy system modelling of carbon-neutral hydrogen as an enabler of sectoral integration within a decarbonization pathway. Energies 12(13), article 2551. https://doi.org/10.3390/en12132551
226. Evans BP, Prokopenko M (2021) A maximum entropy model of bounded rational decision-making with prior beliefs and market feedback. Entropy 23, article 669. https://doi.org/10.3390/e23060669
227. Evans JA (2008) Electronic publication and the narrowing of science and scholarship. Science 321(5887):395–399. https://doi.org/10.1126/science.1150473
228. Evans JA, Foster JG (2011) Metaknowledge. Science 331(6018):721–725. https://doi.org/10.1126/science.1201765

229. Ezbakhe F, Pérez-Foguet A (2021) Decision analysis for sustainable development: the case of renewable energy planning under uncertainty. Eur J Oper Res 291(2):601–613. https://doi.org/10.1016/j.ejor.2020.02.037
230. Farkas JZ, Hoyk E, de Morais MB, Csomós G (2023) A systematic review of urban green space research over the last 30 years: a bibliometric analysis. Heliyon 9, article e13406. https://doi.org/10.1016/j.heliyon.2023.e13406
231. Fazey I, Schäpke N, Caniglia G, Hodgson A, Kendrick I, Lyon C, Page G, Patterson J, Riedy C, Strasser T, Verveen S et al (2020) Transforming knowledge systems for life on Earth: visions of future systems and how to get there. Energy Res Soc Sci 70, article 101724. https://doi.org/10.1016/j.erss.2020.101724
232. Feagan M, Matsler M, Meerow S, Muñoz-Erickson TA, Hobbins R, Gim C, Miller CA (2019) Redesigning knowledge systems for urban resilience. Environ Sci Policy 101:358–363. https://doi.org/10.1016/j.envsci.2019.07.014
233. Feng X, Xiu C, Li J, Zhong Y (2021) Measuring the evolution of urban resilience based on the exposure–connectedness–potential (ECP) approach: a case study of Shenyang City, China. Land 10(12), article 1305. https://doi.org/10.3390/land10121305
234. Fernandez-Bou AS, Ortiz-Partida JP, Classen-Rodriguez LM, Pells C, Dobbin KB, Espinoza V, Rodríguez-Flores JM, Thao C, Hammond Wagner CR, Fencl A, Flores-Landeros H (2021) 3 challenges, 3 errors, and 3 solutions to integrate frontline communities in climate change policy and research: lessons from California. Front Clim. 3, article 717554. https://www.frontiersin.org/articles/https://doi.org/10.3389/fclim.2021.717554/full
235. Ferraro PJ, Sanchirico JN, Smith MD (2019) Causal inference in coupled human and natural systems. Proc Natl Acad Sci 116(12):5311–5318. https://doi.org/10.1073/pnas.1805563115
236. Ferretti V (2016) From stakeholders analysis to cognitive mapping and multi-attribute value theory: an integrated approach for policy support. Eur J Oper Res 253(2):524–541. https://doi.org/10.1016/j.ejor.2016.02.054
237. Ferretti V, Pluchinotta I, Tsoukiàs A (2019) Studying the generation of alternatives in public policy making process. Eur J Oper Res 273:353–363. https://doi.org/10.1016/j.ejor.2018.07.054
238. Finger M, Portmann E (2016) What are cognitive cities?. In Portmann E, Finger M (eds) Towards cognitive cities. Studies in systems, decision and control. Springer, Cham, vol 63. https://doi.org/10.1007/978-3-319-33798-2_1
239. Fisher J, Arora P, Chen S, Rhee S, Blaine T, Simangan D (2021) Four propositions on integrated sustainability: toward a theoretical framework to understand the environment, peace, and sustainability nexus. Sustain Sci 16:1125–1145. https://doi.org/10.1007/s11625-021-00925-y
240. Fistola R, Gargiulo C, La Rocca RA (2020) Rethinking vulnerability in city-systems: a methodological proposal to assess “urban entropy.” Environ Impact Assess Rev 85:106464. https://doi.org/10.1016/j.eiar.2020.106464
241. Flyvbjerg B (2014) What you should know about megaprojects and why: an overview. Proj Manag J 45(2):6–19. https://doi.org/10.1002/pmj.21409
242. Flyvbjerg B (2021) Top ten behavioral biases in project management: an overview. Proj Manag J 52:531–546. https://doi.org/10.1177/87569728211049046
243. Fornell C (1992) A national customer satisfaction barometer: the Swedish experience. J Mark 56(1):6–21. https://doi.org/10.1177/002224299205600
244. Fortunato S, Bergstrom CT, Börner K, Evans JA, Helbing D, Milojević S, Petersen AM, Radicchi F, Sinatra R, Uzzi B, Vespignani A (2018) Science of science. Science 359(6379), article eaao0185. https://doi.org/10.1126/science.aao0185
245. Fossey Y (2022a) Transports: ça roule pour le Tram 13. Department of Yvelynes, France, Yvelines-Infos, 42, p 16. Online on 07 July 2022. https://www.yvelines-infos.fr/transports-ca-roule-pour-le-tram-13/
246. Fossey Y (2022b) Redonner la splendeur du XIXe siècle à la place d’Armes. Department of Yvelynes, France, Yvelines-Infos, 42, p 15. Online on 11 July 2022. https://www.yvelines-infos.fr/le-departement-veut-redonner-la-splendeur-du-19e-a-la-place-darmes/

247. Fox N, Campbell-Arvai V, Lindquist M, Van Berkel D, Serrano-Vergel R (2022) Gamifying decision support systems to promote inclusive and engaged urban resilience planning. Urb Plann 7(2). https://doi.org/10.17645/up.v7i2.4987
248. Fraga AL, Vegetti M, Leone HP (2020) Ontology-based solutions for interoperability among product lifecycle management systems: a systematic literature review. J Ind Inf Integr 20:100176. https://doi.org/10.1016/j.jii.2020.100176
249. François C (1999) Systemics and cybernetics in a historical perspective. Syst Res Behav Sci Off J Int Feder Syst Res 16(3):203–219. https://doi.org/10.1002/(SICI)1099-1743(199905/06)16:3%3C203::AID-SRES210%3E3.0.CO;2-1
250. Freeth R, Caniglia G (2020) Learning to collaborate while collaborating: advancing interdisciplinary sustainability research. Sustain Sci 15(1):247–261. https://doi.org/10.1007/s11625-019-00701-z
251. Frescura FA, Hiley BJ (1980) The implicate order, algebras, and the spinor. Found Phys 10:7–31. https://doi.org/10.1007/BF00709014
252. Frese I, Kolb G, Tiemann D, Krämer PM, Ciumasu IM, Weber CM (2003) Development of an optical detection cell for an automated miniaturized immunochemical device for on-site screening of pesticides residues in water. In: The 10th international trade fair & conference 'sensor 2003', 13–15 May, Nuremberg, Germany
253. Frick V (2023) Understanding the democratic promise of the city. Philos Soc Critic, published online on 7 January. https://doi.org/10.1177/01914537221150456
254. Frostell B (2016) Eco-cycle models as support in urban planning for a circular economy leading to eco-cities. In: Linnaeus Eco-Tech (Kalmar, Sweden, November 21–23), pp 25–25. https://open.lnu.se/index.php/eco-tech/article/view/631
255. Fu F, Nowak MA, Hauert C (2010) Invasion and expansion of cooperators in lattice populations: prisoner's dilemma vs. snowdrift games. J Theor Biol 266(3):358–366. https://doi.org/10.1016/j.jtbi.2010.06.042
256. Fu G, Jin Y, Sun S, Yuan, Butler D (2022) The role of deep learning in urban water management: a critical review. Water Res. 223, article 118973. https://doi.org/10.1016/j.watres.2022.118973
257. Fung A (2006) Varieties of participation in complex governance. Public Adm Rev 66:66–75. https://doi.org/10.1111/j.1540-6210.2006.00667.x
258. Funtowicz S, Ravetz J (1993) Science for the post-normal age. Futures 31(7):735–755. https://doi.org/10.1016/0016-3287(93)90022-L
259. Funtowicz S, Ravetz J (2018) Post-normal science. In: Companion to environmental studies. Routledge, pp 443–447. https://www.taylorfrancis.com/chapters/edit/https://doi.org/10.4324/9781315640051-89/
260. Gaber J (2019) Building "a ladder of citizen participation." J Am Plann Assoc 85(3):188–201. https://doi.org/10.1080/01944363.2019.1612267
261. Galderisi A, Treccozzi E (2017) Green strategies for flood resilient cities: the Benevento case study. Procedia Environ Sci 37:655–666. https://doi.org/10.1016/j.proenv.2017.03.052
262. Galende-Sánchez E, Sorman AH (2021). From consultation toward co-production in science and policy: a critical systematic review of participatory climate and energy initiatives. Energy Res Soc Sci 73, article 101907. https://doi.org/10.1016/j.erss.2020.101907
263. Galli A, Iha K, Pires SM, Mancini MS, Alves A, Zokai G, Lin D, Murthy A, Wackernagel M (2020). Assessing the ecological footprint and biocapacity of Portuguese cities: critical results for environmental awareness and local management. Cities 96, article 102442. https://doi.org/10.1016/j.cities.2019.102442
264. Gbadegeshin SA, Al Natsheh A, Ghafel K, Mohammed O, Koskela A, Rimpiläinen A, Tikkanen J, Kuoppala A (2022) Overcoming the valley of death: a new model for high technology startups. Sustain Fut 4:100077. https://doi.org/10.1016/j.sftr.2022.100077
265. Geacu S, Grigorescu I (2022) Historical changes in urban and peri-urban forests: evidence from the Galaţi Area, Romania. Land 11(11), article 2043. https://doi.org/10.3390/land11112043

266. Gelfond AL, Erofeeva IV, Moiseenko VA (2020) An analytical reading of the planning structure in Munich. IOP Conf Ser Mater Sci Eng 913(3), article 032022). https://doi.org/10.1088/1757-899X/913/3/032022
267. Georgescu-Roegen N (1971) The entropy law and the economic process. Harvard University Press. https://doi.org/10.4159/harvard.9780674281653
268. Ghosh S, Halappanavar M, Tumeo A, Kalyanaraman A, Lu H, Chavarrià-Miranda D, Khan A, Gebremedhin AH (2018) Distributed Louvain Algorithm for graph community detection. In: 2018 IEEE International Parallel and Distributed Processing Symposium, pp 885–895. https://doi.org/10.1109/IPDPS.2018.00098
269. Gibbons M (1999) Science's new social contract with society. Nature 402(Suppl 6761):C81–C84. https://doi.org/10.1016/j.cosust.2020.01.006
270. Gibson R, Bon E, Darius P, Smyth P (2023) Are online political influencers accelerating democratic deconsolidation? Media Commun 11(3):175–186. https://doi.org/10.17645/mac.v11i3.6813
271. Giddings B, Hopwood B, O'Brien G (2002) Environment, economy and society: fitting them together into sustainable development. Sustain Dev 10(4):187–196. https://doi.org/10.1002/sd.199
272. Gielen D, Lyons M (2022) Critical materials for the energy transition: rare earth elements. Int Renew Energy Agency (IRENA, Abu Dhabi). ISBN 978-92-9260-437-0. https://www.irena.org/Technical-Papers/Rare-Earth-Elements
273. Giler-Velásquez G, Marcillo-Delgado B, Vaca-Cardenas M, Vaca-Cardenas L (2021) Software frameworks that improve HCI focused on cognitive cities. A systematic literature review. In Guarda T, Portela F, Santos MF (eds) Advanced research in technologies, information, innovation and sustainability. ARTIIS 2021. Communications in computer and information science, vol 1485. Springer, Cham. https://doi.org/10.1007/978-3-030-90241-4_12
274. Gingras Y (2013) Sociologie des sciences. Presses Universitaires de France, Paris (France), Series Que sais-je?, No. 3950. ISBN 978-2-7154-0485-4. https://www.puf.com/content/Sociologie_des_sciences
275. Giuseppi A, Pietrabissa A (2022) Bellman's principle of optimality and deep reinforcement learning for time-varying tasks. Int J Control 95(9):2448–2459. https://doi.org/10.1080/00207179.2021.1913516
276. Glaeser EL, Kominers SD, Luca M, Naik N (2018) Big data and big cities: the promises and limitations of improved measures of urban life. Econ Inq 56(1):114–137. https://doi.org/10.1111/ecin.12364
277. Godefroid S, Koedam N (2003) Distribution pattern of the flora in a peri-urban forest: an effect of the city–forest ecotone. Landsc Urban Plan 65(4):169–185. https://doi.org/10.1016/S0169-2046(03)00013-6
278. Godschalk DR (2003) Urban hazard mitigation: creating resilient cities. Nat Hazard Rev 4(3):136–143. https://doi.org/10.1061/(ASCE)1527-6988(2003)4:3(136)
279. Goldman D, Assaraf OBZ, Shaharabani D (2013) Influence of a non-formal environmental education programme on junior high-school students' environmental literacy. Int J Sci Educ 35(3):515–545. https://doi.org/10.1080/09500693.2012.749545
280. González A, Donnelly A, Jones M, Chrysoulakis N, Lopes M (2013) A decision-support system for sustainable urban metabolism in Europe. Environ Impact Assess Rev 38:109–119. https://doi.org/10.1016/j.eiar.2012.06.007
281. Gopstein A, Nguyen C, O'Fallon C, Hastings N, Wollman D (2021) NIST framework and roadmap for smart grid interoperability standards, release 4.0. United States of America, Department of Commerce, National Institute of Standards and Technology. https://doi.org/10.6028/NIST.SP.1108r4
282. Gowdy J, Erickson JD (2005) The approach of ecological economics. Camb J Econ 29:207–222. https://doi.org/10.1093/cje/bei033
283. Graziul C, Belikov A, Chattopadyay I, Chen Z, Fang H, Girdhar A, Jia X, Krafft PM, Kleiman-Weiner M, Lewis C, Liang C (2023) Does Big Data serve policy? Not without context. An experiment with in silico social science. Comput Math Organ Theor 29(1):188–219. https://doi.org/10.1007/s10588-022-09362-3

284. Green JF, Sterner T, Wagner G (2014) A balance of bottom-up and top-down in linking climate policies. Nat Clim Chang 4(12):1064–1067. https://doi.org/10.1038/nclimate2429
285. Green JF (2017) Don't link carbon markets. Nature 543:484–486. https://doi.org/10.1038/543484a
286. Green JF (2021) Does carbon pricing reduce emissions? A review of ex-post analyses. Environ Res Lett 16, article 043004. https://doi.org/10.1088/1748-9326/abdae9
287. Grimm NB, Faeth SH, Golubiewski NE, Redman CL, Wu J, Bai X, Briggs JM (2008) Global change and the ecology of cities. Science 319(5864):756–760. https://doi.org/10.1126/science.1150195
288. Gurjar BR, Ciumasu IM, Costica N, Kumar A, Ojha CSP (2006) Overexploitation of ecosystem resources vs. the costs of storms and flooding risk management. In Höppe P, Pielke Jr RA (2006, May) Workshop on climate change and disaster losses. In: Understanding and Attributing Trends and Projections, Hohenkammer, Germany, Final Workshop Report, pp 122–134. https://citeseerx.ist.psu.edu/document?repid=rep1&type=pdf&doi=34b1043e23e1bc313c3efe452733b2e9ac6bf382
289. Guo F, Zhu P, Wang S, Duan D, Jin Y (2017) Improving natural ventilation performance in a high-density urban district: A building morphology method. Procedia Eng 205:952–958. https://doi.org/10.1016/j.proeng.2017.10.149
290. Gusenbauer M (2021) The age of abundant scholarly information and its synthesis—a time when 'just google it' is no longer enough. Res Synth Methods 12(6):684–691. https://doi.org/10.1002/jrsm.1520
291. Gustavsson K (2021) Charlie Dunbar Broad. In: Zalta EN (ed) The Stanford Encyclopedia of Philosophy (Fall 2021 edn). https://plato.stanford.edu/archives/fall2021/entries/broad/
292. Habermas J (1981) Theorie des kommunikativen Handelns. Suhrkamp Verlag. ISBN 978-3-518-28775-0. https://www.suhrkamp.de/buch/juergen-habermas-theorie-des-kommunikativen-handelns-t-9783518287750
293. Habermans J, Bohman J, Rehg W (eds) (1997) Deliberative democracy: essays on reason and politics. MIT Press. ISBN 0-262-02434-9
294. Häggström S, Tivander J, Carlson R (2006) Strategic data acquisition addressed to support implementation of design for environment. Chalmers University of Technology, Gothenburg, Sweden. https://www.lifecyclecenter.se/wp-content/uploads/2006_13.pdf
295. Hall BH, Khan B (2003) Adoption of new technology. National Bureau of Economic Research, NBER Working Paper N. 9730. NBER, Cambridge. https://www.nber.org/papers/w9730; https://doi.org/10.3386/w9730
296. Hämäläinen M (2020) Smart city development with digital twin technology. In: 33rd Bled eConference-Enabling Technology for a Sustainable Society, 28–29 June 2020 (online conference proceedings). University of Maribor. https://urn.fi/URN:ISBN:978-961-286-362-3
297. Hämäläinen M (2021) Urban development with dynamic digital twins in Helsinki city. IET Smart Cities 3(4):201–210. https://doi.org/10.1049/smc2.12015
298. Hannibal B, Meerow S, Woodruff SC, Roy M, Gilbertson PG, Matos M (2022) Who collaborates on urban resilience? An analysis of flood resilience planning networks in four coastal cities. J Urb Affairs (online first). https://doi.org/10.1080/07352166.2022.2137033
299. Hänninen S, Sandberg B (2006) Consumer learning roadmap: a necessary tool for new products. Int J Knowl Learn 2(3–4):298–307. https://doi.org/10.1504/IJKL.2006.010997
300. Harding AR, Belaia M, Keith DW (2022) The value of information about solar geoengineering and the two-sided cost of bias. Clim Policy 23(3):355–365. https://doi.org/10.1080/14693062.2022.2091509
301. Hawas MA (2017) Are we intentionally limiting urban planning and intelligence? A causal evaluative review and methodical redirection for intelligence systems. IEEE Access 5:13253–13259. https://doi.org/10.1109/ACCESS.2017.2725138
302. Haybron DM (2011) Central Park: Nature, context, and human wellbeing. Int J Wellbeing 1(2):235–254. https://doi.org/10.5502/ijw.v1i2.6

303. Hayes AT, Jandaghian Z, Lacasse MA, Gaur A, Lu H, Laouadi A, Ge H, Wang L (2022) Nature-based solutions (nbss) to mitigate urban heat island (UHI) effects in Canadian cities. Buildings 12(7), article 925. https://doi.org/10.3390/buildings12070925
304. Héder M (2017) From NASA to EU: the evolution of the TRL scale in public sector innovation. Innov J: Publ Sect Innov J 22(2), article 3. https://eprints.sztaki.hu/9204/1/Heder_1_3269345_ny.pdf
305. Hedman KM, Emerson TE, Fort MA, Lambert JM, Betzenhauser AM, Pauketat TR (2022) Corn, climate, and the human population of greater Cahokia. In: Following the Mississippian spread: climate change and migration in the Eastern US (ca. AD 1000–1600). Springer International Publishing, Cham, pp 37–63. https://doi.org/10.1007/978-3-030-89082-7_2
306. Heisler GM, Brazel AJ (2010) The urban physical environment: temperature and urban heat islands. Urb Ecosyst Ecol 55:29–56. https://doi.org/10.2134/agronmonogr55.c2
307. Helbing D (2012) Systemic risks in society and economics. In: Social self-organization. Springer, Berlin, pp 261–284. https://doi.org/10.1007/978-3-642-24004-1_14. Also available at: https://webarchiv.ethz.ch/soms/teaching/2010_crowds/material/systemic_risk_society_econ.pdf
308. Helbing D (2013) Globally networked risks and how to respond. Nature 497(7447):51–59. https://doi.org/10.1038/nature12047
309. Helgeson C (2018). Structuring decisions under deep uncertainty. Topoi, 1–13. https://doi.org/10.1007/s11245-018-9584-y
310. Henry M, Schraven D, Bocken N, Frenken K, Hekkert M, Kirchherr J (2021) The battle of the buzzwords: a comparative review of the circular economy and the sharing economy concepts. Environ Innov Soc Trans 38:1–21. https://doi.org/10.1016/j.eist.2020.10.008
311. Herberg J, Schmitz S, Stasiak D, Schmieg G (2021) Boundary speak in sustainability studies: computational reading of a transversal field. Sci Publ Policy 48(3):398–411. https://doi.org/10.1093/scipol/scab006
312. Hermans Y, Le Cun B, Bui A (2012) Individual decisions & schedule planner in a vehi-cle-to-grid context. In: 2012 IEEE International Electric Vehicle Conference, Greenville, US, pp 1–6. https://doi.org/10.1109/IEVC.2012.6183275
313. Herrera H (2017) Resilience for whom? The problem structuring process of the resilience analysis. Sustainability 9, article 1196. https://doi.org/10.3390/su9071196
314. Hinojosa AS, Gardner WL, Walker HJ, Cogliser C, Gullifor D (2017) A review of cognitive dissonance theory in management research: opportunities for further development. J Manag 43(1):170–199. https://doi.org/10.1177/0149206316668236
315. Hjørland B (2016) Knowledge organization (KO). KO Knowl Organ 43(6):475–484. https://www.isko.org/cyclo/knowledge_organization.htm
316. Hjørland B (2018) Data (with big data and database semantics). Knowl Organ 45(8):685–708. https://doi.org/10.5771/0943-7444-2018-8-685
317. Hjørland B (2021) Information retrieval and knowledge organization: a perspective from the philosophy of science. Information 12(3), article 135. https://doi.org/10.3390/info12030135
318. Hodgkinson GP, Burkhard B, Foss NJ, Grichnik D, Sarala RM, Tang Y, Van Essen M (2023) The heuristics and biases of top managers: past, present, and future. J Manage Stud (first online: 29 May 2023). https://doi.org/10.1111/joms.12937
319. Hofmann P, Samp C, Urbach N (2020) Robotic process automation. Electron Mark 30(1):99–106. https://doi.org/10.1007/s12525-019-00365-8
320. Hofstadter DR, Mitchell M (1995) The copycat project: a model of mental fluidity and analogy-making. Adv Connect Neural Comput Theor 2:205–267. https://melaniemitchell.me/PapersContent/HofstadterMitchellCopycat1995.pdf
321. Howell EL, Brossard D (2021) (Mis) informed about what? What it means to be a science-literate citizen in a digital world. Proc Natl Acad Sci 118(15), article e1912436117. https://doi.org/10.1073/pnas.1912436117
322. Hjørland B (2016) Knowledge organization. Knowl Organ 43(6):475–484. https://doi.org/10.5771/0943-7444-2016-6-475

323. Hjørland B (2021) Information retrieval and knowledge organization: a perspective from the philosophy of science. Information 12, article 135. https://doi.org/10.3390/info12030135
324. Hritonenko V, Yatsenko Y (2022) Sustainable adaptation and mitigation in regions and cities: review of decision-support methods. Resour Conserv Recycl Adv 13, article 200066. https://doi.org/10.1016/j.rcradv.2022.200066
325. Huang W, Cui S, Yarime M, Hashimoto S, Managi S (2015) Improving urban metabolism study for sustainable urban transformation. Environ Technol Innov 4:62–72. https://doi.org/10.1016/j.eti.2015.04.004
326. Huang X, Wang H, Shan L, Xiao F (2021) Constructing and optimizing urban ecological network in the context of rapid urbanization for improving landscape connectivity. Ecol Indicat 132, article 108319. https://doi.org/10.1016/j.ecolind.2021.108319
327. Ichikawa JJ, Steup M (2017) The analysis of knowledge. Stanford Encyclopedia of Philosophy. https://plato.stanford.edu/entries/knowledge-analysis/. Retrieved 27 May 2022
328. IEA (2020) Energy technology perspectives: special report on clean energy innovation. IEA (the International Energy Agency), Paris, France. https://iea.blob.core.windows.net/assets/04dc5d08-4e45-447d-a0c1-d76b5ac43987/Energy_Technology_Perspectives_2020_-_Special_Report_on_Clean_Energy_Innovation.pdf
329. IPBES (2019) Global assessment report on biodiversity and ecosystem services of the intergovernmental science-policy platform on biodiversity and ecosystem services. Brondizio ES, Settele J, Díaz S, Ngo HT (eds) IPBES Secretariat, Bonn, Germany. ISBN 9783947851133. https://doi.org/10.5281/zenodo.3831673. https://www.ipbes.net/global-assessment. Accessed on 18 Oct 2023
330. IPCC-WGI (2022) The physical science basis. Working Group I contribution to the sixth assessment report. In: The International Panel for Climate Change. https://www.ipcc.ch/report/sixth-assessment-report-working-group-i/
331. IPCC-WGII (2022) Impacts, adaptation and vulnerability. Working Group II contribution to the sixth assessment report. In: The International Panel for Climate Change. https://www.ipcc.ch/report/sixth-assessment-report-working-group-ii/
332. IPCC-WGIII (2022) Mitigation of climate change. In: Working Group III contribution to the sixth assessment report. The International Panel for Climate Change. https://www.ipcc.ch/report/sixth-assessment-report-working-group-3/
333. Ipsen KL, Zimmermann RK, Nielsen PS, Birkved M (2019) Environmental assessment of Smart City Solutions using a coupled urban metabolism—life cycle impact assessment approach. Int J Life Cycle Assess 24:1239–1253. https://doi.org/10.1007/s11367-018-1453-9
334. Irwin EG, Bockstael NE (2007) The evolution of urban sprawl: evidence of spatial heterogeneity and increasing land fragmentation. Proc Natl Acad Sci 104(52):20672–20677. https://doi.org/10.1073/pnas.0705527105
335. Israilidis J, Odusanya K, Mazhar MU (2021) Exploring knowledge management perspectives in smart city research: a review and future research agenda. Int J Inform Manage 56, article 101989. https://doi.org/10.1016/j.ijinfomgt.2019.07.015
336. Jacquemod C, Nguyen KT, Sevin K, Dekneuvel E, Aguir K, Nicolle B, Lorenzini P, Jacquemod G (2016) Low cost wireless current sensor for NIALM application. Sens Actuat A 252:209–224. https://doi.org/10.1016/j.sna.2016.10.020
337. Jakimowicz A (2020) The role of entropy in the development of economics. Entropy 22(4), article 452. https://doi.org/10.3390/e22040452
338. Janssen M, Van Der Voort H, van Veenstra AF (2015) Failure of large transformation projects from the viewpoint of complex adaptive systems: management principles for dealing with project dynamics. Inf Syst Front 17(1):15–29. https://doi.org/10.1007/s10796-014-9511-8
339. Javied T, Bakakeu J, Gessinger D, Franke J (2018) Strategic energy management in Industry 4.0 environment. In: 2018 annual IEEE international systems conference (SysCon), 1–4. IEEE (2018). https://doi.org/10.1109/SYSCON.2018.8369610
340. Jayakrishna K, Vinodh S, Anish S (2016) A graph theory approach to measure the performance of sustainability enablers in a manufacturing organization. Int J Sustain Eng 9(1):47–58. https://doi.org/10.1080/19397038.2015.1050970

341. Jaynes ET (1957) Information theory and statistical mechanics I. Phys Rev 106:620–630. https://doi.org/10.1103/PhysRev.106.620
342. Jaynes ET (1957) Information theory and statistical mechanics II. Phys Rev 108:171–190. https://doi.org/10.1103/PhysRev.108.171
343. Jennings V, Bamkole O (2019) The relationship between social cohesion and urban green space: An avenue for health promotion. Int J Environ Res Publ Health 16(3), article 452. https://doi.org/10.3390/ijerph16030452
344. Jing R, Hastings A, Guo M (2020) Sustainable design of urban rooftop food-energy-land nexus. Iscience 23(11). https://doi.org/10.1016/j.isci.2020.101743
345. Johns RA, Pontes R (2019) Parks, rhetoric and environmental education: challenges and opportunities for enhancing ecoliteracy. J Outdoor Environ Educ 22(1):1–19. https://doi.org/10.1007/s42322-019-0029-x
346. Jørgensen K (2016) What is landscape architecture about? Sci J Latvia Univ Agric Lands Architect Art 7(7):46–49. https://www.researchgate.net/profile/Karsten-Jorgensen/publication/295547814_What_is_landscape_architecture_about/links/56cb28c808ae5488f0dae4ac/What-is-landscape-architecture-about.pdf
347. Kagermann H, Lukas WD, Wahlster W (2011) Industrie 4.0. Mit dem Internet der Dinge auf dem Weg zur 4. industriellen Revolution. VDI Nachrichten (2011; https://www.vdi-nachrichten.com). https://web.archive.org/web/20130304101009/; http://www.vdi-nachrichten.com/artikel/Industrie-4-0-Mit-dem-Internet-der-Dinge-auf-dem-Weg-zur-4-industriellen-Revolution/52570/1. Archived; Accessed 27 May 2022
348. Kahneman D (2011) Thinking, fast and slow. Farrar, Straus & Giroux, New York. ISBN 978-0374275631
349. Kalogeri C, Spyrou C, Koukoula M, Saviolakis PM, Pappa A, Loupis M, Masouras C, Katsafados P (2023) Modeling the impact of the green roofs as a nature-based solution to mitigate the urban heat island effects over Attica, Greece. Environ Sci Proc 26(1), article 174. https://doi.org/10.3390/environsciproc2023026174
350. Kamiński B, Poulin V, Prałat P, Szufel P, Théberge F (2019) Clustering via hypergraph modularity. PLoS ONE 14(11):e0224307. https://doi.org/10.1371/journal.pone.0224307
351. Kane IO, Vanderlinden JP, Baztan J, Touili N, Claus S (2014) Communicating risk through a DSS: A coastal risk centred empirical analysis. Coast Eng 87:240–248. https://doi.org/10.1016/j.coastaleng.2014.01.007
352. Kang J-N, Wei Y-M, Liu L-C, Han R, Yu B-Y, Wang J-W (2020) Energy systems for climate change mitigation: A systematic review. Appl Energy 263, article 114602. https://doi.org/10.1016/j.apenergy.2020.114602
353. Kao TS, Kao HF, Tsai YJ (2017) The context, status and challenges of environmental education in formal education in Taiwan. Jpn J Environ Educ 26(4):15–20. https://doi.org/10.5647/jsoee.26.4_15
354. Kattel GR, Elkadi H, Meikle H (2013) Developing a complementary framework for urban ecology. Urb Forestr Urb Green 12(4):498–508. https://doi.org/10.1016/j.ufug.2013.07.005
355. Kay CA, Rohnke AT, Sander HA, Stankowich T, Fidino M, Murray MH, Lewis JS, Taves I, Lehrer EW, Zellmer AJ, Schell CJ (2022) Barriers to building wildlife-inclusive cities: Insights from the deliberations of urban ecologists, urban planners and landscape designers. People Nat 4(1):62–70. https://doi.org/10.1002/pan3.10283
356. Kazemi F, Mohorko R (2017) Review on the roles and effects of growing media on plant performance in green roofs in world climates. Urb Forest Urb Green 23:13–26. https://doi.org/10.1016/j.ufug.2017.02.006
357. Kearns M (2012) Experiments in social computation. Commun ACM 55(10):56–67. https://doi.org/10.1145/2347736.2347753
358. Kehler S, Birchall SJ (2021) Social vulnerability and climate change adaptation: the critical importance of moving beyond technocratic policy approaches. Environ Sci Policy 124:471–477. https://doi.org/10.1016/j.envsci.2021.07.025
359. Kennedy C, Pincetl S, Bunje P (2011) The study of urban metabolism and its applications to urban planning and design. Environ Pollut 159(8–9):1965–1973. https://doi.org/10.1016/j.envpol.2010.10.022

360. Kemp R, Loorbach D, Rotmans J (2007) Transition management as a model for managing processes of co-evolution towards sustainable development. Int J Sustain Dev World Ecol 14(1):78–91. https://doi.org/10.1080/13504500709469709
361. Keohane RO, Victor DG (2016) Cooperation and discord in global climate policy. Nat Clim Chang 6(6):570–575. https://doi.org/10.1038/nclimate2937
362. Khan A, Samiullah Ayub M (2022) Impact of soil sealing on the genesis of pluvial flood in Peshawar, Pakistan. Arab J Geosci 15(7), article 575. https://doi.org/10.1007/s12517-022-09816-7
363. Kim J, Ko Y, Kim W, Kim G, Lee J, Eyman OTG, Chowdhury S, Adiwal J, Son Y, Lee WK (2023) Understanding the Impact of the COVID-19 Pandemic on the perception and use of urban green spaces in Korea. Int J Environ Res Publ Health 20(4), article 3018. https://doi.org/10.3390/ijerph20043018
364. Kingsley P, Anderson T (1998) Facing life without the internet. Internet Res 8(4):303–312. https://doi.org/10.1108/10662249810231041
365. Kitzberger T (2012) Ecotones as complex arenas of disturbance, climate, and human impacts: the trans-Andean forest-steppe ecotone of northern Patagonia. In: Ecotones between forest and grassland. Springer, New York, pp 59–88. https://doi.org/10.1007/978-1-4614-3797-0_3
366. Klaus VH, Kiehl K (2021) A conceptual framework for urban ecological restoration and rehabilitation. Basic Appl Ecol 52:82–94. https://doi.org/10.1016/j.baae.2021.02.010
367. Klein JT (2014) Discourses of transdisciplinarity: looking back to the future. Futures 63:68–74. https://doi.org/10.1016/j.futures.2014.08.008
368. Kling R (1999) Can the "next-generation internet" effectively support "ordinary citizens"? Inf Soc 15(1):57–63. https://doi.org/10.1080/019722499128673
369. Klitsie JB, Price RA, De Lille CSH (2019) Overcoming the valley of death: a design innovation perspective. Des Manage J 14(1):28–41. https://doi.org/10.1111/dmj.12052
370. Koberg E, Longoni A (2019) A systematic review of sustainable supply chain management in global supply chains. J Clean Prod 207:1084–1098. https://www.sciencedirect.com/science/article/pii/S0959652618330476
371. Kolb G, Frese I, Hessel V, Jiang F, Löwe H, Tiemann D, Ciumasu IM, Weber C, Krämer P (2003). Entwiklung eines mobilen immunochemischen Fließinjektionssystems für die vor-Ort-Analytik von umweltrelevanten Chemikalien. In: The 6th Dresdner-Sensor-Symposium, 08–10 December, Dresden, Germany
372. Kolb G, Frese I, Hessel V, Ciumasu IM, Krämer PM, Löwe H, Tiemann D (2004) An automated, portable immunochemical flow injection system for on-site analysis of environmentally hazardous chemicals. In: Proceedings of the oral presentations at Lab Automation 2004, 1–5 February, San Jose, CA, USA
373. Koller M, Eckert K, Ferber U, Gräbe G, Verbücheln M, Wendler K (2022) Resource management as part of sustainable urban district development. Sustainability 14(7), article 4224. https://doi.org/10.3390/su14074224
374. Kong F, Yin H, Nakagoshi N, Zong Y (2010) Urban green space network development for biodiversity conservation: identification based on graph theory and gravity modeling. Landsc Urban Plan 95(1–2):16–27. https://doi.org/10.1016/j.landurbplan.2009.11.001
375. Kong L, Mu X, Hu G, Zhang Z (2022). The application of resilience theory in urban development: a literature review. Environ Sci & Pollut Res 29, 49651–49671. https://doi.org/10.1007/s11356-022-20891-x
376. Koos-Morar TA, Ntanou K, Ciumasu IM (2015) Eco-cities of the future through landscape management methods: conciliating cultural heritage and green infrastructure in Versailles, France. In the 4th LE: NOTRE Landscape Forum, 21–25 Apr 2015, Bucharest, Romania. http://forum.ln-institute.org/
377. Kotis KI, Zachila K, Paparidis E (2021) Machine learning meets the semantic web. Artif Intell Adv 3(1):63–70. https://doi.org/10.3233/SW-212892
378. Kourtit K, Nijkamp P, Wahlström MH (2021) How to make cities the home of people–a 'soul and body'analysis of urban attractiveness. Land Use Policy 111, article 104734. https://doi.org/10.1016/j.landusepol.2020.104734

379. Krämer PM (1996) Biosensors for the measurement of pesticide residues in the environment: past, present, and future. J AOAC Int 79(6):1245–1254. https://doi.org/10.1093/jaoac/79.6.1245
380. Krämer PM (1998) A strategy to validate immunoassay test kits for TNT and PAHs as a field screening method for contaminated sites in Germany. Anal Chim Acta 376:3–11. https://doi.org/10.1016/S0003-2670(98)00446-2
381. Krämer PM, Baumann BA, Stoks PG (1997) Prototype of a newly developed immunochemical detection system for the determination of pesticides residues in water. Anal Chim Acta 347(1–2):187–198. https://doi.org/10.1016/S0003-2670(96)00635-6
382. Krämer PM, Ciumasu IM, Weber C-M, Kolb G, Frese I, Werner B, Kettrup AA (2002) Development of an automated miniaturized immunochemical device for on-site screening of pesticide residues in water In: Proceedings of the Tenth IUPAC International Congress on the Chemistry of Crop Protection, 4–9 August, Basel, Switzerland. sub-topic 6a—trends in analytical methods and instrumentation, p 201
383. Krämer PM, Ciumasu IM, Weber CM, Kolb G, Tiemann D, Frese I, Löwe H, Kettrup AA (2004) A new, automated, portable immunochemical system for field screening. In: The International Association of Environmental Analytical Chemistry—IAEAC: the 6th Workshop on Biosensors and Bioanalytical μ-Techniques in Environmental and Clinical Analysis, ENEA—University of Rome "La Sapienza", 08–12 October, Rome, Italy
384. Krämer PM, Kremmer E, Weber CM, Ciumasu IM, Forster S, Kettrup AA (2005) Development of new rat monoclonal antibodies with different selectivities and sensitivities for 2, 4, 6-trinitrotoluene (TNT) and other nitroaromatic compounds. Anal Bioanal Chem 382:1919–1933. https://doi.org/10.1007/s00216-005-3294-1
385. Kretschmer F, Franziskowski S, Huber F, Ertl T (2023) Chances and barriers of building information modelling in wastewater management. Water Sci Technol 87(7):1630–1642. https://doi.org/10.2166/wst.2023.079
386. Krieg RM (1995) Information technology and low-income, inner-city communities. J Urban Technol 3(1):1–17. https://doi.org/10.1080/10630739508724513
387. Kriegler E, Edmonds J, Hallegatte S, Ebi KL, Kram T, Riahi K, Winkler H, Van Vuuren DP (2014) A new scenario framework for climate change research: the concept of shared climate policy assumptions. Clim Change 122(3):401–414. https://doi.org/10.1007/s10584-013-0971-5
388. Krüger EL, Minella FO, Rasia F (2011) Impact of urban geometry on outdoor thermal comfort and air quality from field measurements in Curitiba, Brazil. Build Environ 46:621–634. https://doi.org/10.1016/j.buildenv.2010.09.006
389. Kruse S, Bendig D, Brettel M (2023) How does CEO decision style influence firm performance? The mediating role of speed and innovativeness in New Product Development. J Manage Stud 60(5):1206–1235. https://doi.org/10.1111/joms.12913
390. Kühn I, Klotz S (2006) Urbanization and homogenization—comparing the floras of urban and rural areas in Germany. Biol Cons 127(3):292–300. https://doi.org/10.1016/j.biocon.2005.06.033
391. Kuhn TS (1962) The structure of scientific revolutions. University of Chicago Press, Chicago. https://sciencepolicy.colorado.edu/students/envs_5110/structure_of_scientific_revolutions.pdf
392. Kumazawa T, Saito O, Kozaki K, Matsui T, Mizoguchi R (2009) Toward knowledge structuring of sustainability science based on ontology engineering. Sustain Sci 4(1):99–116. https://doi.org/10.1007/s11625-008-0063-z
393. Kunisch S, Denyer D, Bartunek JM, Menz M, Cardinal LB (2023) Review research as scientific inquiry. Organ Res Methods 26(1):3–45. https://doi.org/10.1177/10944281221127292
394. Kuo T-C, Smith S (2018) A systematic review of technologies involving eco-innovation for enterprises moving towards sustainability. J Clean Prod 192:207–220. https://doi.org/10.1016/j.jclepro.2018.04.212
395. Kutzschenbach MV, Schmid A, Schoenenberger L (2018) Using feedback systems thinking to explore theories of digital business for medtech companies. In: Business information systems

and technology 4.0. Springer, Cham, pp 161–175. https://doi.org/10.1007/978-3-319-74322-6_11
396. Kuznetsova E, Ruiz C, Li YF, Zio E (2013) Reliable microgrid energy management under environmental uncertainty and mechanical failures: an agent-based modeling and robust optimization approach. In: Safety, reliability and risk analysis: beyond the Horizon (ESREL 2013), pp 2873–2882
397. Kuznetsova E, Li YF, Ruiz C, Zio E (2014) An integrated framework of agent-based modelling and robust optimization for microgrid energy management. Appl Energy 129:70–88. https://doi.org/10.1016/j.apenergy.2014.04.024
398. Kwon M, Karamcheti S, Cuellar MF, Sadigh D (2021) Targeted data acquisition for evolving negotiation agents. In: Proceedings of the 38th International Conference on Machine Learning, PMLR 139, pp 5894–5904. http://proceedings.mlr.press/v139/kwon21a.html
399. Kwon YI, Son JK (2018) A case study on the promising product selection indicators for small and medium-sized enterprises (SMEs). J Open Innov Technol Market Complex 4(4):56. https://doi.org/10.3390/joitmc4040056
400. Laermans W, Costica M, Costica N, Koos-Morar TA, Ntanou K, Ciumasu IM (2015) People and urban parks: searching the new spiritus loci. In: Nature & Urban Wellbeing. Nature-based Solutions to Societal Challenges, 18–20 May, Ghent, Belgium. http://www.alter-net.info/outputs/conf-2015
401. Lampinen J, Tuomi M, Fischer LK, Neuenkamp L, Alday JG, Bucharova A, Cancellieri L, Casado-Arzuaga I, Čeplová N, Cerveró L, Deák B (2021) Acceptance of near-natural greenspace management relates to ecological and socio-cultural assigned values among European urbanites. Basic Appl Ecol 50:119–131. https://doi.org/10.1016/j.baae.2020.10.006
402. Landauer M, Juhola S, Klein J (2019) The role of scale in integrating climate change adaptation and mitigation in cities. J Environ Planning Manage 62:741–765. https://doi.org/10.1080/09640568.2018.1430022
403. Landemore H (2017) Deliberative democracy as open, not (just) representative democracy. Dædalus 146(3):51–63. https://doi.org/10.1162/DAED_a_00446
404. Langley WN, Corry RC, Brown RD (2019) Core knowledge domains of landscape architecture. Landsc J 37(1):9–21. https://doi.org/10.3368/lj.37.1.9
405. Lapchak PA, Zhang JH, Noble-Haeusslein LJ (2013) RIGOR guidelines: escalating STAIR and STEPS for effective translational research. Transl Stroke Res 4:279–285. https://doi.org/10.1007/s12975-012-0209-2
406. Larson W, Liu F, Yezer A (2012) Energy footprint of the city: effects of urban land use and transportation policies. J Urban Econ 72(2–3):147–159. https://doi.org/10.1016/j.jue.2012.05.001
407. Lee SJ (2022) Citizen participation as a force for persistent Korean democracy. Taiwan J Democracy 18(2):25–46. https://doi.org/10.29654/TJD
408. Le Moniteur (2022) À Versailles (Yvelines), le Département et l'établissement public du château lancent le projet d'aménagement de la place d'Armes. Le Moniteur, 27 September, 2023. https://www.lemoniteur.fr/article/yvelines-reamenagement-de-la-place-d-armes-a-versailles.2225657
409. Lemonde C, Arsenio E, Henriques R (2021) Integrative analysis of multimodal traffic data: addressing open challenges using big data analytics in the city of Lisbon. Eur Transp Res Rev 13(1):1–22. https://doi.org/10.1186/s12544-021-00520-3
410. Leurs K, Georgiou M (2016) Digital makings of the cosmopolitan city? Young people's urban imaginaries of London. Int J Commun 10:3689–3709. https://ijoc.org/index.php/ijoc/article/view/4425
411. Lewandowski M (2016) Designing the business models for circular economy—towards the conceptual framework. Sustainability 8(1), article 43. https://doi.org/10.3390/su8010043
412. Lewin AY, Zhong X (2013) The evolving diaspora of talent: a perspective on trends and implications for sourcing science and engineering work. J Int Manage 19(1):6–13. https://www.sciencedirect.com/science/article/pii/S107542531200066X

413. Leydesdorff L (2012) The Triple Helix, Quadruple Helix, …, and an N-Tuple of Helices: explanatory models for analyzing the knowledge-based economy? J Knowl Econ 3:25–35. https://doi.org/10.1007/s13132-011-0049-4
414. Leydesdorff L, Etzkowitz H (1996) Emergence of a Triple Helix of university—industry—government relations. Sci Publ Policy 23(5):279–286. https://doi.org/10.1093/spp/23.5.279
415. Lieder M, Rashid A (2016) Towards circular economy implementation: a comprehensive review in context of manufacturing industry. J Clean Prod 115:36–51. https://doi.org/10.1016/j.jclepro.2015.12.042
416. Lim SB, Malek JA, Yussoff MFYM, Yigitcanlar T (2021) Understanding and acceptance of smart city policies: practitioners' perspectives on the Malaysian Smart City Framework. Sustainability 13, article 9559. https://doi.org/10.3390/su13179559
417. Linde L, Sjödin D, Parida V, Wincent J (2021) Dynamic capabilities for ecosystem orchestration: a capability-based framework for smart city innovation initiatives. Technol Forecast Soc Change 166, article 120614. https://doi.org/10.1016/j.techfore.2021.120614
418. Linder M, Williander M (2017) Circular business model innovation: inherent 31 uncertainties. Bus Strateg Environ 26(2):182–196. https://doi.org/10.1002/bse.1906
419. Litavniece L (2014) Risk management in provision of city attractiveness. J Positive Manage 5(3):3–14. https://doi.org/10.12775/JPM.2014.016
420. Liu H, Zhou G, Wennersten R, Frostell B (2014) Analysis of sustainable urban development approaches in China. Habitat Int 41:24–32. https://doi.org/10.1016/j.habitatint.2013.06.005
421. Liu J, Dietz T, Carpenter SR, Alberti M, Folke C, Moran E, Pell AN, Deadman P, Kratz T, Lubchenco J, Ostrom E (2007) Complexity of coupled human and natural systems. Science 317:1513–1516. https://doi.org/10.1126/science.11440
422. Liu J, Dietz T, Carpenter SR, Taylor WW, Alberti M, Deadman P, Redman C, Pell A, Folke C, Ouyang Z, Lubchenco J (2021) Coupled human and natural systems: the evolution and applications of an integrated framework. Ambio 50(10):1778–1783. https://doi.org/10.1007/s13280-020-01488-5
423. Liu J, Hull V, Godfray HCJ, Tilman D, Gleick P, Hoff H, Pahl-Wostl C, Xu Z, Chung MG, Sun J, Li S (2018) Nexus approaches to global sustainable development. Nat Sustain 1(9):466–476. https://doi.org/10.1038/s41893-018-0135-8
424. Liu X, Derudder B (2013) Analyzing urban networks through the lens of corporate networks: a critical review. Cities 31:430–437. https://doi.org/10.1016/j.cities.2012.07.009
425. Lloyd GER (1979; 1999) The development of empirical research. In Lloyd GER (ed) Magic, reason, and experience: studies in the origins and development of Greek science. Hackett Publishing. 978-0-87220-528-4. https://hackettpublishing.com/classics/ancient-philosophy/magic-reason-and-experience. A reprint of the first edition in 1979, Cambridge University Press, Chapter 3
426. Lokatis S, Jeschke JM, Bernard-Verdier M, Buchholz S, Grossart HP, Havemann F, Hölker F, Itescu Y, Kowarik I, Kramer-Schadt S, Mietchen D (2023) Hypotheses in urban ecology: building a common knowledge base. Biol Rev (first online: 18 April). https://doi.org/10.1111/brv.12964
427. Londo DL (2020) The impact of international migration on immigrant's home country: exploring diaspora power and conflict. Am J Int Relat 5(1):14–25. https://doi.org/10.47672/ajir.508
428. Lorenzen M, Mudambi R, Schotter A (2020) International connectedness and local disconnectedness: MNE strategy, city-regions and disruption. J Int Bus Stud 51:1199–1222. https://doi.org/10.1057/s41267-020-00339-5
429. Lu Y (2017) Industry 4.0. A survey on technologies, applications and open research issues. J Ind Inf Integr 6:1–10. https://doi.org/10.1016/j.jii.2017.04.005
430. Lubchenco J (1998) Entering the century of the environment: a new social contract for science. Science 279:491–497. https://doi.org/10.1126/science.279.5350.491
431. Lubchenco J, Rapley C (2020) Our moment of truth: the social contract realized? Environ Res Lett 15(11), article 110201. https://doi.org/10.1088/1748-9326/abba9c

432. Lührmann A (2021) Disrupting the autocratization sequence: towards democratic resilience. Democratization 28(5):1017–1039. https://doi.org/10.1080/13510347.2021.1928080
433. Lundin RA, Söderholm A (1995) A theory of the temporary organization. Scand J Manag 11(4):437–455. https://doi.org/10.1016/0956-5221(95)00036-U
434. Lyons G, Mokhtarian P, Dijst M, Böcker L (2018) The dynamics of urban metabolism in the face of digitalization and changing lifestyles: understanding and influencing our cities. Resour Conserv Recycl 132:246–257. https://doi.org/10.1016/j.resconrec.2017.07.032
435. MacGregor-Fors I, Morales-Perez L, Schondube JE (2011) Does size really matter? Species–area relationships in human settlements. Divers Distrib 17:112–121. https://doi.org/10.1111/j.1472-4642.2010.00714.x
436. Machin J, Batista E, Martínez-Ballesté A, Solanas A (2021) Privacy and security in cognitive cities: a systematic review. Appl Sci 11(10):4471. https://doi.org/10.3390/app11104471
437. Macrorie R, Marvin S, While A (2021) Robotics and automation in the city: a research agenda. Urban Geogr 42(2):197–217. https://doi.org/10.1080/02723638.2019.1698868
438. MacVaugh J, Schiavone F (2010) Limits to the diffusion of innovation: a literature re-view and integrative model. Eur J Innov Manag 13(2):197–221. https://doi.org/10.1108/14601061011040258
439. Magnani L (2015) The eco-cognitive model of abduction: Ἀπαγωγή now: naturalizing the logic of abduction. J Appl Log 13(3):285–315. https://doi.org/10.1016/j.jal.2015.04.003
440. Maidique MA, Zirger BJ (1984) A study of success and failure in product innovation: the case of the US electronics industry. IEEE Trans Eng Manage 4:192–203. https://doi.org/10.1109/TEM.1984.6447537
441. Majoor S (2015) Urban megaprojects in crisis? Ørestad Copenhagen revisited. Eur Plann Stud 23(12):2497–2515. https://doi.org/10.1080/09654313.2015.1014780
442. Majoor S (2018) Ambiguity as an opportunity: coping with change in urban megaprojects. In Salet W (ed) The Routledge Handbook of Institutions and Planning in Action, pp 331–346. Routledge, New York. https://doi.org/10.4324/9781315111230
443. Majoor SJH (2016) De stad als experiment. De organisatie van stedelijke innovatie. (In English: The city as an experiment. The organization of urban innovation.) Hogeschool van Amsterdam, Amsterdam. https://www.amsterdamuas.com/binaries/content/assets/hva/lectorale-redes/boekjes/lectorale-rede---de-stad-als-experiment---dr.-stan-j.h-majoor.pdf
444. Malyshev S, Krötzsch M, González L, Gonsior J, Bielefeldt A (2018) Getting the most out of Wikidata: semantic technology usage in Wikipedia's knowledge graph. In The Semantic Web–ISWC 2018: 17th International Semantic Web Conference, Monterey, CA, USA, October 8–12, Proceedings, Part II 17. Springer, Cham, pp 376–394. https://doi.org/10.1007/978-3-030-00668-6_23
445. Mankins JC (1995) Technology readiness levels: a white paper. National Aeronautics and Space Administration (NASA). Washington, DC. http://www.hq.nasa.gov/office/codeq/trl/trl.pdf
446. Mankins JC (2009) Technology readiness assessments: a retrospective. Acta Astronaut 65(9–10):1216–1223. https://doi.org/10.1016/j.actaastro.2009.03.058
447. Manolescu I, Borza M, Talmaciu M, Costica N, Ciumasu IM (2010) Extracurricular, nonformal undergraduate education for sustainability: the ECO-FUN project. In: International Society for Ecological Economics—ISEE 2010: Advancing Sustainability in a Time of Crisis, 22–25 August, Oldenburg & Bremen, Germany
448. Mariani M, Borghi M (2019) Industry 4.0: a bibliometric review of its managerial intellectual structure and potential evolution in the service industries. Technol Forecast Soc Change 149, article 119752. https://doi.org/10.1016/j.techfore.2019.119752
449. Markkanen S, Anger-Kraavi A (2019) Social impacts of climate change mitigation policies and their implications for inequality. Clim Policy 19:827–844. https://doi.org/10.1080/14693062.2019.1596873
450. Markham SK, Ward SJ, Aiman-Smith L, Kingon AI (2010) The valley of death as context for role theory in product innovation. J Prod Innov Manag 27(3):402–417. https://doi.org/10.1111/j.1540-5885.2010.00724.x

451. Martin P, Le Gal E, Choy DL (2016) Effect of social and institutional fragmentation on collective action in peri-urban settings. In: Maheshwari B, Thoradeniya B, Singh VP (eds) Balanced urban development: options and strategies for liveable cities, pp 465–480. Water Science and Technology Library, vol 72. Springer, Cham. https://doi.org/10.1007/978-3-319-28112-4_28
452. Martín EG, Giordano R, Pagano A, Van Der Keur P, Costa MM (2020) Using a system thinking approach to assess the contribution of nature based solutions to sustainable development goals. Sci Total Environ 738:139693. https://doi.org/10.1016/j.scitotenv.2020.139693
453. Martínez A, Mammola S (2021) Specialized terminology reduces the number of citations of scientific papers. Proc Royal Soc B 288(1948), article 20202581. https://doi.org/10.1098/rspb.2020.2581
454. Martins VWB, Rampasso IS, Anholon R, Quelhas OLG, Leal Filho W (2019) Knowledge management in the context of sustainability: literature review and opportunities for future research. J Clean Prod 229:489–500. https://doi.org/10.1016/j.jclepro.2019.04.354
455. Marvuglia A, Havinga L, Heidrich O, Fonseca J, Gaitani N, Reckien D (2020) Advances and challenges in assessing urban sustainability: an advanced bibliometric review. Renew Sustain Energy Rev 124, article 109788. https://doi.org/10.1016/j.rser.2020.109788
456. Marzluff JM (2005) Island biogeography for an urbanizing world: how extinction and colonization may determine biological diversity in human-dominated landscapes. Urb Ecosyst 8:157–177. https://doi.org/10.1007/s11252-005-4378-6
457. Marzluff JM, Shulenberger E, Endlicher W, Simon U, ZumBrunnen C, Alberti M, Bradley G, Ryan C (2008) An introduction to Urban Ecology as an interaction between humans and nature. In: Marzluff JM, Shulenberger E, Endlicher W, Simon U, ZumBrunnen C, Alberti M, Bradley G, Ryan C (eds) Urban ecology, pp vii–xii. Springer, Cham. ISBN: 978-0-387-73411-8
458. Mauser W, Klepper G, Rice M, Schmalzbauer BS, Hackmann H, Leemans R, Moore H (2013) Transdisciplinary global change research: the co-creation of knowledge for sustainability. Curr Opin Environ Sustain 5(3):420–431. https://doi.org/10.1016/j.cosust.2013.07.001
459. Mayer AL, Donovan RP, Pawlowski CW (2014) Information and entropy theory for the sustainability of coupled human and natural systems. Ecol Soc 19(3). https://www.jstor.org/stable/26269639
460. McDonald RI, Mansur AV, Ascensão F, Colbert ML, Crossman K, Elmqvist T, Gonzalez A, Güneralp B, Haase D, Hamann M, Hillel O (2020) Research gaps in knowledge of the impact of urban growth on biodiversity. Nat Sustain 3(1):16–24. https://doi.org/10.1038/s41893-019-0436-6
461. McDowall W, Geng Y, Huang B, Barteková E, Bleischwitz R, Türkeli S, Kemp R, Doménech T (2017) Circular economy policies in China and Europe. J Ind Ecol 21(3):651–661. https://doi.org/10.1111/jiec.12597
462. McKinney ML (2006) Urbanization as a major cause of biotic homogenization. Biol Cons 126:247–260. https://doi.org/10.1016/j.biocon.2005.09.005
463. Meckling J, Nahm J (2018) The power of process: state capacity and climate policy. Governance 31:741–757. https://doi.org/10.1111/gove.12338
464. Medved P, Kim JI, Ursic M (2020) The urban social sustainability paradigm in Northeast Asia and Europe: a comparative study of sustainable urban areas from South Korea, China, Germany and Sweden. Int Rev Spat Plann Sustain Dev 8(4):16–37. https://doi.org/10.14246/irspsd.8.4_16
465. Meerow S, Newell JP (2015) Resilience and complexity: a bibliometric review and prospects for industrial ecology. J Ind Ecol 19(2):236–251. https://doi.org/10.1111/jiec.12252
466. Meerow S, Newell JP (2016) Urban resilience for whom, what, when, where, and why? Urban Geogr 1–21. https://doi.org/10.1080/02723638.2016.1206395
467. Meerow S, Newell JP, Stults M (2016) Defining urban resilience: a review. Landsc Urban Plan 147:38–49. https://doi.org/10.1016/j.landurbplan.2015.11.011
468. Meerow S, Newell JP (2017) Spatial planning for multifunctional green infrastructure: growing resilience in Detroit. Landsc Urban Plan 159:62–75. https://doi.org/10.1016/j.landurbplan.2016.10.005

469. Meerow S, Newell JP (2019) Urban resilience for whom, what, when, where, and why? Urban Geogr 40(3):309–329. https://doi.org/10.1080/02723638.2016.1206395
470. Mehmood H, Gilman E, Cortes M, Kostakos P, Byrne A, Valta K, Tekes S, Riekki J (2019) Implementing big data lake for heterogeneous data sources. In: 2019 IEEE 35th international conference on data engineering workshops (ICDEW). IEEE, pp 37–44. https://doi.org/10.1109/ICDEW.2019.00-37
471. Mejia C, Kajikawa Y (2021) The academic landscapes of manufacturing enterprise performance and environmental sustainability: a study of commonalities and differences. Int J Environ Res Publ Health 18(7), article 3370. https://doi.org/10.3390/ijerph18073370
472. Midgley G, Cavana RY, Brocklesby J, Foote JL, Wood DR, Ahuriri-Driscoll A (2013) Towards a new framework for evaluating systemic problem structuring methods. Eur J Oper Res 229(1):143–154. https://doi.org/10.1016/j.ejor.2013.01.047
473. Miller GA (2003) The cognitive revolution: a historical perspective. Trends Cogn Sci 7(3):141–144. https://doi.org/10.1016/S1364-6613(03)00029-9
474. Miller KD (2015) Agent-based modeling and organization studies: a critical realist perspective. Organ Stud 36(2):175–196. https://doi.org/10.1177/0170840614556921
475. Miller P (1994) Gender discrimination in training: an Australian perspective. Br J Ind Relat 32(4):539–564. https://doi.org/10.1111/j.1467-8543.1994.tb01050.x
476. Mingers J, Rosenhead J (2004) Problem structuring methods in action. Eur J Oper Res 152(3):530–554. https://doi.org/10.1016/S0377-2217(03)00056-0
477. Mitchell M (1993) Analogy-making as perception: a computer model. MIT Press. ISBN 9780262515443. https://mitpress.mit.edu/9780262132893/. 2nd edition (2003) available at: https://mitpress.mit.edu/9780262515443/analogy-making-as-perception/
478. Mitchell M (2001) Analogy making as a complex adaptive system. In: Santa Fe Institute Studies in the sciences of complexity-proceedings, vol 2001-June. Reading, Massachusetts, Addison-Wesley, pp 335–360
479. Mitchell M (2006) Complex systems: network thinking. Artif Intell 170:1194–1212. https://doi.org/10.1016/j.artint.2006.10.002
480. Mitchell M (2021) Abstraction and analogy-making in Artificial Intelligence. Ann N Y Acad Sci 1505(1):79–101. https://doi.org/10.1111/nyas.1461
481. Mitchell M (2021b) Why AI is harder than we think. Preprint, arXiv preprint arXiv:2104.12871. https://doi.org/10.48550/arXiv.2104.12871
482. Mohammed SB (2022) Re-thinking the global cosmopolis: an analysis of the UN-Habitat "city we need" policies in Helsinki and Sydney. Cogent Arts Hum 9(1), article 2132615. https://doi.org/10.1080/23311983.2022.2132615
483. Monah D (2005) Religie si arta in cultura Cucuteni [Religion and art in Cucuteni culture]. In: Dumitroaia G (ed) Primul muzeu Cucuteni din Romania [The first Cucuteni museum for Romania] (in Romanian). Bibliotheca Memoriae Antiquitatis XV:162–173. Editura Foton, Piatra-Neamţ, Romania
484. Monjardet B (2003) The presence of lattice theory in discrete problems of mathematical social sciences. Why Math Soc Sci 46(2):103–144. https://doi.org/10.1016/S0165-4896(03)00072-6
485. Montoya-Weiss MM, Calantone R (1994) Determinants of new product performance: a review and meta-analysis. J Prod Innov Manag 11(5):397–417. https://doi.org/10.1016/0737-6782(94)90029-9
486. Moore GC, Benbasat I (1991) Development of an instrument to measure the perceptions of adopting an information technology innovation. Inf Syst Res 2(3):192–222. https://doi.org/10.1287/isre.2.3.192
487. Moore J, Kissinger M, Rees WE (2013) An urban metabolism and ecological footprint assessment of Metro Vancouver. J Environ Manage 124:51–61. https://doi.org/10.1016/J.JENVMAN.2013.03.009
488. Moreau CP, Lehmann DR, Markman AB (2001) Entrenched knowledge structures and consumer response to new products. J Mark Res 38(1):14–29. https://doi.org/10.1509/jmkr.38.1.14.18836

489. Morganti M (2021) Spatial metrics to investigate the impact of urban form on microclimate and building energy performance: an essential overview. In: Palme M, Salvati A (eds) Urban microclimate modelling for comfort and energy studies. Springer, Cham. https://doi.org/10.1007/978-3-030-65421-4_18
490. Morris MG, Venkatesh V (2000) Age differences in technology adoption decisions: implications for a changing work force. Pers Psychol 53(2):375–403. https://doi.org/10.1111/j.1744-6570.2000.tb00206.x
491. Morris Z, Weissburg M, Bras B (2018) Towards a biologically-inspired urban-industrial ecosystem. Procedia CIRP 69:861–866. https://doi.org/10.1016/j.procir.2017.11.055
492. Morris ZB, Weissburg M, Bras B (2021) Ecological network analysis of urban–industrial ecosystems. J Ind Ecol 25(1):193–204. https://doi.org/10.1111/jiec.13043
493. Mostashari A, Arnold F, Mansouri M, Finger M (2011) Cognitive cities and intelligent urban governance. Netw Indus Quarter 13(3):4–7. https://web.archive.org/web/20180410022321id_/http://newsletter.epfl.ch/mir/index.php?module=epflfiles&func=getFile&fid=240&inline=1
494. Mostafavi N, Farzinmoghadam M, Hoque S (2017) Urban residential energy consumption modeling in the integrated urban metabolism analysis tool (IUMAT). Build Environ 114:429–444. https://doi.org/10.1016/J.BUILDENV.2016.12.035
495. Moynihan DP (2008) Learning under uncertainty: networks in crisis management. Public Adm Rev 68(2):350–365. https://doi.org/10.1111/j.1540-6210.2007.00867.x
496. Mucunguzi P (1995) A review of non-formal environmental education in Uganda. Environ Educ Res 1(3):337–344. https://doi.org/10.1080/1350462950010308
497. Muñoz-Erickson TA (2014) Co-production of knowledge–action systems in urban sustainable governance: the KASA approach. Environ Sci Policy 37:182–191. https://doi.org/10.1016/j.envsci.2013.09.014
498. Muñoz-Erickson TA, Cutts BB (2016) Structural dimensions of knowledge-action networks for sustainability. Curr Opin Environ Sustain 18:56–64. https://doi.org/10.1016/j.cosust.2015.08.013
499. Murakami Y (2014) Influences of return migration on international collaborative research networks: cases of Japanese scientists returning from the US. J Technol Transf 39(4):616–634. https://link.springer.com/article/https://doi.org/10.1007/s10961-013-9316-9
500. Murray A, Skene K, Haynes K (2017) The circular economy: an interdisciplinary exploration of the concept and application in a global context. J Bus Ethic 140(3):369–380. https://link.springer.com/article/https://doi.org/10.1007/s10551-015-2693-2
501. Myeong S, Kim Y, Ahn MJ (2020) Smart city strategies—technology push or culture pull? A case study exploration of Gimpo and Namyangju, South Korea. Smart Cities 4(1):41–53. https://doi.org/10.3390/smartcities4010003
502. Narechania A, Du F, Sinha AR, Rossi R, Hoffswell J, Guo S, Koh E, Navathe SB, Endert A (2023) DataPilot: utilizing quality and usage information for subset selection during visual data preparation. In: Proceedings of the 2023 CHI conference on human factors in computing systems, article 809. https://doi.org/10.1145/3544548.3581509
503. Natanian J, Aleksandrowicz O, Auer T (2019) A parametric approach to optimizing urban form, energy balance and environmental quality: the case of Mediterranean districts. Appl Energy 254, article 113637. https://doi.org/10.1016/j.apenergy.2019.113637
504. Natarajan L (2017) Socio-spatial learning: a case study of community knowledge in participatory spatial planning. Prog Plan 111:1–23. https://doi.org/10.1016/j.progress.2015.06.002
505. NATO-SC (2022) NATO 2022 strategic concept. The North Atlantic Treaty Organization. www.nato.int/strategic-concept/. Accessed 03 June 2022
506. NATO-IF (2022) NATO innovation fund. The North Atlantic Treaty Organization. https://www.nato.int/cps/en/natohq/news_197494.htm. Accessed: 16 Oct 2023
507. Nativi S, Santoro M, Giuliani G, Mazzetti P (2019) Towards a knowledge base to support global change policy goals. Int J Digit Earth 13:188–216. https://doi.org/10.1080/17538947.2018.1559367

508. Nguyen KT, Dekneuvel E, Nicolle B, Zammit O, Van CN, Jacquemod G (2013) Using FPGA for real time power monitoring in a NIALM system. In: 2013 IEEE international symposium on industrial electronics. IEEE, pp 1–6. https://doi.org/10.1109/ISIE.2013.6563850
509. Nielsen MW, Andersen JP (2021) Global citation inequality is on the rise. Proc Natl Acad Sci 118(7), article e2012208118. https://doi.org/10.1073/pnas.2012208118
510. Niemelä J (1999) Ecology and urban planning. Biodivers Conserv 8:119–131. https://doi.org/10.1023/A:1008817325994
511. Nijhuis S, de Vries J (2020) Design as research in landscape architecture. Landsc J 38(1–2):87–103. https://doi.org/10.3368/lj.38.1-2.87
512. Nola R, Sankey H (2000) A selective survey of theories of scientific method. In: After Popper, Kuhn and Feyerabend: Recent issues in theories of scientific method. Springer Science+Business Media, B.V., pp 1–65. https://doi.org/10.1007/978-94-011-3935-9
513. Noss RF (1995) Ecological integrity and sustainability: buzzwords in conflict? In: Westra L, Lemons J (eds) Perspectives on ecological integrity. Environ Sci Technol Libr 5:60–76. Springer, Dordrecht. https://doi.org/10.1007/978-94-011-0451-7_5
514. Ntanou K, Morar TA, Liamidi H, Dean J, Ciumasu IM (2014) Is it possible to develop a model of sustainable urban water cycle? In: Proceedings of Deltas in Times of Climate Change II, 24–26 Sept 2014, Rotterdam, The Netherlands, p 119. http://archive.northsearegion.eu/files/repository/20151028152236_UNESCO-IHE6.ABSTRACTWEBDITCC2.pdf
515. Núñez R, Allen M, Gao R, Miller Rigoli C, Relaford-Doyle J, Semenuks A (2019) What happened to cognitive science? Nat Hum Behav 3(8):782–791. https://doi.org/10.1038/s41562-019-0626-2
516. Ochoa D, Riano-Briceno G, Quijano N, Ocampo-Martinez C (2019) Control of urban drainage systems: Optimal flow control and deep learning in action. In: 2019 American Control Conference (ACC). IEEE, pp 4826–4831. https://doi.org/10.23919/ACC.2019.8814958
517. Odobleja S (1938; 1939) Psychologie consonantiste, vol 1; 2. Librairie Maloine, Paris. https://www.worldcat.org/oclc/563648574
518. Odobleja S (1978) Diversity and unity in cybernetics. In: Rose J (ed) Current topics in cybernetics and systems. Springer, Berlin. https://doi.org/10.1007/978-3-642-93104-8_98
519. Odobleja S (1983) Cybernétique générale: psychologie consonantiste: science des sciences. Editions Nagard, Paris. https://www.worldcat.org/oclc/955246887
520. OECD (2015) Frascati Manual 2015: guidelines for collecting and reporting data on research and experimental development. The Organisation for Economic Co-operation and Development (OECD), Paris. https://www.oecd.org/sti/frascati-manual-2015-9789264239012-en.htm
521. Olechowski A, Eppinger SD, Joglekar N (2015) Technology readiness levels at 40: a study of state-of-the-art use, challenges, and opportunities. In: 2015 Portland International Conference on Management of Engineering & Technology (PICMET), pp 2084–2094. IEEE. http://web.mit.edu/eppinger/www/pdf/Eppinger_PICMET2015.pdf
522. Oleinik E, Zakharova A (2019) City: economic growth and social attractiveness issues. Entrepreneurship Sustain Iss 7(1), article 454. https://doi.org/10.9770/jesi.2019.7.1(32)
523. Olsen J, Gjerding A (2019) Modalities of abduction: a philosophy of science-based investigation of abduction. Hum Arenas 2:129–152. https://doi.org/10.1007/s42087-018-0044-4
524. Omata N (2023) The role of developmental 'buzzwords' in the international refugee regime: self-reliance, resilience, and economic inclusion. World Dev 167:106248. https://doi.org/10.1016/j.worlddev.2023.106248
525. Omukuti J, Barrett S, White PCL, Marchant R, Averchenkova A (2022) The green climate fund and its shortcomings in local delivery of adaptation finance. Clim Policy (published online: 06 July 2022). https://doi.org/10.1080/14693062.2022.2093152
526. Oppenheimer M, O'Neill BC, Webster M (2008) Negative learning. Clim Change 89(1):155–172. https://doi.org/10.1007/s10584-008-9405-1
527. Opute J (2018) Management of diaspora businesses: issues and learning. In: Hack-Polay D, Siwale J (eds) African diaspora direct investment. Palgrave Studies of Entrepreneurship in Africa. Palgrave Macmillan, Cham. https://doi.org/10.1007/978-3-319-72047-0_9

528. Ortega J, Moslem S (2023) Decision support system for evaluating park & ride system using the analytic hierarchy process (AHP) method. Urb Plann Transp Res 11(1), article 2194362. https://doi.org/10.1080/21650020.2023.2194362
529. Ortega PA, Braun DA (2013) Thermodynamics as a theory of decision-making with information-processing costs. Proc Royal Soc A: Math Phys Eng Sci 469:20120683. https://doi.org/10.1098/rspa.2012.0683
530. Osman K, Pudin S (2009) The adults non-formal environmental education (EE): a scenario in Sabah, Malaysia. Procedia-Soc Behav Sci 1(1):2306–2311. https://doi.org/10.1016/j.sbspro.2009.01.405
531. Osman MA, Noah SAM, Saad S (2022) Ontology-based knowledge management tools for knowledge sharing in organization—a review. IEEE Access 10:43267–43283. https://doi.org/10.1109/ACCESS.2022.3163758
532. Österblom H, Folke C, Rocha J, Bebbington J, Blasiak R, Jouffray JB, Selig ER, Wabnitz CC, Bengtsson F, Crona B, Gupta R (2022) Scientific mobilization of keystone actors for biosphere stewardship. Scientific Rep 12(1), article 3802. https://doi.org/10.1038/s41598-022-07023-8
533. Ostrom E (1999) Coping with tragedies of the commons. Annu Rev Polit Sci 2(1):493–535. https://doi.org/10.1146/annurev.polisci.2.1.493
534. Ostrom E (2010) A multi-scale approach to coping with climate change and other collective action problems. Solutions 1(2):27–36. https://hdl.handle.net/10535/5774
535. Ostrom E (2014) A polycentric approach for coping with climate change. Ann Econ Financ 15(1):97–134. https://doi.org/10.1098/rspa.2012.0683
536. Otero-Cerdeira L, Rodríguez-Martínez FJ, Gómez-Rodríguez A (2015) Ontology matching: a literature review. Expert Syst Appl 42(2):949–971. https://doi.org/10.1016/j.eswa.2014.08.032
537. Otte E, Rousseau R (2002) Social network analysis: a powerful strategy, also for the information sciences. J Inf Sci 28(6):441–453. https://doi.org/10.1177/016555150202800601
538. Paavola S (2021). Practical abduction for research on human practices: enriching rather than testing a hypothesis. In: Shook JR, Paavola S (eds) Abduction in cognition and action. Studies in applied philosophy, epistemology and rational ethics. Springer, Cham, vol 59, pp 31–48. https://doi.org/10.1007/978-3-030-61773-8_2
539. Pahl-Wostl C (2007) The implications of complexity for integrated resources management. Environ Model Softw 22(5):561–569. https://doi.org/10.1016/j.envsoft.2005.12.024
540. Pahl-Wostl C, Vörösmarty C, Bhaduri A, Bogardi J, Rockström J, Alcamo J (2013) Towards a sustainable water future: shaping the next decade of global water research. Curr Opin Environ Sustain 5(6):708–714. https://doi.org/10.1016/j.cosust.2013.10.012
541. Pahl-Wostl C (2017) An evolutionary perspective on water governance: from understanding to transformation. Water Resour Manage Int J Publ Eur Water Resour Assoc (EWRA) 31(10):2917–2932. https://doi.org/10.1007/s11269-017-1727-1
542. Palusci O, Monti P, Cecere C, Montazeri H, Blocken B (2022) Impact of morphological parameters on urban ventilation in compact cities: the case of the Tuscolano-Don Bosco district in Rome. Sci Total Environ 807, article 150490. https://doi.org/10.1016/j.scitotenv.2021.150490
543. Pan C (2020) Enfolding wholes in parts: quantum holography and international relations. Eur J Int Relat 26(1_suppl):14–38. https://doi.org/10.1177/1354066120938844
544. Panda DK, Das S (2021) Smart grid architecture model for control, optimization and data analytics of future power networks with more renewable energy. J Clean Prod 301, article 126877. https://doi.org/10.1016/j.jclepro.2021.126877
545. Papadis E, Tsatsaronis G (2020) Challenges in the decarbonization of the energy sector. Energy 205, article 118025. https://doi.org/10.1016/j.energy.2020.118025
546. Papatheodorou A, Rosselló J, Xiao H (2010) Global economic crisis and tourism: consequences and perspectives. J Travel Res 49(1):39–45. https://journals.sagepub.com/https://doi.org/10.1177/0047287509355327
547. Parnell S, Elmqvist T, McPhearson T, Nagendra H, Sörlin S (2018) Introduction: situating knowledge and action for an urban planet. In: Elmqvist T, Bai X, Frantzeskaki N, Griffith

C, Maddox D, McPhearson T, Parnell S, Roberts D, Romero-Lankao P, Simon D, Watkins M (eds) Urban planet: knowledge towards sustainable cities. Cambridge University Press, Cambridge, pp 1–16. https://doi.org/10.1017/9781316647554.002

548. Partelow S (2018) A review of the social-ecological systems framework: applications, methods, modifications, and challenges. Ecol Soc 23(4), article 36. https://doi.org/10.5751/ES-10594-230436
549. Pascual U, Palomo I, Adams WM, Chan KM, Daw TM, Garmendia E, Gómez-Baggethun E, De Groot RS, Mace GM, Martín-López B, Phelps J (2017) Off-stage ecosystem service burdens: a blind spot for global sustainability. Environ Res Lett 12(7), article 075001. https://iopscience.iop.org/article/https://doi.org/10.1088/1748-9326/aa7392/meta
550. Pauketat TR (2020) When the rains stopped: evapotranspiration and ontology at ancient Cahokia. J Anthropol Res 76(4):410–438. https://doi.org/10.1086/711102
551. Pauleit S, Duhme F (2000) Assessing the environmental performance of land cover types for urban planning. Landsc Urban Plan 52:1–20. https://doi.org/10.1016/S0169-2046(00)00109-2
552. Peel MC, Finlayson BL, McMahon TA (2007) Updated world map of the Köppen-Geiger climate classification. Hydrol Earth Syst Sci 11(5):1633–1644. https://doi.org/10.5194/hess-11-1633-2007
553. Pei Y, Biswas S, Fussell DS, Pingali K (2019) An elementary introduction to Kalman filtering. Commun ACM 62(11):122–133. https://doi.org/10.1145/3363294
554. Pelorosso R, Gobattoni F, Leone A (2017) The low-entropy city: a thermodynamic approach to reconnect urban systems with nature. Landsc Urban Plan 168:22–30. https://doi.org/10.1016/j.landurbplan.2017.10.002
555. Peralta ME, Soltero VM (2020) Analysis of fractal manufacturing systems framework towards Industry 4.0. J Manuf Syst 57:46–60. https://doi.org/10.1016/j.jmsy.2020.08.004
556. Perez C (2010) Technological revolutions and techno-economic paradigms. Camb J Econ 34(1):185–202. https://doi.org/10.1093/cje/bep051
557. Peroni F, Pappalardo SE, Facchinelli F, Crescini E, Munafò M, Hodgson ME, De Marchi M (2022) How to map soil sealing, land take and impervious surfaces? Syst Rev. Environ Res Lett 17(5), article 053005. https://doi.org/10.1088/1748-9326/ac6887
558. Perrotti D, Stremke S (2020) Can urban metabolism models advance green infrastructure planning? Insights from ecosystem services research. Environ Plann B: Urb Anal City Sci 47(4):678–694. https://doi.org/10.1177/2399808318797131
559. Peters HP (2013) Gap between science and media revisited: Scientists as public communicators. Proc Natl Acad Sci 110(supplement_3):14102–14109. https://doi.org/10.1073/pnas.1212745110
560. Phelan PE, Kaloush K, Miner M, Golden J, Phelan B, Silva H III, Taylor RA (2015) Urban heat island: mechanisms, implications, and possible remedies. Annu Rev Environ Resour 40:285–307. https://doi.org/10.1146/annurev-environ-102014-021155
561. Phillips S, Wilson WH (2010) Categorial compositionality: a category theory explanation for the systematicity of human cognition. PLoS Comput Biol 6(7):e100. https://doi.org/10.1371/journal.pcbi.1000858
562. Pickering M (2004) Experience as horizon: Koselleck, expectation and historical time. Cult Stud 18(2–3):271–289. https://doi.org/10.1080/0950238042000201518
563. Pickering J, Bäckstrand K, Schlosberg D (2020) Between environmental and ecological democracy: theory and practice at the democracy-environment nexus. J Environ Plann Policy Manage 22(1):1–15. https://doi.org/10.1080/1523908X.2020.1703276
564. Pickett ST, Cadenasso ML, Grove JM, Nilon CH, Pouyat RV, Zipperer WC, Costanza R (2001) Urban ecological systems: linking terrestrial ecological, physical, and socioeconomic components of metropolitan areas. Annu Rev Ecol Syst 32(1):127–157. https://doi.org/10.1146/annurev.ecolsys.32.081501.114012
565. Pickett ST, Grove JM (2008) Urban ecosystems: what would Tansley do? Urban Ecosyst 12(1):1–8. https://doi.org/10.1007/s11252-008-0079-2

566. Pignatelli M, Moghadam ST, Genta C, Lombardi P (2023) Spatial decision support system for low-carbon sustainable cities development: an interactive storytelling dashboard for the city of Turin. Sustain Cities Soc 89, article 104310. https://doi.org/10.1016/j.scs.2022.104310
567. Pluchinotta I, Giordano R, Zikos D, Krueger T, Tsoukiàs A (2020) Integrating problem structuring methods and concept-knowledge theory for an advanced policy design: lessons from a case study in Cyprus. J Comparative Policy Anal Res Pract 22(6):626–647. https://doi.org/10.1080/13876988.2020.1753512
568. Polk M (2014) Achieving the promise of transdisciplinarity: a critical exploration of the relationship between transdisciplinary research and societal problem solving. Sustain Sci 9(4):439–451. https://doi.org/10.1007/s11625-014-0247-7
569. Polotski V, Kenne JP, Gharbi A (2020) Kalman filter based production control of a failure-prone single-machine single-product manufacturing system with imprecise demand and inventory information. J Manuf Syst 56:558–572. https://doi.org/10.1016/j.jmsy.2020.07.010
570. Polzin F, Egli F, Steffen B, Schmidt TS (2019) How do policies mobilize private finance for renewable energy? A systematic review with an investor perspective. Appl Energy 236:1249–1268. https://doi.org/10.1016/j.apenergy.2018.11.098
571. Pompeani DP, Hillman AL, Finkenbinder MS, Bain DJ, Correa-Metrio A, Pompeani KM, Abbott MB (2018) The environmental impact of a pre-Columbian city based on geochemical insights from lake sediment cores recovered near Cahokia. Quatern Res 91(2):714–728. https://doi.org/10.1017/qua.2018.141
572. Pompeani DP, Bird BW, Wilson JJ, Gilhooly III WP, Hillman AL, Finkenbinder MS, Abbott MB (2021) Severe little ice age drought in the midcontinental United States during the Mississippian abandonment of Cahokia. Sci Rep 11(1), article 13829. https://doi.org/10.1038/s41598-021-92900-x
573. PoR-WGCC (2023) (National) Working Group for Climate Changes (WGCC), President of Romania. https://www.presidency.ro/ro/media/clima-si-sustenabilitate/raportul-limitarea-schimbarilor-climatice-si-a-impactului-lor-o-abordare-integrata-pentru-romania. Accessed on 15 Sept 2022
574. Portisch J, Heist N, Paulheim H (2022) Knowledge graph embedding for data mining vs. knowledge graph embedding for link prediction–two sides of the same coin? Semantic Web 13(3):399–422. https://doi.org/10.3233/SW-212892
575. Portnov BA (1998) Social attractiveness of the urban physical environment: cities of Siberia. Ann Reg Sci 32(4):525–548. https://doi.org/10.1007/s001680050087
576. Prigogine I (1997) The End of Certainty: Time, Chaos and the New Laws of Nature. The Free Press, New York. ISBN 0684837056
577. Puga-Caballero A, MacGregor-Fors I, Ortega-Álvarez R (2014) Birds at the urban fringe: avian community shifts in different peri-urban ecotones of a megacity. Ecol Res 29:619–628. https://doi.org/10.1007/s11284-014-1145-2
578. Puskás N, Abunnasr Y, Naalbandian S (2021) Assessing deeper levels of participation in nature-based solutions in urban landscapes–a literature review of real-world cases. Lands Urb Plann 210, article 104065. https://doi.org/10.1016/j.landurbplan.2021.104065
579. Qin Y (2020) Urban flooding mitigation techniques: A systematic review and future studies. Water 12(12), article 3579. https://doi.org/10.34068/IYS.02.01.03
580. Quan SJ, Li C (2021) Urban form and building energy use: a systematic review of measures, mechanisms, and methodologies. Renew Sustain Energy Rev 139, article 110662. https://doi.org/10.1016/j.rser.2020.110662
581. Radoli LO (2018) Narratives of migration and development as discourses in transnational digital migrant media: the case of Kenyan migration to Europe. Doctoral Dissertation. Technical University Cottbus–Senftenberg. https://opus4.kobv.de/opus4-btu/files/4780/Lydia_OumaRadoli.pdf
582. Rahm E (2011) Towards large-scale schema and ontology matching. In Bellahsene Z, Bonifati A, Rahm E (eds) Schema matching and mapping, data-centric systems and applications, pp 3–27. https://doi.org/10.1007/978-3-642-16518-4_1

583. Ratti RA, Vespignani JL (2016) Oil prices and global factor macroeconomic variables. Ener Econ 59:198–212. https://www.sciencedirect.com/science/article/pii/S0140988316301372
584. Raubo ANNA (2010) City branding and its impact on city's attractiveness for external audiences. Master Thesis, Erasmus University Rotterdam. https://thesis.eur.nl/pub/7923/MA-Anna%20Raubo%20(Romanski)%20331503%20augustus%202010.pdf
585. Raven R, Sengers F, Spaeth P, Xie L, Cheshmehzangi A, de Jong M (2019) Urban experimentation and institutional arrangements. Eur Plan Stud 27(2):258–281. https://doi.org/10.1080/09654313.2017.1393047
586. Ray JK, Sultana R, Bera R, Sil S, Alfred QM (2023) A comprehensive review on artificial intelligence (AI) and robotic process automation (RPA) for the development of smart cities. In: Bhattacharyya S, Banerjee JS, De D (eds) Confluence of artificial intelligence and robotic process automation. Smart innovation, systems and technologies. Springer, Singapore, vol 335. https://doi.org/10.1007/978-981-19-8296-5_12
587. Rebele F (1994) Urban ecology and special features of urban ecosystems. Glob Ecol Biogeogr Lett 4(6):173–187. https://doi.org/10.2307/2997649
588. Reed MS, Evely AC, Cundill G, Fazey I, Glass J, Laing A, Newig J, Parrish B, Prell C, Raymond C, Stringer LC (2010) What is social learning? Ecol Soc 15(4):1–10. https://www.jstor.org/stable/26268235
589. Rees W (1992) Ecological footprints and appropriated carrying capacity: what urban economics leaves out. Environ Urban 4(2):121–130. https://doi.org/10.1177/095624789200400212
590. Rees W (1996) Revisiting carrying capacity: area-based indicators of sustainability. Popul Environ 17(3):195–215. https://doi.org/10.1007/BF02208489
591. Rees W (2012) Cities as dissipative structures: global change and the vulnerability of urban civilization. In: Weinstein MP, Turner RE (eds) Sustainability science: the emerging paradigm and the urban environment. Springer, London. https://doi.org/10.1007/978-1-4614-3188-6_12
592. Rees W, Wackernagel M (1996) Urban ecological footprints: why the cities cannot be sustainable—and why they are a key to sustainability. Environ Impact Assess Rev 16:223–248. https://doi.org/10.1016/S0195-9255(96)00022-4
593. Rees W, Wackernagel M (2008) Urban ecological footprints: why cities cannot be sustainable—and why they are a key to sustainability. In: Marzluff JM, Shulenberger E, Endlicher W, Alberti M, Bradley G, Ryan C, Simon U, ZumBrunnen C (eds) Urban ecology: an international perspective on the interaction between humans and nature, pp 537–555. Springer, Boston. https://doi.org/10.1007/978-0-387-73412-5_35
594. Rega-Brodsky CC, Aronson MF, Piana MR, Carpenter ES, Hahs AK, Herrera-Montes A, Knapp S, Kotze DJ, Lepczyk CA, Moretti M, Salisbury AB (2022) Urban biodiversity: state of the science and future directions. Urb Ecosyst 25(4):1083–1096. https://doi.org/10.1007/s11252-022-01207-w
595. Reis J, Marques PA, Marques PC (2022) Where are smart cities heading? A meta-review and guidelines for future research. Appl Sci 12(16), article 8328. https://doi.org/10.3390/app12168328
596. Ren J, Liang H, Chan FT (2017) Urban sewage sludge, sustainability, and transition for eco-city: multi-criteria sustainability assessment of technologies based on best-worst method. Technol Forecast Soc Chang 116:29–39. https://doi.org/10.1016/j.techfore.2016.10.070
597. Richins ML, Bloch PH (1986) After the new wears off: the temporal context of product involvement. J Consum Res 13(2):280–285. https://doi.org/10.1086/209067
598. Richter M, Lombardi P, Arendarski B, Naumann A, Hoepfner A, Komarnicki P, Pantaleo A (2021) A vision for energy decarbonization: Planning sustainable tertiary sites as net-zero energy systems. Energies 14(17), article 5577. https://doi.org/10.3390/en14175577
599. Robinson J (2002) Global and world cities: a view from off the map. Int J Urban Reg Res 26(3):531–554. https://doi.org/10.1111/1468-2427.00397
600. Robinson J (2005) Urban geography: world cities, or a world of cities. Prog Hum Geogr 29(6):757–765. https://doi.org/10.1191/0309132505ph582pr

601. Rocak M, Hospers GJ, Reverda N (2016) Searching for social sustainability: the case of the shrinking city of Heerlen, The Netherlands. Sustainability 8, article 382. https://doi.org/10.3390/su8040382
602. Rodrigues M, Franco M (2022) Bibliometric review about eco-cites and urban sustainable development: trend topics. Environ Dev Sustain 24:13683–13704. https://doi.org/10.1007/s10668-021-02006-7
603. Rodríguez-Rojas MI, Grindlay Moreno AL (2022) A discussion on the application of terminology for urban soil sealing mitigation practices. Int J Environ Res Publ Health 19(14), article 8713. https://doi.org/10.3390/ijerph19148713
604. Roelich K, Giesekam J (2019) Decision making under uncertainty in climate change mitigation: introducing multiple actor motivations, agency and influence. Climate Policy 19:175–188. https://doi.org/10.1080/14693062.2018.1479238
605. Rogers E (1962) Diffusion of innovations. The Free Press, New York. https://doi.org/10.1002/jps.2600520633
606. Rogers ML (2010) Dewey and his vision of democracy. Contemp Pragmat 7(1):69–91. https://ssrn.com/abstract=1831305
607. Roh K, Bardow A, Bongartz D, Burre J, Chung W, Deutz S, Han D, Heßelmann M, Kohlhaas Y, König A, Lee JS (2020) Early-stage evaluation of emerging CO_2 utilization technologies at low technology readiness levels. Green Chem 22(12):3842–3859. https://doi.org/10.1039/c9gc04440j
608. Rohde M (2020) The English garden in Germany: some late eighteenth-century concepts of the landscape garden. In Finch J, Woudstra J (eds) Capability brown, royal gardener: the business of place-making in Northern Europe, pp 151–164. White Rose University Press, Heslington. https://doi.org/10.22599/CapabilityBrown.k
609. Romão J, Kourtit K, Neuts B, Nijkamp P (2018) The smart city as a common place for tourists and residents: a structural analysis of the determinants of urban attractiveness. Cities 78:67–75. https://doi.org/10.1016/j.cities.2017.11.007
610. Romero-Lankao P, Gnatz DM, Wilhelmi O, Hayden M (2016) Urban sustainability and resilience: from theory to practice. Sustainability 8:1224. https://doi.org/10.3390/su8121224
611. Romero-Lankao P, Qin H, Dickinson K (2012) Urban vulnerability to temperature-related hazards: a meta-analysis and meta-knowledge approach. Glob Environ Chang 22(3):670–683. https://doi.org/10.1016/j.gloenvcha.2012.04.002
612. Rondinel-Oviedo DR, Keena N (2023) Entropy and cities: a bibliographic analysis towards more circular and sustainable urban environments. Entropy 25(3):532. https://doi.org/10.3390/e25030532
613. Rönnebeck S, Bernholt S, Ropohl M (2016) Searching for a common ground—a literature review of empirical research on scientific inquiry activities. Stud Sci Educ 52(2):161–197. https://doi.org/10.1080/03057267.2016.1206351
614. Rook GA (2013) Regulation of the immune system by biodiversity from the natural environment: an ecosystem service essential to health. Proc Natl Acad Sci 110(46):18360–18367. https://doi.org/10.1073/pnas.1313731110
615. Rosado L, Niza S, Ferrão P (2014) A material flow accounting case study of the Lisbon metropolitan area using the urban metabolism analyst model. J Ind Ecol 18(1):84–101. https://doi.org/10.1111/jiec.12083
616. Rosser JB Jr (2021) Econophysics and the entropic foundations of economics. Entropy 23:1286. https://doi.org/10.3390/e23101286
617. Roth M (2007) Review of urban climate research in (sub) tropical regions. Int J Climatol J R Meteorol Soc 27(14):1859–1873. https://doi.org/10.1002/joc.1591
618. Roure JB, Keeley RH (1990) Predictors of success in new technology based ventures. J Bus Ventur 5(4):201–220. https://doi.org/10.1016/0883-9026(90)90017-N
619. Rosso AM (2023) Methods and management in ancient Egyptian and Greek medicine. Hist Philos Med 5(3), article 14. https://doi.org/10.53388/HPM2023014
620. Roussel F, Alexandre F (2021) Landscape ecological enhancement and environmental inequalities in peri-urban areas, using flora as a socio-ecological indicator—the case of the greater Paris area. Landsc Urb Plann 210, article 104062

621. Rovatsos M (2014) Multiagent systems for social computation. In: Proceedings of the 2014 international conference on autonomous agents and multi-agent systems, ACM, Paris, France, pp 1165–1168. http://dl.acm.org/citation.cfm?id=2617388.2617432
622. Ruiz-Mallén I, Satorras M, March H, Baró F (2022) Community climate resilience and environmental education: opportunities and challenges for transformative learning. Environ Educ Res 28(7):1088–1107. https://doi.org/10.1080/13504622.2022.2070602
623. Russel A, Tillement S, Journé B (2022) Building resilience in temporary organizations: lessons from a shipyard. In: Pinheiro, Frogotto ML, Young M (eds) Towards resilient organizations and societies, public sector organizations, pp 91–116. Palgrave Macmillan (Springer), Cham. https://doi.org/10.1007/978-3-030-82072-5_4
624. Sadin SR, Povinelli FP, Rosen R (1989) The NASA technology push towards future space mission systems. Acta Astronaut 20:73–77. https://doi.org/10.1016/0094-5765(89)90054-4
625. Sakr S, Bonifati A, Voigt H, Iosup A, Ammar K, Angles R, Aref W, Arenas M, Besta M, Boncz PA, Daudjee K (2021) The future is big graphs: a community view on graph processing systems. Commun ACM 64(9):62–71. https://doi.org/10.1145/3434642
626. Sartori RA, Gomes A, Narcizo A, Mata S, Cárcamo AT, Matos RMA, Correa AA, Santana MH, Vieira MA, Rajão H (2023) Urban ecology and biological studies in Brazilian cities: a systematic review. Urb Ecosyst 26:547–558. https://doi.org/10.1007/s11252-022-01324-6
627. Sauer J, Pallathadka A, Ajibade I, Berbés-Blázquez M, Chang H, Cook EM, Grimm NB, Iwaniec DM, Lloyd R, Post GC (2023) Relating social, ecological, and technological vulnerability to future flood exposure at two spatial scales in four US cities. Sustain Cities Soc article 104880. https://doi.org/10.1016/j.scs.2023.104880
628. Scalenghe R, Marsan FA (2009) The anthropogenic sealing of soils in urban areas. Landsc Urban Plan 90(1–2):1–10. https://doi.org/10.1016/j.landurbplan.2008.10.011
629. Schäpke N, Stelzer F, Caniglia G, Bergmann M, Wanner M, Singer-Brodowski M, Loorbach D, Olsson P, Baedeker C, Lang DJ (2018) Jointly experimenting for transformation?: shaping real-world laboratories by comparing them. GAIA 27:85–96. https://doi.org/10.14512/gaia.27.S1.16. Also available at: http://refhub.elsevier.com/S2214-6296(20)30299-1/h0405
630. Scheinerman ER, Ullman DH (2008) Fractional graph theory: a rational approach to the theory of graphs. Wiley. http://www.ams.jhu.edu/~ers/fgt
631. Schimel D, Keller M (2015) Big questions, big science: meeting the challenges of global ecology. Oecologia 177(4):925–934. https://doi.org/10.1007/s00442-015-3236-3
632. Schmalensee R, Stavins RN (2017) Lessons learned from three decades of experience with cap and trade. Rev Environ Econ Policy 11:59–79. https://doi.org/10.1093/reep/rew017
633. Schmieg G, Meyer E, Schrickel I, Herberg J, Caniglia G, Vilsmaier U, Laubichler M, Hörl E, Lang D (2018) Modeling normativity in sustainability: a comparison of the sustainable development goals, the Paris Agreement, and the papal encyclical. Sustain Sci 13(3):785–796. https://doi.org/10.1007/s11625-017-0504-7
634. Schmitz B, Satzger G, Gitzel R (2017) More observations, more variables or more quality?—data acquisition strategies to enhance uncertainty analytics for industrial service contracting. In: Za S, Drăgoicea M, Cavallari M (eds) Exploring services science. IESS 2017. Lecture Notes in Business Information Processing, vol 279, pp 159–172. Springer, Cham. https://doi.org/10.1007/978-3-319-56925-3_13
635. Schöggl J-P, Stumpf L, Baumgartner RJ (2020) The narrative of sustainability and circular economy—a longitudinal review of two decades of research. Resour Conserv Recycl 163, article 105073. https://doi.org/10.1016/j.resconrec.2020.105073
636. Scholz G, Dewulf A, Pahl-Wostl C (2014) An analytical framework of social learning facilitated by participatory methods. Syst Pract Action Res 27(6):575–591. https://doi.org/10.1007/s11213-013-9310-z
637. Scholz RW, Spoerri A, Lang DJ (2009) Problem structuring in transitions: the case of Swiss waste management. Futures 41(3):171–181. https://doi.org/10.1016/j.futures.2008.09.013
638. Schulz-Hardt S, Frey D, Luthgens C, Moscovici S (2000) Biased information search in group decision making. J Person Soc Psychol 78(4):655–669. https://doi.org/10.1037/0022-3514.78.4.655

639. Science|Business (2012) How to plan Europe's energy future. A report of a symposium on the energy R&D challenge. https://sciencebusiness.net/report/how-plan-europes-energy-future. Accessed on 6 June 2022
640. Sellberg MM, Wilkinson C, Peterson GD (2015) Resilience assessment: a useful approach to navigate urban sustainability challenges. Ecol Soc 20(1). https://doi.org/10.5751/ES-07258-200143
641. Selwyn N (2003) Apart from technology: understanding people's non-use of information and communication technologies in everyday life. Technol Soc 25(1):99–116. https://doi.org/10.1016/S0160-791X(02)00062-3
642. Seppelt R, Lautenbach S, Volk M (2013) Identifying trade-offs between ecosystem services, land use, and biodiversity: a plea for combining scenario analysis and optimization on different spatial scales. Curr Opin Environ Sustain 5:458–463. https://doi.org/10.1016/j.cosust.2013.05.002
643. Serrano FA (2003) City competitiveness and attractiveness: a new approach to evaluate economic development in Mexican cities. Doctoral dissertation, University of Glasgow. https://eleanor.lib.gla.ac.uk/record=b2142325
644. Šešelja D, Straßer C, Borg A (2020) Formal models of scientific inquiry in a social context: an introduction. J Gen Philos Sci 51:211–217. https://doi.org/10.1007/s10838-020-09502-w
645. Shao H, Kim G (2022) A comprehensive review of different types of green infrastructure to mitigate urban heat Islands: progress, functions, and benefits. Land 11(10), article 1792. https://doi.org/10.3390/land11101792
646. Shapiro C, Varian HR (1999) Information rules: a strategic guide to the network economy. Harvard Business School Press, Boston, MA. ISBN 0-87584-863-X
647. Sharifi A, Roosta M, Javadpoor M (2021) Urban form resilience: a comparative analysis of traditional, semi-planned, and planned neighborhoods in Shiraz, Iran. Urb Sci 5(1), article 18. https://doi.org/10.3390/urbansci5010018
648. Shore J, Johnson R (1980) Axiomatic derivation of the principle of maximum entropy and the principle of minimum cross-entropy. IEEE Trans Inform Theory 26(1):26–37. https://doi.org/ https://doi.org/10.1109/TIT.1980.1056144
649. Shy O (2001) The economics of network industries. Cambridge University Press, Cambridge, UK. ISBN 0521800951
650. Siderska J (2020) Robotic process automation—a driver of digital transformation? Eng Manage Prod Serv 12(2):21–31. https://doi.org/10.2478/emj-2020-0009
651. Sikandar A (2015) John Dewey and his philosophy of education. J Educ Educ Dev 2(2):191–201. http://jmsnew.iobmresearch.com/index.php/joeed/article/view/214
652. Silaban P (2021) What drives city brand attractiveness? An empirical study on some provinces in Indonesia. Geo J Tourism Geosites 38(4):1299–1308. https://doi.org/10.30892/gtg.38438-772
653. Silva BN, Khan M, Han K (2018) Towards sustainable smart cities: a review of trends, architectures, components, and open challenges in smart cities. Sustain Cities Soc 38:697–713. https://doi.org/10.1016/j.scs.2018.01.053
654. Simon D (ed) Rethinking sustainable cities. Accessible, green and fair. Policy Press, University of Bristol, 2016, Bristol, UK. ISBN 978-1-4473-3284-8
655. Simonofski A, Vallé T, Serral E, Wautelet Y (2021) Investigating context factors in citizen participation strategies: a comparative analysis of Swedish and Belgian smart cities. Int J Inform Manage 56, article 102011. https://doi.org/10.1016/j.ijinfomgt.2019.09.007
656. Simwanda M, Ranagalage M, Estoque RC, Murayama Y (2019) Spatial analysis of surface urban heat islands in four rapidly growing African cities. Rem Sens 11(14), article 1645. https://doi.org/10.3390/rs11141645
657. Skanavis C, Sakellari M, Petreniti V (2005) The potential of free-choice learning for environmental participation in Greece. Environ Educ Res 11(3):321–333. https://doi.org/10.1080/13504620500081194
658. Smith S (2015) Epic Fails: Reconceptualizing failure as a catalyst for developing creative persistence within teaching and learning experiences. J Technol Teach Educ 23(3):329–355. https://www.learntechlib.org/primary/p/151569/

659. Snieska V, Zykiene I (2015) City attractiveness for investment: characteristics and underlying factors. Procedia Soc Behav Sci 213:48–54. https://doi.org/10.1016/j.sbspro.2015.11.402
660. Sobolevsky S, Bojic I, Belyi A, Sitko I, Hawelka B, Arias JM, Ratti C (2015) Scaling of city attractiveness for foreign visitors through big data of human economical and social media activity. In 2015 IEEE International Congress on Big Data, pp 600–607. IEEE. https://doi.org/10.1109/BigDataCongress.2015.92
661. Soh TMT, Meerah TSM (2013) Outdoor education: an alternative approach in teaching and learning science. Asian Soc Sci 9(16), article 1. https://doi.org/10.5539/ass.v9n16p1
662. Song XM, Montoya-Weiss MM (1998) Critical development activities for really new versus incremental products. J Prod Innov Manage Int Publ Prod Dev Manage Assoc 15(2):124–135. https://doi.org/10.1111/1540-5885.1520124
663. Speed PJ (2011) The genius in all of us: why everything you've been told about genetics, talent and intelligence is wrong. J Biol Educ 45(3), article 165. https://doi.org/10.1080/00219266.2011.589466
664. Spinosa C, Flores F, Dreyfus H (1995) Disclosing new worlds: entrepreneurship, democratic action, and the cultivation of solidarity. Inquiry 38(1–2):3–63. https://doi.org/10.1080/00201749508602373
665. Stahel W (2016) The circular economy. Nature 531(7595):435–438. https://doi.org/10.1038/531435a
666. Stasavage D (2016) Representation and consent: why they arose in Europe and not elsewhere. Annu Rev Polit Sci 19(1):145–162. https://doi.org/10.1146/annurev-polisci-043014-105648
667. Steeneveld GJ, Klompmaker JO, Groen RJ, Holtslag AA (2018) An urban climate assessment and management tool for combined heat and air quality judgements at neighbourhood scales. Resour Conserv Recycl 132:204–217. https://doi.org/10.1016/j.resconrec.2016.12.002
668. Stell JG (2014) Formal concept analysis over graphs and hypergraphs. In: Graph structures for knowledge representation and reasoning. Springer, Cham, pp 165–179. https://doi.org/10.1007/978-3-319-04534-4_11
669. Stevanović M (1997) The age of clay: the social dynamics of house destruction. J Anthropol Archaeol 16(4):334–395. https://doi.org/10.1006/jaar.1997.0310
670. Stjerne IS, Söderlund J, Minbaeva D (2019) Crossing times: temporal boundary-spanning practices in interorganizational projects. Int J Project Manage 37:347–365. https://doi.org/10.1016/j.ijproman.2018.09.004
671. Stockwell B (2013) Fragticulture: reconfiguring agricultural systems to enhance resilience and sustainability in fragmented coastal peri-urban catchments. Doctoral dissertation, University of the Sunshine Coast, Queensland, Australia. https://doi.org/10.25907/00267
672. Stuart R, Abetti PA (1987) Start-up ventures: towards the prediction of initial success. J Bus Ventur 2(3):215–230. https://doi.org/10.1016/0883-9026(87)90010-3
673. Supriana CE (2021) Designing knowledge sharing systems to support integrated eco-city planning and management. In: IOP conference series: Earth and Environmental Science, vol 737, no 1, paper 012006. IOP Publishing. https://doi.org/10.1088/1755-1315/737/1/012006
674. Sutton PC, Anderson SJ (2016) Holistic valuation of urban ecosystem services in New York City's Central Park. Ecosyst Serv 19:87–91. https://doi.org/10.1016/j.ecoser.2016.04.003
675. Świąder M, Lin D, Szewrański S, Kazak JK, Iha K, van Hoof J, Belčáková I, Altiok S (2020) The application of ecological footprint and biocapacity for environmental carrying capacity assessment: a new approach for European cities. Environ Sci Policy 105:56–74. https://doi.org/10.1016/j.envsci.2019.12.010
676. Syamil A, Doll WJ, Apigian CH (2004) Process performance in product development: measures and impacts. Eur J Innov Manag 7(3):205–217. https://doi.org/10.1108/14601060410549892
677. Sydow J, Braun T (2018) Projects as temporary organizations: an agenda for further theorizing the interorganizational dimension. Int J Project Manage 36:4–11. https://doi.org/10.1016/j.ijproman.2017.04.012
678. Szajnfarber Z, Groover JA, Wei Z, Broniatowski DA, Chernicoff W, Helveston JP (2022) Evolvability analysis framework: adding transition path and stakeholder diversity to infrastructure planning. Syst Eng 25(1):35–50. https://doi.org/10.1002/sys.21600

679. Taleghani M, Kleerekoper L, Tenpierik M, Van Den Dobbelsteen A (2015) Outdoor thermal comfort within five different urban forms in the Netherlands. Build Environ 83:65–78. https://doi.org/10.1016/j.buildenv.2014.03.014
680. Tan X, Huang B, Batty M, Li J (2021) Urban spatial organization, multifractals, and evolutionary patterns in large cities. Ann Am Assoc Geogr 111(5):1539–1558. https://doi.org/10.1080/24694452.2020.1823203
681. Tao Y, Sornette D, Lin L (2021) Emerging social brain: a collective self-motivated Boltzmann machine. Chaos Solit Fract 143, article 110543. https://doi.org/10.1016/j.chaos.2020.110543
682. Taylor DE (1999) Central Park as a model for social control: urban parks, social class and leisure behavior in nineteenth-century America. J Leis Res 31(4):420–477. https://doi.org/10.1080/00222216.1999.11949875
683. Taylor EW, Caldarelli M (2004) Teaching beliefs of non-formal environmental educators: a perspective from state and local parks in the United States. Environ Educ Res 10(4):451–469. https://doi.org/10.1080/1350462042000291001
684. Taylor MR, Rubin ES, Hounshell DA (2003) Effect of government actions on technological innovation for SO_2 control. Environ Sci Technol 37(20):4527–4534. https://doi.org/10.1021/es034223b
685. Taylor P, Derudder B (2004) World city network: a global urban analysis. Routledge. https://doi.org/10.4324/9781315730950
686. Taylor P, Derudder B (2015) World city network: a global urban analysis. Routledge (First edition: 2004). https://doi.org/10.4324/9781315730950
687. Theodorou P (2022) The effects of urbanisation on ecological interactions. Curr Opin Insect Sci 52, article 100922. https://doi.org/10.1016/j.cois.2022.100922
688. The Economist (2017) The world's most valuable resource. The Economist, May 6th–12th, 2017, p 7. https://www.economist.com/leaders/2017/05/06/the-worlds-most-valuable-resource-is-no-longer-oil-but-data
689. Thomas K, Hardy RD, Lazrus H, Mendez M, Orlove B, Rivera-Collazo I, Roberts JT, Rockman M, Warner BP, Winthrop R (2019) Explaining differential vulnerability to climate change: a social science review. Wiley Interdiscip Rev Clim Change 10(2):e565. https://doi.org/10.1002/wcc.565
690. Thomas Muñoz R (2022) Environmental education networks for social empowerment and global citizenship: a case of non-formal education from Mexico. In: Öztürk M (ed) Educational response, inclusion and empowerment for SDGs in emerging economies. Sustainable development goals series. Springer, Cham. https://doi.org/10.1007/978-3-030-98962-0_12
691. Tischer C, Kirjavainen P, Matterne U, Tempes J, Willeke K, Keil T, Apfelbacher C, Täubel M (2022) Interplay between natural environment, human microbiota and immune system: a scoping review of interventions and future perspectives towards allergy prevention. Sci Total Environ 821, article 153422. https://doi.org/10.1016/j.scitotenv.2022.153422
692. Tobias S, Conen F, Duss A, Wenzel LM, Buser C, Alewell C (2018) Soil sealing and unsealing: state of the art and examples. Land Degrad Dev 29(6):2015–2024. https://doi.org/10.1002/ldr.2919
693. Torday JS, Miller Jr WB (2016) Biologic relativity: who is the observer and what is observed? Progr Biophys Molecul Biol 121(1):29–34. https://doi.org/10.1016/j.pbiomolbio.2016.03.001
694. Toyoda Y (2021) Survey paper: achievements and perspectives of community resilience approaches to societal systems. Asia-Pacific J Region Sci 5(3):705–756. https://doi.org/10.1007/s41685-021-00202-x
695. Traoré I, Gavan V, Riffonneau Y, L'Henoret B, Drouet E (2013) Development of a generic and scalable Modelica-based model of a typical French railway station building. In: Proceedings of the 13th Conference of International Building Performance Simulation Association, 26–28 August, Chambéry, France, pp 25–28. https://www.aivc.org/resource/development-generic-and-scalable-modelica-based-model-typical-french-railway-station
696. Treisman D (2023) How great is the current danger to democracy? Assessing the risk with historical data. Compar Polit Stud 56(12), article 00104140231168363. https://doi.org/10.1177/00104140231168363

697. Trencher GP, Yarime M, McCormick KB, Doll NH, Kraines SB (2014) Beyond the third mission: exploring the emerging university function of co-creation for sustainability. Sci Publ Policy 41:151–179. https://doi.org/10.1093/scipol/sct044
698. Tringham R (2005) Weaving house life and death into places: a blueprint for a hypermedia narrative. In: Bailey D, Whittle A, Cummings V (eds) (Un)settling the neolithic, pp 98–111. Oxbow Books, Oxford. https://www.ruthtringham.com/project/weaving-house-life-and-death-2005/
699. Tsoukias A, Montibeller G, Lucertini G, Belton V (2013) Policy analytics: an agenda for research and practice. EURO J Decis Proces 1(1–2):115–134. https://doi.org/10.1007/s40070-013-0008-3
700. Tsuchiya K, Iha K, Murthy A, Lin D, Altiok S, Rupprecht CD, Kiyono H, McGreevy SR (2021) Decentralization & local food: Japan's regional ecological footprints indicate localized sustainability strategies. J Clean Prod 292, article 126043. https://doi.org/10.1016/j.jclepro.2021.126043
701. Tsuchiya N, Taguchi S, Saigo H (2016) Using category theory to assess the relationship between consciousness and integrated information theory. Neurosci Res 107:1–7. https://doi.org/10.1016/j.neures.2015.12.007
702. Turoff M, Hiltz SR, Bañuls VA, Van Den Eede G (2013) Multiple perspectives on planning for emergencies: An introduction to the special issue on planning and foresight for emergency preparedness and management. Technol Forecast Soc Chang 80(9):1647–1656. https://doi.org/10.1016/j.techfore.2013.07.014
703. Tzoulas K, James P (2010) Making biodiversity measures accessible to non-specialists: An innovative method for rapid assessment of urban biodiversity. Urban Ecosyst 13:113–127. https://doi.org/10.1007/s11252-009-0107-x
704. Ulpiani G (2021) On the linkage between urban heat island and urban pollution island: three-decade literature review towards a conceptual framework. Sci Total Environ 751, article 141727. https://doi.org/10.1016/j.scitotenv.2020.141727
705. Unger J (2004) Intra-urban relationship between surface geometry and urban heat island: review and new approach. Clim Res 27(3):253–264. https://www.jstor.org/stable/24868753
706. Urpelainen J (2013) A model of dynamic climate governance: dream big, win small. Int Environ Agreement Polit Law Econ 13(2):107–125. https://doi.org/10.1007/s10784-012-9174-1
707. US-DOD (2019) Indo-Pacific strategy report preparedness, partnerships, and promoting a networked region. United States of America, Department of Defense. https://media.defense.gov/2019/Jul/01/2002152311/-1/-1/1/DEPARTMENT-OF-DEFENSE-INDO-PACIFIC-STRATEGYREPORT-2019.PDF
708. US Department of Energy (2011) DOE G 413.3-4A, Technology readiness assessment guide. Washington D.C. https://www.directives.doe.gov/directives-documents/400-series/0413.3-EGuide-04a
709. US-GAO (2020) Technology readiness assessment guide. Best practices for evaluating the readiness of technology for use in acquisition programs and projects. Report GAO-20-48G. U.S. Government Accountability Office (GAO), Washington DC. https://apps.dtic.mil/sti/pdfs/AD1173994.pdf
710. Uy PD, Nakagoshi N (2008) Application of land suitability analysis and landscape ecology to urban greenspace planning in Hanoi, Vietnam. Urb Forestr Urb Green 7(1):25–40. https://doi.org/10.1016/j.ufug.2007.09.002
711. Vallée R (2009) History of cybernetics. Syst Sci Cybern 3:22–33. http://www.eolss.net/Sample-chapters/C02/E6-46-03-01.pdf
712. Valliere JM (2022) Cultivating scientific literacy and a sense of place through course-based urban ecology research. Ecol Evolut 12, article e8985. https://doi.org/10.1002/ece3.8985
713. Van Cauwenbergh N, Dourojeanni PA, van der Zaag P, Brugnach M, Dartee K, Giordano R, Lopez-Gunn E (2022) Beyond TRL–Understanding institutional readiness for implementation of nature-based solutions. Environ Sci Policy 127:293–302. https://doi.org/10.1016/j.envsci.2021.09.021

714. Van der Aalst WM, Bichler M, Heinzl A (2018) Robotic process automation. Bus Inf Syst Eng 60:269–272. https://doi.org/10.1007/s12599-018-0542-4
715. Van der Wal M, De Kraker J, Offermans A, Kroeze C, Kirschner PA, Van Ittersum M (2014) Measuring social learning in participatory approaches to natural resource management. Environ Policy Gov 24(1):1–15. https://doi.org/10.1002/eet.1627
716. Van der Jagt AP, Buijs A, Dobbs C, van Lierop M, Pauleit S, Randrup TB, Skiba A, Wild T (2023) With the process comes the progress: a systematic review to support governance assessment of urban nature-based solutions. Urb Forestr Urb Green article 128067. https://doi.org/10.1016/j.ufug.2023.128067
717. Vanderlinden JP, Baztan J, Chouinard O, Cordier M, Da Cunha C, Huctin JM, Kane A, Kennedy G, Nikulkina I, Shadrin V, Surette C (2020) Meaning in the face of changing climate risks: Connecting agency, sensemaking and narratives of change through transdisciplinary research. Clim Risk Manage 29, article 100224. https://doi.org/10.1016/j.crm.2020.100224
718. Vanderlinden JP, Baztan J, Touili N, Kane IO, Rulleau B, Simal PD, Pietrantoni L, Prati G, Zagonari F (2017) Coastal flooding, uncertainty and climate change: science as a solution to (mis) perceptions? A qualitative enquiry in three coastal European settings. J Coast Res Spec Iss 77:127–133. https://doi.org/10.2112/SI77-013.1
719. Van Hullebusch ED, Bani A, Carvalho M, Cetecioglu Z, De Gusseme B, Di Lonardo S, Djolic M, Van Eekert M, Griessler Bulc T, Haznedaroglu BZ, Isteniĉ D (2021) Nature-based units as building blocks for resource recovery systems in cities. Water 13(22), article 3153. https://doi.org/10.3390/w13223153
720. Vasantkumar C (2017) From world cities to world sites: Strategic ruralism and the case for an anthropology of actually existing connectivity. Crit Anthropol 37(4):364–382. https://doi.org/10.1177/0308275X17735370
721. Vega KA, Küffer C (2021) Promoting wildflower biodiversity in dense and green cities: the important role of small vegetation patches. Urb Forestr Urban Green 62, article 127165. https://doi.org/10.1016/j.ufug.2021.127165
722. Veerkamp CJ, Schipper AM, Hedlund K, Lazarova T, Nordin A, Hanson HI (2021) A review of studies assessing ecosystem services provided by urban green and blue infrastructure. Ecosyst Serv 52, article 101367. https://doi.org/10.1016/j.ecoser.2021.101367
723. Venkataraman N (2002) The value center. In Ward J, Peppard J (eds) Strategic planning for information systems. Wiley, Chester
724. Verginer L, Riccaboni M (2021) Talent goes to global cities: the world network of scientists' mobility. Res Policy 50(1), article 104127. https://doi.org/10.1016/j.respol.2020.104127
725. Vernay AL, Rahola TS, Ravesteijn W (2010) Growing food, feeding change: towards a holistic and dynamic approach of eco-city planning. In: Next generation infrastructure systems for eco-cities, 11–13 November, pp 1–6. IEEE. https://doi.org/10.1109/INFRA.2010.5679234
726. Victor DG, House JC, Joy S (2005) A Madisonian approach to climate policy. Science 309:1820–1821. https://doi.org/10.1126/science.1113180
727. Vik J, Melås AM, Stræte EP, Søraa RA (2021) Balanced readiness level assessment (BRLa): a tool for exploring new and emerging technologies. Technol Forecast Soc Change 169, article 120854. https://doi.org/10.1016/j.techfore.2021.1
728. Voytenko Y, McCormick K, Evans J, Schliwa G (2016) Urban living labs for sustainability and low carbon cities in Europe: Towards a research agenda. J Clean Prod 123:45–54. https://doi.org/10.1016/j.jclepro.2015.08.053
729. Vu HM (2020) A review of dynamic capabilities, innovation capabilities, entrepreneurial capabilities and their consequences. J Asian Finan Econ Bus 7(8):485–494. https://doi.org/10.13106/jafeb.2020.vol7.no8.485
730. Wackernagel M, Kitzes J, Moran D, Goldfinger S, Thomas M (2006) The ecological footprint of cities and regions: comparing resource availability with resource demand. Environ Urban 18(1):103–112. https://doi.org/10.1177/0956247806063978
731. Wagner N, Velander S, Biber-Freudenberger L, Dietz T (2023) Effectiveness factors and impacts on policymaking of science-policy interfaces in the environmental sustainability context. Environ Sci Policy 140:56–67. https://doi.org/10.1016/j.envsci.2022.11.008

732. Walch C (2015) Expertise and policy-making in disaster risk reduction. Nat Clim Chang 5:706–707. https://doi.org/10.1038/nclimate2680
733. Wang B, Sun S, Duan M (2018) Wind potential evaluation with urban morphology-A case study in Beijing. Energy Procedia 153:62–67. https://doi.org/10.1016/j.egypro.2018.10.078
734. Wang C, Wang ZH, Yang J, Li Q (2018) A backward-Lagrangian-stochastic footprint model for the urban environment. Bound-Layer Meteorol 168:59–80. https://doi.org/10.1007/s10546-018-0338-6
735. Wang K, Yan F, Zhang Y, Xiao Y, Gu L (2020) Supply chain financial risk evaluation of small-and medium-sized enterprises under smart city. J Adv Transp 2020, Article 8849356. https://doi.org/10.1155/2020/8849356
736. Wang J, Fleischmann M, Venerandi A, Romice O, Kuffer M, Porta S (2023) EO+ Morphometrics: understanding cities through urban morphology at large scale. Landsc Urb Plann 233, article 104691. https://doi.org/10.1016/j.landurbplan.2023.104691
737. Waqar A, Othman I, Falqi II, Almujibah HR, Alshehri AM, Alsulamy SH, Benjeddou O (2023) Assessment of barriers to robotics process automation (RPA) implementation in safety management of tall buildings. Buildings 13(7), article 1663. https://doi.org/10.3390/buildings13071663
738. Ward M, Halliday S, Uflewska O, Wong TC (2018) Three dimensions of maturity required to achieve future state, technology-enabled manufacturing supply chains. Proc Instit Mechan Eng Part B J Eng Manuf 232(4):605–620. https://doi.org/10.1177/0954405417710045
739. Waring TM, Kline MA, Brooks JS, Goff SH, Gowdy J, Janssen MA Smaldino PE, Jacquet J (2015) A multilevel evolutionary framework for sustainability analysis. Ecol Soc 20(2), article 34. https://doi.org/10.5751/ES-07634-200234
740. Weil C, Bibri SE, Longchamp R, Golay F, Alahi A (2023) Urban digital twin challenges: a systematic review and perspectives for sustainable smart cities. Sustain Cities Soc 99:104862. https://doi.org/10.1016/j.scs.2023.104862
741. Weltevreden JW, Rietbergen TV (2007) E-shopping versus city centre shopping: the role of perceived city centre attractiveness. Tijdschr Econ Soc Geogr 98(1):68–85. https://doi.org/10.1111/j.1467-9663.2007.00377.x
742. Welzel C (2021) Democratic horizons: what value change reveals about the future of democracy. Democratization 28(5):992–1016. https://doi.org/10.1080/13510347.2021.1883001
743. Welzel C (2021) Why the future is democratic. J Democr 32(2):132–144. https://doi.org/10.1353/jod.2021.0024
744. Wenger E (1998) Communities of practice: learning as a social system. Syst Thinker 9(5):2–3. https://participativelearning.org/pluginfile.php/636/mod_resource/content/3/Learningasasocialsystem.pdf
745. Wentz EA, York AM, Alberti M, Conrow L, Fischer H, Inostroza L, Jantz C, Pickett ST, Seto KC, Taubenböck H (2018) Six fundamental aspects for conceptualizing multidimensional urban form: a spatial mapping perspective. Landsc Urb Plan 179:55–62. https://doi.org/10.1016/j.landurbplan.2018.07.007
746. West JD, Bergstrom CT (2021) Misinformation in and about science. Proc Natl Acad Sci 118(15), article e1912444117. https://doi.org/10.1073/pnas.1912444117
747. West M, Fowler J (1999) Developing high quality data models. The European Process Industries STEP Technical Liaison Executive (EPISTLE). https://sites.google.com/site/drmatthewwest/publications/princ03.pdf
748. Westman L, Moores E, Burch SL (2021) Bridging the governance divide: the role of SMEs in 44 urban sustainability interventions. Cities 108, article 102944. https://doi.org/10.1016/j.cities.2020.102944
749. White AJ, Stevens LR, Lorenzi V, Munoz SE, Schroeder S, Cao A, Bogdanovich T (2019) Fecal stanols show simultaneous floodingand seasonal precipitation change correlate with Cahokia's population decline. Proc Natl Acad Sci 116(12):5461–5466. https://doi.org/10.1073/pnas.1809400116

750. White R, Engelen G (1994) Urban systems dynamics and cellular automata: fractal structures between order and chaos. Chaos, Solitons Fractals 4(4):563–583. https://doi.org/10.1016/0960-0779(94)90066-3
751. White R, Marzano M, Fesenko E, Inman A, Jones G, Agstner B, Mumford R (2022) Technology development for the early detection of plant pests: a framework for assessing Technology Readiness Levels (TRLs) in environmental science. J Plant Dis Prot 129(5):1249–1261. https://doi.org/10.1007/s41348-022-00599-3
752. WHO (2016) Urban green spaces and health. (No. WHO/EURO: 2016-3352-43111-60341). World Health Organization. Regional Office for Europe. https://iris.who.int/handle/10665/345751
753. Wigley TM, Richels R, Edmonds JA (1996) Economic and environmental choices in the stabilization of atmospheric CO_2 concentrations. Nature 379(6562):240. https://www.nature.com/articles/379240a0
754. Wilkinson M (2013) Testing the null hypothesis: the forgotten legacy of Karl Popper? J Sports Sci 31(9):919–920. https://doi.org/10.1080/02640414.2012.753636
755. Williams J (2016) Can low carbon city experiments transform the development regime? Futures 77:80–96. https://doi.org/10.1016/j.futures.2016.02.003
756. Willmoth A (2008) Real science using TASC. Gift Educ Int 24(2–3):249–254. https://doi.org/10.1177/026142940802400314
757. Wittig R (2004) The origin and development of the urban flora of Central Europe. Urb Ecosyst 7:323–339. https://doi.org/10.1007/s11252-005-6833-9
758. Woods WI (2004) Population nucleation, intensive agriculture, and environmental degradation: the Cahokia example. Agric Hum Values 21(2):255–261. https://doi.org/10.1023/B:AHUM.0000029398.01906.5e
759. Workman M, Darch G, Dooley K, Lomax G, Maltby J, Pollitt H (2021) Climate policy decision making in contexts of deep uncertainty-from optimisation to robustness. Environ Sci Policy 120:127–137. https://doi.org/10.1016/j.envsci.2021.03.002
760. Wu L, Wang D, Evans JA (2019) Large teams develop and small teams disrupt science and technology. Nature 566(7744):378–382. https://doi.org/10.1038/s41586-019-0941-9
761. Xu S, He H, Liu Z (2022) New promises of advanced molecular recognition: Bioassays, single cell analysis, cancer therapy, and beyond. Chin J Chem 40(5):635–650. https://doi.org/10.1002/cjoc.202100679
762. Yarime M (2017) Facilitating data-intensive approaches to innovation for sustainability: opportunities and challenges in building smart cities. Sustain Sci 12:881–885. https://doi.org/10.1007/s11625-017-0498-1
763. Zeithaml VA (1988) Consumer perceptions of price, quality, and value: a means-end model and synthesis of evidence. J Mark 52(3):2–22. https://doi.org/10.1177/002224298805200302
764. Zhang Y (2013) Urban metabolism: a review of research methodologies. Environ Pollut 178:463–473. https://doi.org/10.1016/j.envpol.2013.03.052
765. Zheng Y, Capra L, Wolfson O, Yang H (2014) Urban computing: concepts, methodologies, and applications. ACM Trans Intell Syst Technol (TIST) 5(3):1–55. https://doi.org/10.1145/2629592
766. Zhou X, Li S, Wei C (2024) Consensus reaching process for group decision-making based on trust network and ordinal consensus measure. Inform Fus 101, article 101969. https://doi.org/10.1016/j.inffus.2023.101969
767. Zikargae MH, Woldearegay AG, Skjerdal T (2022) Empowering rural society through non-formal environmental education: an empirical study of environment and forest development community projects in Ethiopia. Heliyon 8(3). https://ui.adsabs.harvard.edu/link_gateway/2022Heliy...809127Z/https://doi.org/10.1016/j.heliyon.2022.e09127

Afterword

July 10, 2023, Versailles—France.

A decade ago, at the end of a key project in the Econoving cluster of eco-innovation at the University of Versailles (UVSQ), and based on the knowledge of the time, I asked a long-view question: *Is there a future for eco-cities?* [1]. The same year, as mentioned in the Foreword, I received the suggestion from Springer to consider writing a book on eco-city scenarios, an idea which I then pursued as a direct attempt to answer the apparently simple question above. Today, in this very book, I propose an apparently simple answer: *Yes, and it takes long-term work.*

By its very nature, a city is not only a window to the complexity of our world but is also a practical platform for understanding the objective space of options we have or can generate for addressing the unsustainability crisis realistically.

As I am embarking on other projects, I happily acknowledge that I am constantly finding new ways, people, and ideas, and that every experience in this kind of work is formative—not only through discovery and novelty per se but also by a continual refinement the current big picture, ergo an amplification of the insights that are still coming from past projects and experiences, and in a challenging global context that demands realism, steady development of competence and perseverance. I dare say that involvement in eco-city projects and programs is really about us learning to fly—better and better—over the horizon and into a new era that just started.

As you have noticed, the three autonomous but interconnected modular chapters of this book can be read as stand-alone units—and re-read for further in-depth study, as these modules connect to relevant bodies of literature on sustainability, cities, and related themes, through citations that were carefully chosen to cover both the basics and the domain of emerging themes. This does not and cannot possibly exhaust the immense richness of issues and insights coming out every week, but it is meant to be a fair and solid start basis and support for concrete projects. As you will compare it with other books and expert resources on the matter, I am sure the projects will lead you through taking the best out of each.

I hope the book will be a good companion for you, and that you will be able to share your own experience with others around the world.

I. M. Ciumasu, *Eco-cities*, Green Energy and Technology,
https://doi.org/10.1007/978-3-319-14702-4

If you will care to share your thoughts with me too, I will do my best to respond to all of your inquiries and further ideas (for example, by email, LinkedIn, and X/formerly Twitter). Equally important, if inspired by the references cited in this book, I hope you will contact some of those authors too: we are a worldwide human resource for eco-city projects and programs. If we put our minds together, we can do anything.

Let's get to work!

Reference

1. Ciumasu IM (2013c). Is there a future for eco-cities? Prospective roles of Versailles Chantiers within the Greater Paris: A hub of multi-modal communication. Technical report of the Econoving Integration Seminar (2011–2013). ECONOVING Chair for Generating Eco-Innovation, REEDS Center, University of Versailles Saint-Quentin-en-Yvelines. https://doi.org/10.13140/RG.2.1.3866.8240

MIX
Papier aus verantwortungsvollen Quellen
Paper from responsible sources
FSC® C105338

If you have any concerns about our products,
you can contact us on
ProductSafety@springernature.com

In case Publisher is established outside the EU,
the EU authorized representative is:
Springer Nature Customer Service Center GmbH
Europaplatz 3, 69115 Heidelberg, Germany

Printed by Libri Plureos GmbH
in Hamburg, Germany